Modern Physics
with Modern Computational Methods

Modern Physics

with Modern Computational Methods

Third Edition

John Morrison

ELSEVIER

ACADEMIC PRESS
An imprint of Elsevier

Academic Press is an imprint of Elsevier
125 London Wall, London EC2Y 5AS, United Kingdom
525 B Street, Suite 1650, San Diego, CA 92101, United States
50 Hampshire Street, 5th Floor, Cambridge, MA 02139, United States
The Boulevard, Langford Lane, Kidlington, Oxford OX5 1GB, United Kingdom

Library of Congress Cataloging-in-Publication Data
A catalog record for this book is available from the Library of Congress

British Library Cataloguing-in-Publication Data
A catalogue record for this book is available from the British Library

ISBN: 978-0-12-817790-7

For information on all Academic Press publications
visit our website at https://www.elsevier.com/books-and-journals

Publisher: Katey Birtcher
Editorial Project Manager: Alexandra Ford
Production Project Manager: Radhika Sivalingam
Designer: Maggie Reid

Typeset by VTeX

Working together
to grow libraries in
developing countries

www.elsevier.com • www.bookaid.org

Dedication

About the dedication

In a fractured world consisting of rich and poor countries, physics has tried with varying degrees of success to reach out to the entire world. Graduate programs in the United States and Europe have for many years offered their graduate teaching assistantships to students from poorer countries. A graduate teaching assistantship is a way that a student from a poorer country can get an advanced degree while paying for his/her own education. In the beginning graduate recruitment was done almost exclusively by mail. A big pile of applications was processed by a committee within the science departments with the members of the committees having no contact or understanding of the countries from which the students came. More recently faculty members like myself have traveled to Latin America, the Middle East, and India. The scientists and mathematicians I know who have traveled to these parts of the world have been surprised and felt privileged to teach students who are so committed to their studies. Being forced to live on one's own resources can strengthen the mind.

The graduate students are a strength of our department at University of Louisville. Within recent years, many of our best graduate students have come from central Russia, from Palestine, and from Bengal in India. These students were recruited by faculty members who have themselves taught in those parts of the world. To what I have said previously, I should mention the International Centre of Theoretical Physics founded by Abdus Salam in Trieste, Italy which hosts students in Physics, Mathematics and related topics from the developing countries for a one-year Master course with the possibility of continuing for a PhD degree.

Dedication

This book is dedicated to the physics students from Latin America, Africa, the Middle East, and Asia, to their teachers, to the international scientists who have taught there, and to the International Centre of Theoretical Physics in Trieste, Italy and universities in England, Amsterdam, and Denmark that have offered advance courses to students from developing countries for many years. I am presently working together with my Latin American students and colleagues to prepare a version of this modern physics book in the Spanish language.

John C. Morrison
University of Louisville, Louisville, KY, United States
Birzeit University, Ramallah, Palestine
University of Costa Rica, San Jose, Costa Rica

Contents

Preface xi
Introduction xiii

1. The wave-particle duality

1.1 The particle model of light 1
 1.1.1 The photoelectric effect 1
 1.1.2 The absorption and emission of light by atoms 4
 1.1.3 The Compton effect 11
1.2 The wave model of radiation and matter 12
 1.2.1 X-ray scattering 12
 1.2.2 Electron waves 13
Suggestions for further reading 15
Basic equations 15
Summary 16
Questions 16
Problems 17

2. The Schrödinger wave equation

2.1 The wave equation 19
2.2 Probabilities and average values 23
2.3 The finite potential well 26
2.4 The simple harmonic oscillator 28
2.5 Time evolution of the wave function 30
Suggestion for further reading 33
Basic equations 33
Summary 34
Questions 35
Problems 35

3. Operators and waves

3.1 Observables, operators, and eigenvalues 37
3.2 The finite well and harmonic oscillator using finite differences 40
3.3 The finite well and harmonic oscillator with spline collocation 47
3.4 Electron scattering 56
 3.4.1 Scattering from a potential step 56
 3.4.2 Barrier penetration and tunneling 60
 3.4.3 T-matrices 61
 3.4.4 Scattering from more complex barriers 61
3.5 The Heisenberg uncertainty principle 64

 3.5.1 Wave packets and the uncertainty principle 65
 3.5.2 Average value of the momentum and the energy 68
Suggestion for further reading 76
Basic equations 76
Summary 78
Questions 78
Problems 79

4. The hydrogen atom

4.1 The Gross structure of hydrogen 81
 4.1.1 The Schrödinger equation in three dimensions 81
 4.1.2 The energy levels of hydrogen 82
 4.1.3 The wave functions of hydrogen 84
 4.1.4 Probabilities and average values in three dimensions 89
 4.1.5 The intrinsic spin of the electron 95
4.2 Radiative transitions 95
 4.2.1 The Einstein A and B coefficients 96
 4.2.2 Transition probabilities 97
 4.2.3 Selection rules 100
4.3 The fine structure of hydrogen 102
 4.3.1 The magnetic moment of the electron 102
 4.3.2 The Stern-Gerlach experiment 105
 4.3.3 The spin of the electron 105
 4.3.4 The addition of angular momentum 107
 4.3.5 * The fine structure 108
 4.3.6 * The Zeeman effect 109
Suggestion for further reading 111
Basic equations 111
Summary 113
Questions 113
Problems 114

5. Many-electron atoms

5.1 The independent-particle model 117
 5.1.1 Antisymmetric wave functions and the Pauli exclusion principle 118
 5.1.2 The central-field approximation 119
5.2 Shell structure and the periodic table 120
5.3 The LS term energies 122
5.4 Configurations of two electrons 122

5.4.1 Configurations of equivalent electrons   123
5.4.2 Configurations of two nonequivalent electrons   125
5.5 The Hartree-Fock method   126
5.5.1 The Hartree-Fock applet   127
5.5.2 The size of atoms and the strength of their interactions   131
5.6 Further developments in atomic theory   135
Suggestions for further reading   138
Basic equations   138
Summary   138
Questions   138
Problems   139

6. The emergence of masers and lasers

6.1 Radiative transitions   141
6.2 Laser amplification   142
6.3 Laser cooling   146
6.4 * Magneto-optical traps   147
Suggestions for further reading   150
Basic equations   150
Summary   151
Questions   151
Problems   151

7. Diatomic molecules

7.1 The hydrogen molecular ion   153
7.2 The Hartree-Fock method   163
7.3 Exoplanets   164
References   166
Summary   167
Questions   168

8. Statistical physics

8.1 The nature of statistical laws   169
8.2 An ideal gas   172
8.3 Applications of Maxwell-Boltzmann statistics   174
8.3.1 Maxwell distribution of the speeds of gas particles   174
8.3.2 Black body radiation   180
8.4 Entropy and the laws of thermodynamics   185
8.4.1 The four laws of thermodynamics   187
8.5 A perfect quantum gas   189
8.6 Bose-Einstein condensation   193
8.7 Free-electron theory of metals   194
Suggestions for further reading   199
Basic equations   199
Summary   201
Questions   201
Problems   202

9. Electronic structure of solids

9.1 The Bravais lattice   206
9.2 Additional crystal structures   210
9.2.1 The diamond structure   210
9.2.2 The hexagonal close-packed structure   210
9.2.3 The sodium chloride structure   211
9.3 The reciprocal lattice   212
9.4 Lattice planes   215
9.5 Bloch's theorem   220
9.6 Diffraction of electrons by an ideal crystal   224
9.7 The band gap   226
9.8 Classification of solids   227
9.8.1 The band picture   227
9.8.2 The bond picture   231
Suggestions for further reading   234
Basic equations   235
Summary   236
Questions   236
Problems   237

10. Charge carriers in semiconductors

10.1 Density of charge carriers in semiconductors   241
10.2 Doped crystals   244
10.3 A few simple devices   245
10.3.1 The p-n junction   245
10.3.2 Solar cells   247
10.3.3 Bipolar transistors   248
10.3.4 Junction field-effect transistors (JFET)   249
10.3.5 MOSFETs   250
Suggestions for further reading   250
Summary   251
Questions   251

11. Semiconductor lasers

11.1 Motion of electrons in a crystal   253
11.2 Band structure of semiconductors   255
11.2.1 Conduction bands   255
11.2.2 Valence bands   256
11.2.3 Optical transitions   256
11.3 Heterostructures   258
11.3.1 Properties of heterostructures   258
11.3.2 Experimental methods   259
11.3.3 Theoretical methods   260
11.4 Quantum wells   262
11.5 Quantum barriers   265
11.5.1 Scattering of electrons by potential barriers   265
11.5.2 Light waves   266
11.5.3 Reflection and transmission by an interface   267
11.5.4 The Fabry-Perot laser   268

11.6   **Phenomenological description of diode lasers**   270
   11.6.1 The rate equation   270
   11.6.2 Well below threshold   272
   11.6.3 The laser threshold   273
   11.6.4 Above threshold   274
Suggestions for further reading   275
Basic equations   275
Summary   277
Questions   277
Problems   278

12.   The special theory of relativity

12.1   **Galilean transformations**   279
12.2   **The relative nature of simultaneity**   282
12.3   **Lorentz transformation**   284
   12.3.1 The transformation equations   284
   12.3.2 Lorentz contraction   287
   12.3.3 Time dilation   288
   12.3.4 The invariant space-time interval   291
   12.3.5 Addition of velocities   291
   12.3.6 The Doppler effect   293
12.4   **Space-time diagrams**   295
   12.4.1 Particle motion   295
   12.4.2 Lorentz transformations   298
   12.4.3 The light cone   300
12.5   **Four-vectors**   301
Suggestions for further reading   305
Basic equations   305
Summary   307
Questions   307
Problems   308

13.   The relativistic wave equations and general relativity

13.1   **Momentum and energy**   311
13.2   **Conservation of energy and momentum**   314
13.3   ***The Dirac theory of the electron**   318
   13.3.1 Review of the Schrödinger theory   318
   13.3.2 The Klein-Gordon equation   320
   13.3.3 The Dirac equation   321
   13.3.4 Plane wave solutions of the Dirac equation   324
13.4   ***Field quantization**   327
13.5   **The general theory of relativity**   329
   13.5.1 The principle of equivalence   329
   13.5.2 The path of a freely-falling body in curvilinear coordinates   329
   13.5.3 Relations between partial derivatives of $g_{\mu\nu}$ and $\Gamma^{\lambda}_{\mu\nu}$ 330
   13.5.4 A slow moving particle in a weak gravitational field   331
   13.5.5 Vectors and tensors   332
   13.5.6 Transformation of the affine connection   333
   13.5.7 Covariant differentiation   333

   13.5.8 The parallel transport of a vector along a curve   334
   13.5.9 The curvature tensor   335
   13.5.10 Einstein's field equations   335
Suggestions for further reading   336
Basic equations   336
Summary   338
Questions   338
Problems   339

14.   Particle physics

14.1   **Leptons and quarks**   341
14.2   **Conservation laws**   348
   14.2.1 Energy, momentum, and charge   348
   14.2.2 Lepton number   348
   14.2.3 Baryon number   350
   14.2.4 Strangeness   351
   14.2.5 Charm, beauty, and truth   353
14.3   **Spatial symmetries**   354
   14.3.1 Angular momentum of composite systems   354
   14.3.2 Parity   356
14.4   **Isospin and color**   357
   14.4.1 Isospin   358
   14.4.2 Color   363
14.5   **Feynman diagrams**   364
   14.5.1 Electromagnetic interactions   365
   14.5.2 Weak interactions   366
   14.5.3 Strong interactions   368
14.6   **The $R(3)$ and $SU(3)$ symmetry groups**   369
   14.6.1 The rotation group in three dimensions   369
   14.6.2 The $SU(3)$ symmetry group   372
   14.6.3 The representations of $SU(3)$ 375
14.7   ***Gauge invariance and the electroweak theory**   380
14.8   **Spontaneous symmetry breaking and the discovery of the Higgs**   382
14.9   **Supersymmetry**   385
   14.9.1 Symmetries in physics   385
   14.9.2 The Poincaré algebra   386
   14.9.3 The supersymmetry algebra   387
Suggestions for further reading   388
Basic equations   388
Summary   389
Questions   389
Problems   390

15.   Nuclear physics

15.1   **Properties of nuclei**   393
   15.1.1 Nuclear sizes   394
   15.1.2 Binding energies   397
   15.1.3 The semiempirical mass formula   398
15.2   **Decay processes**   401
   15.2.1 Alpha decay   402
   15.2.2 The β-stability valley   403

15.2.3 Gamma decay   406
15.2.4 Natural radioactivity   407
15.3 The nuclear shell model   408
15.3.1 Nuclear potential wells   408
15.3.2 Nucleon states   409
15.3.3 Magic numbers   411
15.3.4 The spin-orbit interaction   411
15.4 Excited states of nuclei   412
Suggestions for further reading   416
Basic equations   416
Summary   417
Questions   417
Problems   417

Index   421

Online appendices

A. Constants and conversion factors

B. Atomic masses

C. Introduction to MATLAB®

D. Solution of the oscillator equation

E. The average value of the momentum

F. The Hartree-Fock applet

G. Integrals that arise in statistical physics

Preface

This book presents the ideas that have shaped modern physics and provides an introduction to current research in the different fields of physics. Intended as the text for a first course in modern physics following an introductory course in physics with calculus, the book begins with a brief and focused account of experiments that led to the formulation of the new quantum theory, while ensuing chapters go more deeply into the underlying physics.

This book helps prepare engineering students for the upper division courses on devices they will later take and provides engineering students and physics majors an overview of contemporary physics as it is presently understood. The course on modern physics is often the last course in physics most engineering students will ever take and the course occurs before the specialty courses physics majors later take on the various fields of contemporary physics. For this reason, the book covers a few topics that are ordinarily taught at a more advanced level. These subjects are included because they are relevant and interesting. They would ordinarily be unavailable to engineering students and they give physics majors a good sense of the fields of contemporary physics in which they may later specialize. Topics such as Bloch's theorem, matrix methods for calculating the scattering of electrons by complex barriers, and the relativistic Dirac equation broaden the background of both engineering and physics majors. The books I have used to prepare later chapters of this book are just the books used in upper-division courses in the various fields of contemporary physics.

This new third edition

This new edition has provided me the opportunity of presenting the elements of modern quantum theory in the way one would approach computational problems today. Erwin Schrödinger worked in a time without modern computers or the techniques of numerical linear algebra that have been developed in recent years. He began his treatment of the hydrogen atom as we would today by separating the equation that has become associated with his name into an angular and a radial equation but he then solved each of these equations by expanding the solution in an infinite series. The solution of the angular equation for hydrogen leads to Legendre polynomials that are used in quantum chemistry even to this day, while the solution of the radial equation lead to Laguerre polynomials known only by a comparatively fewer number of senior physicists. The angular part of the wave function for hydrogen are products of Legendre polynomials and exponential functions that are eigenfunctions of the z-component of the angular momentum. These functions called *spherical harmonics*, are denoted by $Y_{lm_l}(\theta, \phi)$. The spherical harmonics for $l \leq 2$ are given in Chapter 4 which is devoted to the hydrogen atom. The radial part of the hydrogen wave functions are obtained in Chapter 4 by solving the radial equation numerically.

In this new edition we have converted the differential equations that arise into sets of linear equation or matrix equations by making a finite difference approximation of the derivatives or by using the spline collocation method, and we have solved the resulting sets of equations. MATLAB® programs are described for solving the eigenvalue equations for a particle in a finite well and the simple harmonic oscillator and for solving the radial equation for hydrogen. The lowest-lying solutions of these problems are plotted using MATLAB and the physical significance of these solutions are discussed.

The various fields of contemporary physics continue to develop, and I have attempted to conclude each of the later chapters with a description of modern developments. The chapters on many-electron atomic atoms concludes with a description of accurate many-body calculations and the chapter on diatomic molecules concludes with a new section on recent efforts to find life on exo-planets. Chapter 10 on charge carriers in semiconductors includes a new section on solar cells that are essential for our finding alternatives to the burning of fossil fuels. The current energy consumption of the United States could be supplied by solar power if farmers in Kansas converted just 4% of existing farmland into solar installations. Chapter 14 on particle physics now concludes with a section on supersymmetry and Chapter 13 on relativity now concludes with a section giving the essential elements of general relativity.

Acknowledgments

For the new Third Edition, I would like to acknowledge the help of a number of leading scientists and engineers. I would like to thank Shamus McNamara in the Electrical Engineering Department of University of Louisville for providing me with a first draft of the new section on solar cells and working with me to adopt the section for this book, I would also like to thank Matteo Bertolini of the International School of Advanced Studies in Trieste, Italy who made his lectures on supersymmetry available and worked with me to improve the section I wrote on supersymmetry, Rodrigo Alvarado of the University of Costa Rico who gave me a copy of his lectures on general relativity and corrected my section on general relativity, and my graduate student Allan Lasky-Headrick who assisted me on the section on general relativity. Thanks are also due to John Black of Chalmers University of Technology in Göteborg, Sweden, Jacek Kobus at Nickolas Copernicus University in Torun, Poland, and Carlos Bunge of the Autonomous University of Mexico for helping me understand molecular spectra detected in space. Thanks are also due to Bethsaida Zamora, Hector Saballos, William Porras, Edwin Santiago, and the other engaged physics majors at University of Costa Rica where my teaching of modern physics took on a new computational character.

For the second edition, I would like to thank Keith Ellis of the Theory Group at Fermilab for discussing recent developments in particle physics with me and correcting the two sections I wrote on local gauge invariance and the discovery of the Higgs boson. Thanks are also due to John Wilkins and William Plalmer at Ohio State University, Charlotte Fischer who provided me the first version of the Hartree-Fock applet, Howard Georgi, who allowed me to attend his class on group theory and particle physics at Harvard University, Massimilliano Galeazzi at University of Miami, Mike Santos and Michael Morrison at University of Oklahoma, the Nobel prize winner, Eric Cornel at University of Colorado, and Dirk Walecka at College of William and Mary.

I would also like to give special thanks to Leslie Friesen who drew all of the figures and Ken Hicks who has given me many valuable suggestions and wrote the solution manuals.

Supplements to the text

For purchasers of the book, student resources can be found on this book's companion website. To access these files, please visit https://www.elsevier.com/books-and-journals/book-companion/9780128177907.

For instructors using this book for a course, please visit https://educate.elsevier.com/ to register for access to the ancillary materials.

Introduction

Every physical system can be characterized by its size and the length of time it takes for processes occurring within it to evolve. This is as true of the distribution of electrons circulating about the nucleus of an atom as it is of a chain of mountains rising up over the ages.

Modern physics is a rich field including decisive experiments conducted in the early part of the twentieth century and more recent research that has given us a deeper understanding of fundamental processes in nature. In conjunction with our growing understanding of the physical world, a burgeoning technology has led to the development of lasers, solid state devices, and many other innovations. The introduction and the first three chapters of this book provide an introduction to modern quantum physics and the remaining twelve chapters describe the various fields of contemporary physics as they are presently understood. In recognition of the evolving nature of modern physics, our description of the various fields of modern physics concludes with a section on further developments in the field. A number of distinguished physicists and engineers have participated in the writing of these sections.

I.1 The concepts of particles and waves

While some of the ideas currently used to describe microscopic systems differ considerably from the ideas of classical physics, other important ideas are classical in origin. We begin this chapter by discussing the important concepts of a particle and a wave which have the same meaning in classical and modern physics. A *particle* is an object with a definite mass concentrated at a single location in space, while a *wave* is a disturbance that propagates through space. The first section of this chapter, which discusses the elementary properties of particles and waves, provides a review of some of the fundamental ideas of classical physics and indicates the changes that occur in modern quantum physics. The second section of this chapter describes some of the central ideas of quantum physics and also discusses the size and time scales of the physical systems considered in this book.

I.1.1 The variables of a moving particle

The position and velocity vectors of a particle are illustrated in Fig. I.1. The position vector **r** extends from the origin to the particle, while the velocity vector **v** points in the direction of the particle's motion. Other variables, which are appropriate for describing a moving particle, can be defined in terms of these elementary variables.

The *momentum* **p** of the particle is equal to the product of the mass and velocity **v** of the particle

$$\mathbf{p} = m\mathbf{v}.$$

We shall find that the momentum is useful for describing the motion of electrons in an extended system such as a crystal.

The motion of a particle moving about a center of force can be described using the *angular momentum*, which is defined to be the cross product of the position and momentum vectors

$$\boldsymbol{\ell} = \mathbf{r} \times \mathbf{p}.$$

The cross product of two vectors is a vector having a magnitude equal to the product of the magnitudes of the two vectors times the sine of the angle between them. Denoting the angle between the momentum and position vectors by θ as in Fig. I.1, the magnitude of the angular momentum vector momentum can be written

$$|\boldsymbol{\ell}| = |\mathbf{r}|\,|\mathbf{p}|\sin\theta.$$

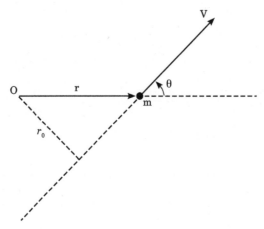

FIGURE I.1 The position **r** and the velocity **v** of a moving particle of mass m. The point O denotes the origin, and r_0 denotes the distance between the line of motion and the origin.

This expression for the angular momentum may be written more simply in terms of the distance between the line of motion of the particle and the origin, which is denoted by r_0 in Fig. I.1. We have

$$|\ell| = r_0 |\mathbf{p}|.$$

The angular momentum is thus equal to the distance between the line of motion of the particle and the origin times the momentum of the particle. The direction of the angular momentum vector is generally taken to be normal to the plane of the particle's motion. For a classical particle moving under the influence of a central force, the angular momentum is conserved. The angular momentum will be used in later chapters to describe the motion of electrons about the nucleus of an atom.

The *kinetic energy* of a particle with mass m and velocity **v** is defined by the equation

$$KE = \frac{1}{2} m v^2,$$

where v is the magnitude of the velocity or the speed of the particle. The concept of *potential energy* is useful for describing the motion of particles under the influence of conservative forces. In order to define the *potential energy* of a particle, we choose a point of reference denoted by R. The potential energy of a particle at a point P is defined as the negative of the work carried out on the particle by the force field as the particle moves from R to P. For a one-dimensional problem described by a variable x, the definition of the potential energy can be written

$$V_P = -\int_R^P F(x)dx. \tag{I.1}$$

As a first example of how the potential energy is defined we consider the harmonic oscillator, which consists of a body of mass m moving under the influence of a linear restoring force

$$F = -kx, \tag{I.2}$$

where x denotes the distance of the body from its equilibrium position. The constant k, which occurs in Eq. (I.2), is called the *force constant*. The restoring force is proportional to the displacement of the body and points in the direction opposite to the displacement. If the body is displaced to the right, for instance, the restoring force points to the left. It is natural to take the reference position R in the definition of the potential energy of the oscillator to be the equilibrium position for which $x = 0$. The definition of the potential energy (I.1) then becomes

$$V(x) = -\int_0^x (-kx')dx' = \frac{1}{2} kx^2. \tag{I.3}$$

Here x' is used within the integration in place of x to distinguish the variable of integration from the limit of integration.

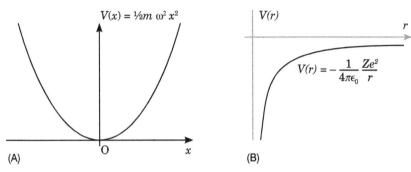

FIGURE I.2 (A) The potential energy $V(x)$ for the simple harmonic operator. (B) The potential energy $V(r)$ of an electron in a hydrogen-like ion with nuclear charge Ze.

If one were to pull the mass from its equilibrium position and release it, the mass would oscillate with a frequency independent of the initial displacement. The angular frequency of the oscillator is related to the force constant of the oscillator and the mass of the particle by the equation

$$\omega = \sqrt{k/m},$$

or

$$k = m\omega^2.$$

Substituting this last expression for k into Eq. (I.3), we obtain the following expression for the potential energy of the oscillator

$$V(x) = \frac{1}{2} m\omega^2 x^2. \tag{I.4}$$

The oscillator potential is illustrated in Fig. I.2(A). The harmonic oscillator provides a useful model for a number of important problems in physics. It may be used, for instance, to describe the vibration of the atoms in a crystal about their equilibrium positions and the radiation field within a cavity.

As a further example of potential energy, we consider the potential energy of a particle with electric charge q moving under the influence of a charge Q. According to Coulomb's law the electromagnetic force between the two charges is equal to

$$F = \frac{1}{4\pi\epsilon_0} \frac{Qq}{r^2},$$

where r is the distance between the two charges and ϵ_0 is the permittivity of free space. The reference point for the potential energy for this problem can be conveniently chosen to be at infinity where $r = \infty$ and the force is equal to zero. Using Eq. (I.1), the potential energy of the particle with charge q at a distance r from the charge Q can be written

$$V(r) = -\frac{Qq}{4\pi\epsilon_0} \int_\infty^r \frac{1}{r'^2} dr'.$$

Evaluating the above integral, one finds that the potential energy of the particle is

$$V(r) = \frac{Qq}{4\pi\epsilon_0} \frac{1}{r}.$$

An application of this last formula will arise when we consider the motion of electrons in an atom. For an electron with charge $-e$ moving in the field of an atomic nucleus having Z protons and hence a nuclear charge of Ze, the formula for the potential energy becomes

$$V(r) = -\frac{Ze^2}{4\pi\epsilon_0} \frac{1}{r}. \tag{I.5}$$

The potential energy of an electron in a hydrogen atom given by Eq. (I.5) is illustrated in Fig. I.2(B).

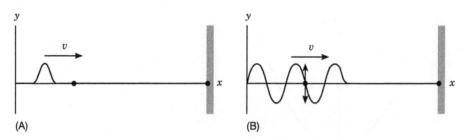

FIGURE I.3 (A) A pulse moving with velocity **v** along a stretched string. (B) An extended sinusoidal wave moving along a string.

In upcoming chapters of this book, we will consider the motion of particles in fields that confines their motion. We will call such a potential field that confines the motion of particles a potential well. Fig. I.2(A) that describes the potential energy of a particle moving under the influence of a linear restoring force and Fig. I.2(B) that describes potential energy of the electron in a hydrogen atom, are examples of such potential wells. A distinctive feature of the new quantum theory is that the bound states of particles moving in a potential well can only have distinct values of the energy. This is unlike anything we have seen in classical physics. The energy of a planet moving about the Sun can be increased or reduced by a small amount, and the energy of the planet need not be one of a number of possible values but depends on the process by which the solar system was formed. In the new quantum theory, a particle in moving in a potential well with one value of the energy can make a transition to another level with less energy and emit a photon with an energy equal to the difference in energy between the two levels.

I.1.2 Elementary properties of waves

We consider now some of the elementary properties of waves. Various kinds of waves arise in classical physics, and we shall encounter other examples of wave motion when we apply the new quantum theory to microscopic systems.

Traveling waves

If one end of a stretched string is moved abruptly up and down, a pulse will move along the string as shown in Fig. I.3(A). A typical element of the string will move up and then down as the pulse passes. If instead the end of the string moves up and down with the time dependence,

$$y = \sin \omega t,$$

an extended sinusoidal wave will travel along the string as shown in Fig. I.3(B). A wave of this kind which moves up and down with the dependence of a sine or cosine is called a *harmonic wave*.

The wavelength of a harmonic wave will be denoted by λ and the speed of the wave by v. The wavelength is the distance from one wave crest to the next. As the wave moves, a particular element of the string which is at the top of a crest will move down as the trough approaches and then move back up again with the next crest. Each element of the string oscillates up and down with a period, T. The frequency of oscillation f is equal to $1/T$. The period can also be thought of as the time for a crest to move a distance of one wavelength. Thus, the wavelength, wave speed, and period are related in the following way

$$\lambda = vT .$$

Using the relation, $T = 1/f$, this equation can be written

$$\lambda f = v. \tag{I.6}$$

The dependence of a harmonic wave upon the space and time coordinates can be represented mathematically using the trigonometric sine or cosine functions. We consider first a harmonic wave moving along the x-axis for which the displacement is

$$y(x,t) = A \sin[2\pi(x/\lambda - t/T)], \tag{I.7}$$

where A is the amplitude of the oscillation. One can see immediately that as the variable x in the sine function increases by an amount λ or the time increases by an amount T, the argument of the sine will change by an amount 2π, and the function

$y(x, t)$ will go through a full oscillation. It is convenient to describe the wave by the *angular wave number*,

$$k = \frac{2\pi}{\lambda},\tag{I.8}$$

and the *angular frequency*,

$$\omega = \frac{2\pi}{T}.\tag{I.9}$$

Using the relation, $T = 1/f$, the second of these two equations can also be written

$$\omega = 2\pi f.\tag{I.10}$$

The angular wave number k, which is defined by Eq. (I.8), has SI units of radians per meter, while ω, which is defined by Eq. (I.9), has SI units of radians per second. Using Eqs. (I.8) and (I.9), the wave function (I.7) can be written simply

$$y(x, t) = A \sin(kx - \omega t).\tag{I.11}$$

Eq. (I.11) describes a traveling wave. We can see this by considering the crest of the wave where the value of the phase of the sine function in Eq. (I.11) is equal to $\pi/2$. The location of the crest is given by the equation

$$kx_{\text{crest}} - \omega t = \frac{\pi}{2}.$$

Solving this last equation for x_{crest}, we get

$$x_{\text{crest}} = \frac{\omega t}{k} + \frac{\pi}{2k}.$$

An expression for the velocity of the wave crest can be obtained by taking the derivative of x_{crest} with respect to time to obtain

$$v = \frac{dx_{\text{crest}}}{dt} = \frac{\omega}{k}.\tag{I.12}$$

Eq. (I.11) thus describes a sinusoidal wave moving in the positive x-direction with a velocity of ω/k. Eq. (I.12) relating the velocity of the wave to the angular wave number k and angular frequency ω can also be obtained by solving Eq. (I.8) for λ and solving Eq. (I.10) for f. Eq. (I.12) is then obtained by substituting these expressions for λ and f into Eq. (I.6).

Using the same approach as that used to understand the significance of Eq. (I.11), one can show that

$$y(x, t) = A \sin(kx + \omega t)\tag{I.13}$$

describes a sinusoidal wave moving in the negative x-direction with a velocity of ω/k.

Fig. I.4(A) illustrates how the harmonic function (I.11) varies with position at a fixed time chosen to be $t = 0$. Setting t equal to zero, Eq. (I.11) becomes

$$y(x, 0) = A \sin kx.\tag{I.14}$$

The wave described by the function $A \sin x$ and illustrated in Fig. I.4(A) does not depend upon the time. Such a wave, which is described by its dependence upon a spatial coordinate, is called a *stationary wave*. As for the traveling wave (I.11), the angular wave number k is related to the wavelength by Eq. (I.8). Similarly, Fig. I.4(B) shows how the function (I.11) varies with time at a fixed position chosen to be $x = 0$. Setting x equal to zero in Eq. (I.11) and using the fact that the sine is an odd function, we obtain

$$y(0, t) = -A \sin \omega t.\tag{I.15}$$

The wave function (I.15) oscillates as the time increases with an angular frequency ω given by Eq. (I.10).

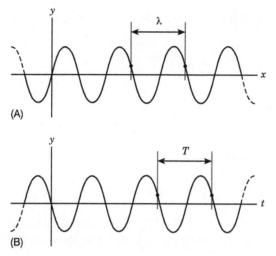

FIGURE I.4 (A) The x-dependence of the sinusoidal function for a fixed time $t = 0$. (B) The time dependence of the sinusoidal function at the fixed point $x = 0$.

Standing waves

Suppose two waves travel simultaneously along the same stretched string. Let $y_1(x, t)$ and $y_2(x, t)$ be the displacements of the string due to the two waves individually. The total displacement of the string is then

$$y(x, t) = y_1(x, t) + y_2(x, t).$$

This is called the *principle of superposition*. The displacement due to two waves is generally the algebraic sum of the displacements due to the two waves separately. Waves that obey the superposition principle are called *linear* waves and waves that do not are called *nonlinear* waves. It is found experimentally that *most* of the waves encountered in nature obey the superposition principle. Shock waves produced by an explosion or a jet moving at supersonic speeds are uncommon examples of waves that do not obey the superposition principle. In this text, only linear waves will be considered. Two harmonic waves reinforce each other or cancel depending upon whether or not they are in phase (in step) with each other. This phenomenon of reinforcement or cancellation is called *interference*.

We consider now two harmonic waves with the same wavelength and frequency moving in opposite directions. A wave moving in the opposite direction would be produced when a wave is reflected by the wall of a potential well. Two waves having equal amplitudes are described by the wave functions

$$y_1(x) = A \sin(kx - \omega t)$$

and

$$y_2(x) = A \sin(kx + \omega t).$$

According to the principle of superposition, the combined wave is described by the wave function

$$y(x, t) = y_1(x) + y_2(x) = A\left[\sin(kx - \omega t) + \sin(kx + \omega t)\right]. \tag{I.16}$$

Using the trigonometric identity,

$$\sin(A \pm B) = \sin A \cos B \pm \cos A \sin B, \tag{I.17}$$

Eq. (I.16) may be written

$$u(x, t) = [2A \cos \omega t] \sin kx. \tag{I.18}$$

This function describes a *standing wave*.

At a particular time the quantity within square brackets in Eq. (I.18) has a constant value and may be thought of as the amplitude of the wave. The amplitude function $2A \cos \omega t$ varies with time having both positive and negative values. The

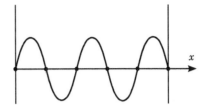

FIGURE I.5 Function describing the spatial form of a standing wave. The nodes, which have zero displacement, are represented by dots.

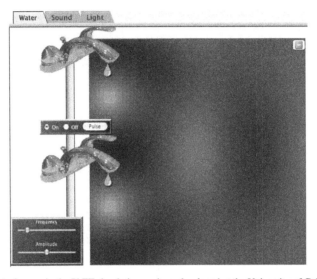

FIGURE I.6 A simulation of wave interference in the PhET simulation package developed at the University of Colorado.

function $\sin kx$ has the spatial form illustrated in Fig. I.5 being zero at the points satisfying the equation

$$kx = n\pi, \quad \text{for } n = 0, 1, 2, \ldots .$$

Substituting $k = 2\pi/\lambda$ into this equation, we get

$$x = n\frac{\lambda}{2}, \quad \text{for } n = 0, 1, 2, \ldots .$$

The function $\sin kx$ is thus equal to zero at points separated by half a wavelength. At these points, which are called *nodes*, the lateral displacement, is *always* equal to zero. An example of a standing wave is provided by the vibrating strings of a guitar. The ends of the guitar strings are fixed and cannot move. In addition to the ends of the strings, other points along the strings separated by half a wavelength have zero displacements.

A novel feature of the quantum theory we will soon consider is that waves are associated with particles. The first examples of the quantum theory we will consider in Chapter 2 of this book will involve particles confined to move in a potential well. We can imagine that a particle confined to move in a potential well would spend its days moving back and forth in the well. So, we should not be surprise that the waves associated with these particles should be standing waves.

One can gain an intuitive understanding of the properties of waves by using the PhET simulation package developed at the University of Colorado. The simulations can be found at the web site: phet.colorado.edu/en/simulations. Choosing the categories "physics" and "sound and waves", one can initiate the simulation called "wave interference". Choosing the tab "water" and the option "one drip", one sees the waves spreading across a body of water when drips from a single faucet strike the water surface. Choosing then the option "two drips", one sees the waves produced by the drips of two faucets striking the water surface. This figure is shown in Fig. I.6. As we have just described the waves from the two disturbances add together and destructively interfere to produce a complex disturbance on the surface of the water. One can observe similar effects with sound and light waves by choosing the tabs "sound" and "light".

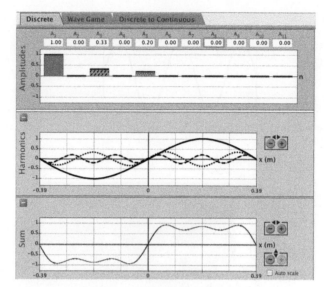

FIGURE I.7 A representation of a square wave function formed by adding harmonic waves together using the PhET simulation package developed at the University of Colorado.

The Fourier theorem

We have thus far considered sinusoidal waves on a string and would now like to consider wave phenomenon when the shape of the initial disturbance is *not* sinusoidal. In the decade of the 1920's, Jean Baptiste Fourier showed that *any* periodic and reasonably continuous function $f(x)$, which is defined in the interval $0 \leq x < L$, can be represented by a series of sinusoidal waves

$$f(x) = \sum_{n=1,2,...} S_n \sin nkx, \quad \text{for} \quad 0 \leq x \leq L, \tag{I.19}$$

where $k = \pi/L$ and

$$S_n = \frac{2}{L} \int_0^L \sin nkx \, f(x) dx. \tag{I.20}$$

A sketch of the derivation of Eq. (I.20) is given in Problem 4.

As an example, we consider a square wave

$$f(x) = \begin{cases} +A & \text{if } x \leq x < \frac{1}{2}L \\ -A & \text{if } \frac{1}{2}L < x \leq L \end{cases} \tag{I.21}$$

Using Eqs. (I.19) and (I.20), the square wave (I.21) can be shown to be equal to the following infinite sum of sinusoidal waves

$$f(x) = A\frac{4}{\pi}\left(\sin kx + \frac{1}{3}\sin 3kx + \frac{1}{5}\sin 5kx + \dots\right), \tag{I.22}$$

where $k = 2\pi/L$ is the angular wave number of the fundamental mode of vibration.

We can gain some insight into how harmonic waves combine to form the square wave function (I.21) by using the simulation package "Fourier: Making Waves" at the web site http://phet.colorado.edu/en/simulation/fourier. A reproduction of the window that comes up is shown in Fig. I.7. With "Preset Function" set to "sine/cosine" and "Graph controls" set at "Function of: space (x)" and "sin", one can begin by setting $A_1 = 1$ and $A_3 = 0.33$ and then gradually adding $A_5 = 0.20$, $A_7 = 0.14$, $A_9 = 0.11$, and $A_{11} = 0.09$. As one adds more and more sine functions of higher frequency, the sum of the waves shown in the lower screen becomes more and more like a square wave. One can understand in qualitative terms how the harmonic waves add up to produce the square wave. Using the window reproduced by Fig. I.7, one can view each sinusoidal wave by setting the amplitude of the wave equal to one and all other amplitudes equal to zero. The amplitude A_1 corresponds to the fundamental wave for which a single wavelength stretches over the whole region. This sinusoidal wave

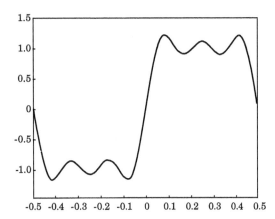

FIGURE I.8 Plot produced by MATLAB Program I.1.

– like the square wave – is zero at the center of the region and assumes negative values to the left of center and positive values to the right of center. The sinusoidal waves with amplitudes A_3, A_5, A_7, A_9, and A_{11} all have these same properties but being sinusoidal waves of higher frequencies they rise more rapidly from zero as one moves to the right from the center of the region. By adding waves with higher frequencies to the fundamental wave, one produces a wave which rises more rapidly as one moves to the right from center and declines more rapidly as one moves to the left from center; however, the sum of the waves oscillate with a higher frequency than the fundamental frequency in the region to the right and left of center. As one adds more and more waves, the oscillations due to the various waves of high frequency destructively interfere and one obtains the square wave.

The above result can also be obtained using the MATLAB® software package. A short introduction to MATLAB can be found in Appendix C and a more extensive presentation in Appendix CC. MATLAB Program I.1 given below adds sinusoidal waves up to the fifth harmonic. The first three lines of the program define the values of A, L, and k, and the next line defines a vector x with elements between -L/2, and +L/2 with equal steps of L/100. The plot of x versus y produced by this MATLAB program is shown in Fig. I.8. This figure is very similar to Fig. I.7 produced by the PhET simulation package.

MATLAB Program I.1
This program adds the Fourier components up to the 5th harmonic to produce a square wave of amplitude 1.0 and width 1.0.

```
A=1;
L=1;
k=2*pi/L;
x = -L/2 : L/100 : L/2;
y = (A*4/pi)*( sin(k*x)+(1/3)*sin(3*k*x)+(1/5)*sin(5*k*x) );
plot(x,y)
```

The Fourier theorem has wide-ranging consequences. No matter what the shape of a disturbance, one can think of the disturbance as being a sum of harmonic waves.

Representation of waves using exponentials

It is often convenient to represent waves using exponential functions. For instance, a stationary wave can be described by the function

$$\psi(x) = Ae^{ikx}, \tag{I.23}$$

where the constant A is a real number. The function $\psi(x)$ can be resolved into its real and imaginary parts using Euler's equation,

$$e^{i\theta} = \cos\theta + i\,\sin\theta, \tag{I.24}$$

to obtain

$$\psi(x) = A\cos(kx) + iA\sin(kx).$$

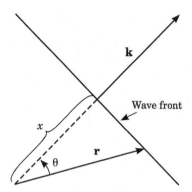

FIGURE I.9 The wave vector **k** and the position vector **r** for a wave traveling in three-dimensions is shown together with a particular wave front. The angle θ appearing in Eq. (I.20) is also shown.

Notice that the imaginary part of the function $\psi(x)$ is equal to $A\sin(kx)$. This function corresponds to the stationary wave shown in Fig. I.4(A). Similarly, the real part of the function $\psi(x)$ is equal to the function $A\cos(kx)$ which can be obtained by shifting the function shown in Fig. I.4(A) to the left by an amount $\pi/2$.

A traveling wave can be described by the exponential function

$$\psi(x,t) = Ae^{i(kx - \omega t)}. \tag{I.25}$$

Using Euler's Eq. (I.24), one may readily show that the imaginary part of the right-hand side of this last equation is equal to the sinusoidal function appearing in Eq. (I.11).

The exponential function has mathematical properties which makes it more convenient to use than the trigonometric functions. For instance, the product of an exponential function e^A and a second exponential function e^B can be evaluated by simply adding up the exponents

$$e^A \cdot e^B = e^{A+B}.$$

We now consider stationary waves for which the direction in which the value of the function changes most rapidly does not coincide with the x-direction, and we consider traveling waves moving in other directions than the positive and negative x-directions. Imagine that in a particular region of space we identify a point where the wave function has a local maximum and we identify other points near our original point that are also local maxima. A surface passing through these points is called a *wave front*. We denote by **k** a vector pointing in a direction perpendicular to the wave fronts with a magnitude

$$|\mathbf{k}| = \frac{2\pi}{\lambda}.$$

The magnitude of **k** will be denoted by k. The vector **k**, which is called the *wave vector*, is shown together with a position vector **r** and a particular wave front in Fig. I.9. The scalar product $\mathbf{k} \cdot \mathbf{r}$ can be written

$$\mathbf{k} \cdot \mathbf{r} = |\mathbf{k}|\,|\mathbf{r}|\cos\theta. \tag{I.26}$$

Notice that the quantity $|\mathbf{r}|\cos\theta$ shown in Fig. I.9 is the projection of the position vector **r** upon the direction of the vector **k**. All of the points on a wave front correspond to the same value of $|\mathbf{r}|\cos\theta$. The quantity $\mathbf{k} \cdot \mathbf{r}$ is the product of k and the distance to a wave front measured along the vector **k**. Hence, $\mathbf{k} \cdot \mathbf{r}$ plays the same role as kx does for waves in one dimension. The wave function for a stationary wave in three dimensions can be written

$$\psi(\mathbf{r}) = Ae^{i\mathbf{k} \cdot \mathbf{r}}.$$

Similarly, a traveling wave in three-dimensions can be described by the function

$$\psi(\mathbf{r}) = Ae^{i(\mathbf{k} \cdot \mathbf{r} - \omega t)}.$$

The wave vector **k** is perpendicular to the wave fronts pointing in the direction the wave propagates.

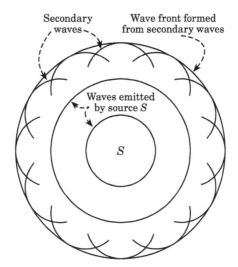

FIGURE I.10 Huygens' principle states that every point on a wave front may be considered as a source of secondary waves.

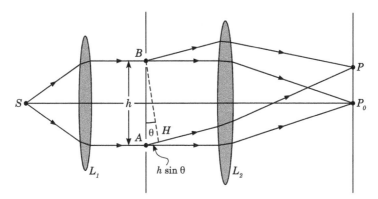

FIGURE I.11 Interference produced by two slits.

I.1.3 Interference and diffraction phenomena

The variation of amplitude and intensity that occur when waves encounter a physical barrier can be understood using Huygens' principle, which states that each point on a wave front may be considered as a source of secondary waves. The position of the wave front at a later time can be found by superimposing these secondary waves. Waves emitted by the wave front thus serve to regenerate the wave and enable us to analyze its propagation in space. This is illustrated in Fig. I.10.

The word *interference* is used to describe the superposition of two waves, while *diffraction* is interference produced by several waves. For both interference and diffraction phenomena, Huygens' principle enables us to reconstruct subsequent wave fronts and to calculate the resulting intensities.

A good example of interference effects is provided by the two slit interference experiment shown schematically in Fig. I.11. In the experiment, the light source S, which lies in the focal plane of the lens L_1, produces a beam of parallel rays falling perpendicularly upon the plane containing the double slit. The interference of secondary waves emitted by the two apertures leads to a variation of the intensity of the transmitted light in the secondary focal plane of the lens L_2. Whether or not constructive interference occurs at the point P depends upon whether the number of waves along the upper path (BP) shown in Fig. I.11 differs from the number of waves along the lower segment (AP) by an integral number of wavelengths. The difference in the length of the two paths is equal to the length of the segment AH. If we denote the distance between the two slits by h, then the length of AH is equal to $h \sin \theta$, and the condition for constructive interference is

$$h \sin \theta = n\lambda. \tag{I.27}$$

Constructive interference occurs when the difference in path lengths is equal to an integral number of wave lengths. The intensity distribution of the light incident upon the screen at the right is illustrated in Fig. I.12(A). A photograph of the interference pattern produced by a double slit is shown in Fig. I.12(B).

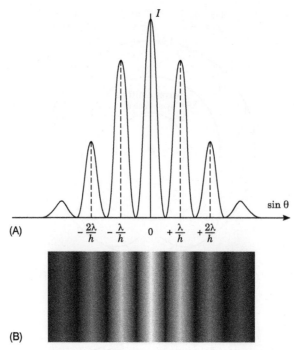

(A)

(B)

FIGURE I.12 (A) Intensity distribution produced on a screen by light passing through a double slit. (B) A photograph of an interference pattern produced by two slits.

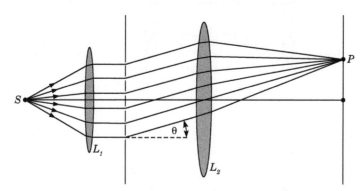

FIGURE I.13 Diffraction by a grating.

The two slit interference experiment which we have discussed clearly illustrates the ideas of constructive and destructive interference. The bright fringes produced in the experiment corresponds to angles at which light traveling through the two slits arrive at the focal plane of the second lens in phase with each other, while the dark fringes correspond to angles for which the distance traveled by light from the two slits differ by an odd number of half wavelengths and the light destructively interferes. In the new quantum theory, radiation and matter have both a particle and a wave nature. The two-slit interference experiment we have just described has been performed using a beam of electrons.

An optical *grating* can be made by forming a large number of parallel equidistant slits. A grating of this kind is illustrated in Fig. I.13. As in the case of a double slit, intensity maxima can be observed in the focal plane of the lens L_2. The brightest maxima occur at points corresponding to the values of θ satisfying Eq. (I.27) where h here represents the distance between the centers of neighboring slits. At such points, light from all of the different slits arrive with the same phase. Eq. (I.27) thus gives all of the angles for which constructive interference occurs for the double slit interference experiment and the angles for which the principal maxima occur for a grating. For a grating, however, a large number of secondary maxima occur separated by a corresponding number of secondary minima. The gratings used in modern spectroscopic experiments consist typically of aluminum or silver-coated glass plates which have thin lines ruled on them by a fine diamond needle. A grating having several hundred thousand lines produces a number of narrow bright lines on a dark background, each line

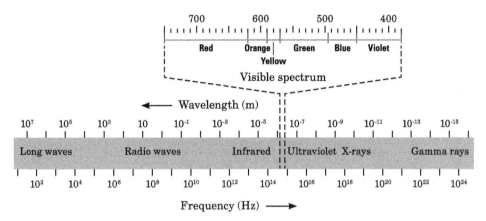

FIGURE I.14 The frequencies and wavelengths of electromagnetic radiation.

corresponding to a different value of n in Eq. (I.27). Using this equation and the measured angles of the maxima, one may readily calculate the wavelength and frequency of the incident light.

Electromagnetic waves

The wave model may be used to describe the propagation of electromagnetic radiation. The frequencies and wavelengths of the most important forms of electromagnetic radiation are shown in Fig. I.14. The human eye can perceive electromagnetic radiation (light) with wavelengths between 400 and 700 nanometers (that is between 400×10^{-9} m and 700×10^{-9} m). The wavelength of light is also commonly given in Angstroms. One Angstrom (Å) is equal to 1×10^{-10} m or one-tenth of a nanometer. When wavelengths are expressed in Angstroms, the wavelength of visible light is between 4000 Å and 7000 Å. Ultraviolet light, x-rays, and γ-rays have wavelengths which are shorter than the wavelength of visible light, while infrared light, microwaves, and radio waves have wavelengths which are longer. We shall denote the speed of light in a vacuum by c. Substituting c for v in Eqs. (I.6), we have

$$\lambda f = c. \tag{I.28}$$

Solving this equation for f, we get

$$f = \frac{c}{\lambda}. \tag{I.29}$$

According to Eq. (I.29), the frequency and the wavelength of light are inversely related to each other. If the wavelength increases, for instance, the frequency will decrease. The speed of light is given together with other physical constants in Appendix A. Using Eq. (I.29) and the value of c given in this appendix, one may easily obtain the values of the frequency given in Table I.1.

TABLE I.1 Wavelength (λ) and Frequency (f) of Light.

Color	λ (nm)	f (Hz)
red	700	4.28×10^{14}
violet	400	7.49×10^{14}

The unit of one cycle per second (s^{-1}) is referred to as a Hertz and abbreviated Hz. In the SI system of units, 10^3 is denoted by kilo (k), 10^6 is denoted by mega (M), 10^9 is denoted by giga (G) and 10^{12} is denoted by tera (T). Red light thus has a frequency of 428 THz and violet light has frequency of 749 THz.

Example I.1

Calculate the wavelength of electromagnetic radiation having a frequency of $f = 100\,\text{MHz} = 100 \times 10^6\,\text{s}^{-1}$.

Solution

Solving Eq. (I.28) for λ, we obtain

$$\lambda = \frac{c}{f}.$$

We then substitute the value of c given in Appendix A and $f = 100 \times 10^6\,\text{s}^{-1}$ into this last equation to obtain

$$\lambda = \frac{2.998 \times 10^8\,\text{m/s}}{100 \times 10^6\,\text{s}^{-1}} = 299.8\,\text{m}.$$

The radiation, which has a wavelength of about three hundred meters, corresponds to radio waves.

I.2   An overview of quantum physics

Microscopic systems differ in a number of ways from macroscopic systems for which the laws of classical physics apply. One of the most striking new features of physical systems on a microscopic level is that they display a wave-particle duality. Certain phenomena can be understood by considering radiation or matter as consisting of particles, while other phenomena demand that we think of radiation or matter as consisting of waves.

At the beginning of the twentieth century, electromagnetic radiation was thought of as a continuous quantity described by waves, while it was thought that matter could be resolved into constituent particles. The first evidence that electromagnetic radiation had a discrete quality appeared in 1900 when Max Planck succeeded in explaining the radiation field within a cavity. In his theory, Planck assumed that the electromagnetic field interchanged energy with the walls of the cavity in integral multiples of hf where h is a physical constant now called Planck's constant and f is the frequency of the radiation. While Planck was careful to confine his assumption to the way the radiation field exchanges energy with its environment, Albert Einstein broke entirely with the tenets of classical physics five years later when he proposed a theory of the photoelectric effect. The photoelectric effect refers to the emission of electrons by a metal surface when light is incident upon the surface. Einstein was able to explain the observed features of the photoelectric effect by supposing that the radiation field associated with the incident light consisted of quanta of energy. In keeping with the earlier work of Planck, Einstein supposed that these quanta have an energy

$$E = hf. \tag{I.30}$$

The theory of the radiation field developed by Planck will be described in Chapter 7, while the theory of the photoelectric effect of Einstein will be described in Chapter 1.

The theories of Planck and Einstein, which have since been confirmed by experiment, were the first indication that electromagnetic radiation has a dual wave-particle quality. While the interference and diffraction phenomena discussed in the previous section require that we think of light as consisting of electromagnetic waves, the phenomena associated with the absorption and emission of radiation demand that we think of light as consisting of quanta of energy, which we call *photons*. In 1923, Louis de Broglie suggested that just as light has both a wave and a particle character, the objects we think of as particles should also display a wave-particle dualism. This remarkable suggestion, which placed the theories of radiation and matter on the same footing, has since been confirmed by experiment. While a beam of electrons passing through a magnetic field is deflected in the way charged particles would be deflected, a beam of electrons, which is reflected by the planes of atoms within a crystal, displays the same interference patterns that we would associate with waves. The electron and the particles that constitute the atomic nucleus all have this dual wave-particle character. The theory of de Broglie and experiments that confirm his theory are described in Chapter 1.

The dual nature of waves and particles determines to a considerable extent the mathematical form of modern theories. The distinctive feature of modern theories is that they are formulated in terms of *probabilities*. The equations of modern quantum theory are not generally used to predict with certainty the outcome of an observation but rather the probability of obtaining a particular possible result. To give some idea of how the concept of probability arises from the wave-particle dualism, we consider again the interference experiment shown in Fig. I.11. In this experiment, light is incident upon the two slits shown in the figure and an interference pattern is formed on the screen to the right. As we have seen, the intensity pattern produced on the screen, which is shown in Fig. I.12(A), can be interpreted in terms of the interference of secondary waves

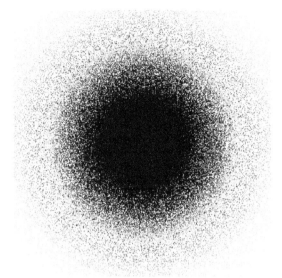

FIGURE I.15   Cloud surrounding the hydrogen nucleus. The density of the cloud is related to the probability of finding an electron.

emitted by the two slits. This intensity pattern is predicted unequivocally by classical optics. The concept of probability enters the picture when we consider the interference experiment from the particle point of view. The beam of light can be thought of not only as a superposition of waves but also as a stream of photons. If the screen were made of a light sensitive material and the intensity of the light were sufficiently low, the impact of each photon could be recorded. The cumulative effect of all of the photons passing through the slits and striking the screen would produce the effect illustrated in Fig. I.12(A). Each photon has an equal probability of passing through either slit. The density of the image produced at a particular point on the screen is proportional to the probability that a photon would strike the screen at that point. We are thus led to use the concept of probability not due to any shortcoming of classical optics, but due to the fact that the incident light can be described both as waves and as particles.

Ideas involving probability play an important role in our description of all microscopic systems. To show how this occurs, we shall conduct a thought experiment on a collection of hydrogen atoms. Hydrogen is the lightest and simplest atom with a single electron moving about a nucleus. Imagine we have a sensitive camera which can record the position of the electron of a hydrogen atom on a photographic plate. If we were to take a large number of pictures of the electrons in different hydrogen atoms superimposed on a single photographic plate, we would get a picture similar to that shown in Fig. I.15 in which the hydrogen nucleus is surrounded by a cloud. The density of the cloud at each point in space is related to the probability of finding an electron at that point. In modern quantum theory, the electron is described by a wave function ψ which is a solution of an equation called the Schrödinger equation. The probability of finding the electron at any point in space is proportional to the absolute value squared of the wave function $|\psi|^2$. The situation is entirely analogous to the two-slit interference experiment we have just considered. The photon in the interference pattern and the electrons surrounding the nucleus of an atom are both particles corresponding to waves which are accurately described by the theory. The concept of probability is required to describe the position of a particle that corresponds to a wave and is thus due to the wave-particle nature of photons and electrons.

The electron in the hydrogen atom moves about a nucleus which has a single proton with a positive charge. As we shall see in Chapter 4, the radius of the cloud surrounding the hydrogen nucleus is equal to $a_0 = 0.529$ Å or 0.529×10^{-10} m, and the diameter of the cloud is thus approximately one Angstrom or one tenth of a nanometer. Atoms of helium and lithium, which follow hydrogen in the Periodic Table, have two and three electrons respectively and their nuclei have corresponding numbers of protons. Since electrons have a negative charge, the cloud surrounding the nucleus of an atom can be interpreted as a charge cloud. As one moves from one atom to the next along a row of the Periodic Table, the nuclear charge increases by one and an additional electron is added to the charge cloud. The positive electric charge of atomic nuclei attracts the negatively charged electrons and draws the electron cloud in toward the nucleus. For this reason, the size of atoms increases only very slowly as the number of electrons increases. Xenon which has 56 electron is only two to three times larger than the helium atom which has two electrons.

Atoms and nuclei emit and absorb radiation in making transitions from one state to another. The basic principles of radiative transitions were given by Niels Bohr in 1913. Bohr proposed that an atom has stationary states in which it has well-defined values of the energy and that it emits or absorbs a photon of light when it makes a transition from one state to another. In an emission process, the atom makes a transition to a state in which it has less energy and emits a single photon, while in an absorption process, the atom absorbs a photon and makes a transition to a state in which it has more energy. These ideas have been found to apply generally to molecules and nuclei as well.

The energies of atoms depend upon the nature of the electron charge cloud. These energies can conveniently be expressed in electron volts (eV). One electron volt is the kinetic energy an electron would have after being accelerated through a potential difference of one volt. The energy of a photon that has been emitted or absorbed is given by Eq. (I.30), which is due to Planck and Einstein. An expression for the energy of a photon in terms of the wavelength of light can be obtained by substituting the expression for the frequency provided by Eq. (I.29) into Eq. (I.30) giving

$$E = \frac{hc}{\lambda}. \tag{I.31}$$

Using the values of Planck's constant and the speed of light given in Appendix A, one may easily show that the product of the constants appearing in the above equation is

$$hc = 1240 \text{ eV} \cdot \text{nm}. \tag{I.32}$$

This is a good number to remember for later reference.

Example I.2

Calculate the energy of the photons for light having a wavelength $\lambda = 400$ nm.

Solution
Substituting the wavelength λ of the light and the product hc given by Eq. (I.32) into Eq. (I.31) gives

$$E = \frac{1240 \text{ eV} \cdot \text{nm}}{400 \text{ nm}} = 3.1 \text{ eV}.$$

The light may thus be thought of as consisting of photons having an energy of 3.1 eV.

Eq. (I.32) for the product hc is convenient for visible light and for atomic transitions. The transitions made by nuclei typically involve millions of electron volts (MeV) of energy. For problems involving nuclear radiation, it is convenient to write the product of the constants h and c as follows

$$hc = 1240 \text{ MeV} \cdot \text{fm}, \tag{I.33}$$

which expresses hc in terms of MeV and Fermi (fm). One MeV is equal to 10^6 eV, and one Fermi (fm), which is the approximate size of an atomic nucleus, is equal to 10^{-15} m or $10^{-6} \times 10^{-9}$ m. The unit MeV appearing in Eq. (I.33) is thus one million times larger than the unit eV appearing in Eq. (I.32), while the unit fm appearing in Eq. (I.33) is one million times smaller than the unit nm appearing in Eq. (I.32).

The following table, which gives a few typical photon energies, is arranged according to decreasing wavelength or increasing photon energy.

TABLE I.2 Typical Photon Energies.

Wavelength (λ)	Photon Energy (E)	Type of Radiation
700 nm	1.77 eV	Red Light
400 nm	3.10 eV	Violet Light
1 Å = 0.1 nm	12.4 keV	X-Rays
1240 fm	1.0 MeV	Gamma Rays

The photons of visible light have energies of a few eV. Red light with a wavelength of 700 nm has the least energetic photons and violet light with a wavelength of 400 nm has the most energetic photons in the visible region. We shall find that the outer electrons of an atom are bound to the atom by a few electron volts of energy. The third row of Table I.2

gives the photon energy for light with a wavelength of one Angstrom, which is a typical distance separating the atoms of a crystal. One Angstrom is equal to 10^{-10} m which is the same as 0.1 nm. Light of this wavelength known as x-rays is commonly used to study crystal structures. As can be seen from the third row of the table, x-ray photons have energies of tens of thousands of electron volts. The gamma rays emitted by nuclei range in energy up to about one MeV. For this reason, we chose the photon energy in the last row of the table to be 1.0 MeV and then used Eqs. (I.31) and (I.33) to obtain the wavelength corresponding to this radiation.

As can be seen from the entries in Table I.2, the electromagnetic radiation emitted by atoms and nuclei has a much longer wavelength than the size of the species that emits the radiation. Violet light has a wavelength of 4000 Å and is thus four thousand times larger than an atom. Very energetic 1 MeV gamma rays have a wavelength 1240 times the size of the nucleus. Less energetic gamma rays, which are commonly emitted by nuclei, have even longer wavelengths. Such considerations will be important in later chapters when we study radiation processes.

We can think about the time scales appropriate for describing atomic processes in the same way as we think about time in our own lives. Our life is a process which begins with our birth and extends on until the day we die. It takes us a certain length of time to overcome different kinds of adversities and to respond to changes in our environment. The same could be said of the motion of the electrons in an atom; however, the time scale is different. As we shall see in Chapter 4, the unit of time that is appropriate for describing an electron in an atom is 2.4×10^{-17} seconds. We can think of this as the time required for an electron in an atom to circulate once about the nuclear center. It is like the pulse rate of our own bodies. Atoms generally decay from excited states to the ground state or readjust to changes in their environment in about a nanosecond which is 10^{-9} seconds. While 10^{-9} seconds is a very short time in our own lives, it is a very long time for electrons circulating about the nucleus of an atom. The coupling between atomic electrons and the outside world is usually sufficiently weak that an electron in an atom has to circulate about the nucleus tens of millions of times before it makes a transition.

We turn our attention now to the much smaller world of the atomic nucleus. The nucleus of an atom consists of positively charged protons and neutrons which are electrically neutral. The masses of the electron, proton, and neutron are given in Appendix A. Denoting the masses of these particles by m_e, m_p, and m_n respectively, one can readily confirm that they are related by the equations

$$m_p = 1836.2 \, m_e$$
$$m_n = 1838.7 \, m_e .$$

The masses of the proton and neutron are about the same, and they are approximately equal to two thousand times the mass of an electron. We shall use the common term *nucleon* to refer to a proton or a neutron. For a neutral atom, the number of protons in the nucleus is equal to the number of electrons in the charge cloud surrounding the nucleus. The nucleus also generally contains a number of neutrons. The basic unit of time for describing a process occurring in the nucleus is the time it would take a nucleon having a kinetic energy of 40 or 50 MeV to traverse a distance of 10^{-15} meters which is the size of the nucleus. This length of time is about 10^{-23} seconds. Nuclear processes evolving over a longer period of time can be thought of as delayed processes.

Modern efforts to understand the elementary constituents of matter can be traced back to the experiments of J.J. Thomson in 1897. Thomson knew that the radiation emitted by hot filaments could be deflected in a magnetic field and therefore probably consisted of particles. By passing a beam of this radiation through crossed electric and magnetic fields and adjusting the field strengths until the deflection of the beam was zero, he was able to determine the charge to mass ratio of these particles. Thomson found that the charge to mass ratio of the particles in this kind of radiation was very much greater than the ratio for any other known ion. This meant that the charge of the particle was very large or that the mass of the particle was very small. He used the word *electron* to denote the charge of the particles, but this term was later applied to the particles themselves. The mass of the electron is about two thousand times smaller than the mass of the hydrogen atom which consists of a proton and an electron.

The basic structure of atoms was determined by Rutherford in 1911. Rutherford and his students, Geiger and Marsden, performed an experiment in which alpha rays were scattered by a gold foil. The alpha rays they used in their experiments were emitted by a radiative radium source. Rutherford was surprised to find that some of the alpha particles were scattered through wide angles and concluded that most of the mass of the gold atom was concentrated in a small region at the center of the atom which he called the nucleus. The wide angle scattering occurred when an alpha particle came very close to a gold nucleus. Rutherford showed that the nuclear model of the atom led to an accurate description of his scattering data. The discovery of the neutron by J. Chadwick in 1932 made it possible to explain the mass of the chemical elements. The nucleus of the most common isotope of carbon, for instance, consists of six protons and six neutrons.

Since the early discoveries of Thomson, Rutherford, and Chadwick, hundreds of new particles have been identified in the energetic beams of particles produced in modern accelerators, and our understanding of the nature of nucleons has also grown. We have learned that all charged particles have an antiparticle with the same mass and the opposite charge. The antiparticle of the negatively-charged electron e^- is the positively charged positron e^+. We have also learned that the proton and neutron are made up of more elementary particles called *quarks*. There are six quarks: *up, down, strange, charmed, bottom,* and *top*. The proton consists of two up-quarks and a down-quark, while the neutron consists of two down-quarks and an up-quark. In conjunction with our growing understanding of the elementary constituents of matter, we have learned more about the fundamental forces of nature. We now know that quarks are held together by a very powerful force called the *strong* force, and there is another fundamental force called the *weak* force, which is responsible for the decay of a number of unstable particles. An example of a decay process that takes place through the weak interaction is provided by the decay of the neutron. Although neutrons may be stable particles within the nucleus, in free space the neutron decays by the weak interaction into a proton, an electron and an antineutrino

$$n \rightarrow p + e^- + \bar{\nu}.$$

All in all, there are four fundamental forces in nature: the electromagnetic force, the strong and weak forces, and the gravitational force. The nature of these four forces are described in the following table.

TABLE I.3 The Four Fundamental Forces.

Force	Quantum	Typical Interaction Times
Electromagnetic	photon	10^{-14}–10^{-20} s
Strong	gluon	$< 10^{-22}$ s
Weak	$W^\pm, Z^0$	10^{-8}–10^{-13} s
Gravitational	graviton	Years

Each entry in the first column of the table gives a fundamental force of nature, while the entries in the second column of the table give the quantum associated with the force. The photon, which we have discussed in conjunction with the emission and absorption of light, can be thought of as the quantum or "carrier" of the electromagnetic force. The quantum of the strong force is called the *gluon*, while the quanta of the weak force are the W^+, W^-, and Z^0 particles. The quantum of the gravitational force is called the *graviton*. Continuing to the right in Table I.3, the entries in the third column give the typical length of time it takes for processes involving the force to run their course. Scattering processes involving the strong force take place within 10^{-22} seconds, while processes involving the weaker electromagnetic force typically take place in 10^{-14} to 10^{-20} seconds. The weak interaction is very much weaker than the electromagnetic interaction. As shown in the table, processes that depend upon the weak interaction generally take between 10^{-8} and 10^{-13} seconds which is much longer than the times associated with strong and electromagnetic processes. The force of gravity is very much smaller than the other forces; however, the gravitational force is always attractive and has an infinite range. Gravity is responsible for the motions of the planets in their orbits about the Sun and, on a larger scale, for the collective motions of stars and galaxies.

In later chapters of this book, we shall find that the fundamental forces of nature are communicated by means of quanta interchanged during the interaction process with the range of the force being related to the mass of the quantum exchanged. The quantum or carrier of the electromagnetic, strong, and gravitational forces have zero mass and the range of these forces are infinite, while the weak force, which is carried by massive particles has a short range. We should note that some forces that are commonly referred to involve composite particles and are not fundamental in nature. The *Van der Waals force* between neutral atoms composed of a nucleus and outer electrons and the *nuclear force* between nucleons composed of quarks are examples of interactions between composite particles. The force between composite particles typically falls off much more rapidly than the force between fundamental particles. The range of the nuclear force is about a fermi (10^{-15} m) which is equal to the size of the atomic nucleus.

The understanding we have today of modern physics has evolved through the interaction between experiments and theories in particle and nuclear physics as well as in atomic physics and the expanding field of research into semiconductors. Each field of physics has played its own role in giving us the view of the world we have today. It is an exciting story that needs to be told.

In the next chapter, we shall consider a few key experiments performed in the latter part of the nineteenth century and the early part of the twentieth century that enabled physicists to characterize the way radiation interacts with matter. The principles of quantum mechanics are introduced in the context of these experiments in Chapters 2 and 3. The quantum theory is then used to describe atoms and diatomic molecules in Chapters 4–7 and to treat condensed matter physics in Chapters 7–10. After our description of relativity theory in Chapters 11 and 12, we will return to the discussion of the fundamental forces and to particle and nuclear physics.

Basic equations

Variables of particles

Momentum

$$\mathbf{p} = m\mathbf{v}$$

Angular momentum

$$\boldsymbol{\ell} = \mathbf{r} \times \mathbf{p}$$

Kinetic energy

$$KE = \frac{1}{2}m\mathbf{v}^2$$

Potential energy

$$V_P = -\int_R^P F(x)dx$$

Potential energy of oscillator

$$V(x) = \frac{1}{2}m\omega^2 x^2$$

Potential energy of an electron due to nucleus

$$V(r) = -\frac{Ze^2}{4\pi\epsilon_0}\frac{1}{r}$$

Properties of waves

Relations involving λ, k, and ω

$$\lambda f = v, \qquad k = \frac{2\pi}{\lambda}, \qquad \omega = 2\pi f$$

Stationary wave

in one dimension $\qquad \psi(x) = Ae^{ikx}$

in three dimensions $\qquad \psi(\mathbf{r}) = Ae^{i\mathbf{k}\cdot\mathbf{r}}$

Traveling waves

in one dimension $\qquad \psi(x,t) = Ae^{i(kx-\omega t)}$

in three dimensions $\qquad \psi(\mathbf{r}) = Ae^{i(\mathbf{k}\cdot\mathbf{r}-\omega t)}$

Electromagnetic radiation

Photon energy

$$E = hf, \qquad E = \frac{hc}{\lambda}$$

Value of hc

$$hc = 1240 \, \text{eV} \cdot \text{nm}$$

Summary

The elementary properties of particles and waves are reviewed, and then an overview is given of the new quantum theory used to describe microscopic systems. A novel feature of physical systems on a microscopic level is that they display a wave-particle duality. Certain phenomena require that we consider radiation or matter as consisting of particles, while other phenomena demand that we think of radiation or matter as consisting of waves.

The wave-particle duality causes us to describe the motion of particles using the concepts of probability. A particle such as an electron is described by a wave function which enables us to determine the probability that the particle can be found at a particular point.

All physical systems can be characterized by their size and by the length of time it takes for processes occurring within them to evolve. Atoms, which have diameters of about an angstrom or 10^{-10} m, make transitions typically in about a nanosecond or 10^{-9} seconds. The atoms in a crystal are separated by about an angstrom, which is equal to the wavelength of x-rays. The basic unit of time for describing a process occurring in the nucleus is the time it would take a proton or neutron having a kinetic energy of 40 or 50 MeV to traverse a distance of 10^{-15} meters which is the size of the nucleus. This length of time is about 10^{-23} seconds.

There are four fundamental forces in nature: the strong interaction, the electromagnetic and weak interactions, and gravitation. A quantum or carrier is associated with each of these interactions and also a characteristic time necessary for processes due to these interactions to take place.

Suggestions for further reading

Thomas A. Moore, *Six Ideas That Shaped Physics*, Second Edition (New York: Mc Graw Hill, 2003).

R. Mills, *Space, Time and Quanta* (New York: W.H. Freeman and Company, 1994).

Gary Zukav, *The Dancing Wu Li Masters* (New York: Bantam Books, 1980).

Questions

1. Write down an equation defining the angular momentum of a particle with position **r** and momentum **p**.
2. Which physical effect or experiment shows that light has a wave nature?
3. Express the kinetic energy KE of a particle in terms of its momentum p.
4. What would be the wavelength of a wave described by the function $u(x, t) = A \sin(2x/\text{cm} - 10\text{s})$?
5. What would be the frequency of a wave described by the function $u(x, t) = A \sin(2x/\text{cm} - 10t/\text{s})$?
6. What would be the phase velocity of a wave described by the function, $u(x, t) = A \sin(2x/\text{cm} - 10t/\text{s})$?
7. Write down the exponential function corresponding to a traveling wave with a wavelength of 10 cm and a frequency of 10 Hz.
8. Write down a trigonometric function describing a stationary wave with a wavelength of 10 cm.
9. What condition must be satisfied by the difference of the two path lengths for constructive interference to occur for the two-slit interference experiment?
10. How would the interference pattern of a two slit interference experiment change if the distance between the two slits were to increase?
11. Which forms of electromagnetic radiation have a wavelength shorter than visible light?
12. Calculate the frequency of electromagnetic radiation having a wavelength of 10 nm.
13. Write down a formula expressing the energy of a photon in terms of the frequency of light.
14. Write down a formula expressing the energy of a photon in terms of the wavelength of light.
15. What is the energy of the photons for light with a wavelength of 0.1 nm?
16. Suppose that it were possible to increase the charge of an atomic nucleus without increasing the number of electrons. How would the probability cloud around the nucleus change as the charge of the nucleus increased?
17. How long does it generally take for an atom to make a transition?
18. Under what circumstance would an α-particles be scattered directly backward in Rutherford's experiment?
19. Use the fact that the proton is composed of two up-quarks and a down-quark and the neutron is composed of two down-quarks and an up-quark to find the charge of the up- and down-quarks.
20. In what sense are protons and neutrons composite particles?

21. The K^+ meson decays in 1.24×10^{-8} s according to the following reaction formula

$$K^+ \rightarrow \mu^+ + \nu_\mu.$$

What interaction is responsible for this decay process?

22. Give the size of a nucleus and an atom in SI units.

Problems

1. Calculate the frequency of light having a wavelength $\lambda = 500$ nm.

2. Calculate the energy of the photons for light having a wavelength $\lambda = 500$ nm.

3. Suppose that a beam of light consists of photons having an energy of 5.4 eV. What is the wavelength of the light?

4. The sinusoidal functions $\sin nkx$ appearing in Eq. (I.19) may be shown to satisfy the identity

$$\int_0^L \sin mkx \sin nkx dx = \frac{L}{2}\delta_{n,m},$$

where $\delta_{n,m}$ is the Kronecker delta function defined to be equal to one if n and m are equal and zero if n and m are not equal. Using this identity, multiply Eq. (I.19) from the left with $\sin mkx$ and integrate from 0 to L, to obtain

$$\int_0^L \sin mkx f(x) dx = \frac{L}{2} S_m.$$

This last equation is equivalent to Eq. (I.20).

5. From an experiment in which x-rays are scattered from a crystal, one finds that the wavelength of the radiation is 1.2 Å. What is the energy of the x-ray photons?

6. A wave is described by the function

$$\psi(x) = A e^{i(\alpha x + \beta t)}.$$

What are the wavelength and frequency of the wave in terms of the constants α and β?

7. Find the length of the smallest standing wave that can be formed with light having a frequency of 600 THz. Recall that 1 THz $= 10^{12}$ Hz.

8. Suppose an atom makes a transition from a state in which it has an energy E_2 to a state having an energy E_1 where $E_2 > E_1$. What is the energy of the quantum of light emitted by the atom? Derive an expression for the wavelength of the emitted light.

9. Consider a light wave with wavelength 400 nm incident on a double slit with distance $h = 1.2$ μm between the slits. What is the angle of the first two diffraction maxima (beyond the central maximum at $0°$)?

10. Using Euler's identity Eq. (I.24) show that adding two waves given by $y_1 = A e^{i(kx - \omega t)}$ and $y_2 = A e^{i(kx + \omega t)}$ gives a new wave with time-dependent amplitude $2A \cos(\omega t)$ and position dependence e^{ikx}.

11. A very sensitive detector measures the energy of a single photon from starlight at 2.5 eV and at the same time measures its wavelength at 495 nm. What is the value of Planck's constant at that far-away star?

12. Write a MATLAB program to plot the function $y = \sqrt{30} x(1 - x)$ between $x = 0$ and $x = 1$.

13. Using the plotting capability of MATLAB, show that the addition of two waves, $y_1 = \sin \theta \cos 3\theta$ and $y_2 = \sin 3\theta \cos \theta$ gives the expected result from the trigonometric identity Eq. (I.17).

14. Following the MATLAB example in the text, plot the first three components of the Fourier series for a square wave together on the same figure. Label each line appropriately.

15. Following the MATLAB example in the text, plot the result of adding together the first three components of the Fourier series for a square wave.

Chapter 1

The wave-particle duality

Contents

1.1 The particle model of light	1	Summary	16
1.2 The wave model of radiation and matter	12	Questions	16
Suggestions for further reading	15	Problems	17
Basic equations	15		

Certain phenomena can be understood by considering radiation or matter to consist of particles, while other phenomena demand that we think of radiation or matter as consisting of waves.

We consider in this chapter a few key experiments that enable us to characterize the possible ways in which radiation interacts with matter. The first section describes experiments in which electromagnetic radiation is absorbed or emitted or is scattered by free particles. The results of these experiments can be understood by supposing that electromagnetic radiation consists of little packets of energy called photons. The second section describes experiments in which a beam of electromagnetic radiation is incident upon a crystal and experiments in which a collimated beam of electrons is incident upon a crystal or upon two-slits. The variation of the intensity of the scattered radiation in these experiments can be interpreted as the interference patterns produced by waves. The experiments described in this chapter taken together show that radiation and matter have a dual particle-wave character. Certain phenomena can be understood by considering radiation or matter to consist of particles, while other phenomena demand that we think of radiation or matter as consisting of waves.

1.1 The particle model of light

At the end of the nineteenth century, light was thought of as a form of electromagnetic waves. We have since found that certain phenomena involving the absorption and the emission of light can only be understood by thinking of a beam of light as a stream of particles. This has lead to a more balanced way of thinking about light as exhibiting both wave and particle properties. In this section, we will consider three kinds of experiments in which the particle-like character of light reveals itself. The first is the photoelectric effect, the second involves the emission and absorption of light by atoms, and the third involves the scattering of light by free electrons.

1.1.1 The photoelectric effect

As mentioned in the introduction, the photoelectric effect refers to the emission of electrons by a metal surface when light is incident upon the surface. Two simple experiments for studying the photoelectric experiment are illustrated in Fig. 1.1. In both of these experiments, light falls on a metal plate called a *cathode* causing electrons to be liberated from the metal surface and collected by a nearby conducting plate called an *anode*. In the first of these experiments illustrated in Fig. 1.1(A), the cathode and the anode are connected by an ammeter having negligible resistance. Electrons collected by the anode can then freely flow through the ammeter, which measures the number of electrons flowing through it per second. While not every electron emitted from the cathode flows through the ammeter, the current measured by the ammeter is proportional to the number of electrons emitted from the cathode. This experiment makes it possible to determine how the number of electrons emitted depends upon the frequency and intensity of light.

In the second experiment illustrated in Fig. 1.1(B), the two plates are connected by a voltmeter that has essentially infinite resistance. The voltmeter measures the voltage difference between the anode and the cathode. As light liberates more and more electrons from the cathode, the cathode becomes more and more positively charged and the anode becomes more negatively charged. As illustrated in Fig. 1.1(B), this creates an electric field and a potential difference between the

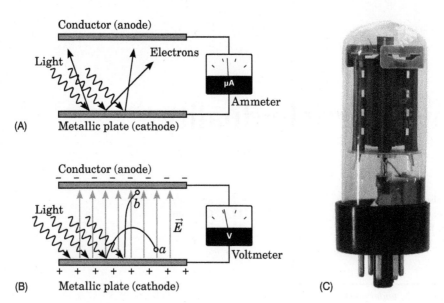

(A) Metallic plate (cathode)

(B) Metallic plate (cathode)   (C)

FIGURE 1.1 An apparatus for observing the photoelectric effect.

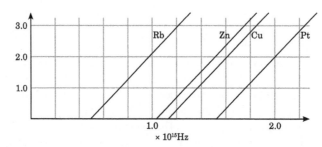

FIGURE 1.2 The maximum kinetic energy versus frequency for a number of metals.

cathode and the anode which resists the further flow of electrons. When the potential difference increases to a certain value called the *stopping potential*, the photoelectric current goes to zero. The potential difference required to stop electrons times the charge of an electron is equal to the maximum kinetic energy of electrons emitted from the metal surface.

Experimental studies of the photoelectric effect show that whether or not electrons are emitted from a metal surface depends on which metal is being studied and on the frequency – not the intensity – of the light. If the frequency is below a certain value called the *cutoff frequency*, no electrons are emitted from the metal surface, no matter how intense the light is. Above the cutoff frequency, the maximum kinetic energy of electrons increases linearly with increasing frequency. Fig. 1.2 shows plots of the maximum kinetic energy versus frequency for a number of metals. For each metal, the maximum kinetic energy is equal to zero for the cutoff frequency and the maximum kinetic energy increases as the frequency increases linearly. The slopes of all the straight lines shown in Fig. 1.2 are equal to Planck's constant h. For any frequency above the cutoff frequency, the number of electrons emitted from the metal surface is proportional to the intensity of the light.

In 1905, Einstein proposed a successful theory of the photoelectric effect in which he applied ideas that Planck had used earlier to describe the radiation field within a cavity. Planck had assumed that the electromagnetic field in a cavity interchanges energy with the walls of the cavity in integral multiples of hf where h is the physical constant we now called Planck's constant and f is the frequency of the radiation. In explaining the photoelectric effect, Einstein supposed that the light incident upon a metal surface consists of particles having an energy

$$E = hf. \tag{1.1}$$

As before, the particles of light will be called quanta or photons. If the light shining upon a metal surface can be thought of as consisting of photons with an energy hf, and if the electrons are bound to the metal with an energy W, then the photon energy hf must be greater than W for electrons to be emitted. The *threshold frequency* f_0 thus satisfies the equation

$$hf_0 = W. \tag{1.2}$$

If the frequency of the incident light is greater than f_0, each of the quanta of light has an energy greater than the work function W, and electrons absorbing a photon may then be emitted from the metal.

The work function W is the energy required to free the least tightly bound electrons from the metal. More energy will have to be supplied to more tightly bound electrons to free them from the metal. For a particular frequency of light, the maximum kinetic energy of the emitted electrons is equal to the energy of the photons given by Eq. (1.1) minus the work function of the metal. We thus have

$$(KE)_{\max} = hf - W. \tag{1.3}$$

The value of Planck's constant h and the work function W of the metal can be calculated by plotting the maximum kinetic energy of the emitted electrons versus the frequency of the light as shown in Fig. 1.2. According to Eq. (1.3), the maximum kinetic energy of the electrons and the frequency of the light are linearly related, and the plot of $(KE)_{\max}$ versus f results in a straight line. Planck's constant h is equal to the slope of the line. For points in the upper right-hand corner of Fig. 1.2, the frequency of the incident light and the maximum kinetic energy of the emitted electron are both large. As the frequency of the incident light decreases, the maximum kinetic energy of the photoelectrons will also decrease. The threshold frequency f_0 may be identified as the value of the frequency for which the maximum kinetic energy of the electrons is equal to zero. One may determine the threshold frequency for a particular metal by finding the value of f for which the straight line corresponding to the metal intersects the horizontal axis. For the threshold frequency f_0, the maximum kinetic energy of the electrons is equal to zero and Eq. (1.3) reduces to Eq. (1.2).

The work function of several metals is given in Table 1.1.

TABLE 1.1 Work Functions of Selected Metals.

Metal	W (eV)
Na	2.28
Al	4.08
Fe	4.50
Cu	4.70
Zn	4.31
Ag	4.73
Pt	6.35
Pb	4.14

Following along the lines of our derivation in the introduction, we may obtain an expression for the energy of photons in terms of the wavelength of light by using the fact that the frequency and the wavelength of light are related by the equation

$$f\lambda = c, \tag{1.4}$$

where c is the speed of light in a vacuum. Eq. (1.4) may be solved for the frequency f giving

$$f = \frac{c}{\lambda}.$$

We may then substitute this last equation into Eq. (1.1) to obtain the following expression for the photon energy in terms of the wavelength

$$E = \frac{hc}{\lambda}, \tag{1.5}$$

where, as seen in the introduction, $hc = 1240$ eV · nm. Photons corresponding to the threshold wavelength λ_0 have just enough energy to free an electron from the metal. We thus have

$$\frac{hc}{\lambda_0} = W.$$

For light with a wavelength λ less than λ_0, the energy of the incident photons given by Eq. (1.5) will be greater than the work function W and electrons will be emitted, while for wavelengths greater than λ_0 the energy of the incident photons will be less than W and electrons will not be emitted. It is very unlikely that an electron in the metal will ever absorb more than a single photon. Again, using the fact that the maximum kinetic energy of the emitted electrons is equal to the energy of the photons minus the work function of the metal, we obtain

$$(KE)_{\text{max}} = \frac{hc}{\lambda} - W. \tag{1.6}$$

Example 1.1

A surface of zinc having a work function of 4.3 eV is illuminated by light with a wavelength of 200 nm. What is the maximum kinetic energy of the emitted photoelectrons?

Solution
Using Eq. (1.5) and the fact that $hc = 1240 \, \text{eV} \cdot \text{nm}$, we obtain the following value of the energy of the incident photons

$$E = \frac{1240 \, \text{eV} \cdot \text{nm}}{200 \, \text{nm}} = 6.2 \, \text{eV}.$$

The incident light may thus be thought of as a stream of photons which have an energy of 6.2 eV. According to Eq. (1.6), the maximum kinetic energy of the emitted electrons is equal to the difference of the photon energy and the work function of the metal. We thus have

$$(KE)_{\text{max}} = 6.2 \, \text{eV} - 4.3 \, \text{eV} = 1.9 \, \text{eV}.$$

Albert Einstein received the Nobel prize for his theory of the photoelectric effect in 1921. His theory, which raised a good deal of controversy in its time, was based on the premise that a radiation field interacting with matter could be considered as a collection of photons. Further evidence that light consists of photons is provided by the optical spectra of atoms.

1.1.2 The absorption and emission of light by atoms

Hot bodies like the Sun emit electromagnetic radiation with the intensity verses the wavelength of the radiation being referred to as the *spectrum* of the body. The first detailed description of the spectrum of the Sun was completed by Fraunhofer in 1815. As Wollaston had discovered in 1802, Fraunhofer found that the solar spectrum was crossed by a series of dark lines. Since these lines were present in every kind of sunlight, whether reflected from terrestrial objects or from the moon or planets, he concluded that these lines depended upon the properties of the Sun. In his efforts to develop precise optical equipment which would resolve light into its spectral components, Fraunhofer produced fine gratings by drawing regular lines on gold films with a diamond needle. He used the wave theory of light to accurately determine the wavelength corresponding to the different features of his spectra. The optical equipment and techniques developed by Fraunhofer and his associates contributed in an important way to the rapid advances in astronomy that occurred in the latter part of the nineteenth century.

The dark lines in the solar spectra were finally explained by Bunsen and Kirchhoff in 1859, who were able to demonstrate that the same frequencies of light that were emitted by a flame containing a metal would be absorbed when the radiation passed through a cooler environment containing the same constituent. Unlike solid materials which emit a continuous range of frequencies, individual atoms emit a number of distinct frequencies which are characteristic of the atoms involved.

As one might expect, hydrogen, which is the lightest atom, emits the simplest spectra. An electric discharge in hydrogen gas produces a visible spectrum consisting of four lines between the red and violet arranged in obvious regularity. The wavelength of these four lines are shown in Fig. 1.3. Once the wavelengths emitted by hydrogen in the visible region had been determined, a number of unsuccessful attempts were made to provide a formula for the wavelengths using the idea that the frequencies should be related like the harmonics of a classical oscillating system. Then, in 1885, Johann Balmer, who was a geometry teacher in a Swiss high school, showed that the wavelengths of hydrogen in the visible region were given by the formula

$$\lambda_n = 3645.6 \, \frac{n^2}{(n^2 - 4)} \times 10^{-8} \text{cm} \, , \text{ where } n = 3, 4, 5, \ldots . \tag{1.7}$$

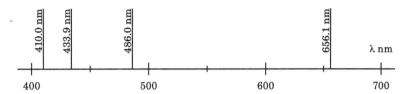

FIGURE 1.3 Spectral lines emitted by hydrogen in the visible region.

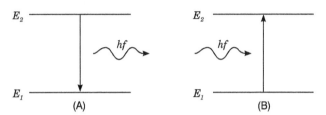

FIGURE 1.4 (A) An emission process in which an atom falls from an upper level with energy E_2 to a lower level with energy E_1 and emits light of frequency f. (B) An absorption process in which an atom with energy E_1 absorbs light of frequency f and makes a transition to a higher level with energy E_2.

From his boyhood, Balmer was a devoted Pythagorean who was convinced that the explanation of the mysteries of the universe depended upon our seeing the correlation between observed phenomena and the appropriate combinations of integers. His empirical formula anticipated by forty years the theoretical work of Heisenberg and Schrödinger who deduced the spectra of hydrogen from physical principles.

In 1890, Rydberg, who was a distinguished Swedish spectroscopist, showed that the Balmer formula could be written in a form that could easily be used to describe other series of spectral lines

$$\frac{1}{\lambda} = R \left(\frac{1}{m^2} - \frac{1}{n^2} \right), \text{ where } m = 2, \text{ and } n = 3, 4, 5, \ldots . \tag{1.8}$$

The constant R, which appears in this equation, is now called the *Rydberg constant*. We leave it as an exercise (Problem 10) to show that Eq. (1.7) can be written in the form (1.8) provided that the Rydberg constant is assigned the value $R = 1.0972 \times 10^5$ cm^{-1}. The formula for other series of lines can be obtained from Eq. (1.8) by replacing the integer m in this equation by integers other than 2. In 1908, Lyman discovered a series of spectral lines in the ultraviolet region of the spectrum with $m = 1$, and Paschen discovered a series of lines in the infrared with $m = 3$.

1.1.2.1 Principles of atomic spectra

While the Rydberg formula (1.8) provides a very accurate description of the spectrum of hydrogen, it was discovered empirically and lacked any kind of fundamental justification. The difficult task of providing a theoretical framework for explaining atomic spectra was initiated by Niels Bohr in 1913. Bohr formulated two principles which have since been shown to have universal validity:

1) Atoms have stationary states in which they have well-defined values of the energy.
2) The transition of an atom from one level to another is accompanied by the emission or absorption of one quantum of light with the frequency

$$f = \frac{E_2 - E_1}{h}, \tag{1.9}$$

where E_2 is the higher of the two energies, E_1 is the lower of the energies, and h is Planck's constant.

In an emission process, the atom falls from an upper level with energy E_2 to a lower level with energy E_1 and emits light corresponding to the frequency given by Eq. (1.9). In an absorption process, an atom, which has an energy E_1 absorbs light with this frequency and makes a transition to the level with energy E_2. These two processes are illustrated in Fig. 1.4. As for the photoelectric effect, the light emitted or absorbed by an atom may be thought of as consisting of particles called photons. Each photon has an energy

$$E_{\text{photon}} = hf, \tag{1.10}$$

where f is the frequency of light. Multiplying Eq. (1.9) by h and using Eq. (1.10), we obtain

$$E_{\text{photon}} = E_2 - E_1.$$

This last equation may be thought of as a statement of the condition that energy be conserved in these processes. For the emission process described in Fig. 1.4(A), the photon carries off the energy which the atom loses, and for the absorption process described in Fig. 1.4(B) the photon gives to the atom the energy necessary for it to make the transition. The energy of a photon is related to the wavelength of the light by Eq. (1.5). In formulating the two principles given above, Bohr was influenced by earlier work of Planck and Einstein.

We will often denote the difference in energy between the upper and lower levels in a transition by

$$\Delta E = E_2 - E_1,$$

where Δ is the upper case form of the Greek letter *delta*. Eq. (1.9) for the transition frequency can then be written simply

$$f = \frac{\Delta E}{h}. \tag{1.11}$$

To obtain an equation for the wavelength, we may solve Eq. (1.4) to obtain

$$\lambda = \frac{c}{f},$$

where c is the speed of light. Substituting Eq. (1.11) into this last equation then gives the following equation for the wavelength of the light

$$\lambda = \frac{hc}{\Delta E}. \tag{1.12}$$

As discussed previously, the product of constants h and c is given in Appendix A and is approximately equal to $1240\,\text{eV} \cdot \text{nm}$.

1.1.2.2 The Bohr model of the atom

In addition to enunciating the principles given above, Bohr proposed a model of the hydrogen atom in which the electron moves in one or another of a discrete set of orbits. We now know that this model contains features which are true and others which are incorrect. Because of its historical importance, we give a brief review of the Bohr model, and we shall then say which features of the model are incompatible with modern theories. We first note that the basic idea that the electron in hydrogen moves in discrete orbits breaks in two fundamental respects with classical physics. The classical theory of orbital motion allows there to be a continuous range of orbits. As mentioned in the Introduction, each of the planets could have an orbit about the Sun which is slightly larger or slightly smaller than the orbit it has. The orbits of the planets depend upon the circumstances in which they were formed and not upon whether they belong to a specific set of allowed orbits. It is also important to take into account the fact that the electron is a charged particle. According to the classical theory of electromagnetism, a charged particle which is accelerating radiates energy. An electron orbiting about a nucleus should radiate energy and spiral in toward the nucleus. While Bohr had a great deal of respect for classical physics, he had become convinced that classical physics could not explain atomic spectra. He simply stated that atoms have stable orbits and that electrons emit radiation only when they make transitions from one stable state of motion to another. Bohr learned that the atom had a nucleus from the experiments of Rutherford with whom he was associated as a young scientist, and he had recently become aware of the Balmer formula. In formulating his model of the atom, he worked back and forth between different possible assumptions and the Balmer formula.

A simple illustration of an electron moving in a circular orbit about the nucleus of the hydrogen atom is given in Fig. 1.5. An electron with charge $-e$ moving in an orbit with radius r about the nucleus with charge $+e$ would be attracted to the nucleus by the Coulomb force

$$F = -\frac{1}{4\pi\epsilon_0} \frac{e^2}{r^2}.$$

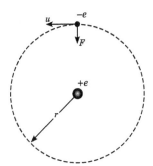

FIGURE 1.5 The Bohr model of the hydrogen atom.

According to Newton's second law, the force attracting the electron to the nucleus should be equal to the product of the mass of the electron m and its radial acceleration v^2/r. This gives

$$\frac{1}{4\pi\epsilon_0}\frac{e^2}{r^2} = m\frac{v^2}{r}. \tag{1.13}$$

The kinetic energy of the electron is given by the equation

$$KE = \frac{1}{2}mv^2.$$

Solving Eq. (1.13) for mv^2 and substituting the result into this last equation, we obtain

$$KE = \frac{1}{2}\frac{1}{4\pi\epsilon_0}\frac{e^2}{r} \tag{1.14}$$

The potential energy of the electron with charge, $-e$, moving in the field of the nucleus with charge, $+e$, is

$$V(r) = -\frac{1}{4\pi\epsilon_0}\frac{e^2}{r}. \tag{1.15}$$

The total energy of the electron is the sum of the kinetic and potential energies. Using Eqs. (1.14) and (1.15), we get

$$E = -\frac{1}{8\pi\epsilon_0}\frac{e^2}{r}. \tag{1.16}$$

This equation for the energy does not imply in any way that the electron is moving in a discrete orbit.

In his early work, Bohr tried two different rules for the stable orbits (quantization rules). The rule which is simplest requires that the orbital angular momentum of the electron be an integral multiple of Planck's constant h divided by 2π. We shall denote $h/2\pi$ simply as $\hbar$. Since the orbital angular momentum of an electron moving in a circle is equal to mvr, the quantization rule may be written

$$mvr = n\hbar, \tag{1.17}$$

where n is a positive integer known as the *principal quantum number*. We would now like to use Eqs. (1.13) and (1.17) to eliminate the velocity of the electron and solve for r. To do this, we first multiply Eq. (1.13) by mr to obtain

$$(mv)^2 = \frac{1}{4\pi\epsilon_0}\frac{me^2}{r}, \tag{1.18}$$

and we then square Eq. (1.17) giving

$$(mv)^2 r^2 = n^2\hbar^2.$$

Substituting the expression for $(mv)^2$ given by Eq. (1.18) into this last equation and solving for r, we obtain the following equation for the radius of the Bohr orbits

$$r = n^2 \frac{4\pi\epsilon_0 \hbar^2}{me^2}.$$ (1.19)

In the Bohr model, the integer n can have the values $n = 1, 2, \ldots$. The radius of the smallest orbit, which corresponds to the ground state of hydrogen, is called the *Bohr radius* and denoted by a_0. Setting $n = 1$ in Eq. (1.19), we obtain

$$a_0 = \frac{4\pi\epsilon_0 \hbar^2}{me^2}.$$ (1.20)

Using the values of the constants given in Appendix A, the Bohr radius can be shown to be approximately equal to 5.29×10^{-11} or 0.529 Å. The radius of the hydrogen atom is thus equal to about a half an Angstrom.

The expression for the energy in the Bohr model can be obtained by substituting Eq. (1.19) into Eq. (1.16) giving

$$E_n = -\frac{me^4}{2(4\pi\epsilon_0)^2 \hbar^2} \frac{1}{n^2},$$

where $n = 1, 2, 3, \ldots$, and, as before, $\hbar$ is equal to Planck's constant divided by 2π. The energies E_n given by this formula can be evaluated using the table of the fundamental constants given in Appendix A. In order to obtain a simpler form of the above equation for E_n, we rewrite it in the following way

$$E_n = -\frac{1}{2}\left(\frac{1}{4\pi\epsilon_0}\frac{e^2}{a_0}\right)\frac{1}{n^2}.$$ (1.21)

The combination of constants within parenthesis can be identified as the magnitude of the potential energy of the electron when it is separated from the hydrogen nucleus by a_0. Using the values of the constants given in Appendix A, this combination of constants may be shown to be approximately equal to 27.2 eV. The equation for the energy levels of hydrogen may thus be written simply

$$E_n = -\frac{13.6\,\text{eV}}{n^2}.$$ (1.22)

Before using the Bohr model to describe the spectra of light emitted or absorbed by the hydrogen atom, we shall review the different assumptions Bohr made to obtain his model. There is now a good deal of experimental evidence which confirms Bohr's hypothesis that the hydrogen atom has stationary states in which it has well-defined values of the energy and the angular momentum. Eq. (1.22) for the energy has been confirmed by experiment, and is also consistent with the predictions of the modern theory called *quantum mechanics*. As we shall find in Chapter 4, the values of the angular momentum can be specified by giving the value of a quantum number l which is an integer in the range between zero and $n - 1$. For the ground state of hydrogen, the integer n in Eq. (1.22) has the value one. The quantum number l must then have the value zero, and the angular momentum must be zero. Thus, while the angular momentum is quantized, it does not have the values given by the Bohr model. We should also note, as Heisenberg has, that important features of the Bohr theory, such as the position and velocity of the electron cannot be determined continuously as the electron moves within the atom.

1.1.2.3 The energy levels and spectra of hydrogen

The energy levels of hydrogen, which are shown in Fig. 1.6, can be obtained by substituting the integer values $n = 1, 2, 3, \ldots$ into Eq. (1.22). For the lowest level with $n = 1$, the energy is $-13.6\,\text{eV}/1^2 = -13.6$ eV. The second level, which corresponds to $n = 2$ has an energy equal to $-13.6\,\text{eV}/2^2 = -3.4$ eV, and so forth. The transitions, which are responsible for the emission lines of the Balmer, Lyman, and Paschen series, are also shown in Fig. 1.6. The Balmer emission lines correspond to transitions from the levels for which n is greater than or equal to 3 down to the level for which $n = 2$. These transitions all produce light in the visible part of the spectra. The first member of the series, which corresponds to a transition from the $n = 3$ level to the $n = 2$ level, is denoted H_α, the second member corresponding to a transition from the $n = 4$ to the $n = 2$ level is denoted H_β, the third member is denoted H_γ, and so forth. The members of the Lyman series correspond to transitions to the $n = 1$ level giving light in the ultraviolet portion of the spectra, while the Paschen series corresponds to transitions to the $n = 3$ level giving light in the infrared. The transitions of these series are referred to in a manner similar to that used for the transitions in the visible part of the spectra. The first members of the Lyman series, for

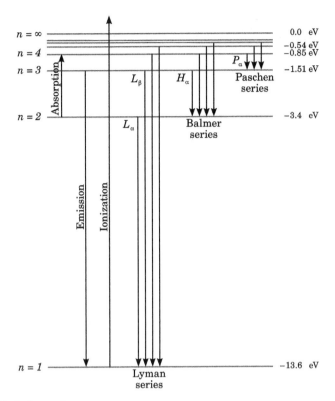

FIGURE 1.6 The energy levels of the hydrogen atom.

instance, corresponds to the transition $n = 2 \rightarrow n = 1$ and is referred to as Lyman-α (L_α), while the first member of the Paschen series corresponds to the transition $n = 4 \rightarrow n = 3$ and is referred to as Paschen-α (P_α). Successive members of these series are referred to as Lyman-β and Paschen-β, and so forth.

Example 1.2

Calculate the wavelength of the second member of the Balmer series.

Solution

For this transition, the n values for the upper and lower levels are 4 and 2, respectively. So, the difference between the energies of the upper and lower states is

$$\Delta E = -\frac{13.6 \text{ eV}}{4^2} - \left(-\frac{13.6 \text{ eV}}{2^2} \right) = 2.55 \text{ eV}.$$

Using Eq. (1.12), the wavelength of the light is

$$\lambda = \frac{hc}{\Delta E} = \frac{1240 \text{ eV} \cdot \text{nm}}{2.55 \text{ eV}} = 486 \text{ nm}.$$

This corresponds to blue light.

An atomic electron, which absorbs a photon, may be emitted from the atom. The incident light is then said to *ionize* the atom. For this to occur, the energy of the photon must be sufficient to raise the electron up from its initial state to a state having an energy greater than or equal to zero. The energy required to ionize an atom in its lowest state is referred to as the *ionization energy* of the atom. Using Eq. (1.22) with $n = 1$ or the position of the lowest level shown in Fig. 1.6, one may easily see that the ionization energy of hydrogen is approximately equal to $+13.6$ eV. A more accurate value of the ionization energy of hydrogen is given in Appendix A. Less energy is required, of course, to ionize an electron from a higher lying level. As for the photoelectric effect considered before, the kinetic energy of an electron, which absorbs light

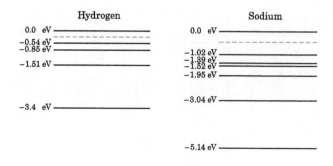

FIGURE 1.7 (A) The energy levels of the hydrogen atom. (B) The experimentally determined energy levels of the sodium atom.

and is emitted from an atom, is equal to the energy of the incident photon minus the energy required to free the electron from the atom.

Example 1.3

A collection of hydrogen atoms in the $n = 3$ state are illuminated with blue light of wavelength 450 nm. Find the kinetic energy of the emitted electrons.

Solution
The energy of the photons for 450 nm light is

$$E = \frac{hc}{\lambda} = \frac{1240 \text{ eV} \cdot \text{nm}}{450 \text{ nm}} = 2.76 \text{ eV}.$$

The kinetic energy of the emitted electrons is equal to the energy of the photons minus the energy necessary to ionize the atom from the $n = 3$ level

$$KE = 2.76 \text{ eV} - \frac{13.6 \text{ eV}}{3^2} = 1.25 \text{ eV}.$$

The success of the principles of Bohr in describing the spectra of hydrogen was an important milestone in the efforts of physicists to understand atomic structure. Niels Bohr received the Nobel Prize in 1922 for his investigations of atomic structure and the radiation emitted by atoms.

As for the photoelectric effect, electrons emitted from atoms following the absorption of light are called photoelectrons. The velocity distribution of photoelectrons emitted by atoms depends upon the frequency of the incident light and the energy level structure of the atoms. The experimentally determined energy levels of the sodium atom are compared with the energy levels of the hydrogen atom in Fig. 1.7. The two level schemes are qualitatively similar. The lowest levels of both atoms are well separated with the levels becoming more closely spaced as one approaches the ionization limit. A dashed line is drawn in each figure to indicate there are other bound levels above the highest bound level shown. If an atom is in a particular energy level and light is incident upon the atom with a photon energy greater than the ionization energy of

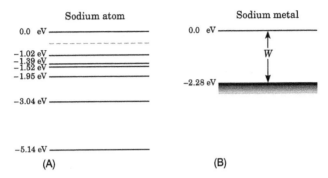

FIGURE 1.8 (A) The energy levels of (A) the sodium atom and (B) sodium metal.

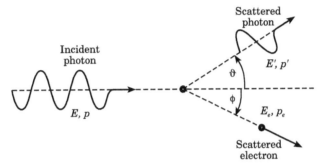

FIGURE 1.9 The scattering of a photon with energy E and momentum p from an electron which is initially at rest. The energy and momentum of the scattered photon are denoted by E' and p', while the energy and momentum of the scattered electron are denoted by E_e and p_e.

that state, an electron with a particular value of the kinetic energy will be emitted from the atom. The kinetic energy of the photoelectron is equal to the difference between the photon energy and the ionization energy of that particular state.

The energy levels of the sodium atom are compared to the highest-lying energy levels of the sodium metal in Fig. 1.8. The novel feature of solids is that the energy levels occur in dense bands with gaps occurring between the bands. For an electron to free itself from a solid, the electron must traverse a gap of a few electron volts between the highest occupied energy level and the unbound spectrum which begins at zero electron volts. This energy gap, which is called the *work function W*, is shown in Fig. 1.8(B). If light with a photon energy greater than the work function is incident upon a solid, electrons with a continuous range of kinetic energies will be emitted from the solid. As we have seen, the maximum kinetic energy of the photoelectrons is equal to the difference between the photon energy and the work function of the solid.

1.1.3 The Compton effect

The experiments of Arthur Compton on the scattering of x-rays by loosely bound electrons had the greatest influence in demonstrating the particle nature of light. In Compton's experiment performed in 1919, a beam of x-rays emanating from a molybdenum electrode were directed against a graphite target and the x-rays emerging from the target at right angles to the direction of the incident beam were carefully analyzed. The x-rays coming from the target were found to have a lower frequency. As Compton pointed out in his paper, this result was in fundamental disagreement with the classical theory of J.J. Thomson which predicted that the electrons should vibrate in unison with the electromagnetic field of the x-rays and emit radiation with the same frequency. Compton was able to explain the lower frequency of the scatter x-rays by assuming that the scattered radiation was produced by the collisions between photons and electrons. Since a recoiling electron would absorb part of the energy of the incident photon, the photons that were scattered by the electrons would have a lower energy and hence a lower frequency.

At the time of his experiment, Compton was aware of relativity theory which was formulated by Einstein in 1905. (Relativity theory will be described in Chapters 11 and 12.) According to relativity theory, the momentum of a massless particle is equal to its energy divided by c. Compton thus assumed that the momentum of a photon, which has a frequency f and an energy hf, is equal to hf/c. A schematic drawing of the scattering process in Compton's experiment is shown in Fig. 1.9. The angle between the line of flight of the scattered photon and the direction of the incident photon is denoted by θ and the direction of the scattered electron is denoted by ϕ. Using the fact that energy and momentum are conserved in collision processes, Compton derived a formula relating the wavelength λ' of the scattered radiation to the wavelength λ of the incident radiation and the angle θ of the scattered radiation. Compton's formula is

$$\lambda' - \lambda = \frac{h}{mc}(1 - \cos\theta), \tag{1.23}$$

where h is Plank's constant, m is the mass of the electron, and c is the velocity of light. This equation accurately describes Compton's data and the data of subsequent experiments on the scattering of x-rays by loosely-bound electrons. Since x-ray photons have energies of tens of thousands of electron volts and the recoiling electrons can thus have a large velocity, the derivation of Compton's equation should be done using relativity theory which is necessary for describing the motion of rapidly moving particles. A derivation of the Compton formula can be found in Chapter 12 together with a description of other high-energy scattering processes.

Arthur Compton received the Nobel prize in 1927 for his discovery of the effect named after him. The description given by Compton of the scattering of x-rays by loosely bound electrons depended upon the particle model of light and is thus consistent with the description we have given earlier of the photoelectric effect and atomic transitions.

1.2 The wave model of radiation and matter

Thus far we have been concerned with inelastic processes in which radiation field acts as though it were concentrated in quanta of energy. We turn our attention now to experiments in which radiation is scattered elastically by material objects. We shall find that the variation of the intensity of the scattered radiation in the experiments, which we shall consider, can be interpreted in terms of the interference of waves reflected by the different parts of the object.

1.2.1 X-ray scattering

The first experiment we will consider involves the scattering of X-rays by crystals. X-rays were discovered by Wilhelm Roentgen at the University of Würzburg, Germany in 1895. Only a few weeks before his discovery, Roentgen had begun working to repeat an experiment involving a cathode ray tube in which electrons accelerate through several thousand volts before striking a metal electrode. Although the tube was entirely covered by a shield of black cardboard, Roentgen noted that a piece of barium-cyanide paper that was sensitive to light was affected by radiation coming from the tube. Apparently, a form of radiation emitted by the tube passed through the dark shielding and illuminated the light-sensitive paper. Roentgen's discovery caused an immediate sensation. The mysterious rays, which passed through solid objects, were called X-rays.

Physicists soon found that X-rays are not deflected by electric or magnetic fields and could therefore only consist of electromagnetic radiation or a beam of neutral particles. We now know that X-rays consist of electromagnetic radiation with very short wavelength which would be emitted by the metal electrode in Roentgen's experiment when the energetic electrons in the cathode ray tube struck the metal electrode. In 1911, C.G. Barkla demonstrated that X-rays are scattered from small particles in suspension and can excite atoms in a body to emit fluorescent radiation. Max von Laue, whose scientific achievements spanned the period of the early rise of quantum theory and the theory of relativity, inferred from the available data that X-rays consisted of a form of electromagnetic radiation having a wavelength much shorter than that of visible light. He reasoned that X-rays would be diffracted by a grating with the lines drawn much closer together than in an ordinary optical grating. This led him to the brilliant idea that the regularly spaced atoms forming a crystal lattice should be used as a diffraction grating. To test these ideas, von Laue had W. Friedrich and P. Knipping perform an experiment in which X-rays were scattered by a crystal. In their experiment, a beam of X-rays was incident upon the face of a thin zinc-sulfide crystal. The pattern of spots formed on a photographic plate placed a few centimeters behind the crystal was successfully accounted for by von Laue on the assumption that the X-rays consisted of waves of short wavelengths diffracted by the atoms of the crystal. The theoretical work of von Laue and the experiments of Friedrich and Knipping opened up vast new areas of research into the properties of X-rays and the structure of crystals. William L. Bragg, who was then a research student with J.J. Thomson at Cambridge, immediately examined the analysis of von Laue. To simplify von Laue's treatment, Bragg assumed as we have in our discussion of Huygens' principle that the atoms of an illuminated crystal radiate secondary waves. The treatment of X-ray diffraction by crystals that we now give follows the treatment of William L. Bragg and his father William H. Bragg who were early pioneers in the field of X-ray crystallography.

Imagine now that X-rays are incident upon a crystal. Two paths of the radiation striking the crystal are illustrated in Fig. 1.10. The upper path is shorter than the lower path by the length of the two segments AB and BC. From the figure, one may show that the length of each of these segments is equal to $d\sin\theta$ where d is the separation between two lattice planes

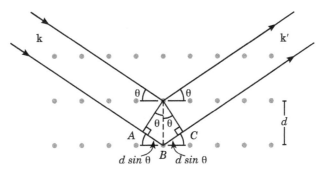

FIGURE 1.10 Scattering of X-rays by the planes of a crystal. The distance between two lattice planes is denoted by d and the scattering angle is denoted by θ.

of the crystal. For constructive interference to occur, it is necessary that the difference in path length be equal to a multiple of the wavelength of the radiation. In this way, one can readily show that the condition for constructive interference is

$$2d\sin\theta = n\lambda, \quad \text{for } n = 1, 2, 3, \ldots . \tag{1.24}$$

This equation is known as *Bragg's law* for X-ray diffraction.

Example 1.4

At what angle would the first maximum appear for x-rays with a wavelength of 1.2×10^{-10} meters if the lattice spacing is 0.8 Å?

Solution

The sine of the angle for the first maximum can be found immediately from Eq. (1.24) by setting $n = 1$ and solving for $\sin\theta$ to obtain

$$\sin\theta = \frac{\lambda}{2d} = \frac{1.2 \times 10^{-10} \text{ m}}{2 \times 0.8 \times 10^{-10} \text{ m}} = 0.75.$$

The desired angle is the inverse sine of 0.75, or $\theta = 48.6°$.

1.2.2 Electron waves

The ideas of constructive and destructive interference that have enabled us to derive Bragg's law depend upon the wave model of electromagnetic radiation. One of the remarkable discoveries of modern physics is that the wave model can also be used to describe the movements of particles such as electrons. This idea, which was originally suggested by Louis de Broglie in 1923, provides a basis for wave mechanics which plays a fundamental role in our understanding of microscopic systems. Prior to de Broglie's work, Einstein had put forward his theory of the photoelectric effect and Bohr had proposed his model of the atom. These theories, each of which depended upon the idea that electromagnetic radiation consists of photons, were the subject of intense debate at the time de Broglie made his suggestion. He argued that just as light has both a wave and a particle character, the objects which we think of as particles should also display this wave-particle dualism. Radiation and matter should be thought of as being at once both particles and waves.

De Broglie turned to the theory of relativity introduced earlier by Einstein to find an equation relating the dynamical variables of a particle to the parameters that specify the properties of the corresponding wave. As will be discussed in Chapter 12, relativity theory provides general expressions for the energy and the momentum of a particle which reduce to the familiar expressions when the velocity of the particle is sufficiently slow. A photon of light has a zero mass and moves with the velocity of light. As we have seen in conjunction with the Compton effect and as we shall consider more fully in Chapter 12, the momentum of a particle with zero mass is related to the energy of the particle by the equation

$$p = \frac{E}{c} .$$

Substituting Eq. (1.1) into this last equation and using Eq. (1.4) gives

$$p = \frac{hf}{c} = \frac{h}{\lambda} . \tag{1.25}$$

The momentum of the particle and the wavelength of the corresponding wave are thus inversely related to each other. As in the earlier quantum relation (1.1), Planck's constant serves as the mediator relating the particle and wave variables. Eq. (1.25) can, of course, be solved for the wavelength associated with the particle to obtain

$$\lambda = \frac{h}{p}. \tag{1.26}$$

The value of the wavelength given by this formula is referred to as the *de Broglie wavelength* of the particle.

The de Broglie relation (1.25) may also be conveniently expressed in terms of the angular wave number k defined by Eq. (I.8) of the introduction. We have

$$p = \frac{h}{\lambda} = \frac{h}{2\pi} \cdot \frac{2\pi}{\lambda},$$

or simply

$$p = \hbar k, \tag{1.27}$$

where again $\hbar$ is equal to $h/2\pi$.

De Broglie was led to his revolutionary hypothesis by his belief in the underlying symmetry in nature. The first experiments showing interference effects involving electrons were carried out independently by Thomson and Reid at the University of Aberdeen in Scotland in 1927 and by Davisson and Germer at the Bell Telephone Laboratories in 1928. Their experiments showed that the scattering of electrons by thin films and by crystals produce interference patterns characteristic of waves having a wavelength equal to the de Broglie wavelength (1.26). Louis de Broglie received the Nobel prize in 1929 for his discovery of the wave nature of electrons. In today's world, the electron microscope and a number of other electrical devices take advantage of the wave nature of the electron.

Example 1.5

Find the de Broglie wavelength of a 10 eV electron.

Solution

We begin by converting the kinetic energy of the electron into SI units

$$(KE) = 10 \text{ eV} \times \frac{1.602 \times 10^{-19} \text{ J}}{1 \text{ eV}} = 1.602 \times 10^{-18} \text{ J}.$$

The kinetic energy of the electron can be written in terms of the momentum

$$KE = \frac{1}{2}mv^2 = \frac{1}{2m}(mv)^2 = \frac{p^2}{2m}.$$

Solving this equation for the momentum gives

$$p = \sqrt{2m(KE)}.$$

Hence, the de Broglie wavelength of the electron is

$$\lambda = \frac{h}{\sqrt{2m(KE)}} = \frac{6.626 \times 10^{-34} \text{ J} \cdot \text{s}}{\sqrt{2 \times 9.11 \times 10^{-31} \text{ kg} \times 1.602 \times 10^{-18} \text{ J}}} = 3.88 \times 10^{-10} \text{ m},$$

about the size of an atom.

The analogy between light waves and the waves associated with particles due to de Broglie suggests that it should be possible to observe two-slit interference effects for beams consisting of electrons. The first double-slit experiments with electrons was performed by Claus Jönsson in 1961 using the apparatus illustrated in Fig. 1.11. In Jönsson's experiment, a beam of electrons produced by a hot filament were accelerated through 50 kV and then passed through a double slit with separation 2.0 μm and width 0.5 μm. The resulting intensity pattern, which is shown in Fig. 1.12, is very similar to the pattern produced by the double-slit interference for light shown in Fig. 1.12(B) of the introduction.

The results of the experiments of Davisson and Germer for the scattering of electrons by crystals and the double-slit interference experiments of Jönsson show that a beam of particles have a wave associated with them just as do beams of

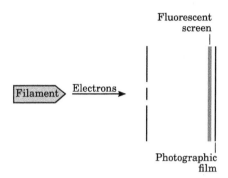

FIGURE 1.11 The two-slit apparatus used to study interference effects involving electrons. Electrons are accelerated from the filament and pass through the double slit producing an interference pattern on the fluorescent screen.

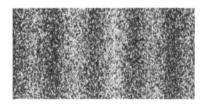

FIGURE 1.12 Photograph of the interference pattern produced by electrons passing through a double slit. The figure was obtained from the web site http://commons.wikimedia.org/wiki.

photons. The momentum of the particles in a beam is related to the wavelength λ of the be Broglie wave by Eq. (1.25), or, equivalently, the momentum is related to the angular wave vector k by Eq. (1.27).

Suggestions for further reading

Edited with Commentaries by H.A. Boorse and L. Motz, *The World of the Atom* (New York: Basic Books, 1966).
Thomas A. Moore, *Six Ideas That Shaped Physics*, Second Edition (New York: Mc Graw Hill, 2003).
A. Pais, *Niels Bohr's Times in Physics, Philosophy and Polity* (Oxford: Clarendon Press, 1991).
H.C. King, *The History of the Telescope* (Cambridge, Massachusetts: Sky Publishing, 1955).

Basic equations

Photoelectric effect

Photon energy

$$E = \frac{hc}{\lambda}, \quad \text{where } hc = 1240 \, \text{eV} \cdot \text{nm}$$

Maximum kinetic energy of electrons

$$(KE)_{\text{max}} = \frac{hc}{\lambda} - W$$

Emission and absorption of radiation by atoms

Transition energy

$$E_{\text{photon}} = E_2 - E_1$$

Energy levels of hydrogen atom

$$E_n = -\frac{13.6 \, \text{eV}}{n^2}, \quad \text{for } n = 1, 2, 3, \ldots$$

Wave properties of radiation and matter

Bragg's law

$$2d \sin \theta = n\lambda$$

de Broglie wavelength

$$\lambda = \frac{h}{p}$$

Summary

Light has both a particle-like and wave-like character. Particles of light have an energy

$$E = \frac{hc}{\lambda},$$

where the product of the constants, h and c, can be conveniently written $hc = 1240 \text{ eV} \cdot \text{nm}$. The theory of light depending on its particle-like nature successfully accounts for the photoelectric effect, the emission and absorption of light by the electrons in an atom, and the scattering of light by loosely bound electrons – the Compton Effect.

The wave character of light is revealed by the interference patterns produced when X-rays are scattered by the regular array of atoms in crystals and by a variety of other interference and diffraction experiments.

In 1923, Louis de Broglie made the remarkable suggestion that particles such as electrons should have wave-like properties. The dual wave-particle character of light should then be a common phenomenon that applies to light and to all microscopic objects. De Broglie proposed that the wave-like and particle-like properties of microscopic objects like electrons are related by the formula

$$\lambda = \frac{h}{p}.$$

Questions

1. How are the threshold frequency f_0 and wavelength λ_0 for the photoelectric effect related to the work function W of a metal?
2. Write down an equation showing how the maximum kinetic energy of photoelectrons depends upon the wavelength of the incident light and the work function of the metal.
3. What would be the work function of a metal with threshold wavelength of 400 nm?
4. Would photoelectrons be emitted from a metal if the wavelength of the incident light were longer than the threshold wavelength?
5. Who was the first person to obtain an accurate spectrum of the Sun?
6. What causes the dark lines in the solar spectra?
7. Suppose that an atom makes a transition from an energy level E_2 to an energy level E_1 emitting light. How is the light's frequency f related to the difference of energy, $E_2 - E_1$?
8. Write down a formula for the energy levels of hydrogen.
9. Suppose that an atom in the $n = 3$ state of hydrogen makes a transition to the $n = 2$ state. What will be the energy of the emitted photon? What would be the wavelength of the light?
10. Give the principle quantum numbers of the initial and final states for the following transitions.
    (a) the first member of the Lyman series
    (b) the third member of the Balmer series
    (c) the second member of the Paschen series
11. What is the energy of the hydrogen atom in the $n = 4$ state?
12. What photon energy would be required to ionize a hydrogen atom in the $n = 5$ state?
13. What common features do the energy levels of the hydrogen atom and the sodium atom have?
14. How do the energy levels of the sodium atom differ from the energy levels of sodium metal?
15. What model of light can be used to describe the *Compton effect*?
16. What is Bragg's law for X-ray diffraction?

17. How are the momentum and the wavelength of a particle related?
18. Express the momentum of an electron in terms of the angular wave vector k.
19. How would the de Broglie wavelength of an electron change if its velocity were to double?
20. Describe the interference pattern is produced when a beam of light having a single wavelength is incident upon two slits?

Problems

1. Calculate the energy of the photons for light having a wavelength $\lambda = 200$ nm.
2. Suppose that a 100 Watt beam of light is incident upon a metal surface. If the light has a wavelength $\lambda = 200$ nm, how many photons strike the surface every second? (*Hint*: Using the power of the beam, find how much energy is incident upon the surface every second.)
3. A helium-neon laser produces red light at a wavelength of 632.8 nm. If the laser output power is 1 milliwatt, what is the rate of photons being emitted from the laser?
4. What will the maximum kinetic energy of the emitted photoelectrons be when ultraviolet light having a wavelength of 200 nm is incident upon the following metal surfaces?
   (a) Na
   (b) Al
   (c) Ag
5. Radiation of a certain wavelength is incident on a cesium surface having a work function of 2.3 eV. Electrons are emitted from the surface with a maximum kinetic energy of 0.9 eV. What is the wavelength of the incident radiation?
6. Light having a wavelength of 460 nm is incident on a cathode, and electrons are emitted from the metal surface. It is observed that the electrons may be prohibited from reaching the anode by applying a stopping potential of 0.72 eV. What is the work function of the metal in the cathode?
7. By how much would the stopping potential in the previous problem increase if the wavelength of the radiation were reduced to 240 nm?
8. The maximum wavelength of radiation that can lead to the emission of electrons from a metal surface is 360 nm. What is the work function of the metal?
9. Electromagnetic radiation having a frequency of 1200 THz is incident upon a surface of sodium metal. What is the maximum kinetic energy of the emitted electrons?
10. Show that Eqs. (1.7) can be written in the form (1.8) provided that the Rydberg constant is assigned the value $R = 1.0972 \times 10^5$ cm^{-1}.
11. An electron in the $n = 5$ state of hydrogen makes a transition to the $n = 2$ state. What are the energy and wavelength of the emitted photon?
12. What must the energy of the photons of light be to cause a transition from the $n = 2$ state of hydrogen to the $n = 4$ state?
13. Light is incident upon the hydrogen atom in the $n = 2$ state and causes a transition to the $n = 3$ state. What is the wavelength of the light?
14. What wavelength of light is required to just ionize a hydrogen atom in the $n = 2$ state?
15. Suppose that light with wavelength 200 nm is absorbed by a hydrogen atom in the $n = 2$ state. What will be the kinetic energy of the emitted electron?
16. Ultraviolet light of wavelength 45.0 nm is incident upon a collection of hydrogen atoms in the ground state. Find the kinetic energy and the velocity of the emitted electrons.
17. What is the wavelength of the following transitions?
    (a) the first member of the Lyman series
    (b) the third member of the Balmer series
    (c) the second member of the Paschen series
18. What is the maximum wavelength of radiation that a hydrogen atom in the $n = 3$ state can absorb?
19. What is the kinetic energy of an electron with a de Broglie wavelength of 0.2 nm?
20. What potential difference must electrons be accelerated through to have the same wavelength as 40 keV X-rays?
21. Electrons, protons, and neutrons have wavelengths of 0.01 nm. Calculate their kinetic energies.
22. Calculate the range of speeds of electrons having wavelengths equal to that of visible light.
23. An electron has a kinetic energy of 40 keV. What kinetic energy must a proton have to have the same de Broglie wavelength as the electron?

Chapter 2

The Schrödinger wave equation

Contents

2.1	The wave equation	19	Suggestion for further reading		33
2.2	Probabilities and average values	23	Basic equations		33
2.3	The finite potential well	26	Summary		34
2.4	The simple harmonic oscillator	28	Questions		35
2.5	Time evolution of the wave function	30	Problems		35

The Schrödinger equation enables one to find the de Broglie wave associated with a particle and thus provides a basic tool for understanding microscopic phenomena.

The wave theory of matter originally proposed by de Broglie was further developed by Erwin Schrödinger who suggested that just as ray optics must be replaced by wave optics when light waves are obstructed by small objects, so classical mechanics must be replaced by wave mechanics in dealing with microscopic systems. Taking advantage of the analogy between optics and classical mechanics, Schrödinger obtained a differential equation which has come to be called the *Schrödinger equation*, and he used the equation to derive an expression for the energy levels of hydrogen obtaining results consistent with the earlier theory of Bohr. In this chapter, we shall obtain the Schrödinger equation and use this equation to study the properties of a number of physical problems in one dimension. Rather than following Schrödinger's original line of argument, we will show that the Schrödinger equation arises naturally in the context of the ideas of de Broglie and the results of electron diffraction experiments.

2.1 The wave equation

The electron diffraction experiments of Davisson and Germer and Thompson and Reid confirmed de Broglie's suggestion that a wave is associated with the motion of a particle. According to Eqs. (1.25) and (1.27) of Chapter 1, the momentum of the particle is related to the wavelength λ and angular wave vector k of the associated wave by the equations

$$p = \frac{h}{\lambda}, \tag{2.1}$$

and

$$p = \hbar k. \tag{2.2}$$

A relation between the kinetic energy of the particle and the angular wave vector can be obtained by writing the kinetic energy in terms of the momentum as follows

$$KE = \frac{1}{2}mv^2 = \frac{1}{2m}(mv)^2.$$

This last equation can be written

$$KE = \frac{1}{2m}p^2, \tag{2.3}$$

where p is the momentum of the particle. Substituting Eq. (2.2) into Eq. (2.3) gives

$$KE = \frac{(\hbar k)^2}{2m}.$$

Modern Physics with Modern Computational Methods. https://doi.org/10.1016/B978-0-12-817790-7.00009-3

The form of the wave function for a stationary wave is given by Eq. (I.23) of the introduction. We have

$$\psi(x) = A e^{ikx}. \tag{2.4}$$

Following de Broglie, we shall suppose that the wave function (2.4) is associated with the motion of a free particle and refer to this function simply as the de Broglie wave function. For a free particle, the potential energy is constant and may be taken to be zero. The energy of the particle is then

$$E = \frac{(\hbar k)^2}{2m}. \tag{2.5}$$

The Davisson-Germer diffraction experiments can be successfully explained by associating with a beam of freely moving electrons the de Broglie wave function (2.4) with the angular wave vector k related to the momentum and the energy of the particles by Eqs. (2.2) and (2.5). Eqs. (2.4), (2.2), and (2.5) together provide a concise summary of the experimental results.

We shall now find differential equations which are satisfied by the de Broglie function (2.4). Taking the derivative of Eq. (2.4) with respect to x, we obtain

$$\frac{d\psi}{dx} = A i k e^{ikx} = ik\psi.$$

We can simplify this result by multiplying the equation through from the left with $-i\hbar$ to get

$$-i\hbar \frac{d\psi}{dx} = \hbar k \, \psi,$$

and then using Eq. (2.2) to identify the term $\hbar k$ appearing on the right as the momentum p. The differential equation satisfied by the de Broglie wave function (2.4) can thus be written

$$-i\hbar \frac{d\psi}{dx} = p \, \psi. \tag{2.6}$$

Another equation satisfied by the de Broglie wave function can be obtained by taking the derivative of Eq. (2.6) with respect to x and multiplying the resulting equation through by $-i\hbar$. We get

$$-\hbar^2 \frac{d^2\psi}{dx^2} = -ip\hbar \frac{d\psi}{dx}.$$

This equation can be simplified using Eq. (2.6) to evaluate the term on the right giving

$$-\hbar^2 \frac{d^2\psi}{dx^2} = p^2 \, \psi.$$

We then divide the equation through by $2m$ and use the relation, $E = p^2/2m$, to replace the coefficient of the wave function on the right by the energy

$$\frac{-\hbar^2}{2m} \frac{d^2\psi}{dx^2} = E \, \psi. \tag{2.7}$$

Since the de Broglie function $\psi(x) = A e^{ikx}$ is a solution of Eq. (2.7) with $E = \hbar^2 k^2/2m$, this equation may be thought of as the equation associated with a free particle with energy E.

To obtain a general equation that would apply to any particle with potential energy $V(x)$, we first note that Eq. (2.7) was obtained by taking the second derivative of the de Broglie wave function (2.4) which is known to describe a free particle. Substituting Eq. (2.4) into Eq. (2.7) gives

$$\frac{\hbar^2 k^2}{2m} A e^{ikx} = E A e^{ikx}.$$

Notice that the coefficient of the de Broglie wave function on the left-hand side of this equation is the kinetic energy of the particle. An obvious way of extending Eq. (2.7) to include the effects of an external force is to add a second term to the

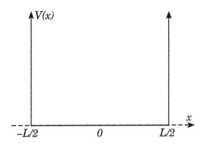

FIGURE 2.1 An infinite potential well extending from $x = -L/2$ to $x = L/2$.

left-hand side of Eq. (2.7) corresponding to the potential energy of the particle. We thus write

$$\frac{-\hbar^2}{2m}\frac{d^2\psi}{dx^2} + V(x)\psi = E\psi. \tag{2.8}$$

This equation, which was originally obtained by Schrödinger in 1926, is called the *Schrödinger time-independent equation*. The Schrödinger equation (2.8) reduces to the free-particle equation (2.7) when the potential energy is equal to zero.

We shall presently solve the Schrödinger equation (2.8) for a number of different physical problems and find that the equation provides a convincing description of microscopic systems. The validity of the Schrödinger equation depends ultimately upon how well the predictions of the theory are confirmed by experiment.

The theory of Schrödinger provides a means for studying the de Broglie waves, which can be thought of as solutions of the Schrödinger equation. The Schrödinger equation also gives the correct energy levels of hydrogen, and thus provides a framework for understanding the earlier results of Balmer and Bohr. When Schrödinger first presented his theory, though, he was unsure how the wave function should be interpreted. To overcome this difficulty, Max Born, who made numerous contributions to the new quantum theory, proposed an interpretation of the wave function in terms of probabilities. Born suggested that the probability of finding the particle in a particular location depended upon the absolute value squared of the wave function. More precisely, the probability dP that the particle is in the infinitesimal interval between x and $x + dx$ is given by the equation

$$dP = |\psi(x)|^2 dx, \tag{2.9}$$

where $|\psi(x)|^2$ is the product of the function $\psi(x)$ and its complex conjugate $\psi^*(x)$

$$|\psi(x)|^2 = \psi^*(x)\psi(x).$$

Eq. (2.9), which defines the relation between the probability and the wave function, can also be written

$$|\psi(x)|^2 = \frac{dP}{dx},$$

and we may thus refer to $|\psi(x)|^2$ as the *probability density*.

The wave mechanical theory of Schrödinger with the probabilistic interpretation of Born is now widely accepted as providing an accurate description of microscopic systems. In this chapter, we shall use the theory to study the properties of a number of physical problems in one dimension.

Example 2.1

Find the wave function and the energy levels of a particle confined within the infinite potential well shown in Fig. 2.1.

Solution
The potential energy function $V(x)$ shown in Fig. 2.1 is equal to zero within the well. Substituting $V(x) = 0$ in Eq. (2.8), we obtain the equation,

$$\frac{-\hbar^2}{2m}\frac{d^2\psi}{dx^2} = E\psi, \tag{2.10}$$

for the wave function $\psi(x)$ within the well with x is in the range $-L/2 \leq x \leq +L/2$. The probability of finding the particle at a particular point may be obtained from the wave function using Eq. (2.9). Since the particle is confined within the interior of the potential well, the wave function is zero beyond the well boundaries at $x = \pm L/2$.

A careful analysis shows that the wave function and its derivative must be continuous at the boundaries of the well. Here we only give the reason in general terms. We will find that in addition to a probability density one can also define a current density which describes the flow of particles within the well. If the wave function or its derivative were discontinuous at $x = \pm L/2$, the boundaries of the well would not simply confine the particles as intended but serve as an emitter or an absorber of particles.

Since the wave function, $\psi(x)$, is continuous and is equal to zero beyond the well boundaries, the wave function must approach zero as x approaches $-L/2$ or $+L/2$. We thus require that the solutions of Eq. (2.10) satisfy the following boundary conditions

$$\psi(-L/2) = 0$$
$$\psi(+L/2) = 0. \tag{2.11}$$

In order to be in a better position to solve the Schrödinger equation (2.10), we bring the term $E\psi$ over to the left-hand side of the equation, and we multiply the entire equation through by $-2m/\hbar^2$ to obtain

$$\frac{d^2\psi}{dx^2} + \left(\frac{2mE}{\hbar^2}\right)\psi = 0.$$

Defining

$$k^2 = \left(\frac{2mE}{\hbar^2}\right), \tag{2.12}$$

the Schrödinger equation becomes

$$\frac{d^2\psi}{dx^2} + k^2\psi = 0. \tag{2.13}$$

One may readily confirm that the functions,

$$\psi(x) = A\cos(kx)$$

and

$$\psi(x) = B\sin(kx),$$

are solutions of Eq. (2.13). For instance, substituting $\psi(x) = A\cos(kx)$ into Eq. (2.13) gives

$$\frac{d^2(A\cos(kx))}{dx^2} + k^2 A\cos(kx) = -Ak^2\cos(kx) + k^2 A\cos(kx) = 0.$$

The cosine function is an even function of x being unchanged when x is replaced by $-x$, while the sine function is an odd function of x changing its sign when x is replaced by $-x$.

We consider first the even solutions. Inside the well, the even solutions have the form

$$\psi(x) = A\cos kx, \quad \text{for } -L/2 \le x \le L/2. \tag{2.14}$$

These solutions satisfy the boundary conditions (2.11) provided that

$$kL/2 = n\pi/2,$$

where n is an odd integer. Similarly, the odd functions,

$$\psi(x) = A\sin kx, \quad \text{for } -L/2 \le x \le L/2, \tag{2.15}$$

satisfy the boundary conditions (2.11) provided that

$$kL/2 = n\pi/2,$$

where n is an even integer. The conditions imposed upon the even and odd solutions both lead to the equation

$$k = \frac{n\pi}{L}, \tag{2.16}$$

where the integer n may be odd or even.

The even solution corresponding $n = 1$ is

$$\psi(x) = A\cos\left(\frac{\pi x}{L}\right),$$

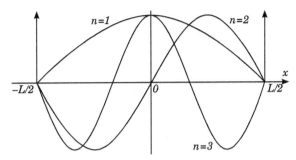

FIGURE 2.2 The solutions of the Schrödinger equation for the infinite well for $n = 1, 2$, and 3. Solutions with n odd are even and solutions with n even are odd.

while the odd solution corresponding to $n = 2$ is

$$\psi(x) = A \sin\left(\frac{2\pi x}{L}\right).$$

One can readily confirm that the solutions of the Schrödinger equation for $n = 1$ and $n = 2$ go to zero on the boundary. For larger values of n, the solutions with n odd are even, and solutions with n even are odd. Fig. 2.2 shows the solutions of the Schrödinger equation for the infinite well corresponding to $n = 1, 2$, and 3.

Solving Eq. (2.12) for the energy, we obtain

$$E = \frac{\hbar^2 k^2}{2m}.$$

An equation for the energy in terms of the integer n can then be obtained by substituting Eq. (2.16) into the above equation giving

$$E = \frac{n^2 \pi^2 \hbar^2}{2m L^2}.$$

Using the fact that $\hbar = h/2\pi$, this last equation may be written more simply

$$E = \frac{n^2 h^2}{8m L^2}. \qquad (2.17)$$

The wave functions and the values of the energy of a particle moving in an infinite well depend upon the positive integer n, which is called a *quantum number*.

The wave functions shown in Fig. 2.2 have the form of standing waves. As for the functions describing the displacement of a guitar string, the wave function of a particle in an infinite well is equal to zero at the ends of the physical region. The displacement of a guitar string must be zero at the ends because the string is tied, while the wave function of a particle in an infinite well must be zero because the particle cannot penetrate the infinite walls of the well.

2.2 Probabilities and average values

The probability dP of finding the particle at a particular point in space is related to the absolute value squared of the wave function. According to Eq. (2.9), the probability that the particle is in the infinitesimal interval between x and $x + dx$ is equal to

$$dP = |\psi(x)|^2 dx.$$

The probability of finding the particle between two points, x_1 and x_2, may be obtained by integrating this last equation for dP over the range between x_1 and x_2. We have

$$P(x_1 \rightarrow x_2) = \int_{x_1}^{x_2} |\psi(x)|^2 dx.$$

In performing the above integration, one adds up the probabilities for the infinitesimal intervals between x_1 and x_2. The requirement that the total probability of finding the particle at any point along the x-axis is equal to one can be written

$$\int_{-\infty}^{\infty} |\psi(x)|^2 dx = 1, \tag{2.18}$$

where the integration is performed over the entire range of the variable x. The condition (2.18) is referred to as the *normalization condition*. It enables us to calculate the value of the constant A in the previous example.

Example 2.2

Use the normalization condition to evaluate the constant A for the wave functions in Example 2.1.

Solution

Since the particle in the infinite well shown in Fig. 2.1 must be located between $-L/2$ and $+L/2$, the probability of finding the particle in the range between $-L/2$ and $+L/2$ must be equal to one. We have

$$\int_{-L/2}^{L/2} |\psi(x)|^2 dx = 1. \tag{2.19}$$

The even solutions for the particle in the infinite well can be obtained using Eqs. (2.14) and (2.16). We obtain

$$\psi(x) = A\cos\left(\frac{n\pi x}{L}\right).$$

Substituting this wave function into the integral in Eq. (2.19) gives

$$A^2 \int_{-L/2}^{L/2} \cos^2\left(\frac{n\pi x}{L}\right) dx = 1.$$

In order to evaluate the above definite integral, we first note that the integral gives the area under the curve of the function $\cos^2(n\pi x/L)$ for a number of half periods. This area is the same as the area under the curve of the function $\sin^2(n\pi x/L)$. We may thus replace $\cos^2(n\pi x/L)$ with $\frac{1}{2}[\sin^2(n\pi x/L) + \cos^2(n\pi x/L)]$ in the above integral to obtain

$$A^2 \int_{-L/2}^{L/2} \frac{1}{2}\left[\sin^2\left(\frac{n\pi x}{L}\right) + \cos^2\left(\frac{n\pi x}{L}\right)\right] dx = 1.$$

Using the identity, $\sin^2(n\pi x/L) + \cos^2(n\pi x/L) = 1$, the last equation may be written simply

$$A^2 \int_{-L/2}^{L/2} \frac{1}{2} dx = 1.$$

Since the above integral has the value $L/2$, we obtain the result

$$A = \sqrt{\frac{2}{L}}.$$

The normalization constant A for the odd functions can be shown to be given by the same equation.

Using this result, the wave function of a particle in an infinite well can be written

$$\psi_n(x) = \begin{cases} \sqrt{2/L}\cos(n\pi x/L), & n \text{ odd} \\ \sqrt{2/L}\sin(n\pi x/L). & n \text{ even} \end{cases} \tag{2.20}$$

The absolute value of the wave function squared gives the probability density of finding the particle at a particular point along the x-axis. For most problems, the probability density will be nonzero over a range of values of x, and a measurement of x may lead to any value within the range. We can still use the wave function, though, to predict the average value of a large number of different measurements of x. In order to motivate the formula that we shall use for the average value of an observable, we consider the problem of determining the average age of the students in a class. Suppose that there are n_1

students having an age a_1, n_2 students having an age a_2, and so forth, and suppose that the total number of students is N. Then the average age of the students in the class will be

$$\text{Average age} = \frac{\sum_i a_i n_i}{N},$$

where the sum runs over all of the different possible ages of the students. This last equation can be written

$$\text{Average age} = \sum_i a_i \left(\frac{n_i}{N}\right),$$

where N has been brought inside the summation. For each value of i, the ratio n_i/N is equal to the probability P_i of a student having the age a_i. So this last result can be written

$$\text{Average age} = \sum_i a_i P_i.$$

In exactly the same way, the average position of a particle can be calculated using the equation

$$<x> = \int_{-\infty}^{\infty} x|\psi(x)|^2 dx, \tag{2.21}$$

where the integration is extended over the entire range of the variable x. The above integration adds up the contributions from all the infinitesimal intervals from $-\infty$ to ∞. For each interval, x is multiplied by the probability that a particle should be in the infinitesimal interval containing x. This formula for calculating the average value of x may be extended so that it enables us to calculate the average value of an arbitrary function of x

$$<f(x)> = \int_{-\infty}^{\infty} f(x)|\psi(x)|^2 dx.$$

We now give the following example of the application of this formula.

Example 2.3

A particle, which is confined to move in one-dimension between 0 and L, is described by the wave function

$$\psi(x) = A x(L - x). \tag{2.22}$$

Use the normalization condition (2.18) to determine the constant A and then derive an expression for the average value of the position of the particle.

Solution

Since the particle is located between 0 and L, the normalization condition is

$$\int_0^L |\psi(x)|^2 dx = 1.$$

Substituting the wave function (2.22) into this equation, we get

$$A^2 \int_0^L x^2(L - x)^2 dx = 1.$$

To evaluate the above integral we multiply the terms in the integrand together to obtain

$$A^2 \int_0^L (L^2 x^2 - 2L x^3 + x^4) dx = 1,$$

and then integrate the polynomial in the integrand term for term to get

$$A^2 \left[\frac{L^2 x^3}{3} - \frac{2L x^4}{4} + \frac{x^5}{5} \right]_0^L = 1.$$

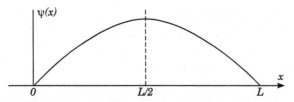

FIGURE 2.3 The function $\psi(x) = Ax(L - x)$, for x in the range, $0 \leq x \leq L$.

The left-hand side of the above equation is evaluated by first substituting $x = L$ into the expression within square brackets and then subtracting the result obtained by substituting $x = 0$ into the expression. We notice that all of the terms obtained by substituting $x = L$ contain a factor of L^5 and all the terms obtained by substituting $x = 0$ give zero. The above equation thus becomes

$$A^2 L^5 \left[\frac{1}{3} - \frac{1}{2} + \frac{1}{5} \right] = A^2 L^5 \frac{1}{30} = 1.$$

Solving for A, we obtain

$$A = \sqrt{\frac{30}{L^5}},$$

and the wave function can be written

$$\psi(x) = \sqrt{\frac{30}{L^5}} x(L - x). \tag{2.23}$$

In order to obtain the average value of the position, we substitute the wave function (2.23) into Eq. (2.21) to obtain

$$<x> = \int_0^L x \frac{30}{L^5} x^2 (L - x)^2 dx = \frac{30}{L^5} \int_0^L (L^2 x^3 - 2L x^4 + x^5) dx.$$

Evaluating the last integral, we obtain

$$<x> = \frac{30}{L^5} \left[\frac{L^2 x^4}{4} - \frac{2L x^5}{5} + \frac{x^6}{6} \right]_0^L.$$

At this point, we note again that all terms obtained by substituting $x = L$ into the expression within square brackets contain a factor of L^6 and all the terms obtained by substituting $x = 0$ give zero. We thus obtain

$$<x> = \frac{30L^6}{L^5} \left[\frac{1}{4} - \frac{2}{5} + \frac{1}{6} \right] = \frac{L}{2}.$$

Notice that the wave function shown in Fig. 2.3 is symmetric with respect to the dotted vertical line passing through the maximum of the function at $x = L/2$ and that the average value of x corresponds to the value of x for which the function attains its maximum. One must be aware, though, that the average value of x does not generally coincide with the maximum value of the wave function.

2.3 The finite potential well

The infinite potential well illustrates how discrete energy levels of a quantum system can arise. A more realistic model of the environment of particles in microscopic systems is provided by the finite well which also has discrete energy levels and provides a realistic description of the active region in semiconductor lasers and is the basis of the nuclear shell model. A finite well is illustrated in Fig. 2.4.

The approach we shall use for the finite well is more complex than the approach we adopted for the infinite well. For the infinite well, we solved the Schrödinger equations inside the well and then required that the wave function goes to zero on the boundaries of the well. In contrast, we shall now solve the Schrödinger equation corresponding to the finite well for values of x inside and outside the well, and we shall require that the solutions be continuous across the boundaries between the regions inside and outside the well. As for the infinite potential well, the Schrödinger equation inside the finite well is

$$\frac{-\hbar^2}{2m} \frac{d^2 \psi}{dx^2} = E\psi, \quad \text{for } -L/2 \leq x \leq L/2, \tag{2.24}$$

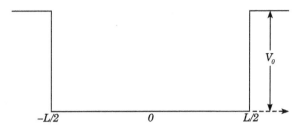

FIGURE 2.4 A finite well of depth V_0 extending from $x = -L/2$ to $x = L/2$.

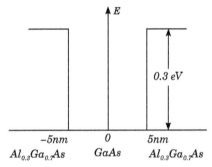

FIGURE 2.5 Energies of conduction electron for a structure in which one layer of $GaAs$ is sandwiched between two layers of $Al_{0.3}Ga_{0.7}As$. Electrons in $GaAs$ are in a quantum well of depth 0.30 eV and width 10 nm.

while the Schrödinger equation outside the well is

$$\frac{-\hbar^2}{2m}\frac{d^2\psi}{dx^2} + V_0\psi = E\psi, \quad \text{for } |x| \geq L/2 . \tag{2.25}$$

As a first example of a finite well, we consider a structure that is formed with layers of the semiconductors, $GaAs$ and $AlGaAs$. $GaAs$ is a semiconductor with Ga at one set of sites in the crystal and As at the other sites. Crystals composed of the alloy, $Al_xGa_{1-x}As$ have Al or Ga at one set of sites in the crystal and As at the other sites. The indices, x and $1 - x$, give the fractional number of the corresponding species in the crystal. As will be discussed in Chapter 10, the alloys of gallium arsenide $GaAs$ and indium phosphide InP are useful for making light-emitting diodes and semiconductor lasers because the conduction electrons of these compounds can make direct optical transitions to unoccupied states in valence band.

Fig. 2.5 shows the energies available to conduction electrons for a structure in which one layer of $GaAs$ is sandwiched between two layers of $Al_{0.3}Ga_{0.7}As$. The $Al_{0.3}Ga_{0.7}As$ crystal has 30% aluminum and 70% gallium. The range of energies available to conduction electrons in $GaAs$ in the center dips down below the levels of $Al_{0.3}Ga_{0.7}As$ on the two sides. Electrons near the bottom of the conduction band of $GaAs$ are not free to move because there are no available states of the same energy in $Al_{0.3}Ga_{0.7}As$. This situation is generally described by saying that the conduction electrons in $GaAs$ are confined within a *quantum well*. As can be seen in Fig. 2.5, the depth of the quantum well created with $GaAs$ sandwiched between two layers of $Al_{0.3}Ga_{0.7}As$ is about 0.3 eV.

The periodic potential due to the atoms in a crystal can enhance or retard the motion of the conduction electrons. As will be discussed in Chapter 9, the screening of the charge of conduction electrons by the ions of the crystal can cause the effective mass of the electrons in a crystal to differ considerably from the mass of free electrons. The effective mass of conduction electrons in $GaAs$ is equal to 0.067 times the free electron mass. This value of the electron mass must be used to find the wave functions of electrons bound in the finite well shown in Fig. 2.5.

As for the infinite well, we multiply Eqs. (2.24) and (2.25) from the left with $2m/\hbar^2$ to obtain

$$-\frac{d^2\psi}{dx^2} = (2mE/\hbar^2)\psi, \quad \text{for } |x| \geq L/2 ,$$

and

$$-\frac{d^2\psi}{dx^2} + (2mV_0/\hbar^2)\psi = (2mE/\hbar^2)\psi, \quad \text{for } |x| \geq L/2 .$$

To be in a position to solve the Schrödinger equations for all values of x, we first introduce dimensionless variables that give the position of the electron in nanometers and the energy and potential energy in electron volts.

$$x = \chi \cdot \text{nm}, \quad E = \epsilon \cdot \text{eV}, \quad V_0 = \mathcal{V}_0 \cdot \text{eV} \tag{2.26}$$

The derivative of ψ with respect to x may be expressed in terms of the derivative with respect to the dimensionless variable χ by using the chain rule. We have

$$\frac{d\psi}{dx} = \frac{d\psi}{d\chi}\frac{d\chi}{dx} = (nm)^{-1}\frac{d\psi}{d\chi},$$

and the second derivative of ψ with respect to x is

$$\frac{d^2\psi}{dx^2} = (nm)^{-2}\frac{d^2\psi}{d\chi^2}.$$

Substituting this expression for the second derivative into the Schrödinger for conduction electrons inside the potential well and multiplying the resulting equations by nm^2, we find that Schrödinger equation for χ inside the well is

$$-\frac{d^2u}{d\chi^2} = E_0\,\epsilon u, \quad \text{for } -5 \le \chi \le 5, \tag{2.27}$$

where the dimensionless number E_0 is given by the equation

$$E_0 = \frac{2m(\text{nn})^2\text{eV}}{\hbar^2}, \tag{2.28}$$

and ϵ is the energy of electrons in electron Volts. Similarly the Schrödinger equation for electrons outside the well is

$$-\frac{d^2u}{d\chi^2} + \mathcal{V}_0 E_0 u = \epsilon\, E_0 u, \quad \text{for } |\chi| \ge 5, \tag{2.29}$$

where $\mathcal{V}_0$ is the depth of the well in electron volts. As we have just discussed, the effective mass of the electrons in a $GaAs$ crystal is 0.067 times the free electron mass. Using this value for the mass of electrons in $GaAs$, the value of E_0 is found to be 1.759.

We postpone the solution of Eqs. (2.27) and (2.29) until Chapter 3 where the wave functions and the energies of electrons in a finite well can be determined within the context of general principles of quantum mechanics.

2.4 The simple harmonic oscillator

Another important application of the wave mechanical ideas we have introduced in this chapter is the simple harmonic oscillator illustrated in Fig. I.2 of the introduction. The classical oscillator consists of a particle of mass m moving under the influence of a linear restoring force. As described in the introduction, the potential energy of the oscillator is given by the function,

$$V(x) = \frac{1}{2}m\omega^2 x^2, \tag{2.30}$$

where ω is the classical frequency.

The Schrödinger equation for the harmonic oscillator may be written

$$\frac{-\hbar^2}{2m}\frac{d^2\psi}{dx^2} + \frac{1}{2}m\omega^2 x^2\psi = E\psi, \tag{2.31}$$

where the first term corresponds to the kinetic energy, $(1/2)m\omega^2 x^2$ is the potential energy and E is the energy of the oscillator. Multiplying the Schrödinger equation of the oscillator from the left by $(2m/\hbar^2)$, we obtain

$$-\frac{d^2\psi}{dx^2} + \left(\frac{m\omega^2 x^2}{\hbar^2}\right)\psi = \left(\frac{2mE}{\hbar^2}\right)\psi. \tag{2.32}$$

This equation can be further simplified by introducing the change of variable

$$\chi = \alpha x,$$

where α is a constant yet to be specified. This last equation can also be written

$$x = \alpha^{-1}\chi. \tag{2.33}$$

As before, the derivate of the wave function can then be expressed in terms of the variable χ by using the chain rule. We have

$$\frac{d\psi}{dx} = \frac{d\chi}{dx}\frac{d\psi}{d\chi} = \alpha\frac{d\psi}{d\chi}.$$

Similarly, the second derivative can be written

$$\frac{d^2\psi}{dx^2} = \alpha^2\frac{d^2\psi}{d\chi^2}. \tag{2.34}$$

Substituting Eqs. (2.33) and (2.34) into Eq. (2.32) and dividing the resulting equation through with α^2, we obtain

$$-\frac{d^2\psi}{d\chi^2} + \frac{1}{\alpha^4}\frac{m^2\omega^2}{\hbar^2}\chi^2\psi = \frac{2mE}{\alpha^2\hbar^2}\psi. \tag{2.35}$$

We may simplify the above equation for the oscillator by choosing

$$\alpha = \sqrt{\frac{m\omega}{\hbar}},$$

and defining

$$\epsilon = \frac{2E}{\hbar\omega}. \tag{2.36}$$

The Schrödinger equation for the oscillator then assumes the following simple form

$$-\frac{d^2\psi}{d\chi^2} + \chi^2\psi = \epsilon\psi, \tag{2.37}$$

where the variable χ is related to the x-coordinate by the equation

$$\chi = \sqrt{\frac{m\omega}{\hbar}}x.$$

Physicist in the early part of the twentieth century routinely solved differential equations such as Eq. (2.37) using power series methods. The classic equations of modern physics that were solved in this way each led to polynomials that were characteristic of the particular physical problem considered. As we shall see, the solution of the harmonic oscillator equation (2.37) leads to Hermite polynomials. The solution of Laplace's equation for problems having spherical symmetry lead to Legendre polynomials, and the solution of the radial equation for hydrogen leads to Laguerre polynomials. However, one should realize that only a very small number of the differential equations that arise in physics can be solved analytically. In the third chapter of this book, we shall develop numerical methods and describe MATLAB® programs for solving the Schrödinger equation for a finite well and for the harmonic oscillator. In Chapter 4, the radial equation for hydrogen will be solved numerically, and the partial differential equations for the hydrogen molecular ion, H_2^+, will be solved numerically in Chapter 7. I tell my students that when they see a differential equation or a partial differential equation, I want them to think of the numerical strategy and the MATLAB program they would write to solve the equation.

Eq. (2.37) is solved in Appendix D using power series methods. Here we give only the results of this calculation, which will serve as a check of the numerical solution obtained in Chapter 3.

The possible energies for the oscillator are

$$E = \hbar\omega(n + 1/2) \tag{2.38}$$

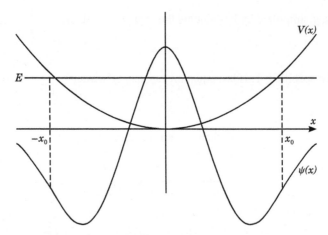

FIGURE 2.6 The potential energy $V(x)$ and the wave function $\psi(x)$ for the $n = 2$ state of the harmonic oscillator.

where n is a nonnegative integer, $n = 0, 1, 2, \ldots$. We note that the lowest state, for which $n = 0$, has a positive energy $\hbar\omega/2$, which is called the *zero point energy*. For a given value of n, the wave function is

$$\psi_n(x) = A_n e^{-y^2/2} H_n(y), \tag{2.39}$$

where y is given by the equations

$$y = \sqrt{\frac{m\omega}{\hbar}} x$$

and the functions $H_n(y)$ are called *Hermite polynomials*. A_n is a normalization constant.

The first few Hermite polynomials are given in Table 2.1.

TABLE 2.1 The first seven Hermite Polynomials.

$H_0(y)$	1
$H_1(y)$	$2y$
$H_2(y)$	$2 - 4y^2$
$H_3(y)$	$12y - 8y^3$
$H_4(y)$	$12 - 48y^2 + 16y^4$
$H_5(y)$	$120y - 160y^3 + 32y^5$
$H_6(y)$	$120 - 120y^2 + 480y^4 - 64y^6$

The wave function for the $n = 2$ state is shown in Fig. 2.6. This function oscillates in the classically allowed region between the two turning points at $x = \pm x_0$ and decreases to zero for $|x| > x_0$. The fact that the wave function is nonzero beyond the classical turning points means that the particle can penetrate into the classically forbidden region. In later chapters, we shall encounter a number of interesting examples of such quantum mechanical tunneling phenomena.

2.5 Time evolution of the wave function

We would now like to consider how the wave function describing a microscopic system evolves in time. We begin as we did when we introduced the Schrödinger time-independent equation by considering first the motion of a free particle. As we discussed in the first section of this chapter, a free particle is described by the wave function

$$\psi(x) = A e^{ikx}. \tag{2.40}$$

The Schrödinger equation for a free particle (2.7) was obtained by finding a second-order differential equation satisfied by the function (2.40).

We recall that the function discussed in the introduction of this book which corresponds to a traveling wave moving in the positive x-direction is

$$\psi(x,t) = A e^{i\,(kx-\omega t)},$$

where ω is the angular frequency. This wave function, which can be written

$$\psi(x,t) = A e^{i\,kx} \cdot e^{-i\,\omega t}, \tag{2.41}$$

has the same dependence upon x as the wave function (2.40) and has, in addition, the time dependence given by the function $e^{-i\,\omega t}$. Since the function for a traveling wave evolves in time as we would expect, we begin our efforts to find an appropriate equation describing the time evolution of the wave function by deriving an equation satisfied by this wave function.

We first note that the function $\psi(x,t)$ depends upon two variables, x and t. The rate a function of two variables changes when one of the variables changes while the other variable remains constant is called a *partial derivative*. The evaluation of a partial derivative of a function is easily accomplished by taking the derivative with respect to one variable while treating the other variable as a constant. The partial derivative of the function (2.41) with respect to time is equal to

$$\frac{\partial \psi}{\partial t} = -i\omega\, A e^{i\,kx} e^{-i\omega t} = -i\omega\,\psi.$$

To obtain this result, we took the derivative of ψ with respect to time in the ordinary way while treating x as a constant.

We may now multiply the above equation through from the left with $i\hbar$, we obtain

$$i\hbar \frac{\partial \psi}{\partial t} = \hbar\omega\,\psi. \tag{2.42}$$

The term $\hbar\omega$ which appears on the right can be identified as the energy. We recall that de Broglie suggested that the equation,

$$E = hf,$$

which was originally proposed by Planck and Einstein to describe electromagnetic radiation, should be regarded as a general equation relating the energy of a particle to the frequency of the corresponding wave. Dividing h in the above equation by 2π and multiplying f by the same factor, the equation becomes

$$E = \frac{h}{2\pi} \cdot (2\pi f),$$

which may be written

$$E = \hbar\omega. \tag{2.43}$$

This last equation is analogous to the equation, $p = \hbar k$, relating the momentum and wave vector of a free particle. Using Eq. (2.43) to replace $\hbar\omega$ with E on the right-hand side of Eq. (2.42), we obtain

$$i\hbar \frac{\partial \psi}{\partial t} = E\,\psi. \tag{2.44}$$

Eq. (2.44), which is satisfied by the wave function of a traveling wave, is similar to Eq. (2.7), which is satisfied by the wave function for a standing wave. Using the notation of the partial derivative introduced recently, Eq. (2.7) may be written

$$\frac{-\hbar^2}{2m} \frac{\partial^2 \psi}{\partial x^2} = E\,\psi. \tag{2.45}$$

One may readily confirm that the wave function (2.41), which describes a traveling wave, satisfies this last equation. We can obtain a second-order equation satisfied by the function describing a traveling wave by equating the left-hand sides of Eqs. (2.45) and (2.44) to obtain

$$\frac{-\hbar^2}{2m} \frac{\partial^2 \psi(x,t)}{\partial x^2} = i\hbar \frac{\partial \psi(x,t)}{\partial t}. \tag{2.46}$$

This equation, which is satisfied by the wave function (2.41), is analogous to Eq. (2.7) satisfied by the wave function (2.40). We can generalize Eq. (2.46) as we generalized Eq. (2.7) by adding a term involving the potential energy to the left-hand side of this last equation giving

$$\left[\frac{-\hbar^2}{2m} \frac{\partial^2}{\partial x^2} + V(x, t) \right] \psi(x, t) = i\hbar \frac{\partial \psi(x, t)}{\partial t}. \tag{2.47}$$

This equation is called the *Schrödinger time-dependent equation*.

The Schrödinger time-dependent equation determines the time development of the wave function. This equation will be used in following chapters to solve dynamical problems involving the absorption of light and the scattering of particles. As for the Schrödinger time-independent equation considered before, the validity of the Schrödinger time-dependent equation depends ultimately upon how well the predictions of the theory are confirmed by experiment. Each new theory has to stand on its own merits.

If the potential energy is independent of time, the Schrödinger time-dependent equation (2.47) may be solved by employing the technique of the *separation of variables*. Using this method, we assume that the wave function can be written as the product of a function which depends only upon the x-coordinate and another function which depends only upon the time

$$\psi(x, t) = u(x)T(t). \tag{2.48}$$

Substituting this form of the wave function into Eq. (2.47) and dividing the resulting equation through with the wave function gives

$$\frac{1}{u(x)T(t)} \left[\frac{-\hbar^2}{2m} \frac{\partial^2}{\partial x^2} + V(x) \right] u(x)T(t) = \frac{1}{u(x)T(t)} i\hbar \frac{\partial(u(x)T(t))}{\partial t}.$$

We note now that the function $u(x)$ can be brought through the partial derivative with respect to time on the right, and the functions $u(x)$ in the numerator and denominator allowed to cancel. Since the function $T(t)$ only depends on the time, the partial derivative with respect to time can then be replaced with the ordinary derivative. Similarly, the functions $T(t)$ occurring on the left can be allowed to cancel and the partial derivatives with respect to x replaced by ordinary derivatives. We thus obtain

$$\frac{1}{u(x)} \left[\frac{-\hbar^2}{2m} \frac{d^2}{dx^2} + V(x) \right] u(x) = \frac{1}{T(t)} i\hbar \frac{dT(t)}{dt}.$$

The right-hand side of this equation now depends only on time, and the left-hand side depends only upon x. The two sides of the equation can only be equal for all values of x and t if they are each equal to a constant. For reasons that will become clear shortly, we denote the constant by E. The equations formed by equating the right- and left-hand sides of this equation with E can be written

$$i\hbar \frac{dT(t)}{dt} = ET(t)$$

$$\left[\frac{-\hbar^2}{2m} \frac{d^2}{dx^2} + V(x) \right] u(x) = Eu(x).$$

The second of these equations is the Schrödinger time-independent equation seen earlier. We may thus identify E as the energy of the system. This justifies our using E to denote the separation constant. The first of the above equations has the solution

$$T(t) = Ae^{-iEt/\hbar}.$$

By substituting the function $T(t)$ given by this last equation into Eq. (2.48), the solution of the Schrödinger time-dependent equation may be written

$$\psi(x, t) = Au(x)e^{-iEt/\hbar},$$

where $u(x)$ is a solution of the Schrödinger time-independent equation. To simplify this result, we note that if $u(x)$ is a solution of the Schrödinger time-independent equation, then $Au(x)$ is also a solution. We may thus absorb the constant A in into the function $u(x)$. Labeling this function by the value of the energy to which it corresponds and using Eq. (2.43) to express the energy E in the exponential function in terms of the angular frequency ω, we obtain finally

$$\psi(x, t) = u_E(x)e^{-i\omega t} , \qquad (2.49)$$

where

$$\omega = \frac{E}{\hbar} .$$

If the potential energy is independent of time, the time evolution of the wave functions is thus due to the multiplicative factor $e^{-i\omega t}$ occurring in Eq. (2.49).

As an example of a solution of the Schrödinger time-dependent equation, we consider again the states of a free particle. Since the potential energy of a free particle does not depend upon time, the wave function of a free particle must be of the general form (2.49) with a temporal part equal to the function $e^{-i\omega t}$ and with a spatial part $u_E(x)$, which is a solution of the Schrödinger time-independent equation. For a free particle, the Schrödinger time-independent equation is identical to Eq. (2.7).

The solutions of Eq. (2.7) are of the general form

$$u_E(x) = Ae^{\pm ikx},$$

where the exponent may be either positive or negative. We consider now separately the solutions associated with the positive and negative exponents. The solution associated with the positive exponent is

$$\psi(x, t) = Ae^{ikx}e^{-i\omega t} = Ae^{i(kx-\omega t)},$$

which represents a traveling wave moving in the positive x-direction. Similarly, the solution associated with the negative exponent is

$$\psi(x, t) = Ae^{-ikx}e^{-i\omega t} = Ae^{-i(kx+\omega t)},$$

which represents a traveling wave moving in the negative x-direction. The solutions of the Schrödinger time-dependent equation for a free particle thus correspond to the traveling waves considered in the introduction. The wave function with an x-dependence equal to e^{ikx} corresponds to a wave moving in the positive x-direction, while wave function with an x-dependence equal to e^{-ikx} corresponds to a wave moving in the negative x-direction.

Suggestion for further reading

W. Moore, *Schrödinger's Life and Thought* (Cambridge University Press, 1989).

Basic equations

The wave equation

Schrödinger time-independent equation

$$\frac{-\hbar^2}{2m}\frac{d^2\psi}{dx^2} + V(x)\psi = E\psi$$

Probability particle is between x and $x + dx$

$$dP = |\psi(x)|^2 dx$$

Average value of x

$$<x> = \int_a^b x|\psi(x)|^2 dx$$

Solutions of Schrödinger time-independent equation

Energy levels of particle in infinite well

$$E = \frac{n^2 h^2}{8mL^2} \quad \text{for } n = 1, 2, 3, \ldots$$

Wave function of particle within infinite well

$$\psi_n(z) = \begin{cases} \sqrt{2/L}\cos(n\pi x/L), & n = 1, 3, 5, \ldots \\ \sqrt{2/L}\sin(n\pi x/L), & n = 2, 4, 6, \ldots \end{cases}$$

where $-L/2 \leq x \leq L/2$. The wave function for a particle in a finite well has the same form within the well and approaches zero exponentially outside the well.

Energy levels of simple harmonic oscillator

$$E = \hbar\omega(n + 1/2) \quad \text{for } n = 0, 1, 2, \ldots$$

Wave function of simple harmonic oscillator

$$\psi_n(x) = A_n e^{-y^2/2} H_n(y),$$

where A_n is a normalization constant, $y = \sqrt{m\omega/\hbar}\, x$, and H_n is a Hermite polynomial.

Time evolution of wave function

Schrödinger time-dependent equation

$$i\hbar\frac{\partial\psi(x,t)}{\partial t} = \left[\frac{-\hbar^2}{2m}\frac{\partial^2}{\partial x^2} + V(x,t)\right]\psi(x,t)$$

Solution when $V(x,t)$ is independent of time

$$\psi(x,t) = u_E(x)e^{-i\omega t}, \quad \text{where} \quad \omega = E/\hbar$$

Summary

The wave function $\psi(x)$ and the energy E of the stationary states of a particle may be obtained by solving the Schrödinger time-independent equation

$$\frac{-\hbar^2}{2m}\frac{d^2\psi}{dx^2} + V(x)\psi = E\psi.$$

The Schrödinger time-independent equation is used in this chapter to study the wave function and the energy of a particle moving in an infinite and a finite potential well and to study the states of the simple harmonic oscillator. In each case, the possible energies of the particle correspond to those values of E for which there is a solution of the Schrödinger time-independent equation that satisfies the boundary conditions. The wave function, which is related to the probability of finding the particle in a particular region in space, may be used to calculate the average value of a function $f(x)$ using the formula

$$< f(x) > = \int_a^b f(x)|\psi(x)|^2 dx.$$

The time evolution of the wave function is described by the Schrödinger time-dependent equation

$$i\hbar\frac{\partial\psi(x,t)}{\partial t} = \left[\frac{-\hbar^2}{2m}\frac{\partial^2}{\partial x^2} + V(x,t)\right]\psi(x,t).$$

For a particle moving in a constant potential, the solutions of the Schrödinger time-dependent equation are of the form

$$\psi(x,t) = u_E(x)e^{-i\omega t},$$

where $u_E(x)$ is a solution of the Schrödinger time-independent equation and $\omega = E/\hbar$.

Questions

1. Write down a function describing a free particle with a momentum p.
2. Write down a first-order and a second-order differential equation satisfied by the wave functions of a free particle.
3. What is the Schrödinger time-independent equation for a particle with a potential energy $V(x)$?
4. Denoting the wave function of a particle by $\psi(x)$, write down an expression for the probability that the particle will be found between a and b.
5. Denoting the wave function of a particle by $\psi(x)$, write down an equation for the average value of x.
6. Suppose that a particle, which is confined to move in one-dimension between 0 and L, is described by the wave function, $\psi(x) = A x(L - x)$. What condition could be imposed upon the wave function $\psi(x)$ to determine the constant A?
7. Suppose that a perfectly elastic ball were bouncing back and forth between two rigid walls with no gravity. Which of the variables, p, $|p|$, E, would have a constant value?
8. Sketch the form of the wave functions corresponding to the three lowest energy levels of a particle confined to an infinite potential well.
9. In what respects is the infinite well an unrealistic model for describing the states of an electron in an atom or in a crystal?
10. What is the value of the kinetic energy of a particle at the classical turning points of an oscillator?
11. Give the energy of the lowest three states of the harmonic oscillator in terms of the oscillator frequency ω.
12. Suppose that a harmonic oscillator made a transition from the $n = 3$ to the $n = 2$ state. What would be the energy of the emitted photon?
13. Describe in qualitative terms the form of the wave functions of the harmonic oscillator between the classical turning points?
14. How does the form of the wave function of the harmonic oscillator change as x increase beyond the classical turning point.
15. Which equation could be used to describe how a physical system evolves when it is placed in an oscillating electric field.
16. Write down the Schrödinger time-dependent equation.
17. What is the general form of the stationary solutions of the Schrödinger time-dependent equation?
18. Describe the wave functions obtained by multiplying the stationary wave Ae^{ikx} by the function $e^{-i\omega t}$.

Problems

1. The lowest energy of an electron confined to a one-dimensional region is 1.0 eV. (a) By describing the electron as a particle in an infinite well, find the size of the region? (b) How much energy must be supplied to the electron to excite it from the ground state to the first level above the ground state?
2. Write down the wave function $\psi(x)$ for an electron in the $n = 3$ state of a 10 nm-wide infinite well with the center of the well being at $x = 0$. What numerical values does the function have at $x = 0, 2, 4, 8$, and 10 nm?
3. An electron in a 10 nm-wide infinite well makes a transition from the $n = 3$ to the $n = 2$ state emitting a photon. Calculate the energy of the photon and the wavelength of the light.
4. Show by direct substitution that the wave function,

$$\psi(x) = Ae^{-m\omega x^2/2\hbar},$$

satisfies Eq. (2.32) for the harmonic oscillator. Calculate the corresponding energy.
5. Determine the constant A in the preceding problem by requiring that the wave function be normalized.
   Hint: For an arbitrary value of the constant a, the integral that arises in doing this problem may be evaluated using the relation

$$\int_0^\infty e^{-ax^2}\, dx = \frac{1}{2}\sqrt{\frac{\pi}{a}}.$$

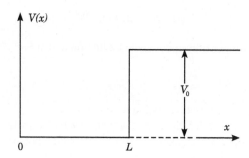

FIGURE 2.7 Potential well for Problem 7.

6. A particle is described by the wave function,

$$\psi(x) = \begin{cases} Ax\,e^{-ax}, & x > 0 \\ 0, & x \leq 0, \end{cases}$$

where A and a are constants.
   (a) Sketch the wave function.
   (b) Use the normalization condition to determine the constant A.
   (c) Find the most probable position of the particle.
   (d) Calculate the average value of the position of the particle.
7. (a) For a particle moving in the potential well shown in Fig. 2.7, write down the Schrödinger equations for the region where $0 \leq x \leq L$ and the region where $x \geq L$. (b) Give the general form of the solution in the two regions. (c) Assuming that the potential is infinite at $x = 0$, impose boundary conditions that are natural for this problem and derive an equation that can be used to find the energy levels for the bound states.
8. Show that the wave function of a traveling wave (2.41) satisfies the time-dependent Schrödinger equation (2.47).
9. Show how the wave function of the even states of a particle in an infinite well extending from $x = -L/2$ to $x = L/2$ evolve in time.

Chapter 3

Operators and waves

Contents

3.1 Observables, operators, and eigenvalues	37	3.5 The Heisenberg uncertainty principle	64
3.2 The finite well and harmonic oscillator using finite differences	40	Suggestion for further reading	76
		Basic equations	76
3.3 The finite well and harmonic oscillator with spline collocation	47	Summary	78
		Questions	78
3.4 Electron scattering	56	Problems	79

Particle variables such as the momentum and the energy can usually have many different values. For each value of a variable, there is a wave function which describes the state of the particle when the variable has that particular value.

We have seen that the energy of microscopic systems can have only certain values. We have also found that a microscopic particle can be described by a wave function which enables us to calculate the probability that the particle is located in a particular region and enables us to calculate average values of functions of the spatial variables.

Taking an inventory of the questions we can answer and the questions we cannot answer, however, we find that there are gaps in our understanding. Suppose, for instance, that we knew that a microscopic system was in a state for which a variable we would use to describe the system had a particular value. How would we calculate the wave function describing the system? What is the connection between the wave functions and the possible values of the physical variables? More generally, what would be our strategy for determining the spectra of observed values of a variable other than the energy?

Questions of this kind can be answered very naturally by using the concept of an operator. The term *operator* is used widely in mathematics to refer to an expression which acts on a function to produce a new function. As an example of the use of the term, consider the following equations

$$x f(x) = g(x)$$
$$\frac{d\, f(x)}{dx} = h(x) .$$

One can say that the coordinate x and the derivative with respect to x have the common feature that they operate on the function, $f(x)$, to produce new functions, $g(x)$ and $h(x)$, and in this respect, x and the derivative with respect to x are operators.

Particle variables such as the momentum and the energy are represented in quantum theory by operators. If a particle were in a state in which a variable describing its motion had a particular value, then the particle would be described by a wave function that satisfies a characteristic equation which contains the operator corresponding to the variable.

3.1 Observables, operators, and eigenvalues

To prepare for our definition of the momentum operator, we shall make use of Eq. (2.6). This equation, which is satisfied by a de Broglie wave, can be written as follows

$$\left(-i\hbar \frac{d}{dx}\right)\psi = p\,\psi. \tag{3.1}$$

Since the product of the operator,

$$-i\hbar\frac{d}{dx},$$

and the wave function ψ appearing on the left-hand side of Eq. (3.1) is equal to

$$-i\hbar\frac{d\psi}{dx},$$

Eq. (3.1) may be regarded as a way of rewriting Eq. (2.6).

The operator associated with the momentum, which we denote by $\hat{p}$, is defined to be

$$\hat{p} = -i\hbar\frac{d}{dx}. \tag{3.2}$$

Eq. (3.1) can thus be written simply

$$\hat{p}\psi = p\,\psi. \tag{3.3}$$

This equation has the general form

$$\text{operator} \times \text{function} = \text{constant} \times \text{function}. \tag{3.4}$$

An equation of this kind is called an *eigenvalue equation*. The constant appearing on the right is referred to as an *eigenvalue*, and the function is called an *eigenfunction*. We shall encounter equations with the general form of Eq. (3.4) often in this chapter and call the function an eigenfunction and the associated constant an eigenvalue.

An operator corresponding to the energy of a particle can be obtained by writing the energy of the particle in terms of the momentum

$$E = \frac{1}{2m}p^2 + V(x) \tag{3.5}$$

and then replacing the momentum with the momentum operator (3.2) to obtain

$$\hat{H} = \frac{-\hbar^2}{2m}\frac{d^2}{dx^2} + V(x). \tag{3.6}$$

Here the energy operator is denoted by $\hat{H}$. Using a terminology which has its origins in classical mechanics, we refer to $\hat{H}$ as the *Hamiltonian* of the system. The eigenvalue equation for the energy is formed by setting the energy operator times the wave function equal to the energy times the wave function. Since we have denoted the energy operator by $\hat{H}$, this gives

$$\hat{H}\psi = E\,\psi. \tag{3.7}$$

The Schrödinger time-independent equation introduced in Chapter 2 can be obtained by substituting the explicit form of $\hat{H}$ given by Eq. (3.6) into Eq. (3.7). Using the terminology we have just introduced, the Schrödinger time-independent equation is thus the eigenvalue equation for the energy.

Example 3.1

Show that the wave function,

$$\psi(x) = Ae^{i\,kx}, \tag{3.8}$$

represents a state for which the momentum of the particle has the value $p = \hbar k$. Find the kinetic energy of the particle in this state.

Solution

To see whether the wave function $\psi(x)$ corresponds to a state for which the particle has a definite value of the momentum, we multiply the momentum operator (3.2) times the wave function (3.8) to obtain

$$\left(-i\hbar\frac{d}{dx}\right)Ae^{i\,kx} = \hbar k\,Ae^{i\,kx}.$$

This last equation may be written more simply in terms of the momentum operator $\hat{p}$ and the wave function $\psi(x)$

$$\hat{p}\psi(x) = \hbar k\,\psi(x),$$

which we may identify as the eigenvalue equation for the momentum (3.3). We may thus identify $\psi(x)$ as an eigenfunction of the momentum corresponding to the eigenvalue $p = \hbar k$.

To see whether the wave function $\psi(x)$ corresponds to a state for which the particle has a definite value of the energy, we use Eq. (3.6) to multiply the energy operator for a free-particle times the wave function (3.8) to obtain

$$\left(\frac{-\hbar^2}{2m}\frac{d^2}{dx^2}\right) Ae^{ikx} = \frac{(\hbar k)^2}{2m} Ae^{ikx}.$$

This last equation may be written more simply in terms of the energy operator $\hat{H}$ and the wave function $\psi(x)$

$$\hat{H}\psi(x) = \frac{(\hbar k)^2}{2m}\psi(x),$$

which we may identify as the eigenvalue equation for the energy (3.7). The wave function $\psi(x)$ is thus an eigenfunction of the energy corresponding to the eigenvalue $E = (\hbar k)^2/2m$. Using the fact that the momentum has the value $p = \hbar k$, the equation for the energy can also be written $E = p^2/2m$ as one would expect.

The function e^{ikx} is an eigenfunction of the momentum operator corresponding to the eigenvalue, $+\hbar k$, and hence describes a particle moving in the positive x-direction. Similarly, the function e^{-ikx} is an eigenfunction of the momentum operator corresponding to the eigenvalue $-\hbar k$ and hence describes a particle moving in the negative x-direction.

The wave functions of a particle in the infinite well are of the form, $\cos kx$ and $\sin kx$. As we have seen in the Introduction, trigonometric functions appear in the expansion of exponential functions by Euler's formula. We have

$$e^{ikx} = \cos(kx) + i\sin(kx).$$

If k is replaced by $-k$, this equation becomes

$$e^{-ikx} = \cos(kx) - i\sin(kx).$$

The last two equations may be solved for $\cos(kx)$ and $\sin(kx)$ giving

$$\cos(kx) = \frac{e^{ikx} + e^{-ikx}}{2} \tag{3.9}$$

$$\sin(kx) = \frac{e^{ikx} - e^{-ikx}}{2i}. \tag{3.10}$$

The trigonometric functions of a particle in an infinite well thus correspond to superpositions of states for which the particle moves in the positive and negative x-directions. One would expect a particle in an infinite well with rigid walls to bounce back and forth between the walls. The trigonometric functions, $\cos kx$ and $\sin kx$, are superpositions of eigenfunctions of the momentum operator corresponding to the eigenvalues, $+\hbar k$ and $-\hbar k$, and they are eigenfunctions of the energy operator corresponding to the eigenvalue $\hbar^2 k^2/2m$. The cosine and sine function given by Eqs. (3.9) and (3.10) can be thought of as standing wave solutions of the Schrödinger time-independent equation.

The idea that physical variables can be represented by operators gives us a unified way of thinking about the results we have obtained thus far and enables us to answer the questions we posed at the beginning of this section. The momentum p is represented in our theory by the momentum operator $\hat{p}$ given by Eq. (3.2), while the position coordinate x is identical to the position operator $\hat{x}$. The operators corresponding to another variable such as the energy can be obtained by writing the variable in terms of the momentum and then replacing the momentum with the momentum operator. This is how we obtained the energy operator denoted by $\hat{H}$.

We can determine whether or not a wave function describes a state of a particle in which a physical variable has a definite value by operating upon the wave function with the operator corresponding to the variable. If the product of the operator and the wave function is equal to a number times the same wave function, then the numerical value obtained is equal to the value of the variable when the particle is in a state described by the wave function. For instance, the product of the momentum operator $\hat{p}$ and the de Broglie wave function e^{ikx} is equal to $\hbar k$ times the function e^{ikx}. This tells us that the wave function e^{ikx} corresponds to a state of the particle in which the momentum has the value, $p = \hbar k$. The product of the energy operator $\hat{H}$ and the wave function $\cos kx$ for a particle in an infinite well is equal $\hbar^2 k^2/2m$ times the function $\cos kx$. The function $\cos kx$ thus describes the state of the particle in which it has the energy $\hbar^2 k^2/2m$.

More generally the possible values of a physical observable are equal to the eigenvalues of the operator corresponding to the observable. We can determine the possible results of measuring a particular variable by setting up the eigenvalue equation for the variable and determining the values of the constant on the right-hand side of the equation for which the equation has a solution satisfying the appropriate boundary conditions.

3.2 The finite well and harmonic oscillator using finite differences

We would now like to consider the finite well again using the concepts of operators and eigenvalue equations described in the previous section. The first strategy, which we shall use to solve the Schrödinger equation for the finite well, will be to use finite-difference approximations of the second derivative to convert the Schrödinger equation for the well into a set of linear equations which can easily be solved with MATLAB®.

The basic idea of the finite-difference approximation is to use the points adjacent to a particular grid point to approximate the derivatives at that point. Because we have chosen the origin of the finite well shown in Fig. 2.5 of Chapter 2 to be in the center of the well, the problem has a reflection symmetry and the solutions of the Schrödinger equations are even or odd. This is important when using finite difference approximation of the derivatives near the center of the well because we know that the values of the wave function for small negative values of x are equal to the wave function or the negative of the wave function for the corresponding positive values of x.

As discussed in Chapter 2, we shall use dimensionless variables that give the position of the particle in nanometers and the energy and potential energy in electron volts.

$$x = \chi \cdot \text{nm}, \quad E = \epsilon \cdot \text{eV}, \quad V_0 = \mathcal{V}_0 \cdot \text{eV} \tag{3.11}$$

According to Eq. (2.27) of Chapter 2, the eigenvalue equation for points inside the well is given by

$$-\frac{d^2\psi}{d\chi^2} = E_0 \epsilon \psi, \quad \text{for } -5 \le \chi \le 5, \tag{3.12}$$

and according to Eq. (2.29) of Chapter 2, the eigenvalue equation for points outside the well is given by

$$-\frac{d^2\psi}{d\chi^2} + \mathcal{V}_0 E_0 \psi = \epsilon E_0 \psi, \quad \text{for } |\chi| \ge 5, \tag{3.13}$$

where $\psi(x)$ is the wave function and E_0 is a dimensionless number given by the equation

$$E_0 = \frac{2m(\text{nm})^2 \text{eV}}{\hbar^2}. \tag{3.14}$$

The finite well described in Chapter 2, consist of a layer of $GaAs$ sandwiched between two layers of $Al_{0.3}Ga_{0.7}As$. The well extends from $x = -5$ nm to $x = +5$ nm and has the depth $V_0 = 0.3$ eV. Because we want to solve the Schrödinger equation for values inside and outside the well, the dimensionless parameter χ extends beyond the edge to the well. In the MATLAB program, for finding the eigenvalues and eigenfunctions of a particle in the well, the value of χ for the edge of the well will be denoted by L and the edge of the entire physical region will be denoted by $xmax$. With n grid points, the step size δ is then equal to $xmax/n$. The parameter $\mathcal{V}_0$ for the depth of the well has the value 0.3. Using the effective mass of conduction electrons in $GaAs$, which is equal to 0.067 times the free electron mass, the constant E_0 can be shown to be 1.759.

We introduce the uniform grid,

$$\chi_i = (i - 1) * \delta, \quad \text{with } i = 1, 2, \ldots, n,$$

where δ is the grid spacing. As we shall see, only the points, $\chi_1, \ldots, \chi_n$ will play a role in the actual computation. The first point, $\chi_1 = 0$ is at the center of the well, and the point χ_n is at $xmax - \delta$. The boundary conditions require that the function ψ be equal to zero at $\chi_{n+1} = \delta * n$. The value of $\psi(\chi)$ corresponding to the grid point χ_i will be denoted by ψ_i. Using a coarse grid with only five points, the Schrödinger equation at the five grid points are

$$-\psi_1'' = E_0 \epsilon \psi_1$$
$$-\psi_2'' = E_0 \epsilon \psi_2$$

$$-\psi_3'' + V_0 E_0 \psi_3 = E_0 \epsilon \psi_3$$
$$-\psi_4'' + V_0 E_0 \psi_4 = E_0 \epsilon \psi_4$$
$$-\psi_5'' + V_0 E_0 \psi_5 = E_0 \epsilon \psi_5$$

We can convert these equations into a matrix equation using the following five-point difference formula to approximate the second derivative

$$\psi_i'' = \frac{-\psi_{i-2} + 16\psi_{i-1} - 30\psi_i + 16\psi_{i+1} - \psi_{i+2}}{12\delta^2}. \tag{3.15}$$

We now use Eqs. (3.15) to evaluate the second derivatives in the above equations, and we multiply each of the resulting equations by $12\delta^2$. We obtain

$$\psi_{-1} - 16\psi_0 + 30\psi_1 - 16\psi_2 + \psi_3 = 12\delta^2 \epsilon E_0 \psi_1$$
$$\psi_0 - 16\psi_1 + 30\psi_2 - 16\psi_3 + \psi_4 = 12\delta^2 \epsilon E_0 \psi_2$$
$$\psi_1 - 16\psi_2 + (30 + 12\delta^2 V_0 E_0)\psi_3 - 16\psi_4 + \psi_5 = 12\delta^2 \epsilon E_0 \psi_3 \tag{3.16}$$
$$\psi_2 - 16\psi_3 + (30 + 12\delta^2 V_0 E_0)\psi_4 - 16\psi_5 = 12\delta^2 \epsilon E_0 \psi_4$$
$$\psi_3 - 16\psi_4 + (30 + 12\delta^2 V_0 E_0)\psi_5 = 12\delta^2 \epsilon E_0 \psi_5$$

In the last two equations, we have taken advantage of the fact that ϕ_6 and ϕ_7 that lie on and beyond the outer boundary are equal to zero.

There are even and odd solutions of the above equations. For the even solutions, the values of the solution to the left of the origin are equal to the values at the corresponding points to the right of the origin. We have

$$\psi_0 = \psi_2 \tag{3.17}$$
$$\psi_{-1} = \psi_3. \tag{3.18}$$

Substituting these values into Eqs. (3.16), we obtain

$$30\psi_1 - 32\psi_2 + 2\psi_3 = 12\delta^2 \epsilon E_0 \psi_1$$
$$-16\psi_1 + 31\psi_2 - 16\psi_3 + \psi_4 = 12\delta^2 \epsilon E_0 \psi_2$$
$$\psi_1 - 16\psi_2 + (30 + 12\delta^2 V_0 E_0)\psi_3 - 16\psi_4 + \psi_5 = 12\delta^2 \epsilon E_0 \psi_3 \tag{3.19}$$
$$\psi_2 - 16\psi_3 + (30 + 12\delta^2 V_0 E_0)\psi_4 - 16\psi_5 = 12\delta^2 \epsilon E_0 \psi_4$$
$$\psi_3 - 16\psi_4 + (30 + 12\delta^2 V_0 E_0)\psi_5 = 12\delta^2 \epsilon E_0 \psi_5$$

Similarly, for the odd solutions of the finite well, we have

$$\psi_{-1} = -\psi_3 \tag{3.20}$$
$$\psi_0 = -\psi_2 \tag{3.21}$$

For the odd solutions, the value of the solution at the origin, ψ_1, is equal to zero. In the first equation, the value of ψ_{-1} to the left of the origin cancels with ψ_3, and the terms involving ψ_0 and ψ_2 cancel. The first equation of Eqs. (3.16) may thus be deleted leaving $n-1$ equations for $n-1$ unknowns,

$$29\psi_2 - 16\psi_3 + \psi_4 = 12\delta^2 \epsilon E_0 \psi_2$$
$$-16\psi_2 + (30 + 12\delta^2 V_0 E_0)\psi_3 - 16\psi_4 + \psi_5 = 12\delta^2 \epsilon E_0 \psi_3 \tag{3.22}$$
$$\psi_2 - 16\psi_3 + (30 + 12\delta^2 V_0 E_0)\psi_4 - 16\psi_5 = 12\delta^2 \epsilon E_0 \psi_4$$
$$\psi_3 - 16\psi_4 + (30 + 12\delta^2 V_0 E_0)\psi_5 = 12\delta^2 \epsilon E_0 \psi_5$$

The equations for the even and odd solutions can each be written in matrix form. For the even solutions of the finite well, we define the matrix

$$
\mathbf{A}_E = \begin{bmatrix}
30 & -32 & +2 & 0 & 0 \\
-16 & 31 & -16 & 1 & 0 \\
1 & -16 & (30 + 12\delta^2 V_0 E_0) & -16 & 1 \\
0 & 1 & -16 & (30 + 12\delta^2 V_0 E_0) & -16 \\
0 & 0 & 1 & -16 & (30 + 12\delta^2 V_0 E_0)
\end{bmatrix}
\tag{3.23}
$$

With this notation, the even solutions of the equations that arise when the differential equation for the finite well is represented by the matrix $\mathbf{A}_E$ are obtained by solving the set of equations

$$
\mathbf{A}_E \psi = 12\delta^2 \epsilon E_0 \psi
\tag{3.24}
$$

with the corresponding eigenvector

$$
\psi = \begin{bmatrix}
\psi_1 \\
\psi_2 \\
\psi_3 \\
\psi_4 \\
\psi_5
\end{bmatrix}
$$

As we have noted previously, the odd solutions are zero at the center of the well. Setting ψ_1 equal to zero, the equations for ψ_2, ψ_3, ψ_4, and ψ_5 can be expressed in terms of a matrix

$$
\mathbf{A}_O = \begin{bmatrix}
29 & -16 & 1 & 0 \\
-16 & (30 + 12\delta^2 V_0 E_0) & -16 & 1 \\
1 & -16 & (30 + 12\delta^2 V_0 E_0) & -16 \\
0 & 1 & -16 & (30 + 12\delta^2 V_0 E_0)
\end{bmatrix},
\tag{3.25}
$$

and the vector ψ by the equation

$$
\psi = \begin{bmatrix}
\psi_2 \\
\psi_3 \\
\psi_4 \\
\psi_5
\end{bmatrix}.
\tag{3.26}
$$

With this notation, the equations for the even solutions with components, ψ_1, ψ_2, ψ_3, ψ_4, and ψ_5, can be written simply

$$
\mathbf{A}_E \psi = \delta^2 E_0 \epsilon \psi,
\tag{3.27}
$$

and the equations for odd solutions with components, ψ_2, ψ_3, ψ_4, and ψ_5 can be written

$$
\mathbf{A}_O \psi = \delta^2 E_0 \epsilon \psi
\tag{3.28}
$$

We have thus converted the eigenvalue problem for the finite well into a matrix problem. A MATLAB program that makes use of special features of MATLAB to construct the matrix and solve the resulting matrix eigenvalue problem is given below.

MATLAB Program FiniteWell.m

The MATLAB program FiniteWell.m for finding the eigenvalues and eigenvectors for an electron moving in a finite well.

```
xmax=20;
L=5;
n=5;
```

```
delta=xmax/n;
n1=fix(L/delta)+1;
n2=n-n1;
e0=1.759;
v0=0.3;
v1=ones(n-1,1);
v2=ones(n-2,1);
v3=ones(n-3,1);
d=[30*ones(n1,1);(30+12*v0*e0*delta^2)*ones(n2,1)];
AE=-16*diag(v1,-1)-16*diag(v1,1)+diag(v2,-2)+diag(v2,2)+diag(d);
% Even Solutions
AE(1,2)=-32;
AE(1,3)=2;
AE(2,2)=31;
AE
% [E V]=eig(AE);
[V,E]=eig(AE);
[EvenStateEnergies,index] = sort(diag(E)/(12.0*e0*delta^2));
EvenStateEnergies(1:2)

% Create grid over entire range between -xmax and xmax
x = linspace(-xmax,xmax,2*n-1)

% Plot first even function
EFW1 = V(:,index(1))
EFW1m0 =EFW1;
EFW1m0(1) = []
plot(x,[flipud(EFW1m0);EFW1])
shg
saveas (gcf,'EFW1.pdf')

% Plot second even function
EFW2 = V(:,index(2))
EFW2m0 =EFW2;
EFW2m0(1) = []
plot(x,[flipud(EFW2m0);EFW2])
shg
saveas (gcf,'EFW2.pdf')

% Odd Solutions
d=[30*ones(n1-1,1);(30+12*v0*e0*delta^2)*ones(n2,1)];
AO=-16*diag(v2,-1)-16*diag(v2,1)+diag(v3,-2)+diag(v3,2)+diag(d);
AO(1,1)=29;
AO
%
[V,E]=eig(AO);
[OddStateEnergies,index] = sort(diag(E)/(12.0*e0*delta^2));
OddStateEnergies(1)

% Plot odd function
OFW1 = V(:,index(1))
plot(x,[-flipud(OFW1);0;OFW1])
shg
saveas (gcf,'OFW1.pdf')
```

In the first line of the program, xmax= 20 defines the length of the physical region and L = 5 is the χ coordinate of the edge of the well. As can be seen in Fig. 2.5 of Chapter 2, the well extends from −5 nm to 5 nm. The wave functions calculated by the program are defined over the region from −20 nm to 20 nm. The variable n is the number of grid points. We have set n equal to 5 so that we can compare the matrices produced by the MATLAB program with the AE matrix given by Eq. (3.23) and the AO matrix defined by Eq. (3.25). The variable n1 defined in the program is the number of grid points inside the well and n2 is the number of grid points beyond the edge of the well. For the coarse five-point grid we have chosen, n1 = 2 and n2 = 3. The value of the diagonal elements inside the well are denoted by $d1$ and the diagonal elements outside the well are denoted by $d2$. Before taking into account the special effect of the boundary conditions, which will be added afterwards, the diagonal elements inside the well are $d1 = 30$, and the diagonal elements outside the well are $d2 = 30 + 12\delta^2 \mathcal{V}_0$. For the well with depth $\mathcal{V}_0 = 0.3$, d2 = 2 + 0.3 ∗ E_0 ∗ δ^2.

The **AE** matrix is given by Eq. (3.23). We shall first construct the basic structure of this matrix and then add the effect of the boundary conditions. Before including the effect of the boundary conditions, the first n1 elements along the diagonal are equal to d1 and the remaining n2 elements are all equal to d2. Also, all of the elements of the matrix **AE** located on either side of the diagonal are equal to -16, and the elements twice removed from the diagonal are +1. The statements, v1 =ones(n-1), v2 =ones(n-2), and v3 = ones(n-3) create vectors of length n-1, n-2, and n-3, respectively with all elements equal to one. The term, -16*diag(v1,-1), places the elements, -16, one unit below the diagonal of the matrix and the term, -16*diag(v1,1), places the elements, -16, one unit above the diagonal. Similarly, the terms, diag(v2,-2) & diag(v2,2), place +1 two units below and above the diagonal of the matrix, while the command diag(d) places the elements of d along the diagonal. After the diagonal elements of **AE** and the elements one unit and two units below and above the diagonal are defined, the next statements defines the special A(1,2), A(1,3), and A(2,2) matrix elements. A MATLAB program suppresses the output of any line ending in a semicolon. To have the AE and AO matrices printed, we wrote AE and AO lines without semicolons so that the program prints out the AE and AO matrices. One can readily confirm that the AE and AO-matrices produced by the program are identical to the matrices given by (3.23) and (3.25). The line of the MATLAB program

```
[V,E]=eig(AE);
[EvenStateEnergies,index] = sort(diag(E)/(12.0*e0*delta^2))
```

produces the eigenvalues of the **AE** with the index giving the number of the corresponding eigenvector. One can see that only the first two eigenvalues, 0.0207 and 0.1517 are less than 0.3 and thus correspond to bound states. Similarly, the lines

```
[V,E]=eig(AO);
[OddStateEnergies,index] = sort(diag(E)/(12.0*e0*delta^2));
```

produce the eigenvalues of the **AO** of which only the eigenvalue 0.786 is less that 0.3. One can obtain more accurate eigenvalues by increasing the number of points (n) in the MATLAB program. With $n = 100$, the lowest two even eigenvalues produced by the program are 0.0342 and 0.2715, and the lowest odd eigenvalue is 0.1326.

The statement

```
x = linspace(-xmax,xmax,2*n-1)
```

produces a grid over the entire range of x between -xmax and +xmax. Before discussing the commands in the program FiniteWell.m that plots the first even eigenfunction over the entire range between $x = -xmax$ and $xmax$, there is a small detail we must mention. The first element of each even eigenfunctions is the value the function has at $x = 0$. We can produce the values of the even functions for negative values of x by using the MATLAB command "flipud" which gives the values of the vector in reverse order, but before issuing this command we must first remove the first element of the vector so that the value at the origin is not included twice.

The MATLAB commands

```
EFW1 = V(:,index(1))\\
EFW1m0 =EFW1; \\
EFW1m0(1) = [] \\
```

define the first even eigenvector for the finite well (EFW1) and then define a vector with the value at the origin removed (EFW1m0). The vector representing the first even eigenfunction over the entire range between $-xmax$ and $xmax$ is then [flipud(EFW1m0);EFW1] which the program then plots with the command

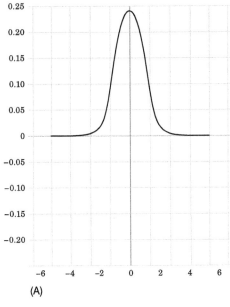
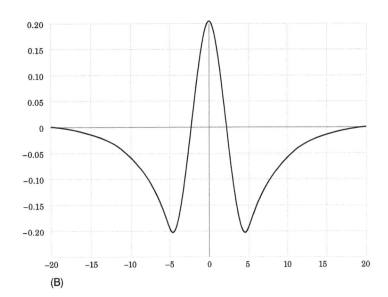

(A)   (B)

FIGURE 3.1 The eigenfunctions corresponding to the two lowest even eigenvalues of an electron moving in a finite well.

```
plot(x,[flipud(EFW1m0);EFW1])\\
shg\\
saveas (gcf,'EFW1.pdf')\\
```

the command shg (short for "show graph") shows the graph of the first even eigenvectors, and the saveas command saves the plot of EFW1 as a pdf file. The second even eigenvector EFW is plotted in a similar fashion. A plot of the eigenfunctions for the two lowest eigenvalues are plotted in Fig. 3.1.

The odd eigenvector OFW1 is plotted and saved with the commands

```
OFW1 = V(:,index(1))\\
plot(x,[-flipud(OFW1);0;OFW1]) \\
shg \\
saveas (gcf,'OFW1.pdf')
```

Recall that the value of the odd functions at the origin were omitted from the function. So, zero must be included between the negative values represented by flipud(OFW1) and the positive values represented by OFW1. A plot of the lowest odd eigenfunction is shown in Fig. 3.2.

We shall next consider the simple harmonic oscillator considered in the fourth section of Chapter 2 where we found that the Schrödinger equation for the oscillator can be written

$$-\frac{d^2\psi}{d\chi^2} + \chi^2\psi = \epsilon\psi . \tag{3.29}$$

The dimensionless variable χ in this equation is related to the x-coordinate by the equation

$$\chi = \sqrt{\frac{m\omega}{\hbar}}\, x, \tag{3.30}$$

and the energy of the oscillator E is related to the eigenvalue ϵ in Eq. (3.29) by the equation

$$E = \frac{1}{2}(\hbar\omega)\epsilon. \tag{3.31}$$

Notice that the term involving the second derivative is the same for the oscillator as it was for the finite well. This implies that the off-diagonal terms in the matrices for the oscillator will be the same as the off-diagonal terms for the finite

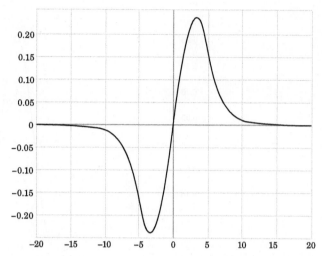

FIGURE 3.2 The eigenfunction corresponding to the lowest odd eigenvalue of a particle moving in a finite well.

well. Only the diagonal terms in the matrices for the even and odd solutions will be different. The MATLAB program HarmonicOscillator.m given below finds the lowest even and odd solutions for the harmonic oscillator.

MATLAB Program HarmonicOscillator.m

The MATLAB program HarmonicOscillator.m for finding the eigenvalues of a particle moving under the influence of a linear restoring force.

```
n = 50
ymax=10;
delta=ymax/n;
deltas=delta*delta;
y=0:delta:ymax;
d = zeros(n,1);
d(1) = 30+12*deltas*y(1)*y(1);
d(2) = 31+12*deltas*y(2)*y(2);
for i = 3:n
    d(i) = 30+12*deltas*y(i)*y(i);
end
v1= ones(n-1,1);
v2 = ones(n-2,1);

% Even Solution
AE = zeros(n,n);
AE = -16*diag(v1,-1)-16*diag(v1,1)+diag(v2,-2)+diag(v2,+2)+diag(d);
AE(1,2) = -32;
AE(1,3) = 2;
E1 = eig(AE);
GroundStateEven = sort(E1./(12.0*deltas));

% Odd Solutions
dod(1) = 29+12*deltas*y(2)^2;
for i = 2:n-1
    dod(i) = 30+12*deltas*y(i+1)*y(i+1);
end
v1 = ones(n-2,1);
v2 = ones(n-3,1);
```

```
AO = zeros(n-1,n-1);
AO = -16*diag(v1,-1)-16*diag(v1,1)+diag(v2,-2)+diag(v2,+2)+diag(dod)
E2 = eig(AO);
GroundStateOdd = sort(E2./(12.0*deltas));

GroundStateEven(1:3)
GroundStateOdd(1:3)
```

Running the program HarmonicOscillator.m, we find that the lowest even eigenvalues ϵ for the harmonic oscillator are approximately 1, 5, 9 which correspond to the energies $E = (1/2)\hbar\omega$, $(5/2)\hbar\omega$, and $(9/2)\hbar\omega$, and the lowest odd eigenvalues for the oscillator are approximately 3, 7, 11 which correspond to the energies $E = (3/2)\hbar\omega$, $(7/2)\hbar\omega$, and $(11/2)\hbar\omega$. This is consistent with the result cited in Chapter 2 that the possible energy levels of the simple harmonic oscillator are

$$E = \hbar\omega(n + 1/2) \quad \text{for } n = 0, 1, 2, \dots . \tag{3.32}$$

Notice that eigenvalues of the even solutions correspond to those energies given by Eq. (3.32) for which $n = 0, 2, 4, 6, \dots$, and the odd solutions correspond those energies for which $n = 1, 3, 5, 7, \dots$.

We will give a more complete solution of the eigenvalue problem for the harmonic oscillator in the next section in which we shall solve the eigenvalue equations for the finite well and the harmonic oscillator using spline collocation. We shall then use the eigenfunctions of the oscillator to calculate the average values of x^2 and the kinetic energy for the several eigenstates of the oscillator.

3.3 The finite well and harmonic oscillator with spline collocation

In the last section, we have used a finite differences approximation of the second derivative to solve the Schrödinger equation for the finite well and the harmonic oscillator. The process of converting a continuous differential equation into a set of linear equations or equivalently into equations involving matrices is called *discretizing* the equation. In this section, we shall convert differential equations into equations involving matrices by developing an approach called the spline collocation method.

For solving the Schrödinger equation using spline collocation, we shall use a continuous differentiable basis of piecewise Hermite cubic splines. To define the spline functions for a single variable x, we introduce the grid

$$x_i = (i - 1) * \delta, \quad \text{with } i = 1, 2, \dots, N$$

where δ is again the grid spacing. The points x_i are called *nodes*.

A spline is a piecewise polynomial that is defined in each interval between adjacent grid points. The basic spines $v_i(x)$ and $s_i(x)$ are defined for $1 \le i \le N - 1$ by the equations

$$v_i(x) = \begin{cases} \frac{1}{\delta^3}(x - x_{i-1})^2[\delta + 2(x_i - x)], & x_{i-1} \le x \le x_i \\ \frac{1}{\delta^3}(x_{i+1} - x)^2[\delta + 2(x - x_i)], & x_i \le x \le x_{i+1} \\ 0, & \text{otherwise} \end{cases} \tag{3.33}$$

$$s_i(x) = \begin{cases} \frac{1}{\delta^3}(x - x_{i-1})^2(x - x_i), & x_{i-1} \le x \le x_i \\ \frac{1}{\delta^3}(x_{i+1} - x)^2(x - x_i), & x_i \le x \le x_{i+1} \\ 0, & \text{otherwise.} \end{cases} \tag{3.34}$$

The spline functions $v_i(x)$ and $s(x)$ together with the special functions at the ends of the entire region are shown in Fig. 3.3.

The values of these functions and their first derivatives at the nodal points follow immediately from Eqs. (3.33) and (3.34)

$$v_i(x_j) = \delta_{ij}, \quad v_i'(x_j) = 0, \quad s_i(x_j) = 0, \quad s_i'(x_j) = \delta_{ij}\frac{1}{h}, \tag{3.35}$$

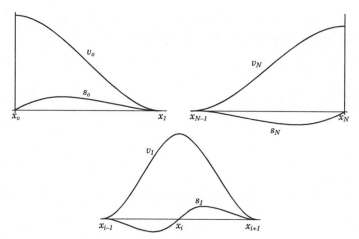

FIGURE 3.3 In the lower part of this figure the v_i and s_i splines are shown with $1 \le i \le n - 1$. The special functions v_0, s_0, v_n, s_n are used to represent functions at the ends of the inter.

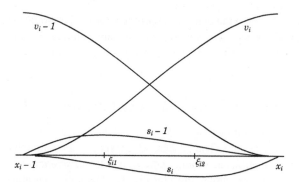

FIGURE 3.4 Four of the splines are nonzero at the Gauss points within the interval $[x_{i-1}, x_i]$.

where δ_{ij} is the Kronecker delta function, which is equal to one if i is equal to j and zero if $i \ne j$. These conditions are sufficient to determine the polynomials within each interval.

An approximate solution of an ordinary differential equation can be expressed as a linear combination of these Hermite splines

$$u(x) = \sum_{i=0}^{N} [\alpha_i v_i(x) + \beta_i s_i(x)]. \tag{3.36}$$

At this point, we introduce the Gauss quadrature points which play an important role in numerical schemes for evaluating integrals. The Gauss points for cubic polynomials are given by the formulas

$$\xi_{i1} = x_{i-1} + \frac{3 - \sqrt{3}}{6}h, \qquad \xi_{i2} = x_{i-1} + \frac{3 + \sqrt{3}}{6}h \tag{3.37}$$

The value of an integral of a function defined at the Gauss quadrature points is equal to the sum of the values of the function at the Gauss points multiplied by the step size δ divided by two. For polynomials of third degree and lower, this expression for the integral is exact.

For the Gauss quadrature points, ξ_{i1} and ξ_{i2}, within the i-th interval, four functions of the basis, v_{i-1}, s_{i-1}, v_i, and s_i, have nonzero values. These functions are illustrated in Fig. 3.4. The values of the function (3.36) at the two Gauss points are

$$u(\xi_{i1}) = \alpha_{i-1} v_{i-1}(\xi_{i1}) + \beta_{i-1} s_{i-1}(\xi_{i1}) + \alpha_i v_i(\xi_{i1}) + \beta_{i-1} s(\xi_{i1}) \tag{3.38}$$

$$u(\xi_{i2}) = \alpha_{i-1} v_{i-1}(\xi_{i2}) + \beta_{i-1} s_{i-1}(\xi_{i2}) + \alpha_i v_i(\xi_{i2}) + \beta_{i-1} s(\xi_{i2}) \tag{3.39}$$

Denoting the values of the spline functions at the first Gauss point (ξ_{i1}) by b_{11}, b_{12}, b_{13}, and b_{14} and the values of the spline functions at the second Gauss point (ξ_{i2}) by b_{21}, b_{22}, b_{23}, and b_{24}, these last two equation can be written simply

$$u(\xi_{i1}) = b_{11}\alpha_{i-1} + b_{12}\beta_{i-1} + b_{13}\alpha_i + b_{14}\beta_i \tag{3.40}$$

$$u(\xi_{i2}) = b_{21}\alpha_{i-1} + b_{22}\beta_{i-1} + b_{23}\alpha_i + b_{24}\beta_i, \tag{3.41}$$

or in matrix form

$$\begin{bmatrix} u(\xi_{i1}) \\ u(\xi_{i2}) \end{bmatrix} = [\mathbf{B}_i] \times \begin{bmatrix} \alpha_{i-1} \\ \beta_{i-1} \\ \alpha_i \\ \beta_i \end{bmatrix} \tag{3.42}$$

where

$$[\mathbf{B}_i] = \begin{bmatrix} b_{11} & b_{12} & b_{13} & b_{14} \\ b_{21} & b_{22} & b_{23} & b_{24} \end{bmatrix} \tag{3.43}$$

The elements of matrix $[\mathbf{B}_i]$, which correspond to the values of the spline basis functions at the collocations points, can be evaluated using Eqs. (3.33) and (3.34). These matrix elements can conveniently expressed in terms of the parameters

$$p1 = (9 - 4*\sqrt{3})/18;$$
$$p2 = (9 + 4*\sqrt{3})/18;$$
$$p3 = (3 - \sqrt{3})/36;$$
$$p4 = (3 + \sqrt{3})/36;$$

with this notation the B-matrix for each interval can be written

$$[\mathbf{B}_i] = \begin{bmatrix} p2 & p4 & p1 & -p3 \\ p1 & p3 & p2 & -p4 \end{bmatrix} \tag{3.44}$$

Treating each of the subintervals in a similar manner, the vector $\mathbf{u_G}$ consisting of the values of the approximate solution at the Gauss points can be written as the product of a matrix times a vector

$$\mathbf{u_G} = \mathbf{Bu}, \tag{3.45}$$

where

$$\mathbf{u} = [\alpha_0, \beta_0, \alpha_1, \beta_1, ..., \alpha_N, \beta_N]^T \tag{3.46}$$

$$\mathbf{u_G} = [u(\xi_{11}), u(\xi_{12}), u(\xi_{21}), u(\xi_{22}), ..., u(\xi_{N1}), u(\xi_{N2})]^T \tag{3.47}$$

The wave function (3.46), which is defined by of the values of the spline coefficients, is called the nodal form of the wave function and Eq. (3.47) gives the values of the wave function at the Gauss quadrature points. $\mathbf{B}$ is a rectangular matrix having $2N + 2$ columns and $2N$ rows. $\mathbf{B}$ has the structure

$$\mathbf{B} = \begin{bmatrix} \boxed{\mathbf{B}_1} & & & & \\ & \boxed{\mathbf{B}_2} & & & \\ & & \ddots & & \\ & & & \boxed{\mathbf{B}_{N-1}} & \\ & & & & \boxed{\mathbf{B}_N} \end{bmatrix}. \tag{3.48}$$

Two adjacent blocks $\mathbf{B}_i$ and $\mathbf{B}_{i+1}$ overlap in two columns.

The first and second derivatives of the approximate solution can also be represented by matrices with the same block structure. We express the vector $\mathbf{u}'_\mathbf{G}$ consisting of the values of the first derivative of the solution at the Gauss points as

$$\mathbf{u}'_\mathbf{G} = \mathbf{C}\mathbf{u}, \tag{3.49}$$

where

$$\mathbf{u} = [\alpha_0, \beta_0, \alpha_1, \beta_1, ..., \alpha_N, \beta_N]^T \tag{3.50}$$

$$\mathbf{u}'_\mathbf{G} = [u'(\xi_{11}), u'(\xi_{12}), u'(\xi_{21}), u'(\xi_{22}), ..., u'(\xi_{N1}), u'(\xi_{N2})]^T. \tag{3.51}$$

The negative of the second derivative of the function at the Gauss points can be written

$$-\mathbf{u}''_\mathbf{G} = \mathbf{A}\mathbf{u}, \tag{3.52}$$

where

$$\mathbf{u} = [\alpha_0, \beta_0, \alpha_1, \beta_1, ..., \alpha_N, \beta_N]^T \tag{3.53}$$

$$\mathbf{u}''_\mathbf{G} = [u''(\xi_{11}), u''(\xi_{12}), u''(\xi_{21}), u''(\xi_{22}), ..., u''(\xi_{N1}), u''(\xi_{N2})]^T. \tag{3.54}$$

The elements of the $[\mathbf{C}_i]$, and $[\mathbf{A}_i]$ can be expressed in terms of the parameters

$$p5 = 2\sqrt{3}$$
$$p6 = \sqrt{3} - 1$$
$$p7 = \sqrt{3} + 1$$

with this notation, the C-matrix for each interval can be written

$$[\mathbf{C}_i] = \begin{bmatrix} 1 & -1/p5 & -1 & 1/p5 \\ 1 & 1/p5 & -1 & -1/p5 \end{bmatrix} / \delta \tag{3.55}$$

and the A-matrix is

$$[\mathbf{A}_i] = \begin{bmatrix} p5 & p7 & -p5 & p6 \\ -p5 & -p6 & p5 & -p7 \end{bmatrix} / \delta^2 \tag{3.56}$$

As for the matrix $\mathbf{B}$, the matrices $\mathbf{C}$ and $\mathbf{A}$ have $2N + 2$ columns and $2N$ rows, and the Hamiltonian of the system will also have this property. The matrices may be converted into square matrices by adding a single row to the top and bottom of these matrices. These additional rows may be chosen to impose the boundary conditions. In the case of homogeneous Dirichlet boundary conditions for which the wave function is zero at the boundary or Neumann boundary conditions for which the derivative of the wave function is zero at a boundary, the boundary conditions can also be imposed by removing from the matrices $\mathbf{B}$, $\mathbf{C}$, and $\mathbf{A}$ the columns corresponding to the zero value of the functions or its derivatives. For the even solutions, the derivative of the wave function at the origin is zero and the wave function will be zero at the edge of the physical region. The boundary conditions can then be imposed by deleting the second column and the second to last column of the matrices. For the odd solutions, the wave function is zero at the origin and the wave function is again zero at the edge of the physical regions. The boundary conditions can then be imposed by deleting the first and the second to last columns. After imposing the boundary conditions in either case, the resulting matrices have the same number of rows and columns.

Eq. (3.12) is the eigenvalue equation for points inside the finite well. For points inside the well, the matrix on the left-hands side of the eigenvalue equation, which we denote by $\mathbf{C}_1$, is equal to the $\mathbf{A}$ matrix representing the negative of the second derivative

$$\mathbf{C}_1 = \mathbf{A}.$$

Outside the well, the terms depend upon both the $\mathbf{A}$ and $\mathbf{B}$ and will be denoted

$$\mathbf{C}_2 = \mathbf{A} + \mathcal{V}_0 * E_0 \mathbf{B}.$$

In the MATLAB program to be described shortly, the $\mathbf{C}_1$ and $\mathbf{C}_2$ matrices are brought together to form the matrix of the Hamiltonian operator, which will be called the collocation matrix. We write

$$\mathbf{C}_{mat} = [\mathbf{C}_1; \mathbf{C}_2]. \tag{3.57}$$

According to Eq. (3.45) a vector consisting of the values of the wave function at the collocation points is related to the vector consisting of the spline coefficient by the equation

$$\mathbf{v} = \mathbf{B}_{mat}\mathbf{u}.$$

Similarly, the values of the negative of the second derivative of the wave function at the Gauss points plus the potential energy times the wave function at the Gauss points is represented by the following equation

$$\mathbf{C}_{mat}\mathbf{u} = \mathbf{A}\mathbf{u} + \mathcal{V}_0 * E_0 \mathbf{B}\mathbf{u}. \tag{3.58}$$

Drawing together these last two equations, the eigenvalue equation for the finite well can be written

$$\mathbf{C}_{mat}\mathbf{u} = E_0\epsilon\mathbf{B}_{mat}\mathbf{u}. \tag{3.59}$$

To obtain a standard eigenvalue problem, we first recall that the boundary conditions may be imposed by deleting two columns from the matrices. After removing the two appropriate columns, $\mathbf{B}_{mat}$ then has the same number of rows and columns and may be inverted. We may then multiply (3.59) from the left with $\mathbf{B}_{mat}^{-1}$ and write the resulting equation as

$$\mathbf{L}_{mat}\mathbf{u} = E_0\epsilon\mathbf{u}, \tag{3.60}$$

where

$$\mathbf{L}_{mat} = \mathbf{B}_{mat}^{-1}\mathbf{C}_{mat}. \tag{3.61}$$

Eq. (3.60) is an ordinary eigenvalue equation.

A matrix eigenvalue equation may thus be written in two equivalent forms. Eq. (3.59) with matrices on both sides of the equation is a generalized eigenvalue equation, and Eq. (3.60) is an ordinary eigenvalue equation. MATLAB has routines that can be used to solve either of these two kinds of eigenvalue problems. In the following, we shall find the eigenvalues and the eigenfunctions of an electron in a finite well by solving a generalized eigenvalue problem, and we shall find the eigenvalues and eigenfunction of the simple harmonic oscillator by solving an ordinary eigenvalue problem. The MATLAB program FiniteWellcoll.m given below uses spline collocation to solve the generalized eigenvalue problem for an electron in a finite well with depth 0.3 eV. This program produces the same bound states with energy less than 0.3 eV as does the program FiniteWell.m described earlier that solved the eigenvalue problem for a particle in a finite well using a five-point finite difference approximation of the second derivative.

MATLAB Program FiniteWellcoll.m

A MATLAB program for finding the eigenvalues and eigenvectors for an electron moving in a finite well using the spline collocation method.

```
xmax=20;
L=5;
n=5;
delta=xmax/n;
deltas=delta*delta;
n1=fix(L/delta);
n2=n-n1;
e0=1.759;
v0=0.3;

% Construct B matrix
p1=(9 -4*sqrt(3))/18;
p2=(9+4*sqrt(3))/18;
```

```
p3=(3-sqrt(3))/36;
p4=(3+sqrt(3))/36;

B=[p2 p4 p1 -p3; p1 p3 p2 -p4]

% Construct A matrix
p5=2*sqrt(3);
p6=sqrt(3)-1;
p7=sqrt(3)+1;

A=[p5 p7 -p5 p6;-p5 -p6 p5 -p7]/deltas

% Construct Full B matrix
Bmat=zeros(2*n, 2*n+2);
for row=1:2:2*n
  Bmat(row:row+1, row:row+3)=B;
end

% Construct Collocation matrix
C1=zeros(2*n1, 2*n+2);
for row=1:2:2*n1
  C1(row:row+1, row:row+3)=A;
end
C2=zeros(2*n2,2*n+2);
for row=1:2:2*n2
  C2(row:row+1,row+2*n1:row+2*n1+3)=A+v0*e0*B;
end
Colmat=[C1;C2]
% Even Solutions
EBmat = Bmat;
EBmat(:,2)=[];
EBmat(:,2*n)=[];
EColmat = Colmat;
EColmat(:,2)=[];
EColmat(:,2*n)=[];
% Even Solutions of Generalized Eigenvalue Equation
V =eig(EColmat,EBmat);
GroundStateEven = sort(V)/e0
% Odd Solutions
OBmat = Bmat;
OBmat(:,1)=[];
OBmat(:,2*n)=[];
OColmat = Colmat;
OColmat(:,1)=[];
OColmat(:,2*n)=[];
% ODD Solutions of Generalized Eigenvalue Problem
V =eig(OColmat,OBmat);
GroundStateOdd = sort(V)/e0
```

As before, the first four lines of the MATLAB Program FiniteWellcoll.m define the length of the physical region (xmax), the x coordinate of the edge of the well (L), the number of intervals (n), and the step size (delta). The MATLAB function *fix* in the next lines of the program rounds the ratio "L/delta" to the integer toward zero. The integer $n1$ is the number of intervals within the well, and the integer $n2$ is the number of intervals outside the well. Each interval contains two collocation points. The lines of the code in which n, $n1$, and $n2$ are defined do not have a semicolon following them so that these variables are printed when the program is executed. After defining the constant E_0, the program then defines the two by four A and B matrices, which are used to construct the collocation matrix $Colmat$. The lines

```
EBmat = Bmat;
EBmat(:,2)=[];
EBmat(:,2*n)=[];
EColmat = Colmat;
EColmat(:,2)=[];
EColmat(:,2*n)=[];
```

define the B-matrix and $Colmat$ for the even solutions and impose the boundary conditions by removing the second and the second to last columns of the matrices. The lines in which the matrices A and B matrices are defined and the separate line with Colmat do not end with a semicolon. So, one can easily check that the matrices produced by the program agree with the equations we have given above. We note in particular that the first columns of Colmat differ from the first columns of $\mathbf{A}$ because the second column of Colmat has been deleted to satisfy the boundary condition on the left. Similarly, the lines of the program

```
OBmat = Bmat;
OBmat(:,1)=[];
OBmat(:,2*n)=[];
OColmat = Colmat;
OColmat(:,1)=[];
OColmat(:,2*n)=[];
```

define the B-matrix and the $Colmat$ for the odd solutions and impose the boundary conditions.

With the number of intervals, n = 100, the lowest even eigenvalues produced by FiniteWellcoll.m are 0.0342 and 0.2715 and the lowest odd eigenvalue is 0.1326 as produced by FiniteWell.m with $n = 100$.

We would now like to describe the MATLAB Program Oscillatorcoll.m which solves the eigenvalue problem of the simple harmonic oscillator using spine collocation. As we have noted earlier, we shall solve the eigenvalue problem for the oscillator by constructing the matrix $\mathbf{B}_{mat}^{-1}$ and multiplying the generalized eigenvalue equation (3.59) from the left with $\mathbf{B}_{mat}^{-1}$ to obtain the ordinary eigenvalue equation (3.60). We can obtain the jth column of a matrix by multiplying the matrix times a unit vector with the jth element being one and all of the other elements being zero. Denoting such a vector by u_j and denoting the jth column of $\mathbf{B}_{mat}^{-1}$ by v^j, we have

$$\mathbf{B}_{mat}^{-1} u_j = v^j. \tag{3.62}$$

This last equation is equivalent to the equation

$$\mathbf{B}_{mat} v^j = u_j. \tag{3.63}$$

In the MATLAB program given below, we solve the linear system (3.63) for each column of $\mathbf{B}_{mat}^{-1}$ and then confirm the inverse of $\mathbf{B}_{mat}$ so formed times $\mathbf{B}_{mat}$ is equal to the identity matrix.

MATLAB Program Oscillatorcoll.m

A MATLAB program for finding the eigenvalues and eigenvectors for a simple harmonic Oscillator.

```
xmax=10;
n=100;
delta=xmax/n;
deltas=delta*delta;

% Gauss Points
xi1=(3-sqrt(3))/6;
xi2=(3+sqrt(3))/6;

% Construct B matrix
p1=(9 -4*sqrt(3))/18;
```

```
p2=(9+4*sqrt(3))/18;
p3=(3-sqrt(3))/36;
p4=(3+sqrt(3))/36;

B=[p2 p4 p1 -p3; p1 p3 p2 -p4];

% Construct A matrix
p5=2*sqrt(3);
p6=sqrt(3)-1;
p7=sqrt(3)+1;

A=[p5 p7 -p5 p6;-p5 -p6 p5 -p7]/deltas;

% Construct Full Matrix Bmat
Bmat=zeros(2*n, 2*n+2);
for row=1:2:2*n
  Bmat(row:row+1, row:row+3)=B;
end

% For Even Solutions
EBmat = Bmat;
EBmat(:,2)=[];
EBmat(:,2*n)=[];

% For Odd Solutions
OBmat = Bmat;
OBmat(:,1)=[];
OBmat(:,2*n)=[];

% Construct Full Matrix Amat
Amat=zeros(2*n, 2*n+2);
for row=1:2:2*n
  Amat(row:row+1, row:row+3)=A;
end
% For Even Solutions
EAmat = Amat;
EAmat(:,2)=[];
EAmat(:,2*n)=[];

% For Odd Solutions
OAmat = Amat;
OAmat(:,1)=[];
OAmat(:,2*n)=[];

% Construct vectors for x  and x squared at the Gauss points
xcol = zeros(2*n,1);
xcol2 = zeros(2*n,1);
ii = 0;
x = 0.0;
for i = 1:n
   x1 = x + xi1*delta;
   x2 = x + xi2*delta;
   xcol(ii+1) = x1;
   xcol(ii+2) = x2;
```

```
    xcol2(ii+1) = x1*x1;
    xcol2(ii+2) = x2*x2;
    ii = ii + 2;
    x = x + delta;
end

% Construct D matrix
dmat = diag(xcol2);

% Colmat
% For Even Solutions
EColmat = zeros(2*n,2*n+2);
EColmat = EAmat + dmat*EBmat;

% For Odd Solutions
OColmat = zeros(2*n,2*n+2);
OColmat = OAmat + dmat*OBmat;


% Construct Inverse of Bmat
% For Even Solutions
EBinv = zeros(2*n,2*n);
EBBinv = zeros(2*n,2*n);
rvec = zeros(2*n);
wf = zeros(2*n);
for j = 1:2*n
    rvec(j) = 1.0;
    wf = EBmat\rvec;
    for i = 1:2*n
        EBinv(i,j) = wf(i);
    end
    rvec(j) =0.0;
end
EBinv;
EBBinv=EBmat*EBinv;
EBBinv;

% For Odd Solutions
OBinv = zeros(2*n,2*n);
OBBinv = zeros(2*n,2*n);
rvec = zeros(2*n);
wf = zeros(2*n);
for j = 1:2*n
    rvec(j) = 1.0;
    wf = OBmat\rvec;
    for i = 1:2*n
        OBinv(i,j) = wf(i);
    end
    rvec(j) =0.0;
end
OBinv;
OBBinv=OBmat*OBinv;
OBBinv;
```

```
% Even Solutions
ELmat = EBinv*EColmat;
[V, D] = eig(ELmat);
[EvenStateEnergies, index] = sort(diag(D));
EvenStateEnergies(1:3)
EvenStateVectors=V(:,index(1:3));

% Odd Sotutions
OLmat = OBinv*OColmat;
[V, D] = eig(OLmat);
[OddStateEnergies, index] = sort(diag(D));
OddStateEnergies(1:3)
OddStateVectors=V(:,index(1:3));

save("oscillator.mat","EvenStateVectors","OddStateVectors");
```

In the MATLAB program, we first give the number of intervals n and the grid spacing $delta$ and then give the coefficients, ξ_{i1} and ξ_{I2} to define the Gauss quadrature points. The 2 x 4 B and A matrices are then defined and these matrices are then used to form the full B and A matrices (Bmat and Amat) with $2*n$ rows and $2*n+2$ columns. Because the boundary conditions are different for even and odd functions, the even and odd B and A matrices must be defined separately. The even B matrices (EBmat), are formed by deleting the second column and the second to last columns of Bmat, while the odd B matrices (OBmat) are formed by deleting the first and second to last columns. As just described, the jth column of $Binv$ is found by multiplying $Binv$ times a unit vector r^j with 1 in the jth row and all other elements equal to zero.

For each column of EBinv, the product Binv*rvec is evaluated by solving the linear system,

```
B wf= rvec
```

with the MATLAB command

```
wf = EBmat\rvec;
```

As before using a finite difference approximation of the second derivative, the program Oscillatorcoll.m produces the even eigenvalues, 1, 5, and 9 (times $(1/2)\hbar\omega$) and the odd eigenvalues, 3, 7, and 11 (times $(1/2)\hbar\omega$). The statement

```
EvenStateVectors=V(:,index(1:3));
```

together with the statements

```
OddStateVectors=V(:,index(1:3));
save("oscillator.mat","EvenStateVectors","OddStateVectors");
```

define the lowest three even and odd eigenvectors and save them in the file "Oscillator.mat", which will be used in a later section of this chapter to plot the three lowest even and odd eigenfunctions of the harmonic oscillator and to calculate the average values of χ^2 and the average value of the kinetic energy for these eigenfunctions.

3.4   Electron scattering

We now consider the important class of problems in which electrons scatter from potential steps and barriers. Such problems provide useful models for describing the processes that occur when electrons are incident upon the interface between different semiconductor materials.

3.4.1   Scattering from a potential step

In order to attain a qualitative understanding of scattering phenomena, we first consider the problem of an electron incident upon the simple potential step shown in Fig. 3.5. For this problem, we first suppose that the energy E is greater than the step height. We shall call the region to the left of the step region 1, and the region to the right of the step region 2. In the first region, for which $x < 0$, the potential energy is $V(x) = 0$, and the Schrödinger equation becomes

$$\frac{-\hbar^2}{2m}\frac{d^2\psi_1}{dx^2} = E\psi_1 .$$

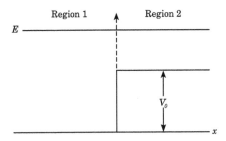

FIGURE 3.5 *Potential step with the energy E greater than the step height V_0.*

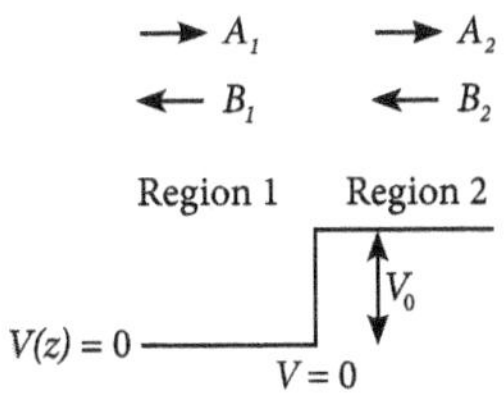

FIGURE 3.6 A potential step at $x = 0$ with the potential energy $V = 0$ for negative x and $V = V_0$ for positive x. The arrows directed towards and away from the step correspond to waves moving in the positive and negative directions.

Defining

$$k_1 = \sqrt{\frac{2mE}{\hbar^2}}, \tag{3.64}$$

this equation can be written

$$\frac{d^2\psi_1}{dx^2} + k_1{}^2\psi_1 = 0.$$

Since for this problem the electron can move freely in either direction, the solution of the wave equation in region 1 has the following form

$$\psi_1(x) = A_1 e^{ik_1 x} + B_1 e^{-ik_1 x}. \tag{3.65}$$

The first term in this equation for the wave function corresponds to an electron which moves in the positive x direction and is incident upon the barrier from the left, while the second term corresponds to an electron which is reflected by the step and moves in the negative x direction.

Similarly, in the second region, for which $x > 0$, the potential energy is $V(x) = V_0$, and the Schrödinger equation is

$$\frac{-\hbar^2}{2m}\frac{d^2\psi_2}{dx^2} + V_0\psi_2 = E\psi_2 .$$

Defining

$$k_2 = \sqrt{\frac{2m(E - V_0)}{\hbar^2}}, \tag{3.66}$$

the equation can be written

$$\frac{d^2\psi_2}{dx^2} + k_2{}^2\psi_2 = 0.$$

The solution for the wave function in the second region is

$$\psi_2(x) = A_2 e^{ik_2 x} + B_2 e^{-ik_2 x} . \tag{3.67}$$

The first term here corresponds to the transmitted wave, which passes over the barrier in the positive x-direction. The second term corresponds to a wave which approaches the step from the right. The amplitudes in the two regions are illustrated in Fig. 3.6.

As we discussed when we considered the example of a particle confined to a finite well, the wave function and its first derivative must be continuous at the boundary between the two regions. We thus require that at $x = 0$ the wave functions satisfy the following boundary conditions

$$\psi_1(0) = \psi_2(0)$$
$$\psi'_1(0) = \psi'_2(0).$$

Substituting the expressions for the wave function in the first and second regions into these equations, we get

$$A_1 + B_1 = A_2 + B_2$$
$$ik_1(A_1 - B_1) = ik_2(A_2 - B_2)$$

Dividing the second equation through by ik_1, these equations become

$$A_1 + B_1 = A_2 + B_2$$
$$A_1 - B_1 = \frac{k_2}{k_1}(A_2 - B_2)$$

We now solve for A_1 in terms of A_2 and B_2 by adding the two equations together, and we solve for B_1 in terms of A_2 and B_2 by subtracting the two equations. The resulting equations can be written

$$A_1 = \left(\frac{k_1 + k_2}{2k_1}\right) A_2 + \left(\frac{k_1 - k_2}{2k_1}\right) B_2 \tag{3.68}$$

$$B_1 = \left(\frac{k_1 - k_2}{2k_1}\right) A_2 + \left(\frac{k_1 + k_2}{2k_1}\right) B_2 \tag{3.69}$$

As a first application of these equations, we consider the simple physical problem for which electrons are incident upon the barrier from the left and are either transmitted or reflected. For that problem, there will not be a wave moving to the left in the second region, and we therefore set the coefficient B_2 equal to zero. Substituting $B_2 = 0$ in the above two equations, we obtain

$$A_1 = \left(\frac{k_1 + k_2}{2k_1}\right) A_2 \tag{3.70}$$

$$B_1 = \left(\frac{k_1 - k_2}{2k_1}\right) A_2. \tag{3.71}$$

Eq. (3.70) provides a relation between the A_1 and A_2 coefficients. An equation relating the A_1 and B_1 coefficients can be obtained by dividing Eq. (3.71) by Eq. (3.70) giving

$$\frac{B_1}{A_1} = \left(\frac{k_1 - k_2}{k_1 + k_2}\right).$$

In a scattering experiment, in which a beam of electrons is incident upon a potential step, the role of the theory is to calculate the probability that an electron will be transmitted or reflected. If we denote the velocity of an electron in the first region by v_1, then an incident electron will travel a distance $v_1 dt$ in an infinitesimal length of time dt. In the length of time dt, an electron located in the region between $x = -v_1 dt$ and $x = 0$ will pass across the point at $x = 0$ on the edge of the step. We have found previously that the probability of finding an electron in an interval is equal to the absolute value of the wave function squared times the length of the interval. The probability that the incident electron, which is described by the wave function

$$\psi_i(x) = A e^{ik_1 x},$$

will be in the interval between $x = -v_1 dt$ and $x = 0$ and thus hit the step in the length of time dt is $|\psi_i|^2 v_1 dt = |A|^2 v_1 dt$. For a one-dimensional problem of this kind, the intensity of the beam at a particular point is equal to the probability per unit time that a particle passes the point. Hence, the intensity of the incident beam of electrons is $|A_1|^2 v_1$. Similarly, the intensity of the transmitted beam is $|A_2|^2 v_2$ and the intensity of the reflected beam is $|B_1|^2 v_1$. The ratio of the intensity of the

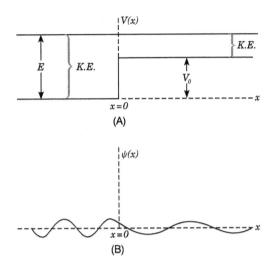

FIGURE 3.7 The wave function for an electron incident upon the potential step.

transmitted beam to the intensity of the incident beam is called the *transmission coefficient* T. It is equal to the probability that an electron, which is incident upon the step, will be transmitted. Using the above expressions for the intensities of the transmitted and incident waves, we obtain

$$T = \frac{|A_2|^2 v_2}{|A_1|^2 v_1}.$$ (3.72)

Similarly, the ratio of the intensity of the reflected electrons to the intensity of the incident electrons, which is called the *reflection coefficient* R, is equal to the probability that an incident electron will be reflected. R is given by the formula

$$R = \frac{|B_1|^2 v_1}{|A_1|^2 v_1} = \frac{|B_1|^2}{|A_1|^2}.$$

The velocity of an electron can be written

$$v = \frac{p}{m} = \frac{\hbar k}{m}.$$

Substituting the expression for the incident amplitude A_1 given by Eq. (3.70) into Eq. (3.72) and using the last equation to evaluate the velocities, we obtain

$$T = \frac{4}{(1 + k_2/k_1)^2} \cdot \frac{k_2}{k_1} = \frac{4 k_1 k_2}{(k_1 + k_2)^2}$$ (3.73)

The corresponding expression for the reflection coefficient is

$$R = \frac{(1 - k_2/k_1)^2}{(1 + k_2/k_1)^2} = \frac{(k_1 - k_2)^2}{(k_1 + k_2)^2}.$$ (3.74)

One may readily confirm that the sum of the T and R coefficients is equal to one.

The potential step is illustrated in Fig. 3.7(A), and the wave function in the two regions is illustrated in Fig. 3.7(B). The kinetic energy in each region corresponds to the difference between the line representing the total energy and the line representing the potential energy. From the figure, it is apparent that the kinetic energy is larger in the region to the left. The kinetic energy and the wavelength of the waves are related to k by the equations

$$KE = \frac{(\hbar k)^2}{2m}$$

and

$$\lambda = \frac{2\pi}{k}.$$

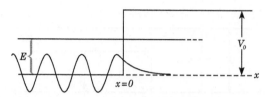

FIGURE 3.8 The wave function for an electron incident upon a potential step when the energy E is less than the step height V_0.

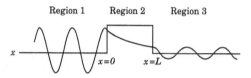

FIGURE 3.9 The wave function for an electron incident upon a potential barrier.

Since the kinetic energy is larger in the region to the left, k is also larger in that region and the wavelength of the wave is smaller. The wave function corresponding to the more energetic incident electron oscillates more rapidly. To be consistent with the boundary conditions, the waves in the two regions must join smoothly together at $x = 0$.

3.4.2 Barrier penetration and tunneling

Thus far we have only considered the case for which the energy E is larger than the step height. If the energy were less than the step height, then, in place of Eq. (3.66), we should define k_2 by the equation

$$k_2 = \sqrt{\frac{2m(V_0 - E)}{\hbar^2}}. \tag{3.75}$$

The equation for the wave function in the second region then becomes

$$\frac{d^2\psi_2}{dx^2} - k_2{}^2\psi_2 = 0,$$

which has the solution

$$\psi_2(x) = Ce^{k_2 x} + De^{-k_2 x}.$$

Since the first term becomes infinite as x increases, we must set $C = 0$. The expression for the wave function in the second region can then be written

$$\psi_2(x) = De^{-k_2 x}. \tag{3.76}$$

The continuity conditions still apply, and they lead as before to linear relations between the coefficients A, B, and D. We leave as an exercise to derive the ratios A/D and B/D in this case and show that $R = 1$ (Problem 9). As for the classical theory, all electrons are reflected from a potential step having a greater height than the kinetic energy of the electrons. An important difference between the wave theory and classical physics follows as a consequence of Eq. (3.76). As for the harmonic oscillator, the wave function decreases monotonically to zero in the forbidden region where the potential energy is greater than the total energy. The wave function for this case is illustrated in Fig. 3.8. Notice that the wave function penetrates into the classically forbidden region. Since the potential energy is greater than the total energy in this region, the kinetic energy of the electron is negative. We leave as an exercise to calculate the probability that an electron with $E < V_0$ penetrates into the barrier. (Problem 10)

Finally, we consider the case of electrons incident upon the potential barrier illustrated in Fig. 3.9. For this problem, the Schrödinger equation must be solved independently in the three regions shown. In regions 1 and 3 for which $x \leq 0$ and $x \geq L$ respectively, the wave functions for the waves are

$$\psi_1(x) = A_1 e^{ik_1 x} + B_1 e^{-ik_1 x}, \quad x \leq 0$$
$$\psi_3(x) = A_3 e^{ik_1 x}, \quad x \geq L,$$

where k_1 is given by Eq. (3.64). In the region within the barrier for which $0 \leq x \leq L$, the wave function is

$$\psi_2(x) = A_2 e^{k_2 x} + B_2 e^{-k_2 x}, \quad 0 \leq x \leq L,$$

where k_2 is given by Eq. (3.75). The conditions that the wave function and its derivative be continuous at $x = 0$ and $x = L$, lead to a total of four boundary conditions imposed upon the five amplitudes, A_1, B_1, A_2, B_2, and A_3. These equations may be solved for the ratio A_1/A_3 of the incident and transmitted amplitudes. (Problem 11) The wave function for this scattering problem is shown in Fig. 3.9. Even though the energy chosen is less than the barrier height, the probability that the incident electron will tunnel through the barrier is nonzero. This possibility, which is an entirely new feature of quantum mechanics, has many important applications.

3.4.3 *T*-matrices

The relations (3.68) and (3.69) between the coefficients, A_1 and B_1, and the coefficients, A_2 and B_2, for the potential step shown in Fig. 3.5 can be expressed in terms of matrices. The advantage of expressing the relation between the incoming and outgoing amplitudes in matrix form is that the transmission and reflection coefficients for complex systems can then be calculated by multiplying the matrices for the individual parts. A review of the elementary properties of matrices is given in Appendix CC.

The *T*-matrix may be defined by the equation

$$\begin{bmatrix} A_1 \\ B_1 \end{bmatrix} = \begin{bmatrix} T_{11} & T_{12} \\ T_{21} & T_{22} \end{bmatrix} \begin{bmatrix} A_2 \\ B_2 \end{bmatrix}. \tag{3.77}$$

Equations for the coefficients, A_1 and B_1, in terms of the elements of the *T*-matrix and the coefficients, A_2 and B_2 can be obtained by multiplying the matrix in Eq. (3.77) times the column vector to its right and equating each component of the resulting vector with the corresponding component of the vector on the left. We obtain

$$A_1 = T_{11} A_2 + T_{12} B_2 \tag{3.78}$$

$$B_1 = T_{21} A_2 + T_{22} B_2 \tag{3.79}$$

With the definition (3.77) of the *T*-matrix, the order of the matrix-vector operation is the same as the order of the corresponding quantities in Fig. 3.6. The incident amplitudes are to the left in both Fig. 3.6 and Eq. (3.77), and the out-going amplitudes are to the right in both the figure and the equation. This is a convenient convention for describing complex scattering processes.

The elements of the *T*-matrix for the potential step can be obtained by identifying the matrix elements on the right-hand side of Eqs. (3.78) and (3.79) with the corresponding terms in Eqs. (3.68) and (3.69). We get

$$T_{11} = \left(\frac{k_1 + k_2}{2k_1} \right), \quad T_{12} = \left(\frac{k_1 - k_2}{2k_1} \right)$$
$$T_{21} = \left(\frac{k_1 - k_2}{2k_1} \right), \quad T_{22} = \left(\frac{k_1 + k_2}{2k_1} \right)$$

where k_1 and k_2 are given by Eqs. (3.64) and (3.75). The *T*-matrix for the potential step, which we denote by $T(k_1, k_2)$, can be written out explicitly in matrix form. We have

$$T(k_1, k_2) = \frac{1}{2k_1} \begin{bmatrix} k_1 + k_2 & k_1 - k_2 \\ k_1 - k_2 & k_1 + k_2 \end{bmatrix}. \tag{3.80}$$

The matrix $T(k_1, k_2)$ determines how the incoming wave is modified by the potential step. According to Eq. (3.77), the vector with components, A_1 and B_1, is equal to the product of $T(k_1, k_2)$ times the vector with components, A_2 and B_2.

3.4.4 Scattering from more complex barriers

The potential step can be thought of as a building block that can be used to build more complicated barriers. Consider, for instance, the potential barrier shown in Fig. 3.10. As for later chapters devoted to semiconductors the z coordinate measures

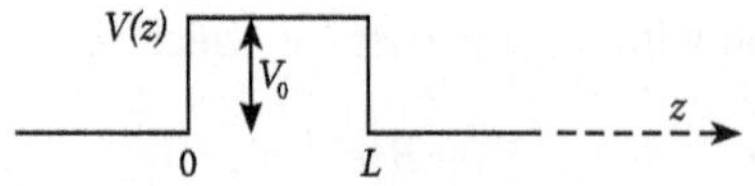

FIGURE 3.10 Potential barrier with $V(z) = V_0$ for $0 \leq z \leq L$ and $V(z) = 0$ elsewhere.

the distance perpendicular to the sides of the well. This barrier has a potential step at $z = 0$, then a level region where the wave oscillates from $z = 0$ to $z = L$, and finally a potential drop. The potential drop is a potential step with k_1 and k_2 interchanged.

To find the effect upon the wave function of a translation by a distance L, we substitute $z + L$ for x in Eq. (3.67) to obtain

$$\psi_2(z + L) = \left(A_2 e^{ik_2 L} \right) e^{ik_2 z} + \left(B_2 e^{-ik_2 L} \right) e^{-ik_2 z}.$$

The translated wave function has $A_2 e^{ik_2 L}$ in place of A_2 and $B_2 e^{-ik_2 L}$ in place of B_2. The effect of the translation upon the wave function can thus be described by the diagonal matrix

$$T_L(k_2) = \begin{bmatrix} e^{ik_2 L} & 0 \\ 0 & e^{-ik_2 L} \end{bmatrix}. \tag{3.81}$$

One can readily confirm that multiplying matrix $T_L(k_2)$ times the vector with components, A_2 and B_2, gives a vector with components, $A_2 e^{ik_2 L}$ and $B_2 e^{-ik_2 L}$.

In the scattering process in which electrons strike the square barrier shown in Fig. 3.10, the wave corresponding to the electrons first approaches the left edge of the barrier. The scattered waves then pass through the region in which the barrier is located and are incident upon the right edge of the barrier. The T-matrix describing the motion of the wave from left to right is the product of the T-matrices for the individual steps. We have

$$T = T(k_1, k_2) T_L(k_2) T(k_2, k_1). \tag{3.82}$$

The order of k_1 and k_2 in the matrix $T(k_1, k_2)$ on the left is the same as the order in k_1 and k_2 in Eq. (3.80). Moving from left to right in the product of matrices, the next matrix translates the solution along the barrier a distance L, and matrix $T(k_2, k_1)$, in which the order of k_1 and k_2 have been interchanged, represents the scattering of the waves by the right edge of the barrier.

The T-matrix for the entire process can be obtained using Eq. (3.80) and Eq. (3.81). We have

$$T = \frac{1}{2k_1} \begin{bmatrix} k_1 + k_2 & k_1 - k_2 \\ k_1 - k_2 & k_1 + k_2 \end{bmatrix} \begin{bmatrix} e^{ik_2 L} & 0 \\ 0 & e^{-ik_2 L} \end{bmatrix} \frac{1}{2k_2} \begin{bmatrix} k_1 + k_2 & k_2 - k_1 \\ k_2 - k_1 & k_1 + k_2 \end{bmatrix}. \tag{3.83}$$

The problem of obtaining the T-matrix thus reduces itself to multiplying the matrices (see Problem 9).

The transmission coefficient for the barrier can be expressed in terms of the elements of the T-matrix. Just as for the potential step, we denote the amplitudes of the incoming and outgoing waves on the left by A_1 and B_1, and we denote the amplitudes of the waves leaving and approaching the barrier on the right by A_2 and B_2. To obtain the transmission and reflection coefficients, we set the incoming amplitude on the right B_2 equal to zero in Eqs. (3.78) and (3.79) to obtain

$$A_1 = T_{11} A_2, \tag{3.84}$$
$$B_1 = T_{21} A_2. \tag{3.85}$$

Eq. (3.84) may be used to obtain an equation for the transmission amplitude t, which is the ratio of the transmitted and incident amplitudes. We have

$$t = \frac{A_2}{A_1} \Big|_{B_2 = 0} = \frac{1}{T_{11}}. \tag{3.86}$$

An equation for the reflection amplitude r can be obtained by dividing Eq. (3.85) by Eq. (3.84) to get

$$r = \frac{B_1}{A_1} \Big|_{B_2 = 0} = \frac{T_{21}}{T_{11}} \tag{3.87}$$

Eqs. (3.86)) and (3.87) give the transmission and reflection amplitudes for a barrier in terms of the elements of the T matrix.

We now apply these ideas to calculate the transmission coefficient as a function of energy for a square potential barrier of $GaAs$ with a height $V_0 = 0.3$ eV and thickness $L = 10$ nm. For a particular value of the energy, we must first use Eqs. (3.64) and (3.75) to calculate the values of k_1 and k_2 and use these values to calculate the transmission amplitude t. As we have seen before, the effective mass of conduction electrons in $GaAs$ is equal to 0.067 times the free electron mass. Using the value of the free electron mass and the value of $\hbar$ given in Appendix A, the value of $2m/\hbar^2$ in Eqs. (3.64) and (3.75) can be conveniently written

$$\frac{2m}{\hbar^2} = 1.7585 \text{ nm}^{-1}\text{eV}^{-1}. \tag{3.88}$$

The transmission coefficient for different values of the energy can easily be obtained using the following MATLAB function.

MATLAB Program FiniteWell.m

function T = Onebarrier(E) This MATLAB function calculates the transmission coefficient for a $GaAs$ barrier of height $V0 = 0.3$ eV and length $L = 10$ nm.

```
function T = Onebarrier(E)
L=10.0;
V0=0.3;
k1=sqrt(1.7585*E);
k2=sqrt(1.7585*(E-V0));
A=[(k1+k2) (k1-k2); (k1-k2) (k1+k2)];
B=[(k1+k2) (k2-k1); (k2-k1) (k1+k2)];
C=[exp(i*k2*L) 0; 0 exp(-i*k2*L)];
M=A*(C*B);
t=(4*k1*k2)/M(1,1);
T = abs(t)^2;
```

The first four lines of the MATLAB program define the size L and depth V_0 of the well and the constants k_1 and k_2. The program then calculates the matrix A, which determines how the incoming wave is modified by the front edge of the barrier, the matrix B, which determines how the wave is effected by the right edge of the barrier, and the matrix C, which translates the wave along the length of the barrier. The MATLAB program then multiplies the three matrices and uses the matrix product M to evaluate the transmission amplitude t and the transmission coefficient. The following table gives a sequence of MATLAB commands and responses using the MATLAB Function onebarrier(E).

MATLAB Output 3.1

A sequence of MATLAB commands and responses using the MATLAB Function Onebarrier.m

```
>> Onebarrier(0.40)

ans =

    0.7020

>> Onebarrier(0.50)

ans =

    0.9740

>> Onebarrier(0.60)

ans =

    0.9205
```

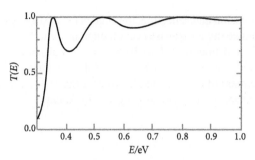

FIGURE 3.11 The transmission coefficient $T(E)$ as a function of the energy E for a square potential barrier of height $V_0 = 0.3$ eV and thickness 10 nm in GaAs.

One can see that the transmission coefficient T is equal to 0.7020 for an energy 0.4 just above the height of the well, rises to 0.9740 for an energy 0.50, and then declines to 0.9205 for an energy 0.60. The transmission coefficient oscillates approaching one for high energies. A plot of the transmission coefficient as a function of energy is produced by the following MATLAB program. The program produces the plot shown in Fig. 3.11.

MATLAB Program OnebarrierPlot.m

This MATLAB program plots the transmission coefficient for a *GaAs* barrier of height $V0 = 0.3$ eV and length $L = 10$ nm for energies from 0.3 eV to 1.0 eV.

```
OnebarrierPlot.m
for n=1:140
    en=0.3+n/200.;
    E(n)=en;
    T(n)=onebarrier(en);
end
plot(E,T);
```

The transmission coefficient of a classical particle would be zero below the barrier height and equal to one for any energy above the barrier height. As one can see from Fig. 3.11, the transmission coefficient for a quantum barrier is more complex, rising from zero to substantial values, oscillating for energies above the barrier height, and approaching a stable value of one only for high energies.

With MATLAB, one can explore more complex structures with multiple interfaces. We leave as an exercise to write a MATLAB function to calculate the transmission coefficient as a function of the energy for two *GaAs* barriers of height 0.3 eV and length $L = 10$ nm separated by a distance $L = 10$ nm and an associate MATLAB program to plot the transmission coefficient from $E = 0.3$ eV to 1.0 eV. (Problem.) One can write other MATLAB functions and programs for barriers separated by different distances, and one can easily extend this approach to more than two barriers. The mirrors of semiconducting lasers are typically made up of several layers of semiconducting materials.

3.5 The Heisenberg uncertainty principle

Modern quantum theory began to emerge in the mid 1920's with the publication of a number of seemingly unrelated articles which approached microscopic systems from radically different points of view. The idea of de Broglie that the motion of a particle such as an electron in an atom could be explained by associating a wave with the particle led Schrödinger to a wave formulation of quantum theory. The first article by Schrödinger on wave mechanics was published in 1926. The success of this theory in describing the spectra of hydrogen seemed to indicate that the electron should be treated as a wave rather than a particle. While the wave theory of Schrödinger was yet to appear, Heisenberg, Born, and Jordan proposed a new theory in 1925. This theory, called matrix mechanics, was the first complete theory presented to the scientific community as a method for solving atomic problems. Instead of questioning the particle nature of the electron, they called into question the classical concept of the measuring process. Because microscopic systems are comparable in size to the smallest means available for measuring them, Heisenberg showed how the measurement of one variable of a microscopic system would disturb the values of other variables of the system and lead to an inherent uncertainty in the results of the measurement.

Using qualitative arguments of this kind, Heisenberg argued that the inherent uncertainties of measurements of the position and momentum of an electron satisfy the equation

$$\Delta x \cdot \Delta p_x \geq \frac{\hbar}{2}.$$

This is known as the *Heisenberg Uncertainty Principle*. Similar relations apply to the other coordinates and the corresponding components of the momenta and to the time and the energy

$$\Delta t \cdot \Delta E \geq \frac{\hbar}{2}.$$

The mathematical formulation of this idea, which has come to be called the *Heisenberg Uncertainty Principle*, is central to the modern interpretation of quantum mechanics. Rather than derive the uncertainty principle as Heisenberg did, by considering an idealized measurement of the position of an electron, we shall consider the prospect of locating particles in space using the wave description of particle motion that we have used thus far.

3.5.1   Wave packets and the uncertainty principle

How an uncertainty relation might arise can be understood by considering functions formed by superimposing waves. The simplest function of this kind consists of a single plane wave. As discussed in the introduction, a plane wave can be represented as the real part of the exponential function

$$\psi(x) = Ae^{ikx}. \tag{3.89}$$

The wave oscillates with a constant amplitude from $-\infty$ to $+\infty$. Since the momentum and the angular wave vector k are related by the equation, $p = \hbar k$, the wave function (3.89) corresponds to a particle having a definite value of the momentum. The probability of finding the particle in the interval between x and $x + dx$ is equal to the product of the absolute value squared of the wave function and the interval length

$$dP = |\psi(x)|^2 dx = |A|^2 dx.$$

Since the probability per interval length is constant, the particle has an equal probability of being found anywhere along the x-axis. While the uncertainty of the momentum of the particle is zero, the uncertainty of the position of the particle is infinite.

Functions, which are localized in space and thus have less uncertainty associated with their position, can be formed by adding together plane waves of this kind. To illustrate how this might occur we consider the following MATLAB program that plots the sum of three sinusoidal waves.

MATLAB Program Sin3.m

This program which plots the sum of three sinusoidal functions.

```
x = -4.5:0.1:4.5;\\
y = sin(x)+sin(2*x)+0.43*sin(3*x)^;\\
plot(x,y);\\
```

The first line of MATLAB Program Sin3 creates a vector x with elements equal to the points of a grid extending from -4.5 to 4.5 with equal steps of 0.1. The program then adds the sine functions, $\sin(x)$, $\sin(2x)$, and $0.43\sin(3x)$, and generates the plot of the sum shown in Fig. 3.12. One sees immediately that the wave packet shown in Fig. 3.12 is more localized than a single sine function that extends over all space.

More well defined wave packets can be formed by making use of Fourier analysis described in the Introduction. We shall do things a little differently here superimposing exponential functions of the form (3.89) rather than sinusoidal functions, and we shall use a Fourier transform $g(k)$ which depends upon a continuous variable k. Any function, which is sufficiently smooth and absolutely integrable, may be represented by the *Fourier integral*

$$\psi(x) = \frac{1}{\sqrt{2\pi}} \int_{-\infty}^{+\infty} g(k)e^{ikx} dk. \tag{3.90}$$

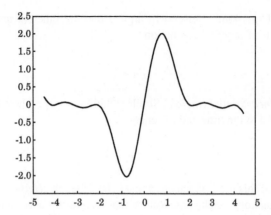

FIGURE 3.12 A wave equal to the sum of the three sinusoidal functions produced by MATLAB Program Sin3.m.

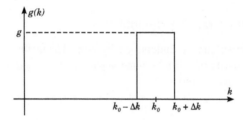

FIGURE 3.13 Fourier transform $g(k)$ corresponding to a packet of waves having k-values near a central value denoted by k_0.

This equation represents the function $\psi(x)$ as a combination of harmonic waves. The function $g(k)$, which specifies the amplitude of the wave with angular wave number k, is called the *Fourier transform* of $\psi(x)$. It is given by the equation

$$g(k) = \frac{1}{\sqrt{2\pi}} \int_{-\infty}^{+\infty} \psi(x) e^{-ikx} dx. \tag{3.91}$$

These equations have many important applications to communication theory because they enable one to represent an arbitrary signal as a superposition of harmonic waves and to study the response of the system to each wave separately.

We now use Eq. (3.90) to form linear combinations of waves corresponding to functions that are localized in space. We consider a packet consisting of waves having k-values near a certain central value that we denote by k_0. The superposition of a packet of waves of this kind can be formed using Eq. (3.90) by choosing a Fourier transform which is nonzero only in a small interval about k_0.

Example 3.2

Find the function corresponding to the Fourier transform $g(k)$ shown in Fig. 3.13.

Solution

The transform $g(k)$ in the figure is equal to g in the interval $k_0 - \Delta k \le k \le k_0 + \Delta k$, and it is zero otherwise. Using Eq. (3.90), we get

$$\psi(x) = \frac{g}{\sqrt{2\pi}} \int_{k_0-\Delta k}^{k_0+\Delta k} e^{ikx} dk.$$

The integral in this equation may be evaluated by defining the change of variable

$$u = ikx.$$

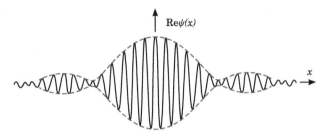

FIGURE 3.14 The dotted line in this figure represents the term within square brackets in Eq. (3.92), while the solid line represents the real part of $\psi(x)$ corresponding to the Fourier transform $g(k)$ shown in Fig. 3.12.

Solving this last equation for k in terms of u and taking the differential dk, we obtain

$$k = \frac{u}{ix}, \quad dk = \frac{du}{ix}.$$

The expression for $\psi(x)$ can thus be written

$$\psi(x) = \frac{g}{\sqrt{2\pi}} \cdot \frac{1}{ix} \int_{ix(k_0-\Delta k)}^{ix(k_0+\Delta k)} e^u \, du = \frac{g}{\sqrt{2\pi}} \cdot \frac{1}{ix} \left[e^{ix(k_0+\Delta k)} - e^{ix(k_0-\Delta k)} \right].$$

The exponential function $e^{ik_0 x}$ may be factored from the term within square brackets giving

$$\psi(x) = \frac{g}{\sqrt{2\pi}} \cdot \frac{1}{ix} e^{ik_0 x} \left[e^{i\Delta k x} - e^{-i\Delta k x} \right].$$

Finally, using Eq. (3.10) which expresses the sine function in terms of exponentials, we obtain

$$\psi(x) = \left[\frac{2g}{\sqrt{2\pi}} \frac{\sin \Delta k \, x}{x} \right] e^{ik_0 x} \tag{3.92}$$

The function given by Eq. (3.92) represents the sum of the group of waves with angular wave vectors near k_0. We shall refer to the quantity within square brackets in Eq. (3.92) as the amplitude function. It is represented by the dotted line in Fig. 3.14. In drawing the figure, we have supposed that Δk is much smaller than k_0, and, therefore, the real part of the exponential function, which corresponds to the solid line in the figure, oscillates much more rapidly than the amplitude function. We use Δx to denote the value of x for which the amplitude function has its first zero. Since the sine has its first zero when its argument is equal to π, we have

$$\Delta k \, \Delta x = \pi.$$

Multiplying this equation through by $\hbar$ and using the fact that $\hbar k = p$, gives

$$\Delta x \, \Delta p = \frac{h}{2}. \tag{3.93}$$

We can thus see that the precisions with which the position and the momentum of the wave packet are defined are complimentary. For a single wave, the momentum has a clearly defined value, while the position of the particle is entirely undefined. By superimposing waves corresponding to different values of k (and hence to different values of the momentum) we can produce a well-defined packet, but then the resulting function corresponds to a range of momenta.

Eq. (3.93) is quite similar to relation between the uncertainty of the position and momentum in the Heisenberg Uncertainty Principle. The differences in the manner in which the two results were obtained are due to differences in the way that the electron is described. In Heisenberg's original line of argument leading to the Uncertainty Principle, the electron is treated as a particle, whereas in our derivation of Eq. (3.93), the state of the electron is represented by a superposition of waves. The underlying fact common to both descriptions is that the precision with which we can simultaneously specify the position and the momentum of a particle is limited.

Example 3.3

An electron is confined to a region of the size of an atom (0.1 nm). What is the minimum uncertainty of the momentum of the electron? What is the kinetic energy of an electron with a momentum equal to this uncertainty?

Solution

Using the Uncertainty Principle, the minimum uncertainty of the momentum of an electron in the atom can be written

$$\Delta p = \frac{\hbar}{2\Delta x} = \frac{1.054573 \times 10^{-34}\,\text{J}\cdot\text{s}}{2.0 \times 10^{-10}\,\text{m}} = 5.27 \times 10^{-25}\,\text{kg m/s}\,.$$

In order to calculate the kinetic energy of an electron with this momentum, we write the kinetic energy in terms of the momentum as before to obtain

$$KE = \frac{1}{2m}p^2\,.$$

Substituting the value of the uncertainty of the momentum in place of the momentum in the above equation gives

$$KE = \frac{(5.27 \times 10^{-25}\,\text{kg m/s})^2}{2 \cdot 9.11 \times 10^{-31}\,\text{kg}} \times \frac{1\,\text{eV}}{1.602 \times 10^{-19}\,\text{J}} = 0.95\,\text{eV}\,.$$

This is a reasonable estimate of the energy of the outer electrons in an atom. The inner electrons are more tightly bound having smaller uncertainties in their position and corresponding greater uncertainties in their momentum.

We have seen how the uncertainty of the position and the uncertainty of the momentum are inversely related with a small value of the uncertainty of the momentum associated with a large uncertainty of the position, and we now want to consider the relation between the uncertainty of the time and the uncertainty of energy. Since the metastable states of an atom decay over a very long period of time, saying that an atom is in a metastable energy level says very little about the time; however, these metastable levels typically have very well-defined values of the energy. By contrast, shorter-lived resonant levels are much broader having smaller lifetimes but much larger uncertainties of the energy of the states. We shall find that this inverse relation between the uncertainty of the time and the energy persists in the processes that occur in particle physics to be considered in Chapter 14. The masses of particles that decay by the strong interaction in 10^{-22} seconds are poorly defined while particles that decay by the electromagnetic or weak interactions over longer periods of time have masses that are more clearly defined.

3.5.2 Average value of the momentum and the energy

The idea that the wave function of a particle can be represented by a superposition of plane waves may be used to obtain an equation for the average momentum of a particle. Recall that the average value of the position of a particle is given by the equation

$$<x> = \int_{-\infty}^{\infty} x|\psi(x)|^2 dx,$$

where the integration is extended over the range, $-\infty \le x \le \infty$. Each value of x occurring in the above integral is multiplied by the probability that a particle should have that value of x. In the same way, we define the average value of the momentum of a particle is given by the equation

$$<p> = \int_{-\infty}^{\infty} \hbar k|g(k)|^2 dk. \tag{3.94}$$

For each value of k in this last integral, the value of the momentum is equal to $p = \hbar k$ and the probability that the particle should have that value of k is equal to $|g(k)|^2 dk$. Eq. (3.94) is thus analogous to the definition of the average value of the position of the particle.

The equation for the average momentum of a particle may be cast into a more convenient form by substituting Eq. (3.91) into Eq. (3.94) and rearranging the terms inside the integral to obtain

$$<p> = \int_{-\infty}^{\infty} \psi^*(x)\left(-i\hbar\frac{d}{dx}\right)\psi(x)dx.$$

A derivation of this last equation can be found in Appendix E. Notice that the quantity appearing within parentheses in the above integral is the momentum operator $\hat{p}$ defined by Eq. (3.2). The above equation may thus be written

$$<p> = \int_{-\infty}^{\infty} \psi^*(x)\, \hat{p}\, \psi(x)dx. \tag{3.95}$$

Eq. (3.95) is a special case of a more general definition of the average value. The average value of the observable Q is given by the following equation

$$<Q> = \int_{-\infty}^{\infty} \psi^*(x)\hat{Q}\psi(x)dx, \tag{3.96}$$

where $\hat{Q}$ is the operator corresponding to Q. The average value of an observable is obtained by sandwiching the operator corresponding to the observable between $\psi^*(x)$ and $\psi(x)$ and integrating over the entire range of the variable x. We note that the expression for the average value of x given by Eq. (2.21) of the second chapter is a special case of Eq. (3.96) since the variable x can be interchanged with $\psi^*(x)$ to rewrite $\psi^*(x)x\psi(x)$ as $x|\psi(x)|^2$.

Example 3.4

Find the average value of χ^2 and the average value of the kinetic energy for the three smallest even and odd eigenvalues of the harmonic oscillator.

Solution

According to Eq. (3.29), the Schrödinger equation for the harmonic oscillator can be written

$$-\frac{d^2\psi}{d\chi^2} + \chi^2\psi = \epsilon\psi, \tag{3.97}$$

where the dimensionless variable χ is related to the x coordinate by the equation

$$\chi = \sqrt{\frac{m\omega}{\hbar}}\, x,$$

and the energy of the oscillator E is related to the eigenvalue ϵ by the equation

$$E = \frac{1}{2}(\hbar\omega)\epsilon.$$

From the Schrödinger equation for the harmonic oscillator (3.97), it is apparent that the kinetic energy operators in these new dimensionless coordinates is equal to the second derivative with respect to χ

$$\hat{KE} = \frac{d^2}{d\chi^2},$$

and the potential energy is equal to

$$\hat{P.E.} = \chi^2.$$

For negative values of χ, the values of even wave functions are equal to the values for the positive values of χ in reverse order, and the odd wave functions are equal to the negative of the wave functions for positive values of χ in reverse order. We shall use the following equation to normalize the even and odd eigenfunctions.

$$\int_0^{\infty} |\psi(\chi)|^2 d\chi = 1 \tag{3.98}$$

Using Eq. (3.96), the average value of the kinetic energy for the harmonic oscillator can then be written

$$<KE> = \int_0^{\infty} \psi^*(\chi)\left[\frac{-d^2\psi(\chi)}{d\chi^2}\right]d\chi$$

and the average value of χ^2 is

$$<\chi^2> = \int_0^{\infty} \psi^*(\chi)\chi^2\psi(\chi)d\chi$$

In Section 3.3 of this chapter, the three lowest even and odd eigenvalues of the harmonic operator were found by the Program Oscillatorcoll.m, and the corresponding eigenvectors were saved in the file "oscillator.mat". The MATLAB program Oscoll_load_vectors.m, which we will now describe, opens the file Oscillator.mat, plots the three lowest even and odd eigenfunctions and use these function to calculate the average value of χ^2 and the average value of the kinetic energy for the three lowest even and odd eigenstates of the harmonic oscillator.

MATLAB Program Oscoll_load_vectors.m

A MATLAB program that plots the eigenfunctions for the three lowest even and odd eigenvalues of the harmonic Oscillator and uses these eigenfunctions to calculate the average value of the kinetic energy and χ^2 for these eigenfunctions.

```
load("oscillator.mat")
xmax=10
n=200
delta=xmax/n
deltas=delta*delta;
% Gauss Points
xi1=(3-sqrt(3))/6;
xi2=(3+sqrt(3))/6;
% Construct B matrix
p1=(9 -4*sqrt(3))/18;
p2=(9+4*sqrt(3))/18;
p3=(3-sqrt(3))/36;
p4=(3+sqrt(3))/36;
B=[p2 p4 p1 -p3; p1 p3 p2 -p4];
% Construct A matrix
p5=2*sqrt(3);
p6=sqrt(3)-1;
p7=sqrt(3)+1;
A=[p5 p7 -p5 p6;-p5 -p6 p5 -p7]/deltas;
% Construct Full Matrix Bmat
for row=1:2:2*n
  Bmat(row:row+1, row:row+3)=B;
end
% For Even Solutions
EBmat = Bmat;
EBmat(:,2)=[];
EBmat(:,2*n)=[];
% For Odd Solutions
OBmat = Bmat;
OBmat(:,1)=[];
OBmat(:,2*n)=[];
% Construct Full Matrix Amat
for row=1i1:2:2*n
  Amat(row:row+1, row:row+3)=A;
end
% For Even Solutions
EAmat = Amat;
EAmat(:,2)=[];
EAmat(:,2*n)=[];
% For Odd Solutions
OAmat = Amat;
OAmat(:,1)=[];
OAmat(:,2*n)=[];
```

```
% Construct vectors for x  and x squared at the Gauss points
ii = 0;
x = 0.0;
for i = 1:n
   x1 = x + xi1*delta;
   x2 = x + xi2*delta;
   xcol(ii+1) = x1;
   xcol(ii+2) = x2;
   xcol2(ii+1) = x1*x1;
   xcol2(ii+2) = x2*x2;
   ii = ii + 2;
   x = x + delta;
end

% First Even Eigenvector.
Ev1nodal = -EvenStateVectors(:,1);
Ev1col = EBmat*Ev1nodal;
% Normalize Ev1nodal & Ev1col
Ev1norm2 = dot(Ev1col,Ev1col)*delta/2;
Ev1nf = sqrt(Ev1norm2);
Ev1 = Ev1col/Ev1nf;
Ev1n = Ev1nodal/Ev1nf;
plot([-flipud(xcol); xcol],[flipud(Ev1); Ev1])
shg
saveas (gcf,'Ev1.pdf')
% Calculate Average Value of x^2 for Ev1
Avx2Ev1 = dot(Ev1,xcol2.*Ev1)*delta/2
% Calculate the Average Value of the Kinetic Energy
AvKEEv1 = dot(Ev1,EAmat*Ev1n)*delta/2


% Second Even Eigenvector.
Ev2nodal= EvenStateVectors(:,2);
Ev2col= EBmat*Ev2nodal;
% Normalize Ev2nodal & Ev2col
Ev2norm2 = dot(Ev2col,Ev2col)*delta/2;
Ev2nf = sqrt(Ev2norm2)
Ev2 = Ev2col/Ev2nf;
Ev2n = Ev2nodal/Ev2nf;
plot([-flipud(xcol); xcol],[flipud(Ev2); Ev2])
shg
saveas (gcf,'Ev2.pdf')
% Calculate Average Value of x^2 for Ev2
Avx2Ev2 = dot(Ev2,xcol2.*Ev2)*delta/2
% Calculate the Average Value of the Kinetic Energy
AvKEEv2 = dot(Ev2,EAmat*Ev2n)*delta/2

% Third Even Eigenvector
Ev3nodal = EvenStateVectors(:,3);
Ev3col = EBmat*Ev3nodal;
% Normalize Ev3
Ev3norm2 = dot(Ev3col,Ev3col)*delta/2;
Ev3nf = sqrt(Ev3norm2)
Ev3 = Ev3col/Ev3nf;
```

```
Ev3n = Ev3nodal/Ev3nf;
plot([-flipud(xcol); xcol],[flipud(Ev3); Ev3])
shg
saveas (gcf,'Ev3.pdf')
% Calculate Average Value of x^2 for Ev3
Avx2Ev3 = dot(Ev3,xcol2.*Ev3)*delta/2
% Calculate the Average Value of the Kinetic Energy
AvKEEv3 = dot(Ev3,EAmat*Ev3n)*delta/2

% First Odd Eigenvector
Ov1nodal = OddStateVectors(:,1);
Ov1col = OBmat*Ov1nodal;
% Normalize Ov1
Ov1norm2 = dot(Ov1col,Ov1col)*delta/2;
Ov1nf = sqrt(Ov1norm2)
Ov1 =  Ov1col/Ov1nf;
Ov1n =  Ov1nodal/Ov1nf;
plot([-flipud(xcol); xcol],[-flipud(Ov1); Ov1])
shg
saveas (gcf, 'Ov1.pdf')
% Calculate Average Value of x^2 for Ov1
Avx2Ov1 = dot(Ov1,xcol2.*Ov1)*delta/2
% Calculate the Average Value of the Kinetic Energy
AvKEOv1 = dot(Ov1,EAmat*Ov1n)*delta/2

% Second Oddd Eigenvector
Ov2nodal= OddStateVectors(:,2);
Ov2col= OBmat*Ov2nodal;
% Normalize Ov2
Ov2norm2 = dot(Ov2col,Ov2col)*delta/2;
Ov2nf = sqrt(Ov2norm2);
Ov2 =  Ov2col/Ov2nf;
Ov2n = Ov2nodal/Ov2nf;
plot([-flipud(xcol); xcol],[-flipud(Ov2); Ov2])
shg
saveas (gcf, 'Ov2.pdf')
% Calculate Average Value of x^2 for Ov2
Avx2Ov2 = dot(Ov2,xcol2.*Ov2)*delta/2
% Calculate the Average Value of the Kinetic Energy
AvKEOv2 = dot(Ov2,OAmat*Ov2n)*delta/2

% Third Odd Eigenvector
Ov3nodal = OddStateVectors(:,3);
Ov3col = OBmat*Ov3nodal;
% Normalize Ov3
Ov3norm2 = dot(Ov3col,Ov3col)*delta/2;
Ov3nf = sqrt(Ov3norm2);
Ov3 = Ov3col/Ov3nf;
Ov3n = Ov3nodal/Ov3nf;
plot([-flipud(xcol); xcol],[-flipud(Ov3); Ov3])
shg
saveas (gcf, 'Ov3.pdf')
% Calculate Average Value of x^2 for Ov3
Avx2Ov3 = dot(Ov3,xcol2.*Ov3)*delta/2
```

```
% Calculate the Average Value of the Kinetic Energy
AvKEOv3 = dot(Ov3,EAmat*Ov3n)*delta/2
```

As before, the first lines of the MATLAB Program Oscoll_load_vectors.m define the length of the physical region (xmax), the number of intervals (n), and the step size (delta). The square of the step size is denoted by deltas. The coefficients $xi1$ and $xi2$ are needed to construct the vector $xcol$ consisting of the values of χ at the Gauss quadrature points. The program then constructs the 2×4 B and A matrices and construct the full B and A matrices with $2n + 2$ columns and $2n$ rows.

As described in Section 3.3, the boundary conditions are different for the even and odd solutions. The even solutions have zero derivatives at the origin and they are zero at the outer boundary, while the odd solutions are zero at the origin and zero at the outer boundary. The full A and B matrices for the even solution are denoted by $EAmat$ and $EBmat$, and the A and B matrices for the odd solutions are denoted by $OAmat$ ans $OBmat$. The boundary conditions for the even solutions are imposed by deleting the second and second to last columns of the $EAmat$ and $EBmat$ matrices, while the boundary conditions for the odd solutions are imposed by deleting the first and second to last columns of the $OAmat$ and $OBmat$ matrices. Once these matrices are constructed, the program finds the values of χ and χ^2 for the Gauss points. The vectors with these components are denoted $xcol$ and $xcol2$.

Before describing how the Program Oscoll_load_vectors.m normalizes the even and odd eigenfunctions of the harmonic oscillator, we recall how MATLAB defines and manipulates vectors. As discussed in Appendix C, a row vector can be created by writing the elements on a single line separated by spaces

```
>> A = [ 1 2 3 4]
```

MATLAB responds to this command by writing

```
A =

    1    2    3    4
```

A column vector can be produced by separating the elements by semicolons

```
>> A = [ 1; 2:  3; 4]
```

with the MATLAB response

```
A =

    1
    2
    3
    4
```

Since a column vector may be regarded as the transpose of a row vector, the above column vector may also be generated by writing

```
>> A = [ 1 2 3 4]'
```

The single quote(') is the *transpose operator* in MATLAB. For illustrating the commands we shall now consider, we introduce a second column vector with the command

```
>> B= [ 2 2 1 1]'
```

The function $dot(A, B)$ returns the dot product of two vectors which is the sum of the products of the individual components. With the A and B vectors just defined

```
dot(A,B) = 13.
```

The program Oscillatorcoll.m finds the nodal solutions of the eigenvalue equations. The nodal solution is a vector of the α_i and β_i spline coefficients (3.46). We defined the nodal solution corresponding to the first even solution by the statement

```
Ev1nodal = -EvenStateVectors(:,1);
```

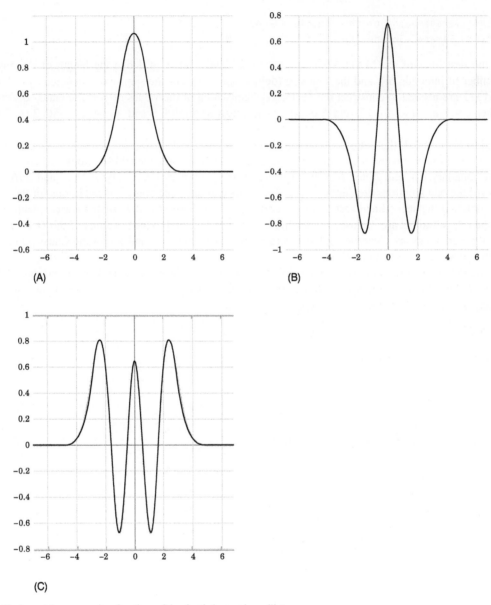

FIGURE 3.15 The lowest three even eigenfunctions of the simple harmonic oscillator.

In quantum mechanics the square of the wave function is related to the probability of finding the particle at a particular point, but the sign of the wave function has no physical significance. We put the minus sign in the above definition of Ev1nodal to get the plot shown in Fig. 3.15(A) rather than the upside-down variant of this plot. The vector consisting of the values of the first eigenfunction at the Gauss points, which is denoted by the vector Ev1col, is obtained by multiplying $EBmat$ matrix times the first nodal eigenvector, Ev1nodal. As described in Section 3.3 of this chapter, the square of the normalization integral of the eigenfunction for the first even state is found by adding up the squares of the values of the eigenfunction at the Gauss points and multiplying by delta/2. This is accomplished by the MATLAB command

```
Ev1norm2 = dot(Ev1col,Ev1col)*delta/2;
```

The square root of $Ev1norm2$ is then used to normalize the vector $Ev1col$ consisting of the values of the first even eigenfunction at the Gauss points and also the first nodal eigenfunction with the commands

```
Ev1nf = sqrt(Ev1norm2);
Ev1 = Ev1col/Ev1nf;
Ev1n = Ev1nodal/Ev1nf;
```

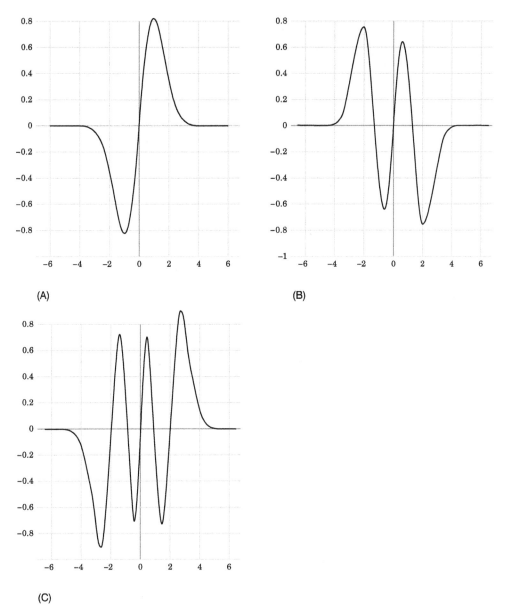

FIGURE 3.16 The lowest three odd eigenfunctions of the simple harmonic oscillator.

The square root of the square of the norm, which is denoted by $Ev1nf$, is thus used to normalize the collocation and the nodal forms of the first eigenfunction. The normalized vector consisting of the values of the first eigenfunction at the Gauss points is denoted by $Ev1$ and normalized nodal solution is denoted by $Ev1n$. For the plot command,

```
plot([-flipud(xcol); xcol],[flipud(Ev1); Ev1])
```

-flipud(xcol) gives the negative of the values of χ at the Gauss quadrature points in reverse order, and flipud(Ev1) gives the values of the vector consisting of the values of the first eigenfunction at the Gauss points in reverse order. The command shg (which is short for show graph) displays the first even function and the following command saveas save the figure in pdf format.

The three lowest even eigenfunctions for the harmonic oscillator are shown in Fig. 3.15 and the three lowest odd eigenfunctions are given in Fig. 3.16. After plotting the first even eigenfunction, the program then calculates the average value of χ^2 for this eigenfunction. The statement

```
Av2Ev1 = dot(Ev1,xcol2.*Ev1)*delta/2
```

multiplies each value of the vector χ^2 at the Gauss quadrature points times the corresponding value of Ev1 at the Gauss points and then forms the dot product of the resulting vector with the first even eigenvector to get the average value of χ^2. With the A and the B vectors used recently to illustrate MATLAB commands,

```
B.*A  = [2 4 3 4]'
```

and

```
dot(A,B.*A) = 2 +8 +9+16= 35
```

The average value of any function of χ would be calculated in the same way.

To calculate the average the kinetic energy we use the fact that the kinetic energy operator in these coordinates is the negative of the second derivative operator represented by Amat. The statement

```
AvKEEv1 = dot(Ev1,EAmat*Ev1n)*delta/2
```

multiplies the A matrix for the even eigenfunctions times the nodal form of the first eigenfunction to get the value of the negative of the second derivative at the Gauss points. The average value of the kinetic energy is then obtained by taking the dot product of this vector with the vector consisting of the values of the first even eigenvector at the Gauss points.

The average value of χ^2 and the average value of the kinetic energy of the three lowest even and odd eigenfunctions of the harmonic oscillator are given in the following table.

TABLE 3.1 Average value of χ^2 and Kinetic Energy for the Three lowest Even and Odd Eigenfunctions of the Harmonic Oscillator.

Eigenfunction	$< \chi^2 >$	$< KE >$	ϵ
Ev1	0.5000	0.5000	1.0000
Ev2	2.5000	2.5000	5.0000
Ev3	4.5000	4.5000	9.0000
Ov1	1.5000	1.3873	3.0000
Ov2	3.5000	3.5000	7.0000
Ov3	5.5000	5.2891	11.0000

Before interpreting these results, we first recall that the energy of the oscillator E is equal to $(1/2)\hbar\omega$ times ϵ. So, the energy eigenvalues shown in the fourth column of Table 3.1 are consistent with the energy eigenvalues of the oscillator given in Chapter 2 by the formula

$$E = \hbar\omega(n + 1/2),$$

where n a nonnegative integer, $n = 0, 1, 2, \ldots$. Since the potential energy of the oscillator is equal to the average value of χ^2, one sees immediately the sum of the potential and kinetic energy for the three lowest even states and the second to the lowest odd states are equal to the corresponding energy eigenvalue of the oscillator. The discrepancies that occur for the first and third odd eigenfunctions are due to a numerical error. The first odd function is located close to the origin and is not properly represented by the grid that expends out to $xmax = 10$ while the third odd eigenfunction extends beyond $xmax$. More accurate values of the average values of the kinetic energy for the odd functions can be obtained by changing the value of $xmax$ and/or increasing the number of grid points.

Suggestion for further reading

Rudra Pratap, *Getting Started with MATLAB, A Quick Introduction for Scientists and Engineers*, Seventh Edition (Oxford University Press, 2017).

Basic equations

Observables, operators, and eigenvalues

Momentum operator

$$\hat{p} = -i\hbar\frac{d}{dx}$$

Energy operator

$$\hat{H} = \frac{-\hbar^2}{2m} \frac{d^2}{dx^2} + V(x)$$

Eigenvalue equation

$$\text{operator} \times \text{function} = \text{constant} \times \text{function}$$

Electron scattering

$E > V_0$ step

The wave equation in region 1 to the left of the step shown in Fig. 3.6 has the form

$$\psi_1(x) = A_1 e^{ik_1 x} + B_1 e^{-ik_1 x}$$

and the function to the right of the step has the form step

$$\psi_2(x) = A_2 e^{ik_2 x} + B_2 e^{-ik_2 x},$$

where

$$k_1 = \sqrt{\frac{2mE}{\hbar^2}},$$

$$k_2 = \sqrt{\frac{2m(E - V_0)}{\hbar^2}}$$

Transmission matrix for the step

$$T(k_1, k_2) = \frac{1}{2k_1} \begin{bmatrix} k_1 + k_2 & k_1 - k_2 \\ k_1 - k_2 & k_1 + k_2 \end{bmatrix}.$$

The effect of a translation upon the wave function in a region with angular wave number k is described by the matrix

$$T_L(k) = \begin{bmatrix} e^{ikL} & 0 \\ 0 & e^{-iktL} \end{bmatrix}.$$

One can obtain the transmission matrix for complex systems by multiplying the transmission matrices for each on the component parts.

The Heisenberg uncertainty principle

Momentum-position uncertainty relation

$$\Delta x \cdot \Delta p_x \geq \frac{\hbar}{2}$$

Energy-time uncertainty relation

$$\Delta t \cdot \Delta E \geq \frac{\hbar}{2}$$

Fourier integral

$$\psi(x) = \frac{1}{\sqrt{2\pi}} \int_{-\infty}^{+\infty} g(k) e^{ikx} dk$$

Fourier transform

$$g(k) = \frac{1}{\sqrt{2\pi}} \int_{-\infty}^{+\infty} \psi(k) e^{-ikx} dx$$

Average value of observable Q

$$< Q > = \int_a^b \psi^*(x) \hat{Q} \psi(x) dx,$$

where $\hat{Q}$ is the operator corresponding to Q.

Summary

Particle variables such as the momentum and the energy are represented in quantum theory by operators. The operator corresponding to the momentum is

$$\hat{p} = -i\hbar \frac{d}{dx}.$$

Operators corresponding to other particle variables such as the energy are obtained by writing the variable in terms of the momentum and then replacing the momentum with the momentum operator. The operator corresponding to the energy is called the Hamiltonian and denoted by H.

The possible results of measuring a variable can be obtained by forming the eigenvalue equation for the operator corresponding to the variable

$$\text{operator} \times \text{function} = \text{constant} \times \text{function}.$$

The results of a measurement are the values of the constant on the right-hand side of the eigenvalue equation for which the equation has a solution satisfying the boundary conditions.

Eigenvalue equations are solved by using a finite difference approximation of the derivatives and by using the spline collocation method. The eigenfunctions obtained are plotted and used to calculate average values of spacial functions and the average value of the kinetic energy.

The wave function for particles incident upon barriers can be obtained by setting up the Schrödinger equation in each region and requiring that the wave function and its derivative be continuous across each interface. Quantum theory makes the remarkable prediction that particles can tunnel through barriers of finite width. Matrix methods are developed the makes if possible to calculate the transmission coefficients for complex systems by multiplying the matrices of the component parts.

The position and the momentum of microscopic objects cannot be precisely known simultaneously. The product of the uncertainty in the position and the uncertainty in the x-component of the momentum satisfy the following relation

$$\Delta x \cdot \Delta p_x \geq \frac{\hbar}{2},$$

which is known as the Heisenberg Uncertainty Principle. Similar relations apply to the other coordinates and the corresponding components of the momentum and to the time and the energy

$$\Delta t \cdot \Delta E \geq \frac{\hbar}{2}.$$

The uncertainty relations may be understood as a fundamental consequence of the wave-particle duality.

The idea that waves are associated with particles leads to the following general equation for the average value of an observable

$$< Q > = \int_a^b \psi^*(x) \hat{Q} \psi(x) dx,$$

where Q is an observable and $\hat{Q}$ is the corresponding operator. This last formula applies to the position, as well as to the momentum and the energy.

Questions

1. Give the operators corresponding to the momentum and energy of a particle.
2. Give the general form of an eigenvalue equation.

3. What significance do the eigenvalues have?

4. What significance do the eigenfunctions have?

5. Write down the momentum eigenvalue equation.

6. Is the function $\cos kx$ an eigenfunction of the momentum operator?

7. Is the function $\cos kx$ an eigenfunction of the operator corresponding to the kinetic energy?

8. An electron is described by the wave function

$$\psi(x) = Ae^{i\alpha x},$$

where α denotes the Greek letter alpha. What is the momentum of the electron?

9. How would you determine whether or not a particular wave function represented a state of the system having a definite value of the momentum?

10. Calculate the dot product of the two vectors

```
>> A = [ 4; 3; 2; 1]
>> B = [ 1; 1; 2; 2]].
```

11. For the two vectors, A and B, given in the preceding question, evaluate B.*A and dot(A,B.*A).

12. Suppose a function $f(x)$ is defined in an interval between 0 and 10 and an equally space grid is created by the command,

```
>> x = linspace(0.0,10.0,10)
```

and suppose the values of the function at the two Gauss quadrature points, ξ_{i1} and ξ_{i2}, within each interval are known. Write down a formula for determining the value of the following integral in terms of the values of the function at the Gauss points

$$\int_0^{10} f(x)dx.$$

13. Suppose a differential equation is solved using spline collocation on a grid created by the command

```
>> x = linspace(0.0,10.0,10)
```

How many rows and columns do the A, B, and C matrices have?

14. Sketch the wave function for an electron incident upon a potential step when the energy E of the electron is greater than the step height V_0.

15. Sketch the wave function for an electron incident upon a potential step when the energy E of the electron is less than the step height V_0.

16. Is the wavelength of a particle that has passed over a potential barrier greater than or less than the wavelength of the incident particle?

17. Sketch the wave function of an electron which tunnels through a potential barrier located between 0 and L.

18. Write down the Heisenberg uncertainty relations for the position and the momentum and for the time and the energy.

19. Use the Heisenberg relation for the time and the energy to describe how the energy profile of an excited atomic state depends upon the lifetime of the state.

20. How would the wave function $\psi(x)$ of a particle change as the width of the Fourier transform increases?

21. Suppose that a particle localized between a and b is described by the wave function $\psi(x)$. Write down an equation for the average value of the kinetic energy of the particle.

Problems

1. Evaluate the product of the momentum operator and each of the two functions

$$\phi_1(x) = \cos kx$$
$$\phi_2(x) = \sin kx.$$

Are these functions eigenfunctions of the momentum operator?

2. Find a linear combination of the functions, $\phi_1(x)$ and $\phi_2(x)$, defined in the previous problem, which is an eigenfunction of the momentum operator.

3. An electron moves in a one-dimensional lattice with the separation between adjacent atoms being equal to a.
   (a) Write down the momentum eigenvalue equation for the electron.
   (b) Find the general form of the solutions of the eigenvalue equation.
   (c) By requiring that the electron's wave function $\psi(x)$ satisfies the periodic boundary condition,

$$\psi(a) = \psi(0),$$

   determine the possible values of the momentum of the electron.
4. Write a MATLAB program to plot the function $y = xe^{-x}$ between 0 and 10.
5. Using a five-point finite difference approximation of the second derivative, develop a MATLAB program similar to the Program FiniteWell.m to find the lowest energy eigenvalue and the corresponding eigenfunction of an electron in a finite well with a depth of 0.2 eV and thickness of $L = 10$ nm in $GaAs$ using the fact that conduction electrons in $GaAs$ have an effective mass equal to 0.067 times the electron mass.
6. For a equally spaced grid with 100 intervals that may be generated by the MATLAB command $x = linspace(0, 10, 100)$, write a program that finds the Gauss quadrature points and uses these Gauss points with the step size ($\delta = 0.1$) to evaluates the following integrals

$$a) \int_0^{10} x^2 dx \quad b) \int_0^{10} x^4 dx \quad c) \int_0^{10} xe^{-x} dx.$$

Compare the values of these integrals so obtained with the values of the integral obtained by explicitly evaluating the integrals. The last integral can be evaluated effectively using integration by parts.
7. Using spline collocation, develop a MATLAB program similar to the Program FiniteWellcoll.m to find the lowest energy eigenvalue and the corresponding eigenfunction of an electron in a finite well with a depth of 0.2 eV and thickness of $L = 10$ nm in $GaAs$ using the fact that conduction electrons in $GaAs$ have an effective mass equal to 0.067 times the electron mass.
8. For the scattering problem illustrated in Fig. 3.5, a particle is incident upon a potential step with the energy of the incident particle being less than the step height. Using the notation for this problem given in the text, derive expressions the ratios A/C and B/C and show that $R = 1$.
9. Using T-matrices, calculate the transmission coefficient as a function of the energy for two $GaAs$ barriers of height 0.3 eV and length $L = 10$ nm separated by a distance $L = 10$ nm and an associate MATLAB program to plot the transmission coefficient from $E = 0.3$ eV to 1.0 eV. One can write other MATLAB functions and programs for barriers separated by different distances, and one can easily extend this approach to more than two barriers.
10. Using the Heisenberg uncertainty principle, estimate the momentum of an electron confined to a 1.0 nm well.
11. Using the Heisenberg uncertainty principle, find the natural line width (ΔE) of an energy level that decays in 4.0×10^{-10} seconds.
12. A particle, which is confined to move in one-dimension between 0 and 1, is described by the wave function

$$\psi(x) = Be^{-x}.$$

Use the normalization condition to determine the constant B.

Chapter 4

The hydrogen atom

Contents

4.1 The Gross structure of hydrogen	81	Basic equations	111
4.2 Radiative transitions	95	Summary	113
4.3 The fine structure of hydrogen	102	Questions	113
Suggestion for further reading	111	Problems	114

The hydrogen atom is the simplest atom imaginable. It has only one electron. Surely, this is an atom we should be able to understand!

4.1 The Gross structure of hydrogen

The hydrogen atom consists of a proton and an electron. Without introducing any difficulty, however, the problem can be generalized to include a nucleus of charge Ze and a single electron. Since the charge of the electron is $-e$, the resulting hydrogen-like ion has a total positive charge equal to $(Z - 1)e$. The possible energies of the electron moving in the field of the nucleus can be determined using the principles of wave mechanics described in the previous chapter.

4.1.1 The Schrödinger equation in three dimensions

In the previous chapters, we found that a particle moving in one dimension may be described by a wave function. The wave function ψ, which may be used to find the average values of variables describing the particle, satisfies the Schrödinger equation

$$\frac{-\hbar^2}{2m} \frac{d^2\psi}{dx^2} + V(x)\psi = E\psi.$$

These ideas may easily be generalized to more realistic problems. In three dimensions, the wave function of a particle satisfies the Schrödinger equation

$$\frac{-\hbar^2}{2m} \left(\frac{\partial^2\psi}{\partial x^2} + \frac{\partial^2\psi}{\partial y^2} + \frac{\partial^2\psi}{\partial z^2} \right) + V(x, y, z)\psi = E\psi, \tag{4.1}$$

where, as before, the symbols ∂ denote the partial derivative. The partial derivative of a function with respect to x, for instance, gives the rate the function changes with respect to x while the spatial coordinates y and z are treated as constants.

In the Schrödinger equation (4.1), the term $V(x, y, z)$, which corresponds to the potential energy of an electron in the field of the nucleus, is described by Eq. (I.5) of the introduction and illustrated in Fig. 4.1. The first term in Eq. (4.1) may be written more compactly by using a special notation to denote the term within parentheses. The *Laplacian* of the function ψ, which is defined by the equation,

$$\nabla^2\psi = \frac{\partial^2\psi}{\partial x^2} + \frac{\partial^2\psi}{\partial y^2} + \frac{\partial^2\psi}{\partial z^2},$$

is the natural generalization of the second derivative to three dimensions. With this special notation, the Schrödinger equation in three dimensions becomes

$$\left[-\frac{\hbar^2}{2m}\nabla^2 - \frac{1}{4\pi\epsilon_0} \frac{Ze^2}{r} \right] \psi(\mathbf{r}) = E\psi(\mathbf{r}), \tag{4.2}$$

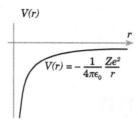

FIGURE 4.1 Potential energy $V(r)$ of electron in a hydrogen-like ion with nuclear charge Z. For the hydrogen atom, $Z = 1$.

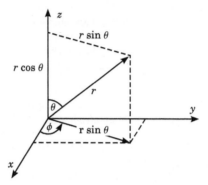

FIGURE 4.2 Spherical polar coordinates.

where $\mathbf{r}$ is the position vector of the electron. The second term in the Schrödinger equation, which corresponds to the potential energy of the electron, only depends upon the distance of the electron from the nucleus.

Systems like the hydrogen atom having spherical symmetry can most easily be studied using a spherical polar coordinate frame such as that illustrated in Fig. 4.2. The spherical coordinates (r, θ, ϕ) are related to the Cartesian coordinates (x, y, z) by the equations

$$
\begin{aligned}
x &= r \sin\theta \cos\phi \\
y &= r \sin\theta \sin\phi \\
z &= r \cos\theta .
\end{aligned}
\tag{4.3}
$$

It is easy to understand the structure of these equations. The term $r \sin\theta$ is equal to the projection of the vector $\mathbf{r}$ upon the $x - y$ plane. It follows that $r \sin\theta \cos\phi$ and $r \sin\theta \sin\phi$ are the x- and y-coordinates of $\mathbf{r}$. Similarly, $r \cos\theta$ is the projection of $\mathbf{r}$ on the z-axis. The Laplacian operator in spherical polar coordinates is given in Appendix AA.

The Schrödinger equation (4.2) in spherical coordinates can be solved by the method of the separation of variables discussed in Chapter 2. To solve the Schrödinger equation in this way, one expresses the wave function as a product of functions of the r, θ, and ϕ coordinates and derives independent equations for the functions of these variables. As we have seen in Chapter 2, the independent equations obtained using the separation of variables technique contain additional constants. The additional constant that arose in the second chapter when we solved the Schrödinger time-dependent equation by the separation of variables corresponded to the energy of the particle. The additional constants that arise when one separates the Schrödinger equation for hydrogen are related to the angular momentum of the electron. This result has a precedent in classical physics. The angular momentum of a classical particle moving in a central field is conserved.

4.1.2 The energy levels of hydrogen

The energy levels of hydrogen and the wave length of light emitted when an atom makes a transition from one level to another were discussed in Chapter 1. We have found that the energy of an electron in the hydrogen atom is given by the

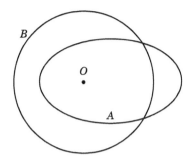

FIGURE 4.3 The motion of two particles moving in elliptical orbits about a center of force located at the point O. The two orbits are denoted by A and B.

equation

$$E_n = -\frac{13.6 \text{ eV}}{n^2},$$

where the principal quantum number n has the possible values, $n = 1, 2, 3, \ldots$. The energy levels of hydrogen are illustrated in Fig. 1.6.

While the energy of an electron in a hydrogen atom can be specified by giving the quantum number n, the motion of an electron moving in three dimensions is not completely specified by giving the value of its energy. The fact that one needs the value of other variables to characterize the motion of an electron in three dimensions can be understood within the framework of classical physics. Fig. 4.3 shows the motion of two particles moving in elliptical orbits about a center of force located at the point O. The particle with the orbit denoted by A in Fig. 4.3 comes closer to the origin and has a smaller value of the angular momentum than the particle with orbit B, and, yet, the particles with orbits A and B can have the same energy. One can specify the state of an electron moving about the nucleus of an atom by giving the value of its energy and by giving also the value of the square of the angular momentum $\mathbf{l}^2$ and the z-component of the angular momentum l_z.

The possible values of the angular momentum of a particle moving about a center of force can be found by solving the Schrödinger equation (4.2) using the separation of variables technique or by taking advantage of the properties of the operators corresponding to the angular momentum. The solution of the Schrödinger equation using the separation of variables is described in Appendix BB, while the determination of the possible values of the angular momentum using operator methods is described in Appendix DD. Using either of these two approaches, the possible values of the angular momentum squared $\mathbf{l}^2$ may be shown to be of the form $l(l + 1)\hbar^2$, where the integer l is the quantum number of the orbital angular momentum. For a given value of l, the z-component of the angular momentum, which we denote by l_z, has the values $m_l \hbar$, where $m_l = -l, -l + 1, ..., l$. The quantum number m_l is the quantum number of the z or *azimuthal component of the angular momentum* – often referred to as the *magnetic quantum number*.

Example 4.1

Compute the value of $\mathbf{l}^2$ for a wave function with $l = 2$. What are the possible values of l_z for this function?

Solution
If $l = 2$, the value of $\mathbf{l}^2$ is $2 \cdot 3\hbar^2$, and the possible values of l_z are $m_l \hbar$, where $m_l = -2, -1, 0, 1, 2$. This can be described by saying that the length of the $\mathbf{l}$ vector is $\sqrt{6}\hbar$ and $\mathbf{l}$ has one of the following components in the z-direction: $l_z = -2\hbar, -1\hbar, 0, 1\hbar, 2\hbar$. The values of l_z, which are illustrated in Fig. 4.4, are equally spaced and differ by $\hbar$.

The orientations in space of the vector $\mathbf{l}$ in Fig. 4.4 correspond to the different values of $m_l \hbar$. From Fig. 4.4, one can see that the cosine of the angle between $\mathbf{l}$ and the z-axis is $l_z/|\mathbf{l}|$, which is equal to $m_l/\sqrt{l(l + 1)}$. Quantum theory thus predicts that for a given value of l only certain orientations of the angular momentum vector in space are allowed. This is called *space quantization*.

Fig. 4.5 shows the energy levels of hydrogen with both the principal quantum number n and angular momentum quantum number l given for each level. The angular momentum quantum numbers are given using the spectroscopic notation in which the letters, $s, p, d, f, g, \ldots$, stand for the l-values, 0, 1, 2, 3, 4, $\ldots$. The lowest energy level is the $1s$ for which $n = 1$ and $l = 0$, while the next higher energy levels are the $2s$- and $2p$-levels with $n = 2$ and l equal to 0 and 1, respectively. There

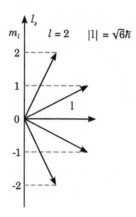

FIGURE 4.4 The orbital angular momentum vector **l** for $l = 2$. The z-component of **l** has the possible values $m_l \hbar$, where $m_l = 2, 1, 0, -1, -2$.

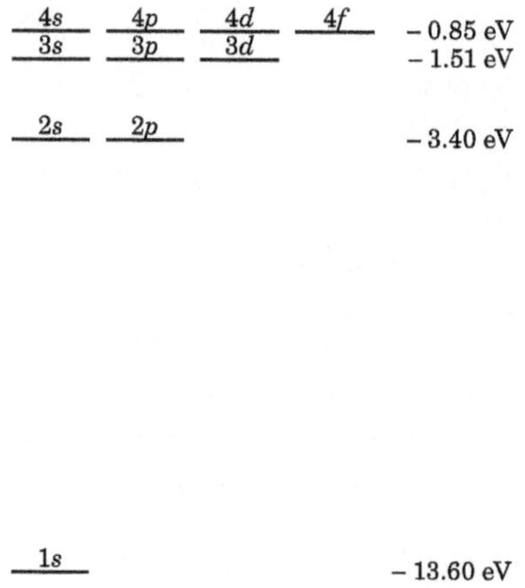

4s	4p	4d	4f	— 0.85 eV
3s	3p	3d		— 1.51 eV

2s	2p		— 3.40 eV

1s	— 13.60 eV

FIGURE 4.5 The energy levels of hydrogen. The principal quantum number n is given for each level and the angular momentum quantum number l is given using the spectroscopic notation.

are three levels for $n = 3$: the $3s$ with $l = 0$, the $3p$ with $l = 1$, and the $3d$ with $l = 2$. The value of the energy for each value of n is the same as those given in Fig. 1.6, while the angular momentum quantum number l has the possible values, $l = 0, 1, \ldots, n - 1$.

4.1.3 The wave functions of hydrogen

The quantum numbers n, l, and m_l may be used to label the radial and angular parts of the hydrogen wave function. We shall write the wave function

$$\psi(r, \theta, \phi) = \frac{P_{nl}(r)}{r} \Theta_{lm_l}(\theta) \Phi_{m_l}(\phi) \,, \tag{4.4}$$

where the radial part of the wave function is expressed as a function $P_{nl}(r)$ divided by r. The angular part of the wave function is called a *spherical harmonic* and denoted $Y_{lm_l}(\theta, \phi)$. It is equal to the product of the functions of θ and ϕ

$$Y_{lm_l}(\theta, \phi) = \Theta_{lm_l}(\theta) \Phi_{m_l}(\phi).$$

The spherical harmonic $Y_{lm_l}(\theta, \phi)$ is an eigenfunction of the operator $\hat{\mathbf{l}}^2$ corresponding to the eigenvalue $l(l + 1)\hbar^2$ and an eigenfunction of l_z corresponding to the eigenvalue $m_l \hbar$. The functions, $\Theta_{lm_l}(\theta)$, $\Phi_{m_l}(\phi)$, and $Y_{lm_l}(\theta, \phi)$ are given in Table 4.1 for a few values of l and m_l.

The spherical harmonics $Y_{lm_l}(\theta, \phi)$ are the product of the two functions, $\Theta_{lm_l}(\theta)$ and $\Phi_{m_l}(\phi)$. One may readily confirm that each entry in the last column of Table 4.1 is equal to the product of the entries from the previous two columns. The

TABLE 4.1 The functions $\Theta_{lm_l}(\theta)$, $\Phi_{m_l}(\phi)$, and $Y_{lm_l}(\theta, \phi)$.

l	m_l	$\Theta_{lm_l}(\theta)$	$\Phi_{m_l}(\phi)$	$Y_{lm_l}(\theta, \phi)$
0	0	$\frac{1}{\sqrt{2}}$	$\frac{1}{\sqrt{2\pi}}$	$\frac{1}{\sqrt{4\pi}}$
1	± 1	$\mp \frac{\sqrt{3}}{2} \sin\theta$	$\frac{1}{\sqrt{2\pi}} e^{\pm i\phi}$	$\mp \sqrt{\frac{3}{8\pi}} \sin\theta e^{\pm i\phi}$
1	0	$\sqrt{\frac{3}{2}} \cos\theta$	$\frac{1}{\sqrt{2\pi}}$	$\sqrt{\frac{3}{4\pi}} \cos\theta$
2	± 2	$\frac{\sqrt{15}}{4} \sin^2\theta$	$\frac{1}{\sqrt{2\pi}} e^{\pm 2i\phi}$	$\sqrt{\frac{15}{32\pi}} \sin^2\theta e^{\pm 2i\phi}$
2	± 1	$\mp \frac{\sqrt{15}}{2} \sin\theta \cos\theta$	$\frac{1}{\sqrt{2\pi}} e^{\pm i\phi}$	$\mp \sqrt{\frac{15}{8\pi}} \sin\theta \cos\theta e^{\pm i\phi}$
2	0	$\sqrt{\frac{5}{2}} \left(\frac{3}{2} \cos^2\theta - \frac{1}{2} \right)$	$\frac{1}{\sqrt{2\pi}}$	$\sqrt{\frac{5}{4\pi}} \left(\frac{3}{2} \cos^2\theta - \frac{1}{2} \right)$

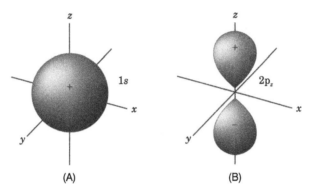

FIGURE 4.6 The spherical harmonics $Y_{lm_l}(\theta, \phi)$ for (A) $l = 0$, $m_l = 0$ and (B) $l = 1$, $m_l = 0$.

spherical harmonic for l equal to zero is spherically symmetric, while the function for $l = 1$ and $m_l = 0$ is directed along the z-axis. These two functions are illustrated in Fig. 4.6. There are linear combinations of the spherical harmonic for $l = 1$, $m_l = 1$ and $l = 1$, $m_l = -1$ that are directed along the x- and y-axes.

The radial functions $P_{nl}(r)$ satisfy the following equation

$$\left[-\frac{\hbar^2}{2m} \frac{d^2}{dr^2} + \frac{\hbar^2 l(l+1)}{2mr^2} - \frac{1}{4\pi\epsilon_0} \frac{Ze^2}{r} \right] P_{nl}(r) = E \, P_{nl}(r). \tag{4.5}$$

A derivation of this equation is given in Appendix EE. The first term in Eq. (4.7) corresponds to the radial part of the kinetic energy of the electron, the second term corresponds to the angular or centrifugal part of the kinetic energy and the third term is the potential energy of the electron in the field of the nucleus. The potential energy term is negative and draws the wave function of the electron toward the nucleus, while the centrifugal term being positive pushes the maxima and zeros of the wave function away from the nucleus. The states having higher values of the angular momentum are affected most by the repulsive centrifugal term.

Calculations in atomic physics are traditionally carried out using *atomic units* for which $\hbar$, the mass (m) and charge (e) of the electron and $4\pi\epsilon_0$ are all equal to one. According to Eq. (1.20) of Chapter 1, the Bohr radius a_0 of hydrogen is given by the formula

$$a_0 = \frac{4\pi\epsilon_0 \hbar^2}{me^2}.$$

Since the Bohr radius is made up entirely of quantities that are equal one in atomic units, the Bohr radius is equal to one. In atomic units, the unit of length is the Bohr radius. As we found in Chapter 1, the Bohr model of the atom gives the following expression for the energy levels of hydrogen

$$E_n = -\frac{me^4}{2(4\pi\epsilon_0)^2 \hbar^2} \frac{1}{n^2}.$$

Again, in atomic units with m, e, $\hbar$, and $4\pi\epsilon_0$ all equal to one, this formula for the energy levels of hydrogen is

$$E_n = -\frac{1}{2}\frac{1}{n^2}. \tag{4.6}$$

The binding energy of hydrogen is thus equal to one half of atomic units. The atomic unit of energy is thus twice the binding energy of hydrogen or 27.2 eV.

To obtain the radial equation for hydrogen in atomic units, we set $\hbar$, the mass (m) and charge (e) of the electron and $4\pi\epsilon_0$ are all equal to one in Eq. (4.5) to obtain

$$\left[-\frac{1}{2}\frac{d^2}{dr^2} + \frac{l(l+1)}{2r^2} - \frac{Z}{r}\right]P_{nl}(r) = E\,P_{nl}(r). \tag{4.7}$$

In this section, we shall solve the eigenvalue problem of an electron in a hydrogen-like ion using the spline collocation method described in Chapter 3. According to Eq. (3.45) of Chapter 3, a vector consisting of the values of the wave function at the collocation points is related to the vector consisting of the spline coefficient by the equation

$$\mathbf{v} = \mathbf{B}_{mat}\mathbf{u}.$$

Similarly, the values of the negative of the second derivative and the potential energy times the wave function at the collocation points is represented by the following equation

$$\mathbf{C}_{mat}\mathbf{u} = (\mathbf{A} + \mathbf{D}_l * \mathbf{B})\mathbf{u},$$

where $\mathbf{D}_l$ is a diagonal matrix having the values of the angular part of the kinetic energy and the attractive potential energy due to the nucleus at the collocation points. Drawing together these last two equations together, the eigenvalue equation for the hydrogen ion can be written

$$\mathbf{C}_{mat}\mathbf{u} = E\mathbf{B}_{mat}\mathbf{u}. \tag{4.8}$$

This equation with matrices on the left- and right-hand side of the equation defines a generalized eigenvalue problem

To obtain a standard eigenvalue problem, we first recall that the boundary conditions may be imposed by deleting two columns from the matrices. After removing the two appropriate columns, $\mathbf{B}_{mat}$ then has the same number of rows and columns and may be inverted. We may then multiply (4.8) from the left with $\mathbf{B}_{mat}^{-1}$ and write the resulting equation as

$$\mathbf{L}_{mat}\mathbf{u} = E_0\epsilon\mathbf{u}, \tag{4.9}$$

where

$$\mathbf{L}_{mat} = \mathbf{B}_{mat}^{-1}\mathbf{C}_{mat}. \tag{4.10}$$

Eq. (4.9) is an ordinary eigenvalue equation.

We shall use this approach to solve the radial equation for hydrogen. The MATLAB® program Hydrogen.m to be described shortly produces the same eigenvalues as those produced by the Bohr model. The lowest three eigenvalues for s-electrons are found to be -0.5, -0.125, and -0.0556 atomic units which correspond to the energies -13.6 eV, -3.4 eV, and -1.51 eV. Similarly, the two lowest eigenvalues for p-electrons are found to be -0.125, and -0.0556, and the lowest eigenvalue for a single d-electron is found to be -0.0556. This shouldn't be too surprising because the Bohr model was deliberately developed by Niels Bohr to reproduce the results of the empirical formula of Balmer.

MATLAB Program Hydrogen.m

A MATLAB program for finding the eigenvalues and eigenvectors for a hydrogen-like ion using the spline collocation method.

```
rrmax=40;
n=400;
delta=rmax/n;
deltas=delta*delta;
```

```matlab
%  Gauss quadrature points
xi1 = (3.0 - sqrt(3.0))/6.0;
xi2 = (3.0 + sqrt(3.0))/6.0;

% Construct [2 x 4] B matrix
p1=(9 -4*sqrt(3))/18;
p2=(9+4*sqrt(3))/18;
p3=(3-sqrt(3))/36;
p4=(3+sqrt(3))/36;

B=[p2 p4 p1 -p3; p1 p3 p2 -p4];

% Construct [2 x 4] A matrix
p5=2*sqrt(3);
p6=sqrt(3)-1;
p7=sqrt(3)+1;

A=[p5 p7 -p5 p6;-p5 -p6 p5 -p7]/deltas;

% Construct Full A matrix
Amat=zeros(2*n, 2*n+2);
for row=1:2:2*n-1
  Amat(row:row+1, row:row+3)=A;
end
Amat(:,1)=[];
Amat(:,2*n)=[];
Amat;
% Construct Full  B matrix
Bmat=zeros(2*n, 2*n+2);
for row=1:2:2*n-1
  Bmat(row:row+1, row:row+3)=B;
end
Bmat(:,1)=[];
Bmat(:,2*n)=[];
Bmat;

% Construct Inverse of Bmat
Binv = zeros(2*n,2*n);
rvec = zeros(2*n);
wf = zeros(2*n);
for j = 1:2*n
    rvec(j) = 1.0;
    wf = Bmat\rvec;
    for i = 1:2*n
      Binv(i,j) = wf(i);
    end
    rvec(j) =0.0;
end
Binv;

% Construct xcol
z = 1
x  = 0.0;
xcol = zeros(2*n,1);
```

```
pot = zeros(2*n,1);
for i = 1:n;
   x1 = x + xi1*delta;
   x2 = x  + xi2*delta;
   xcol(2*i-1) = x1;
   xcol(2*i) = x2;
   x = x + delta;
end
xcol;

% For s States with l=0
for i = 1:2*n;
   pot(i) = -z/xcol(i);
end
D1 = diag(pot);

% Construct Collocation matrix

Colmat = (1/2)*Amat + D1*Bmat;

% Energies and Eigenvectors for s States

Lmat = Binv*Colmat;
[V,D]=eig(Lmat);
[sEnergies,index] = sort(diag(D));
sEnergies(1:3)
sVectors = V(:,index(1:3));

% For p States with l=1
fl = 1;
for i = 1:2*n;
   pot(i) = -z/xcol(i) + fl*(fl+1)/(2.0*xcol(i)*xcol(i));
end
D1 = diag(pot);

% Construct Collocation matrix

Colmat = (1/2)*Amat + D1*Bmat;

% Energies and Eigenvectors for p States

Lmat = Binv*Colmat;
[V,D]=eig(Lmat);
[pEnergies,index] = sort(diag(D));
pEnergies(1:2)
pVectors = V(:,index(1:2));

% For d States with l=2
fl = 2;
for i = 1:2*n;
   pot(i) = -z/xcol(i) + fl*(fl+1)/(2.0*xcol(i)*xcol(i));
end
D1 = diag(pot);
```

```
% Construct Collocation matrix

Colmat = (1/2)*Amat + D1*Bmat;

% Energies and Eigenvectors for d States

Lmat = Binv*Colmat;
[V,D]=eig(Lmat);
[dEnergies,index] = sort(diag(D));
dEnergies(1:1)
dVectors = V(:,index(1));
save("hydrogen.mat","sVectors","pVectors","dVectors")
```

The program begins by giving the length of the physical region (rmax) and the number of intervals (n). The physical region is chosen to be large enough to include the 3s, 3p, and 3d functions that extend far beyond the 1s function which reaches its maximum at one atomic unit. After defining the grid, we give the constants xi1 and xi2 that define the Gauss quadrature points within each interval and define the B and A matrices. Because the wave functions are zero at the origin and at the outer boundary, the boundary conditions are imposed by deleting the first and second to last columns of the B matrix and the collocation matrix Colmat.

As described in the last chapter, a generalize eigenvalue problem can be converted into a standard eigenvalue problem by multiplying by the generalized eigenvalue equation by $\mathbf{B}_{mat}^{-1}$. We can obtain the jth column of $\mathbf{B}_{mat}^{-1}$ by multiplying the matrix times a unit vector with the jth element being one and all of the other elements being zero. Denoting such a vector by u_j and denoting the jth column of $\mathbf{B}_{mat}^{-1}$ by v_j, we have

$$\mathbf{B}_{mat}^{-1} u_j = v_j. \tag{4.11}$$

This last equation is equivalent to the equation

$$\mathbf{B}_{mat} v_j = u_j, \tag{4.12}$$

which is solved by the MATLAB command

```
vj=Bmat \uj
```

In the MATLAB program Hydrogen.m, we construct $\mathbf{B}_{mat}^{-1}$ matrix by solving the linear system (4.12) for each column of $\mathbf{B}_{mat}^{-1}$. The matrix Linv used to solve the eigenvalue problems is then produced by multiplying $\mathbf{B}_{mat}^{-1}$ times the collocation matrix **Colmat**.

The Program Hydrogen.m finds the eigenvalues and eigenvectors of the three lowest-lying s-states with the eigenvectors being in the array "sVectors". Similarly, the program finds the eigenvalues and eigenvectors of the two lowest-lying p-states with the eigenvectors being in the array "pVectors" and the eigenvalues and eigenvector for the lowest lying d-sate with the eigenvectors being in the array "dVectors". The eigenvectors are saved in the file "hydrogen.mat".

We have found that the value of the wave function is related to the probability of finding the particle at a particular point in space. We would now like to generalize this result to three dimensions. In one-dimension, the probability dP of finding the particle in the infinitesimal interval between x and $x + dx$ is given by the equation

$$dP = |\psi(x)|^2 dx.$$

The appropriate generalization of this result to three dimensions is that the probability of finding a particle in the infinitesimal volume element dV at $\mathbf{r}$ is

$$dP = |\psi(\mathbf{r})|^2 dV. \tag{4.13}$$

4.1.4 Probabilities and average values in three dimensions

The volume element in polar coordinates is illustrated in Fig. 4.7. As the variable r increases by dr, the coordinate point moves a distance dr. A change in the angle θ by an amount $d\theta$ with the coordinates r and ϕ fixed, causes the vector $\mathbf{r}$ to rotate about the origin and the coordinate point moves a distance $r d\theta$ in a direction perpendicular to $\mathbf{r}$. Similarly, a change

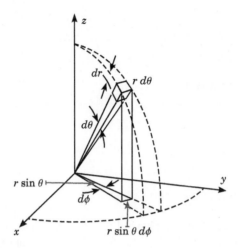

FIGURE 4.7 The volume element in three dimensions.

in the polar angle ϕ by $d\phi$ causes the vector $\mathbf{r}$ to swing around the z-axis and the tip of $\mathbf{r}$ moves a distance $r\sin\theta d\phi$. The volume element in polar coordinates is the product of these three displacements

$$dV = dr \cdot rd\theta \cdot r\sin\theta d\phi$$
$$= r^2 \sin\theta dr d\theta d\phi. \tag{4.14}$$

Substituting Eqs. (4.4) and (4.14) into Eq. (4.13), we obtain the following equation for the probability of finding the particle in dV

$$dP = |\psi(r,\theta,\phi)|^2 dV = P(r)^2 |dr\Theta_{lm_l}(\theta)|^2 |\Phi_{m_l}(\phi)|^2 \sin\theta dr d\theta d\phi.$$

This equation for dP enables us to compute the probability of finding the electron in any given region of space. The probability that the electron is in a spherical shell of inner radius r and thickness dr can be obtained by integrating over the angles θ and ϕ. This gives

$$dP = P(r)^2 dr \int_0^\pi |\Theta_{lm_l}(\theta)|^2 \sin\theta d\theta \int_0^{2\pi} |\Phi_{m_l}(\phi)|^2 d\phi.$$

The θ and ϕ integrals above are each equal to one, since the functions $P(r)$, Θ, and Φ are each normalized. We thus have

$$dP = P(r)^2 dr. \tag{4.15}$$

The probability that the electron is in a shell of thickness dr is thus equal to the square of the function P(r) times dr.

As we have argued for problems in one dimension, the average value of a function $f(r)$ can be evaluated by adding up contributions from all the infinitesimal intervals dr. For a shell between r and $r + dr$, $f(r)$ is multiplied times the probability of the electron being in the shell. Including the contributions from all of the shells by integrating from 0 to ∞, we obtain

$$< f(r) > = \int_0^\infty f(r) dP = \int_0^\infty f(r) P(r)^2 dr. \tag{4.16}$$

The integral that appears in Eq. (4.16) can be evaluated by multiplying the sum of the product of the terms that appear in the integrand at the Gauss quadrature points times the step size delta divided by two. For instance, denoting the vector consisting of the magnitude of the $1s$ function at the Gauss quadrature points by $sV1$ and the vector consisting of the value of r at the Gauss quadrature points by $rcol$, the above integral for the $1s$ state of hydrogen would be evaluated by the MATLAB command

```
AvrsV1 = dot(sV1,rcol.*sV1)*delta/2
```

Here the product rcol.*svi with a dot produces the vector consisting of the values of the product of the elements of rcol and sV1 at the Gauss points and the command "dot" then calculates the sum of the product of each of the elements of this vector with the corresponding element of sV1. We shall use commands of the kind to evaluate the average value of r for the lowest s, p, and d states of hydrogen in the following MATLAB program.

MATLAB Program Hydrogen_load_vectors.m

A MATLAB program for plotting the wave functions for the lowest three s-states of hydrogen, the lowest two p-states, and the lowest d state. The program also evaluates the average value of r for each of these states and evaluate the radial part of the electric dipole integral for the transition $2p \rightarrow 1s$, which we will need in the next section on transition probabilities.

```
load("hydrogen.mat")

xmax=40
n=400
delta=xmax/n

% Gauss Points
xi1=(3-sqrt(3))/6;
xi2=(3+sqrt(3))/6;

% Construct B matrix
p1=(9 -4*sqrt(3))/18;
p2=(9+4*sqrt(3))/18;
p3=(3-sqrt(3))/36;
p4=(3+sqrt(3))/36;

B=[p2 p4 p1 -p3; p1 p3 p2 -p4];

% Construct Full Matrix Bmat
Bmat=zeros(2*n, 2*n+2);
for row=1:2:2*n
  Bmat(row:row+1, row:row+3)=B;
end

% Boundary Conditions
Bmat(:,2)=[];
Bmat(:,2*n)=[];

% Construct vector rccol consisting of the values of r at the Gauss points
ii = 0;
r = 0.0;
rcol = zeros(2*n,1);
for i = 1:n
   r1 = r + xi1*delta;
   r2 = r + xi2*delta;
   rcol(ii+1) = r1;
   rcol(ii+2) = r2;
   ii = ii + 2;
   r = r + delta;
end
rcol;

% First sVector
sV1nodal = -sVectors(:,1);
sV1col = Bmat*sV1nodal;
```

```
% Normalize sV1nodal & sV1col
sV1norm2 = dot(sV1col,sV1col)*delta/2;
sV1nf = sqrt(sV1norm2);
sV1 = sV1col/sV1nf;
plot(rcol, sV1)
saveas (gcf,'sV1.pdf')
shg

% Calculate Average Value of r
AvrsV1 = dot(sV1,rcol.*sV1)*delta/2

% Second sVector
sV2nodal = -sVectors(:,2);
sV2col = Bmat*sV2nodal;
% Normalize sV2nodal & sV2col
sV2norm2 = dot(sV2col,sV2col)*delta/2;
sV2nf = sqrt(sV2norm2);
sV2 = sV2col/sV2nf;
plot(rcol, sV2)
saveas (gcf,'sV2.pdf')
shg


% Calculate Average Value of r
AvrsV2 = dot(sV2,rcol.*sV2)*delta/2

% Third sVector
sV3nodal = -sVectors(:,3);
sV3col = Bmat*sV3nodal;
% Normalize sV3nodal & sV3col
sV3norm2 = dot(sV3col,sV3col)*delta/2;
sV3nf = sqrt(sV3norm2);
sV3 = sV3col/sV3nf;
sV3n = sV3nodal/sV3nf;
plot(rcol, sV3)
saveas (gcf,'sV3.pdf')


% Calculate Average Value of r
AvrsV3 = dot(sV3,rcol.*sV3)*delta/2

% First pVector
pV1nodal = pVectors(:,1);
pV1col = Bmat*pV1nodal;
% Normalize pV1nodal & pV1col
pV1norm2 = dot(pV1col,pV1col)*delta/2;
pV1nf = sqrt(pV1norm2);
pV1 = pV1col/pV1nf;
plot(rcol, pV1)
saveas (gcf,'pV1.pdf')

% Calculate Average Value of r
AvrpV1 = dot(pV1,rcol.*pV1)*delta/2
```

```
% Calculate Electric Dipole
Dipole = dot(sV1,rcol.*pV1)*delta/2

% Second pVector
pV2nodal = pVectors(:,2);
pV2col = Bmat*pV2nodal;
% Normalize pV2nodal & pV2col
pV2norm2 = dot(pV2col,pV2col)*delta/2;
pV2nf = sqrt(pV2norm2);
pV2 = pV2col/pV2nf;
plot(rcol, pV2)
saveas (gcf,'pV2.pdf')

% Calculate Average Value of r
AvrpV2 = dot(pV2,rcol.*pV2)*delta/2

% First dVector
dV1nodal = dVectors(:,1);
dV1col = Bmat*dV1nodal;
% Normalize dV1nodal & dV1col
dV1norm2 = dot(dV1col,dV1col)*delta/2;
dV1nf = sqrt(dV1norm2);
dV1 = -dV1col/dV1nf;
plot(rcol, dV1)
saveas (gcf,'dV1.pdf')

f
% Calculate Average Value of r
AvrdV1 = dot(dV1,rcol.*dV1)*delta/2
```

The program begins by defining the grid and the coefficients, $xi1$ and $xi2$ and generates the B matrix ($Bmat$). Since the wave function is zero at the origin and at the outer boundary, the boundary conditions may be imposed by deleting the first and second to last columns of the $Bmat$.

The program then uses the coefficients $xi1$ and $xi2$ defined earlier to constructs the vector $rcol$ consisting of the values of r at the Gauss quadrature points and defines the nodal form of the $1s$ wave function with the MATLAB command

```
sV1nodal = -sVectors(:,1);
```

As we have noted previously, the sign of the wave functions does not have any physical significance. We have put a minus sign in this last equation so that the wave function is positive for small positive values of r. The vector consisting of the values of the $1s$ wave function at the Gauss points is then produced by multiplying the nodal form of the $1s$ wave function by the B matrix with the command

```
sV1col = Bmat*sV1nodal;
```

The vector $sV1col$ is then normalized with the normalized function denoted $sV1$. At this point, $rcol$ is a vector with elements equal to the value of r at the collocation points and $sV1$ is the normalized P_{1s} wave function at the collocation points. The command

```
plot(rcol, sV1)
```

then produces the plot of the $sV1$ wave function shown in Fig. 4.8(A). The statement

```
AvrsV1 = dot(sV1,rcol.*sV1)*delta/2plot(rcol, sV1)
```

calculates the average value of r for the $1s$ state. In a similar way, the wave function of the $2s$, $3s$, $2p$, $3p$, and $3d$ states are plotted and used to calculate the average value of r for each of these states.

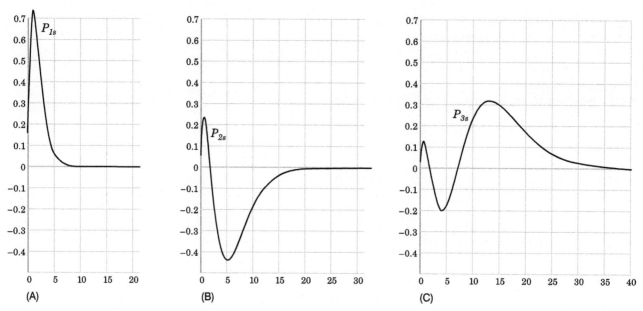

FIGURE 4.8 The lowest lying s-wave functions of hydrogen. The wave functions are as follows: (A) P_{1s}, (B) P_{2s}, and (C) P_{3s}.

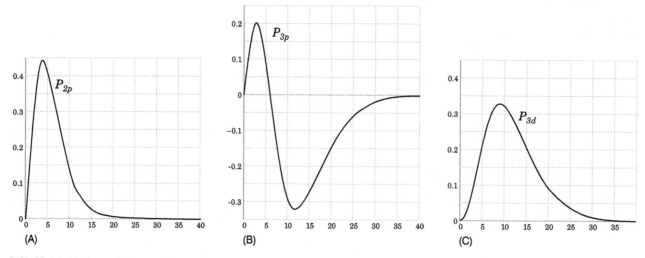

FIGURE 4.9 The lowest lying p and d-wave functions of hydrogen. The wave functions are as follows: (A) P_{2p}, (B) P_{3p}, and (C) P_{3d}.

The energy eigenvalues and the average values r for all of these orbitals are given in Table 4.2.
The program Hydrogen_load_vectors.m also calculates

$$\int_0^\infty P_{1s} r P_{2p} dp$$

which is the radial part of the dipole matrix element to be used in the next section on radiative transitions. The program produces the value

$$\int_0^\infty P_{1s} r P_{2p} dp = 1.290. \tag{4.17}$$

Notice that the $2s$ and $2p$ functions and also the $3s$, $3p$, and $3d$ functions shown in Fig. 4.8 and Fig. 4.9 have similar spatial extents and similar values of the average value of r. This is important – on a more advanced level – for understanding the departures from the single configuration model of many-electron atoms to be described in the next chapter. The $1s^2 2p^2$ configuration of beryllium atom for which two electrons are excited from the $2s$ shell strongly mixes with the ground

TABLE 4.2 The Energy Eigenvalues in both Atomic Units (A.U.) and Electron Volts (eV) and the Average Values of r for the lowest *s*, *p*, and *d* States of Hydrogen.

Electron	E (A.U.)	E (eV)	< r >
1s	−0.5	−13.6	1.5
2s	−0.125	−3.4	6.0
3s	−0.0556	−1.51	13.5
2p	−0.125	−3.4	5.0
3p	−0.0556	−1.51	10.5
3d	−0.0556	−1.51	10.5

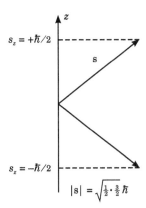

FIGURE 4.10 The spin vector **s**, and the possible *z*-components of **s** with $s_z = \pm \hbar/2$.

$1s^2 2s^2$ configuration even though the $1s^2 2p^2$ configuration lies far above the ground configuration. The reason for this is that the $2s$ and $2p$ wave functions have very similar spatial properties.

4.1.5 The intrinsic spin of the electron

In 1925, Samuel Goudsmit and George Uhlenbeck suggested that the electron has an inherent spin angular momentum with component $s_z = \pm \hbar/2$. As we shall discuss in the third section of this chapter, this hypothesis provides a basis for understanding the fine structure of the energy levels of hydrogen. We shall assign the quantum number s to the spin just as we assigned the quantum number l to the orbital angular momentum. However, unlike l which can be zero or an integer up to $n-1$, the spin quantum number of the electron, only has the value $1/2$. The eigenvalue of $\mathbf{s}^2$ is $s(s+1)\hbar^2 = (1/2)(3/2)\hbar^2$, and the possible eigenvalues of s_z are $m_s \hbar$, where m_s may equal $-1/2$ or $1/2$. These values are illustrated in Fig. 4.10. As with the components of l_z illustrated in Fig. 4.4, the values of s_z are separated by $\hbar$.

The spin of the electron will be described by adding a spin component χ_{m_s} to the orbital wave function (4.4) to obtain the following single electron wave function

$$\psi(r, \theta, \phi) = \frac{P(r)}{r} \Theta_{lm_l}(\theta) \, \Phi_{m_l}(\phi) \chi_{m_s} \, .$$

Each level of the hydrogen atom shown in Fig. 4.5 corresponds to definite values of n and l. We note that for each choice of n and l, there are $2l + 1$ values of m_l and two values of m_s. Each level thus corresponds to a total of $4l + 2$ different sets of quantum numbers. An energy level that corresponds to a number of independent states is said to be *degenerate*. Using this terminology, the energy levels of the hydrogen atom may be said to be $(4l + 2)$-fold degenerate.

4.2 Radiative transitions

In this section, we shall describe the radiative processes by which atoms make transitions from one energy level to another and emit or absorb radiation. Our aim will be to provide a basis for understanding the equations governing transition rates

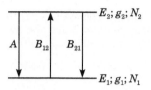

FIGURE 4.11 Two energy levels between which transitions can occur.

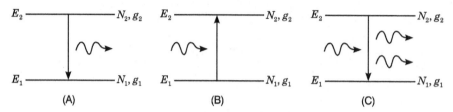

FIGURE 4.12 Radiative processes including: (A) spontaneous emission, (B) absorption, and (C) stimulated emission.

and how these equations can be used in practical examples. The concepts and equations, which we shall obtain, will form the basis for our description of atomic transitions and lasers in following chapters.

4.2.1 The Einstein A and B coefficients

We begin by considering transitions between two energy levels as those shown in Fig. 4.11. The energy of the upper level is denoted by E_2 and the energy of the lower level is denoted by E_1. As shown in the figure, we shall suppose that there are N_2 atoms in the upper level and the degeneracy of the level is g_2, while there are N_1 atoms in the lower level and the degeneracy is g_1. The frequency of light absorbed or emitted by the atom as it makes transitions between the two levels is given by the rule of Niels Bohr given in Chapter 1. According to this rule the frequency of the light is given by the following equation

$$f = \frac{E_2 - E_1}{h},\qquad(4.18)$$

where h is Planck's constant.

Three kinds of radiative processes are shown in Fig. 4.12: *spontaneous emission* in which an atom makes a transition from the upper level E_2 to the lower level E_1 independent of the radiation field, *absorption* in which light is absorbed from the radiation field and the atom makes a transition from the lower to the upper level, and *induced* or *stimulated emission* in which the atom is induced by the radiation field to make a transition from the upper to the lower level. These three processes are described by the coefficients A_{21}, B_{12}, and B_{21}, respectively. They are called the *Einstein coefficients*. The number of transitions per second that atoms make from one level to another depends upon the corresponding coefficient and upon the number of atoms that are in the level from which the transition occurs. For instance, the number of transitions per second that atoms make spontaneously from the energy level E_2 to the level E_1 is equal to A_{21} times N_2. The number of transitions per second that atoms make by absorbing light or by being stimulated to make a transition by a radiation field also depends upon the energy density of the radiation $\rho(f)$. The number of atoms per second that absorb light and make a transition from the energy level E_1 to the level E_2 is equal to $B_{12}N_1\rho(f)df$, while the number of atoms per second that make an induced transition from the level E_2 to the level E_1 is $B_{21}N_2\rho(f)df$.

The rate of change of the populations of the two levels can thus be expressed in terms of the coefficients A_{21}, B_{12}, and B_{21} for spontaneous emission, absorption and induced emission as follows

$$\frac{dN_2}{dt} = -\frac{dN_1}{dt} = -A_{21}N_2 + B_{12}\rho(f)df\,N_1 - B_{21}\rho(f)df\,N_2\,.\qquad(4.19)$$

The A_{21} coefficient gives the number of spontaneous transitions per second a single atom would make from the upper level E_2 to the lower level E_1. Similarly, the coefficients, B_{12} and B_{21}, give the number of absorptions and stimulated emissions that a single atom would make per second and per unit density of the radiation field. For a two-level system, the rate at which the number of atoms in the upper level N_2 increases is equal to the rate at which the number of atoms in the lower level N_1 decreases.

As shown by Einstein, the A and B coefficients are related by the equations

$$A_{21} = \frac{4f^2}{c^3} hf B_{21} \tag{4.20}$$

and

$$g_1 B_{12} = g_2 B_{21} . \tag{4.21}$$

Here g_1 and g_2 are the degeneracies of the states 1 and 2. A proof of these equations is given in the book by Woodgate cited at the end of this chapter.

If an atom is not subjected to radiation, then $\rho(f) = 0$, and the last two terms in Eq. (4.19) do not contribute. We then have

$$\frac{dN_2}{dt} = -A_{21} N_2 .$$

This equation can be solved for the number of atoms in the upper level as a function of time. Assuming there are $N_2(0)$ atoms in the upper level at $t = 0$, we obtain

$$N_2(t) = N_2(0) e^{-A_{21}t} .$$

In the absence of a radiation field, the number of atoms in the upper level thus decreases exponentially with time.

4.2.2 Transition probabilities

In three dimensions, the Schrödinger time-dependent equation described in Chapter 2 can be written

$$\left[\frac{-\hbar^2}{2m} \nabla^2 + V(\mathbf{r}, t) \right] \psi = i\hbar \frac{\partial \psi}{\partial t}, \tag{4.22}$$

where ∇^2 is the Laplacian operator described earlier in this chapter and the wave function depends upon three spatial coordinates and the time. The potential energy $V(\mathbf{r}, t)$ represents the interaction of the electron with the nucleus and with any external field.

If the potential energy is independent of time, the Schrödinger time-dependent equation is separable and the solutions of the equation can be written

$$\psi(\mathbf{r}, t) = \phi(\mathbf{r}) e^{-i\omega t} , \tag{4.23}$$

where the function $\phi(\mathbf{r})$ is a solution of the Schrödinger time-independent equation, and ω is related to the energy by the equation $E = \hbar\omega$. The proof of this result for three dimensions follows along the lines of the derivation for one dimension given in Section 2.5 of Chapter 2.

The stationary states (4.23) represent the possible ways the state of the hydrogen atom can evolve in a constant external field. At a particular time t, the wave function of a hydrogen atom in an oscillating electric field can be expressed as a linear combination of these wave functions as follows

$$\psi(\mathbf{r}, t) = \sum_n c_n(t) \phi_n(\mathbf{r}) e^{-iE_n t/\hbar} , \tag{4.24}$$

where the expansion coefficients $c_n(t)$ depend on time. In order to calculate the number of transitions per second atoms make from the level i to the level j, we suppose that at time, $t = 0$, an atom is in level i. This assumption can be described mathematically by setting the coefficient $c_i(0)$ in Eq. (4.24) equal to one and all the other coefficients $c_j(0)$ equal to zero. Due to the effect of the oscillating radiation field upon the atom, the coefficients $c_j(t)$ will change with time assuming nonzero values. The probability that at a later time t the atom is in the state j is equal to $|c_j(t)|^2$. To derive an expression for the transition probability, we substitute Eq. (4.24) into the Schrödinger time-dependent equation (4.22) and solve for the coefficients $c_j(t)$. In this way, one obtains the following expression for the probability per time that an atom is stimulated to emit light polarized along the z-axis

$$|c_j(t)|^2 / t = \frac{\pi}{\epsilon_0 \hbar^2} \left| \int \phi_1^*(\mathbf{r})(-ez) \phi_2(\mathbf{r}) dV \right|^2 \rho(f) df, \tag{4.25}$$

where ϕ_2 is the wave function of the initial state from which the transition occurs, ϕ_1 is the wave function of the final state, and f is the transition frequency. As before, $\rho(f)$ is the energy density per unit frequency range of the radiation. A derivation of Eq. (4.25) is given in Appendix FF. Our choosing the value of the coefficient of the initial state $c_i(0)$ equal to one corresponds to setting the population of the level N_2 equal to one. By comparing Eq. (4.25) with Eq. (4.19), we can identify the entire factor multiplying $\rho(f)df$ in Eq. (4.25) as the Einstein coefficient for stimulated emission

$$B_{21} = \frac{\pi}{\epsilon_0 \hbar^2} \left| \int \phi_1^*(\mathbf{r})(-ez)\phi_2(\mathbf{r}) dV \right|^2. \tag{4.26}$$

The operator $-ez$, which joins the initial state ϕ_2 to the final state ϕ_1 in the integral on the right, is the z-component of $-e\mathbf{r}$, which is called the *electric dipole operator*. One can describe in simple terms the physical process which has led to the transition. The electric field of the incident radiation, which points in the z-direction, polarizes the atom creating an electric dipole moment. This induced dipole moment of the atom interacts with the changing electromagnetic field causing the atom to make a transition from the initial state described by the wave function ϕ_2 to the final state described by the wave function ϕ_1. Eq. (4.26) gives the transition rate in transitions per atom and per second for stimulated emission. The transition rate for spontaneous emission can be found using Eq. (4.20). We get

$$A_{21} = \frac{16\pi^3 f^3}{\epsilon_0 h c^3} \left| \int \phi_1^*(\mathbf{r})(-ez)\phi_2(\mathbf{r}) dV \right|^2.$$

In calculating the rate at which transitions occur between two energy levels, we shall include the contributions of all the m_s and m_l values of the final state, and we shall assume that the atom has equal probability of being in any one of the initial states. We shall thus sum over all of the m_s and m_l values of the final state and average over all the possible m_s and m_l values of the initial state. The average over the m_s and m_l values of the initial state is obtained by summing over these quantum numbers and dividing through by the degeneracy of the initial state g_2. The sign of the term $-ez$ does not affect the absolute value squared of the integral. We thus obtain the following expression for the transition rate for spontaneous emission

$$A_{21} = \frac{16\pi^3 e^2 a_0^2 f^3}{\epsilon_0 h c^3} \frac{I_{12}}{g_2}, \tag{4.27}$$

where

$$I_{12} = 2 \sum_{m_l^1, m_l^2} \left| \int \phi_{n_1 l_1 m_l^1}^* (z/a_0) \phi_{n_2 l_2 m_l^2} dV \right|^2. \tag{4.28}$$

As before, we have divided the z coordinate in the I_{12} integral by a_0 to express distances in units of the Bohr radius a_0, and we have added a corresponding factor of a_0^2 to the numerator of the expression for A_{12}. Since the operator (z/a_0) is independent of spin, it will not join wave functions having different values of m_s. The two spin states thus contribute independently leading to the factor of 2 that occurs in Eq. (4.28).

Eq. (4.27) for the transition rate can be expressed more simply in terms of the wavelength for the transition. Using the equation $c/f = \lambda$, Eq. (4.27) can be written

$$A_{21} = \frac{64\pi^4}{4\pi\epsilon_0} \frac{e^2 a_0^2}{h} \frac{1}{\lambda^3} \frac{I_{12}}{g_2}.$$

The values of the physical constants in this last equation are given in Appendix A. We thus obtain the following equation for A_{21} giving the number of spontaneous transitions per atom per second

$$A_{21} = \frac{6.078 \times 10^{15}}{\lambda^3} \frac{I_{12}}{g_2} \ (\mathrm{nm}^3\,\mathrm{s}^{-1}), \tag{4.29}$$

where g_2 is the degeneracy of the initial state and I_{12} is given by Eq. (4.28). It is understood that wavelengths are to be measured in nm and the variable r occurring in the transition integral should be expressed in units of a_0. Of course, similar formulas apply for light polarized in the x- and y-directions.

Example 4.2

Calculate the coefficient A_{21} for the hydrogen atom to emit z-polarized light in making the transition $2p \rightarrow 1s$.

Solution

We first calculate the wavelength for the transition. The approach we shall use is the same as that used for working problems of this kind in Chapter 1. Since the n value for the upper level is 2 and the n value for the lower level is 1, the difference in energy of the two levels involved in the transition is

$$\Delta E = E_2 - E_1 = \frac{-13.6 \text{ eV}}{2^2} - \frac{-13.6 \text{ eV}}{1^2} = 10.2 \text{ eV}.$$

The wavelength of the light can be calculated using Eq. (1.12) giving

$$\lambda = \frac{hc}{\Delta E} = \frac{1240 \text{ eV} \cdot \text{nm}}{10.2 \text{ eV}} = 121.6 \text{ nm}. \tag{4.30}$$

We shall now evaluate the transition integrals and use them to calculate the factor I_{12} given by Eq. (4.28). We first consider a transition from a state with $m_l = 0$ to a state with the same value of m_l and denote the initial and final states by $2p_0$ and $1s_0$, respectively. The angular part of the appropriate wave functions are given in Tables 4.1. The nuclear charge Z is also equal to one for hydrogen. Using Eq. (4.3) for the z-coordinate and Eq. (4.14) for the volume element, the transition integral can be written

$$\int \phi_{1s_0}^* \, z \, \phi_{2p_0} \, dV = \int_0^\infty P_{1s}(r) \cdot r \cdot P_{2p}(r) dr \int_0^\pi \frac{1}{\sqrt{2}} \cdot \cos\theta \cdot \sqrt{\frac{3}{2}} \cos\theta \sin\theta \, d\theta \int_0^{2\pi} \frac{1}{\sqrt{2\pi}} \cdot 1 \cdot \frac{1}{\sqrt{2\pi}} \, d\phi. \tag{4.31}$$

The integrations over the r, θ, and ϕ coordinates that appear in the integrals on the right-hand side of this equation are carried out independently. The angular part of wave functions in these integrals are taken from Tables 4.1. For each integral, there is a factor that comes from the wave function of the $1s$, a term that comes from the operator z, another term coming from the wave function of the $2p$, and finally a factor due to the volume element. We note that the radial functions for the two electrons are of the form, $P(r)/r$. The factor of $1/r$ associated with the two functions of r has canceled with the factor of r^2 in the volume element.

The integral over ϕ in Eq. (4.31) gives 1, while the integral over θ reduces to

$$(\sqrt{3}/2) \int_0^\pi \cos^2\theta \sin\theta d\theta.$$

In order to evaluate this integral, we introduce the change of variable $u = \cos\theta$. The differential du is then $-\sin\theta d\theta$ and the integral becomes

$$-(\sqrt{3}/2) \int_1^{-1} u^2 du = 1/\sqrt{3}.$$

Since all of the wave functions corresponding to a particular choice of n and l have the same radial function, there will be a single radial integral for the transitions from the $2p$ level to the $1s$ level. This radial integral was evaluated by the Program Hydrogen_load_Vectors and given by Eq. (4.17). Drawing these contributions together, we get

$$\text{Transition Integral} = \frac{1}{\sqrt{3}} \times 1.290.$$

The factor I_{12}, which is given by Eq. (4.28), is equal twice the square of the transition integral

$$I_{12} = \frac{2}{3}(1.290)^2 = 1.109. \tag{4.32}$$

Finally, substituting Eqs. (4.30) and (4.32) into Eq. (4.29), we obtain the following value for the transition rate

$$A_{21} = \frac{6.078 \times 10^{15}}{(121.6)^3} \frac{1.109}{6} = 6.25 \times 10^8 \text{ per atom per second}. \tag{4.33}$$

Before going further, we would like to assess qualitatively the physical significance of this result. The constant A_{21} given by Eq. (4.33) is the number of transitions per second per atom. The inverse of this constant is the length of time it would take for a single atom to make a transition. Taking the inverse of the term on the right-hand side of Eq. (4.33), we obtain

$$1/A_{21} = 1.6 \times 10^{-9} \text{ sec}. \tag{4.34}$$

The average time it takes a hydrogen atom to make a transition from the $2p$ to the $1s$ level is equal to 1.6 nanoseconds. This is consistent with our statement in the introduction that atoms generally make transitions from one energy level to another in about a nanosecond.

As we have mentioned before, atomic physicist use a system of units called *atomic units* in which e (the charge of the electron), m (the mass of the electron), $\hbar$ (Planck's constant divided by 2π) and $4\pi\epsilon_0$ (4π times the permittivity of the vacuum) are all equal to one. The Bohr radius, (a_0) is the unit of length in atomic units.

The unit of energy in atomic units may be defined as the potential energy of two electrons (units of charge) separated by the Bohr radius a_0, which is the unit of distance. The unit of energy is

$$\text{unit of energy} = \frac{1}{4\pi\epsilon_0} \frac{e^2}{a_0} \tag{4.35}$$

We have seen earlier that the unit of energy in atomic units is equal to twice the binding energy of the hydrogen atom, or about 27.2 eV. In order to identify the atomic unit of time, we first write Schrödinger time-dependent equation (4.22) in the following form

$$H\psi = i\hbar\frac{\partial \psi}{\partial t}, \tag{4.36}$$

where H, which is defined

$$H = \frac{-\hbar^2}{2m}\nabla^2 + V(\mathbf{r}, t), \tag{4.37}$$

is the energy operator for the electron in the hydrogen atom. Dividing Eq. (4.36) by the unit of energy, we obtain

$$\frac{H}{(1/4\pi\epsilon_0)(e^2/a_0)} \psi(\mathbf{r}, t) = i\left(\frac{\hbar}{(1/4\pi\epsilon_0)(e^2/a_0)}\right)\frac{\partial \psi(\mathbf{r}, t)}{\partial t}.$$

The factor multiplying the wave function on the left-hand side of this equation is the energy operator divided by the unit of energy. Since this factor is dimensionless, the left-hand side of this equation must be dimensionless. Consequently, the right-hand side of the equation also has this property, and we may identify the factor appearing within the parenthesis on the right-hand side of this equation as the unit of time in atomic units

$$\text{unit of time} = \frac{\hbar}{(1/4\pi\epsilon_0)(e^2/a_0)}.$$

The magnitude of the unit of time, which can be obtained by substituting the values of the appropriate constants given in Appendix A into the above equation, is equal to 2.4×10^{-17} sec. This is of the same order of magnitude as the time it takes for the electron of the hydrogen atom to circulate once about the nucleus. The length of time required for the $2p$ electron to decay to the $1s$ state, which is given by Eq. (4.34), may thus be understood to be a very long time in the tiny world of the atom. The electron of hydrogen circulates hundreds of millions of times about the nucleus before the electron makes a transition. This justifies the perturbative approach we have used in deriving our basic equation for transition rates in Appendix FF. The electric and magnetic fields of light are usually much weaker than the fields within the atom.

In this section, we have been concerned with calculating transition rates for light which is polarized along the z-axis. Formulas for calculating transition rates for circularly polarized light and for light polarized along the x- and y-axes are given in Appendix GG.

4.2.3 Selection rules

The radiative transitions that can occur from the $n = 2$ and 3 levels of hydrogen are shown in Fig. 4.13. We note, for instance, that an electron in the $3p$ level can make a transition to the $2s$ and $1s$ states, and a $3s$ electron can decay to the $2p$. There are, though, some transitions that do not occur. There is not a transition from $2s$ level to the $1s$, nor does the $3p$ go to the $2p$ or the $3d$ go to the $2s$. Fortunately, simple rules can be formulated that determine which transitions are allowed

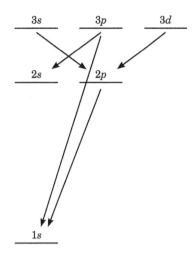

FIGURE 4.13 The electric dipole transitions that can occur from the $n = 2$ and $n = 3$ levels of hydrogen.

and which are forbidden. These rules are called *selection rules*. They determine for which changes in the l and m_l quantum numbers the dipole transition integrals are nonzero and, hence, an electric-dipole transition can occur.

In Example 4.2, we noted that certain integrals of the dipole operators are zero. For instance, while the operator z joins the $2p_0$ state to the $1s_0$ state, it does not join the $2p_{+1}$ or $2p_{-1}$ states to the $1s_0$. Using Eq. (4.4) for the wave functions, Eq. (4.3) for the z-coordinate and Eq. (4.14) for the volume element, the transition integral for the operator z can be written

$$\int \phi^*_{n'l'm'_l} z \phi_{nlm_l} \, dV = \int_0^\infty P_{n'l'}(r) \cdot r \cdot P_{nl}(r) dr$$

$$\int_0^\pi \Theta_{l'm'_l}(\theta) \cdot \cos\theta \cdot \Theta_{lm_l}(\theta) \sin\theta d\theta \int_0^{2\pi} \frac{1}{\sqrt{2\pi}} e^{-im'_l\phi} \frac{1}{\sqrt{2\pi}} e^{im_l\phi} \, d\phi. \quad (4.38)$$

For light polarized in the z-direction, the integral involving the ϕ-coordinate will be zero unless $m'_l = m_l$. Transitions for which x- and y-polarized light is emitted are discussed in Appendix GG. For light polarized in the x- and y-directions, the transition integrals vanish unless the azimuthal quantum number of the final state differs from the initial state by ± 1. The selection rules governing the change in the azimuthal quantum number m_l for radiative transitions may be summarized: $\Delta m_l = 0$, for light polarized along the z-axis; $\Delta m_l = \pm 1$, for the light polarized along the x- or y-axes.

The selection rules for l are more difficult to prove. We first note that the transition integrals are of the general form

$$\int \phi^*_{n'l'm'_l} e x_k \phi_{nlm_l} \, dV, \quad (4.39)$$

where x_k may be any one of the coordinates x, y, or z. Each of the coordinates is odd with respect to inversion through the origin. For the transition integral (4.39) to be nonzero, the initial and final wave functions must thus have opposite parity with respect to inversion. The parity of the spherical harmonic Y_{lm_l} is even or odd with respect to spatial inversion according to whether the angular momentum quantum number l is an even or an odd number. The transition integral will thus be nonzero only if the l values of the initial and final states differ in parity.

To obtain a more precise statement of the selection rules for the l quantum number, one must examine more closely the functions $\Theta_{lm_l}(\theta)$. One may show that the product of $\cos\theta$ or $\sin\theta$ with the functions $\Theta_{lm_l}(\theta)$ gives a linear combination of the functions $\Theta_{l-1m_l}(\theta)$ and $\Theta_{l+1m_l}(\theta)$. This property of the functions $\Theta_{lm_l}(\theta)$ may be used to derive the following selection rule for radiative transitions: $\Delta l = \pm 1$. The selection rules for the m_l and l quantum numbers are summarized in Table 4.3.

The selection rules shown in Table 4.3 give the conditions that must be satisfied for the electric dipole operator to cause radiative transitions. We should note, however, that weaker electronic transitions can occur even when the dipole integrals are zero. These transitions called *forbidden transitions*, which occur with rates that are many of orders of magnitude smaller than electric dipole transitions, still have important applications to astrophysics and to other areas of applied physics.

TABLE 4.3 Selection Rules for Radiative Transitions.

Δm_l	0	for z-polarized light
	± 1	for x- and y-polarized light
Δl	± 1	for all polarizations

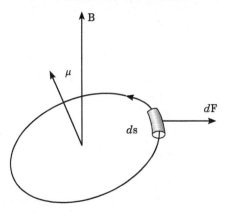

FIGURE 4.14 A current loop with a magnetic moment μ in a magnetic field **B**.

4.3   The fine structure of hydrogen

The appearance of the energy levels of hydrogen is due mainly to the kinetic energy of the electrons and the Coulomb interaction of each electron with the nucleus and the other electrons. In this section, we shall study the effect of the magnetic interactions, which lead to a fine splitting of the energy levels of the atom. The magnetic interactions also have an important effect upon the way that the atom interacts with its environment.

We begin this section by studying the interaction of the hydrogen atom with an external magnetic field.

4.3.1   The magnetic moment of the electron

An electron circulating about the nucleus of an atom can be expected to behave like a current loop which interacts with an external magnetic field. A current loop of this kind is illustrated in Fig. 4.14. As shown in this figure, the element of the loop denoted by **ds** experiences a force **dF** given by the equation

$$\mathbf{dF} = i(\mathbf{ds} \times \mathbf{B}).$$

This force provides a torque which tends to align the loop so that its plane is perpendicular to the direction of the magnetic field. The magnitude of the torque and its effect upon the current loop can best be described in terms of the magnetic moment μ of the loop. The magnitude of the magnetic moment is equal to the product of the current and the area of the loop

$$|\mu| = iA.$$

As shown in Fig. 4.14, the direction of the magnetic moment is perpendicular to the plane of the loop. The torque on the loop is related to the magnetic moment of the loop μ and the magnetic field vector **B** by the equation

$$\tau = \mu \times \mathbf{B}. \tag{4.40}$$

To estimate the magnitude of the magnetic moment of the atom, we shall consider the hydrogen atom as a classical charge distribution. We imagine that the charge cloud of hydrogen consists of filaments of current circulating about the nucleus. The current is due to the motion of the electron which has a charge and which has the dynamical properties of a particle moving about a center of attraction. In order to obtain an expression for the magnetic moment of the atom, we consider a thin ring of charge rotating about the z-axis. This ring is shown in Fig. 4.15. Each of the segment of the rotating ring constitutes a current which contributes to the magnetic moment of the atom. Since the magnetic moment of a current

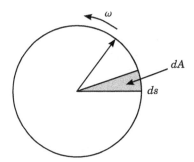

FIGURE 4.15 A thin ring of charge rotating about the z-axis.

loop is equal to the product of the current and the area of the loop, the magnetic moment of the charge ring can be written

$$\mu = \int i\, dA. \tag{4.41}$$

The element of area dA, which is shown in Fig. 4.14, can be expressed in terms of the radius of the ring r and the infinitesimal length of the ring ds using the formula for the area of a triangle. We have

$$dA = \frac{1}{2} r\, ds.$$

Denoting the charge of the element by dq and an infinitesimal element of time by dt, the current due to the element is

$$i = \frac{dq}{dt}.$$

Substituting the above equations for dA and i into Eq. (4.41), we get

$$\mu = \int \frac{dq}{dt} \frac{1}{2} r\, ds = \int dq \cdot \frac{1}{2} r \frac{ds}{dt}.$$

The derivative of s with respect to time is the velocity of the element of the ring. Expressing the velocity of the charge element in terms of its momentum, we have

$$\mu = \frac{1}{2m} \int r p\, dq.$$

The product of r and p is equal to the orbital angular momentum l of the charge element. Carrying out the above integration over the entire charge cloud, we thus obtain the following expression for the magnetic moment of the atom

$$\mu = \frac{-e}{2m} |\mathbf{l}|. \tag{4.42}$$

Here $(-e)$ is the total charge of the atom and $\mathbf{l}$ is the orbital angular momentum of the atom. The ratio of a magnetic moment to the angular momentum giving rise to the magnetic moment is called the *gyromagnetic ratio*. According to Eq. (4.42), the gyromagnetic ratio associated with the orbital motion of the electron is $-e/2m$. Since the charge of the electron is negative, the magnetic moment of the atom points in the opposite direction to the angular momentum. Eq. (4.42) enables us to express the magnetic moment vector in terms of the angular momentum vector. We have

$$\mu = \frac{-e}{2m} \mathbf{l}. \tag{4.43}$$

Using this last equation, the z-component of the magnetic moment can be written

$$\mu_z = \frac{-e}{2m} l_z.$$

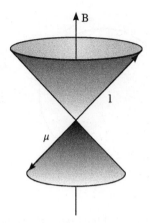

FIGURE 4.16 The orbital angular momentum **l** precessing about a magnetic field **B**.

For an atom in a state for which l_z has the value $\hbar m_l$, the z-component of the magnetic moment has the value

$$\mu_z = -\frac{e}{2m}\,\hbar m_l\,.$$

The constants appearing in this equation for μ_z can be collected together and a new constant defined called the *Bohr magneton* which is given by the following equation

$$\mu_B = \frac{e\hbar}{2m} = 9.2732 \times 10^{-24}\,\mathrm{J\,T^{-1}}\,. \tag{4.44}$$

The equation for μ_z can then be written

$$\mu_z = -\mu_B\,m_l. \tag{4.45}$$

The Bohr magneton is a convenient unit for expressing atomic magnetic moments.

An expression for the torque on an atom due to the magnetic field can be obtained by substituting Eq. (4.43) into Eq. (4.40) giving

$$\tau = \frac{-e}{2m}\,\mathbf{l} \times \mathbf{B}. \tag{4.46}$$

The torque causes the orbital angular momentum of the atom to precess about the magnetic field as illustrated in Fig. 4.16. We shall generally choose the direction of the magnetic field to be the z-axis. The component l_z then remains constant as the atom precesses, while the values of l_x and l_y continually change. The precession of the angular momentum vector of an atom in the magnetic field is analogous to the motion of the axis of a spinning top in a gravitational field. The atom precesses in the field with a frequency ω_L called the *Larmor frequency*. As indicated in Problem 20, the Larmor frequency is given by the equation

$$\omega_L = \frac{eB}{2m}\,.$$

The potential energy of the magnetic moment of the atom in the magnetic field is

$$V = -\mu \cdot \mathbf{B}.$$

If the direction of the magnetic field coincides with the z-axis, this expression for the energy becomes

$$V = -\mu_z\,B.$$

Substituting Eq. (4.45) into this last equation, we get

$$V = \mu_B\,m_l\,B. \tag{4.47}$$

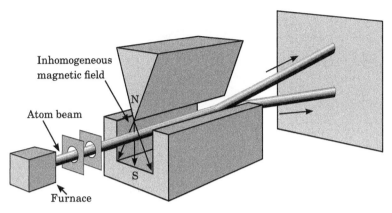

Inhomogeneous magnetic field

Atom beam

N

S

Furnace

FIGURE 4.17 The Stern-Gerlach experiment.

4.3.2 The Stern-Gerlach experiment

In order to study how atoms interact with magnetic fields, we consider the experiment illustrated in Fig. 4.17. In this experiment, a beam of atoms passes through an inhomogeneous magnetic field and is incident upon a screen. Each of the atoms experiences a force which can be obtained by taking the negative gradient of the potential energy (4.47) giving

$$F = -\mu_B m_l \frac{dB}{dz}. \tag{4.48}$$

An atom in a state for which m_l has a particular value will thus experience a definite force. To be concrete, we suppose that the derivative of B with respect to z is positive. As an example of the effect of the magnetic field upon a beam of atoms, we consider a beam of hydrogen atoms prepared in the $n = 2, l = 1$ state. The beam then consists of atoms with $m_l = -1, 0$, and $+1$. According to Eq. (4.48), the atoms with $m_l = +1$ will experience a downward force and be deflected downward, while the atoms with $m_l = -1$ will be deflected upwards. The atoms with $m_l = 0$ will not be deflected. We thus expect atoms with $l = 1$ to produce three lines on the screen. The number of lines on the screen corresponds to the different possible values of m_l. For a particular value of l, m_l can have the values, $m_l = -l, -l + 1, \ldots, l$. There will thus be $2l + 1$ different values of m_l for each value of l, and each value of m_l should produce a distinct image on the screen. We thus expect that atoms with $l = 0$ should produce one line, atoms with $l = 1$ should produce three lines, atoms with $l = 2$ should produce five lines, and so forth.

The first experiment of this kind was performed by Otto Stern and Walter Gerlach in 1921 using a beam of silver atoms. As can be shown using the atomic shell model described in the next chapter, silver has filled shells of electrons and a single outer electron. Since the electrons in filled shells are magnetically neutral and the outer electron is in the $l = 0$ state, one would expect that the beam of atoms would produce a single line on the screen. Stern and Gerlach found instead that the beam of silver atoms produced *two* lines on the screen. Their pioneering experiment provided the first evidence of space quantization which we have discussed earlier. The fact that an $l = 0$ state of silver produced two distinct lines could be explained naturally when Goudsmit and Uhlenbeck proposed in 1925 that the electron has an intrinsic angular momentum or spin. As we have seen, the spin of the electron can be described by the quantum numbers s and m_s. The quantum number s is equal to $1/2$ for the electron, while m_s has the values $\pm 1/2$. The two lines formed by the Stern-Gerlach experiment can thus be attributed to the two possible m_s values of the $l = 0$ electron of silver.

4.3.3 The spin of the electron

An atomic electron has a magnetic moment due to its spin just as the electron has a magnetic moment due to its orbital motion. The magnetic moment associated with the spin of the electron is given by the following equation

$$\mu_s = \frac{-e}{2m} g_s \mathbf{s}. \tag{4.49}$$

Here the constant g_s is called the *g-value* of the spin. According to Eq. (4.49), the ratio of the magnetic moment μ_s to the spin angular momentum $\mathbf{s}$ is equal to $-eg_s/2m$, which is the gyromagnetic ratio associated with the spin of the electron. We

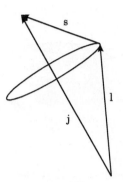

FIGURE 4.18 The spin angular momentum **s** and the orbital angular momentum **l** precessing about the total angular momentum **j**.

can obtain an expression for the total magnetic moment of the electron by adding together the orbital and spin contributions to the magnetic moment given by Eqs. (4.43) and (4.49) giving

$$\mu = \frac{-e}{2m}(\mathbf{l} + g_s \mathbf{s}).$$
(4.50)

We have seen earlier that the spin of the electron can be incorporated into atomic theory by including an additional term χ_{m_s} in the wave function. From the Stern-Gerlach experiment, we know that the spin quantum number m_s can have the values $\pm 1/2$. While the spin of the electron can be accommodated into quantum theory, quantum theory itself does not provide us any guidance as to what the spin might be. In Chapters 11 and 12, we shall study the special theory of relativity. Modern relativity theory began with Einstein who insisted that physical laws should have the same form in coordinate frames moving with respect to each other. Relativity theory has caused us to reevaluate our concepts of space and time and leads to relativistic wave equations which are the relativistic analogues of the Schrödinger equation. The relativistic wave equation for the electron, which is called the Dirac equation, will be studied in Chapter 12. One of the special features of the Dirac theory is that the spin of the electron comes out as a natural consequence of the theory. The Dirac theory of the electron gives a value of g_s equal to two, which is close to the experimental value $g_s = 2.0023192$. Modern theory has shown that the small discrepancy between the Dirac and experimental values of g_s is due to the interaction of the electron with its own radiation field. The value of g_s has recently been measured and calculated to 13 decimal places with excellent agreement between theory and experiment.

Thus far we have only considered the effect of an external magnetic field upon the electron. Since the electron and the nucleus are charged particles, their relative motion gives rise to magnetic fields which are a property of the atom itself. The motion of the nucleus with respect to the electron gives rise to a magnetic field which interacts with the magnetic moment associated with the spin of the electron. This interaction, which is called the *spin-orbit interaction*, can most easily be understood by studying the solutions of the relativistic Dirac equation for an electron moving in an electromagnetic field. If one takes g_s equal to two, the spin-orbit interaction of an electron with a potential energy $V(r)$ is described by the following Hamiltonian

$$h_{s-o} = \frac{1}{2m^2 c^2} \frac{1}{r} \frac{dV}{dr} \mathbf{s} \cdot \mathbf{l}.$$
(4.51)

Just as the spin and orbital magnetic moments of the electron precess about an external magnetic field, the spin-orbit interaction causes the spin and orbital angular momentum vectors to precess about each other. This may easily be understood in physical terms. The relative motion of the nucleus produces a magnetic field at the site of the electron. This magnetic field, which points in the direction opposite to the angular momentum vector **l**, exerts a torque upon the spin of the electron causing the vector **s** to process around **l**. The nature of this motion can be described in a natural way by introducing the total angular momentum **j** which is the sum of **s** and **l**

$$\mathbf{j} = \mathbf{l} + \mathbf{s}.$$

As we have illustrated in Fig. 4.18, the spin-orbit interaction causes the vectors **s** and **l** to process about **j**. Since the spin-orbit interaction is internal to the atom involving the spin and the orbital motions, the total angular momentum vector **j** is unaffected by the interaction.

The effect of the spin-orbit interaction upon atomic states can be calculated using wave functions having definite values of the total angular momentum. Just how angular momenta can be combined to form a total angular momentum is an important problem in atomic physics because even relatively simple atoms have several different angular momenta. The helium atom, for instance, has four angular momenta corresponding to the orbital and spin angular momenta of the two electrons.

4.3.4 The addition of angular momentum

We have found that the orbital angular momentum of an electron is described by the quantum numbers l and m_l, while the spin of an electron is described by the quantum numbers s and m_s. The angular momenta of a single electron are the building blocks for more complex systems having two or more angular momenta.

In order to discuss the problem of combining angular momenta in general terms, we consider a system composed of two angular momenta described by the quantum numbers j_1, m_1 and j_2, m_2 respectively. For example, the quantum numbers j_1, m_1 could be the quantum numbers of the spin of one electron and j_2, m_2 could be the quantum numbers of the spin of another electron, or j_1, m_1 and j_2, m_2 could refer to the spin and orbital angular momenta of a single electron. The composite system can be described by a quantum number J corresponding to the total angular momentum and a quantum number M corresponding to the z-component of the total angular momentum. The possible values of the quantum numbers J and M are given by the following rule.

Rule for addition of angular momenta

For given values of the quantum numbers j_1 and j_2, the quantum number J of the total angular momentum has the values

$$J = j_1 + j_2, j_1 + j_2 - 1, \ldots |j_1 - j_2|.$$

For each value of J, the azimuthal quantum number M can be

$$M = -J, -J + 1, \ldots, J.$$

We now give two examples to show how this rule can be applied.

Example 4.3

What are the possible values of the total spin of two electrons with $s = 1/2$?

Solution

For a system consisting of two spins, j_1 and j_2 are both equal to one half. We denote the total angular momentum quantum number of this two-spin system by S. According to the above rule, the maximum value of S is

$$S_{max} = \frac{1}{2} + \frac{1}{2} = 1,$$

while the minimum value of S is

$$S_{min} = \left| \frac{1}{2} - \frac{1}{2} \right| = 0.$$

Since these two values differ by one, they are the only values of S. The z-component of the total spin has the value $\hbar M_S$ where M_S is an integer in the range $-S \leq M_S \leq S$. For $S = 1$, M_S may be equal to $1, 0,$ or -1. For $S = 0$, M_S is equal to 0.

Example 4.4

What are the possible values of the total angular momentum quantum number j of a single electron having an orbital angular momentum l and a spin $s = 1/2$?

Solution

For this problem, j_1 is equal to l and j_2 is equal to one half. The total angular momentum quantum number of a single electron is denoted by j. According to the rule for adding angular momenta, the maximum value of j is $l + 1/2$ and the minimum value is $|l - 1/2|$. If $l = 0$, these two values are equal to $1/2$, which is then the only possible value of j. If $l > 0$, the maximum and minimum values of j are equal to $l + 1/2$ and $l - 1/2$. Since these two values differ by one, they are the only values of j. The azimuthal quantum number m in each case has values in the range $-j \leq m \leq j$.

The orbital motion of an electron is described by the spherical harmonics $Y_{lm_l}(\theta, \phi)$ where m_l can have the values $l, l - 1, \ldots, -l$, while the spin of an electron is described by the functions χ_{m_s} where m_s can have the values $\pm 1/2$. A composite system having one orbital and one spin angular momentum can be described by product functions consisting of a spherical harmonic and a spin function

$$Y_{lm_l}(\theta, \phi)\chi_{m_s}. \tag{4.52}$$

As we have seen earlier, the spherical harmonic $Y_{lm_l}(\theta, \phi)$ is an eigenfunction of the operator $\mathbf{l}^2$ corresponding to the eigenvalue $l(l + 1)\hbar^2$, and the spin function χ_{m_s} is an eigenfunction of $\mathbf{s}^2$ corresponding to the eigenvalue $\frac{1}{2}(\frac{1}{2} + 1)\hbar^2$. One can form states having definite values of the quantum numbers of the total angular momentum j and m by taking linear combinations of the states (4.52). These states, which will be denoted $|(l\frac{1}{2})jm >$, are called *coupled states*. They are eigenfunctions of the operators $\mathbf{j}^2$ and j_z corresponding to the eigenvalues $j(j + 1)\hbar^2$ and $m\hbar$, and they are eigenfunctions of the operators $\mathbf{l}^2$ and $\mathbf{s}^2$ corresponding to the eigenvalues $l(l + 1)\hbar^2$ and $\frac{1}{2}(\frac{1}{2} + 1)\hbar^2$.

4.3.5   * The fine structure

Using Eq. (I.5) of the introduction for the potential energy $V(r)$, the expression (4.51) for the spin-orbit interaction can be written

$$h_{s-o} = \frac{1}{2m^2c^2} \frac{1}{4\pi\epsilon_0} \frac{Ze^2}{r^3} \mathbf{s} \cdot \mathbf{l}.$$

For a hydrogen atom described by the wave function (4.4), the function $1/r^3$ has the average value

$$\left\langle \frac{1}{r^3} \right\rangle = \int_0^\infty \frac{1}{r^3} P_{nl}(r)^2 \, dr.$$

With this value of $1/r^3$, the spin-orbit interaction can be written

$$h_{s-o} = \zeta \, \mathbf{s} \cdot \mathbf{l}, \tag{4.53}$$

where

$$\zeta = \frac{1}{2m^2c^2} \frac{1}{4\pi\epsilon_0} Ze^2 \left\langle \frac{1}{r^3} \right\rangle. \tag{4.54}$$

ζ is called the *spin-orbit constant*.

In order to evaluate the effect of the operator $\mathbf{s} \cdot \mathbf{l}$, we make use of the identity

$$\mathbf{j}^2 = (\mathbf{s} + \mathbf{l})^2 = \mathbf{s}^2 + \mathbf{l}^2 + 2\mathbf{s} \cdot \mathbf{l}.$$

This equation may be solved for $\mathbf{s} \cdot \mathbf{l}$ to obtain

$$\mathbf{s} \cdot \mathbf{l} = \frac{1}{2}(\mathbf{j}^2 - \mathbf{s}^2 - \mathbf{l}^2),$$

and the spin-orbit operator (4.53) assumes the following form

$$h_{s-o} = \frac{\zeta}{2}(\mathbf{j}^2 - \mathbf{s}^2 - \mathbf{l}^2). \tag{4.55}$$

We now consider the effect of the spin-orbit operator upon the coupled states $|(l\frac{1}{2})jm >$ having a definite value of the spin and orbital angular momentum and also a definite value of the total angular momentum. We recall that these states are eigenfunctions of the operators $\mathbf{l}^2$ and $\mathbf{s}^2$ corresponding to the eigenvalues $l(l + 1)\hbar^2$ and $\frac{1}{2}(\frac{1}{2} + 1)\hbar^2$, and they are also

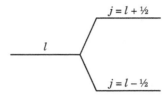

FIGURE 4.19 The spin-orbit splitting of the energy levels of a single electron.

eigenfunctions of the operators $\mathbf{j}^2$ and j_z corresponding to the eigenvalues $j(j+1)\hbar^2$ and $m\hbar$. Hence, the effect of the spin-orbit operator (4.55) upon the coupled states is

$$h_{s-o}|(l\tfrac{1}{2})jm> = \frac{\zeta\hbar^2}{2}\left[j(j+1) - \tfrac{1}{2}(\tfrac{1}{2}+1) - l(l+1)\right]|(l\tfrac{1}{2})jm>. \tag{4.56}$$

The addition of the orbital and the spin angular momentum of a single electron has been considered in Example 4.4. If the orbital quantum number l is greater than zero, the quantum number of the total angular momentum may have the values $j = l \pm 1/2$. Substituting these two values of j into Eq. (4.56), we obtain the following equations for the spin-orbit energy E_{s-o} of the two levels with definite values of the total angular momentum

$$E_{s-o} = \frac{\zeta\hbar^2}{2}l \quad \text{for } j = l+1/2$$

and

$$E_{s-o} = -\frac{\zeta\hbar^2}{2}(l+1) \quad \text{for } j = l-1/2.$$

The splitting of these two levels due to the spin-orbit interaction is illustrated in Fig. 4.19.

In order to get some sense of the magnitude of the spin-orbit interaction h_{s-o}, we rewrite Eq. (4.54) for the spin-orbit constant as follows

$$\zeta = \frac{\alpha^2}{2}Z^4\left(\frac{1}{4\pi\epsilon_0}\frac{e^2}{a_0}\right)\left\langle\left(\frac{a_0}{Zr}\right)^3\right\rangle. \tag{4.57}$$

Expressions for the physical constants appearing in this equation and their numerical values are given in Appendix A. The *fine structure constant* α, which appears in the above expression for the spin-orbit constant, is a dimensionless constant which is approximately equal to $1/137$. The quantity appearing within parentheses is the unit of energy in the atomic system of units. As we have seen before, the unit of energy is approximately equal to 27.2 electron volts. We have grouped the variable r together with the Bohr radius a_0 and the atomic number Z in the combination Zr/a_0 because the radial distance scales as Zr/a_0 with increasing nuclear charge. One can easily confirm this fact by consulting Table 4.2 for the radial wave functions. Due to the appearance of α^2 in the expression for ζ, the spin-orbit interaction is quite small for the lightest atoms; however, because ζ varies as Z^4, the relative importance of the spin-orbit interaction increases with increasing atomic number.

4.3.6 * The Zeeman effect

We consider now the interaction of the hydrogen atom with an external magnetic field **B**. This gives rise to a splitting of the energy levels known as the *Zeeman effect*. The interaction of the magnetic moment of the electron with the magnetic field may be described by the magnetic potential energy

$$V_{\text{mag}} = -\mu \cdot \mathbf{B}, \tag{4.58}$$

where μ is the total magnetic moment of the electron. The contribution to the energy of the atom due to its interaction with the magnetic field can be obtained by taking the average value of V_{mag} for the coupled states having definite values of the quantum numbers j and m. The total magnetic moment of the electron is given by Eq. (4.50). Substituting Eq. (4.50) into Eq. (4.58) gives

$$\begin{aligned} V_{\text{mag}} &= (e/2m)\left(\mathbf{l} + g_s\mathbf{s}\right)\cdot\mathbf{B} \\ &= (e/2m)\,B(l_z + g_s s_z). \end{aligned} \tag{4.59}$$

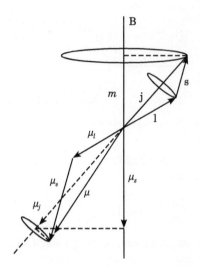

FIGURE 4.20 The precession of the magnetic moments and the angular momentum vectors of the atom.

The interaction of the electron with the magnetic field depends upon the components of the angular momentum vectors in the direction of the magnetic field. As before, we have taken the direction of **B** to correspond with the z-axis. The angular momentum vectors and the magnetic moments associated with the spin and orbital motions of the atom are illustrated in Fig. 4.20.

As we have mentioned earlier, the electric and magnetic fields within the atom are usually much stronger than the fields due to the external environment of the atom. When an atom is in a weak magnetic field, the orbital and spin angular momentum vectors **l** and **s** precess rapidly about the total angular momentum vector **j**, while **j** precesses more slowly about the magnetic field vector **B**. It is thus reasonable to suppose that the effect of the components of **l** and **s** perpendicular to the direction of **j** cancel out over periods of time that include many revolutions. We may then replace the angular momentum vectors with their projection along **j**. We consider first the orbital angular momentum vector. The projection of the orbital angular momentum vector **l** along the total angular momentum vector **j** can be written $(\mathbf{l} \cdot \hat{\mathbf{j}})\hat{\mathbf{j}}$, where $\hat{\mathbf{j}}$ is a unit vector pointing in the direction of **j**. The component of the vector $(\mathbf{l} \cdot \hat{\mathbf{j}})\hat{\mathbf{j}}$ in the direction the magnetic field can be written $(\mathbf{l} \cdot \hat{\mathbf{j}})\hat{j}_z$. For the purpose of evaluating the average value of l_z for coupled states having definite values of the quantum numbers j and m, we may thus make the following replacement

$$l_z \rightarrow (\mathbf{l} \cdot \hat{\mathbf{j}})\hat{j}_z = \frac{(\mathbf{l} \cdot \mathbf{j})j_z}{\mathbf{j}^2} = \frac{(\mathbf{l} \cdot \mathbf{j})j_z}{j(j+1)\hbar^2} . \tag{4.60}$$

In order to evaluate $\mathbf{l} \cdot \mathbf{j}$, we write

$$\mathbf{s} \cdot \mathbf{s} = (\mathbf{j} - \mathbf{l}) \cdot (\mathbf{j} - \mathbf{l}) = \mathbf{j}^2 + \mathbf{l}^2 - 2\mathbf{l} \cdot \mathbf{j}.$$

Solving for $\mathbf{l} \cdot \mathbf{j}$, we obtain

$$\mathbf{l} \cdot \mathbf{j} = \frac{1}{2}(\mathbf{j}^2 + \mathbf{l}^2 - \mathbf{s}^2)$$

and substituting this expression for $\mathbf{l} \cdot \mathbf{j}$ into (4.60), gives the following replacement

$$l_z \rightarrow \frac{(\mathbf{j}^2 + \mathbf{l}^2 - \mathbf{s}^2)j_z}{2j(j+1)\hbar^2} = \frac{j(j+1) + l(l+1) - s(s+1)}{2j(j+1)} j_z . \tag{4.61}$$

The corresponding replacement for the spin angular momentum operator s_z may be shown to be

$$s_z \rightarrow = \frac{j(j+1) - l(l+1) + s(s+1)}{2j(j+1)} j_z . \tag{4.62}$$

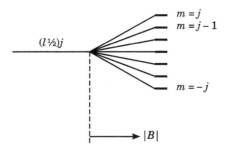

FIGURE 4.21 The splitting of the energy levels of an atom due to the Zeeman effect.

Substituting the expressions (4.61) and (4.62) into Eq. (4.59) gives the following equation for the magnetic potential energy

$$V_{\text{mag}} = \left[\frac{j(j+1) + l(l+1) - s(s+1)}{2j(j+1)} + g_s \frac{j(j+1) - l(l+1) + s(s+1)}{2j(j+1)} \right] (e/2m) \, B j_z \,. \tag{4.63}$$

This equation may be written simply

$$V_{\text{mag}} = g_j \, (e/2m) \, B j_z \,,$$

where

$$g_j = \frac{j(j+1) + l(l+1) - s(s+1)}{2j(j+1)} + g_s \frac{j(j+1) - l(l+1) + s(s+1)}{2j(j+1)} \,.$$

Since the coupled states are eigenfunctions of the operator j_z corresponding to the eigenvalue $m\hbar$, the potential energy V_{mag} leads to the following splitting of the energy levels

$$\Delta E = g_j \, (e/2m) \, B m \hbar \,.$$

This last equation can be written conveniently using the Bohr magneton μ_B, which is defined by Eq. (4.44). We obtain finally

$$\Delta E = g_j \mu_B B m \,.$$

The interaction of the orbital and spin magnetic moments of the electron with an external magnetic field thus produces a splitting of the different m-levels. The splitting of the energy levels of the atom is illustrated in Fig. 4.21. This effect has many important applications to physics. By analyzing the spectra from sunspots, for instance, physicists have been able to determine the strength of the magnetic fields surrounding sunspots.

Suggestion for further reading

G.K. Woodgate, *Elementary Atomic Structure*, Second Edition (Oxford University Press, 1980).

Basic equations

Wave function for hydrogen

Schrödinger equation for hydrogen

$$\left[\frac{-\hbar^2}{2m} \nabla^2 - \frac{1}{4\pi\epsilon_0} \frac{Ze^2}{r} \right] \psi = E\psi$$

Relation between polar and Cartesian coordinates

$$x = r \sin\theta \cos\phi$$
$$y = r \sin\theta \sin\phi$$
$$z = r \cos\theta$$

Wave function in polar coordinates

$$\psi(r, \theta, \phi) = \frac{P(r)}{r} Y_{lm_l}(\theta, \phi)),$$

where

$$Y_{lm_l}(\theta, \phi) = \Theta_{lm_l}(\theta) \Phi_{m_l}(\phi)$$

Volume element in polar coordinates

$$dV = r^2 \sin\theta \, dr \, d\theta \, d\phi$$

Radial equation for hydrogen in atomic units

$$\left[-\frac{1}{2} \frac{d^2}{dr^2} + \frac{l(l+1)}{2r^2} - \frac{Z}{r} \right] P_{nl}(r) = E \, P_{nl}(r)$$

Probabilities and average values

Probability of particle being in shell of thickness dr

$$dP = P(r)^2 \, dr$$

Average value of function $f(r)$

$$< f(r) > = \int_0^\infty f(r) \, dP = \int_0^\infty f(r) P(r)^2 \, dr$$

Transition probabilities

Transition rate for spontaneous emission

$$A_{21} = \frac{6.078 \times 10^{15}}{\lambda^3} \frac{I_{12}}{g_2} \; (\text{nm}^3 \, \text{s}^{-1}),$$

where

$$I_{12} = 2 \sum_{m_l^1, m_l^2} | \int \phi_{n_1 l_1 m_l^1}^* \, ez \, \phi_{n_2 l_2 m_l^2} \, dV |^2$$

Selection rules

$$\Delta m_l \quad 0 \quad \text{for } z\text{-polarized light}$$
$$\Delta m_l \quad \pm 1 \quad \text{for } x\text{- and } y\text{-polarized light}$$
$$\Delta l \quad \pm 1 \quad \text{for all polarizations}$$

The fine structure of hydrogen

Magnetic moment of electron

$$\mu = \frac{-e}{2m} (\mathbf{l} + g_s \mathbf{s})$$

Total angular momentum

$$J = j_1 + j_2, j_1 + j_2 - 1, \ldots |j_1 - j_2|$$

Magnetic potential energy

$$V_{\text{mag}} = -\mu \cdot \mathbf{B}$$

The Zeeman splitting

$$\Delta E = g_j \mu_B B m$$

Summary

The wave function $\psi(\mathbf{r})$ of the electron in the hydrogen atom satisfies the Schrödinger equation

$$\left[\frac{-\hbar^2}{2m}\nabla^2 - \frac{1}{4\pi\epsilon_0}\frac{Ze^2}{r}\right]\psi = E\psi.$$

In polar coordinates, the wave function is of the general form

$$\psi(r,\theta,\phi) = \frac{P(r)}{r}\Theta_{lm_l}(\theta)\Phi_{m_l}(\phi),$$

where the radial part of the wave function is expressed as a function $P(r)$ divided by r. The angular part of the wave function, $\Theta_{lm_l}(\theta)\Phi_{m_l}(\phi)$, is called a spherical harmonic and denoted $Y_{lm_l}(\theta,\phi)$.

The radial function P_{nl} satisfies the equation

$$\left[-\frac{1}{2}\frac{d^2}{dr^2} + \frac{l(l+1)}{2r^2} - \frac{Z}{r}\right]P_{nl}(r) = E\,P_{nl}(r)$$

in atomic units with $\hbar = e = m_e = 4\pi\epsilon_0 = 1$.

The probability that the electron is in the shell between r and $r + dr$ is

$$dP = P(r)^2\,dr.$$

The radial probability density, which is equal to the probability per unit interval in the radial direction, is thus equal to the square of the function P(r). The average value of a function $f(r)$ can be evaluated by multiplying the value of the function at each value of r times the probability $P(r)^2\,dr$ and integrating from 0 to ∞ to obtain

$$<f(r)> = \int_0^\infty f(r)P(r)^2\,dr.$$

The number of transitions per second that atoms make from one level to another depends upon the Einstein coefficients, A_{21}, B_{12}, and B_{21}. The number of transition per second that atoms make spontaneously from the energy level E_2 to the energy level E_1 is equal to A_{21} times N_2. The number of atoms per second that absorb light and make a transition from the energy level E_1 to the level E_2 is equal to $B_{12}N_1\rho(f)df$, while the number of atoms per second that make a stimulated transition from the level E_2 to the level E_1 is $B_{21}N_2\rho(f)df$. Simple rules can be formulated that determine which transitions are allowed and which are forbidden. These rules, which are called selection rules, determine which changes in the values of the quantum numbers, l and m_l, can occur for transitions.

The orbital and spin motions of the electron causes magnetic fields that can be described by the magnetic moments,

$$\mu_l = \frac{-e}{2m}\mathbf{l} \quad \text{and} \quad \mu_s = \frac{-eg_s}{2m}\mathbf{s}.$$

The spin-orbit interaction, which is the interaction of the magnetic moment associated with the spin of the electron with the magnetic field due to the relative motion of the nucleus, leads to a fine splitting of the energy levels of hydrogen. An external magnetic field further splits each level with quantum number j into $2j+1$ sublevels having definite values of the quantum number m. The splitting of the energy levels of an atom due to an external magnetic field is called the Zeeman effect.

Questions

1. Write down the Schrödinger equation for the electron in the hydrogen atom.
2. Write down a general expression for the wave function of the electron in polar coordinates.
3. What are the possible values of $\mathbf{l}^2$ and l_z for an electron with $l = 3$?
4. Make a drawing of the spherical harmonic for l equal to zero and for $l = 1$ and $m_l = 0$.
5. Write a MATLAB program for calculating the vector $rcol$ consisting of the values of r at the Gauss quadrature points.
6. Suppose the nodal form of the $2p$ radial wave function is given by the array $pV1nodal$. What MATLAB statement would produce the vector consisting of the values of the $2p$-radial function at the Gauss quadrature points.

7. Suppose $rcol$ is a vector consisting of the values of r at the Gauss quadrature points and the vector $pV1$ consists of the values of the $2p$ radial function at the Gauss quadrature points. What MATLAB statement would produce a plot of the $2p$ radial function.

8. Make a drawing of the radial function $P(r)$ for the $1s$, $2s$, $2p$, and $3p$ states.

9. Suppose $rcol$ is a vector consisting of the values of r at the Gauss quadrature points, $sV1$ consists of the values of the P_{1s} radial function at the Gauss quadrature points, and the vector $pV1$ consists of the values of the P_{2p} radial function at the Gauss quadrature points. What MATLAB statement would produce the value of the dipole integral, $\int_0^\infty P_{1s} rcol P_{2p} dr$.

10. What is the probability that the electron in hydrogen is between r and $r + dr$?

11. Write down a general formula for the average value of the function $1/r$.

12. What are the possible values of $\mathbf{s}^2$ and s_z for an electron?

13. Which Einstein coefficient would be used to find the number of spontaneous transitions an atom would make between the energy level E_2 and the energy level E_1?

14. Using the appropriate Einstein coefficient, write down an expression for the number of absorptions that would occur causing an atom to make a transition from the energy level E_1 to the energy level E_2.

15. In the absence of radiation, how does the population of atoms in an excited state vary with time?

16. About how long would it take a hydrogen atom to decay from the $2p$ state to the $1s$ state?

17. How large are the units of distance, energy, and time in atomic units?

18. How does the length of time it takes for the electron to make a transition from the $2p$ to the $1s$ state of hydrogen compare with the average length of time for the electron to circulate once about the nucleus.

19. By how much does the orbital angular momentum number l change in an atomic transition?

20. By how much does the azimuthal momentum number m_l change in an atomic transition in which z-polarized light is emitted?

21. By how much does the azimuthal momentum number m_l change in an atomic transition in which x- or y-polarized light is emitted?

22. Write down an equation relating the magnetic moment of the hydrogen atom μ to the orbital angular momentum $\mathbf{l}$ of the atom.

23. How is the Bohr magneton defined?

24. In the Stern-Gerlach experiment, how many lines would be produced on the screen when a beam of atoms in the $l = 0$ state passed through the apparatus?

25. How is the magnetic moment associated with the spin of the electron μ_s related to the spin operator $\mathbf{s}$?

26. Give a qualitative description of the spin-orbit interaction.

27. What are the possible values of the spin quantum number S of two electrons with $s = 1/2$?

28. What are the possible values of the orbital angular momentum L of two electrons with $l = 1$?

29. What are the possible values of the total angular momentum j of a single d-electron with $l = 2$?

30. Describe how the angular momentum vectors, $\mathbf{s}$ and $\mathbf{l}$, evolve in time when a hydrogen atom is placed in a weak magnetic field.

31. What is the separation between adjacent energy levels when a hydrogen atom in a state with a definite value of the total angular momentum j is placed in a weak magnetic field.

32. The electric dipole operator is independent of spin. What should be the selection rule for the quantum number S for the transitions of many-electron atoms?

Problems

1. Modify the MATLAB program Hydrogen.m so that in addition to the functions produced previously it also produces a second d eigenfunction.

2. Write a MATLAB program that inputs the eigenfunctions produced by program described in the previous Problem and plots the $3d$ and $4d$-functions.

3. Write a MATLAB program that inputs the eigenfunctions produced by the program Hydrogen.m and finds the average value of $1/r^3$ for the $2p$ electron of hydrogen.

4. Use the average value of $1/r^3$ for the $2p$ electron of hydrogen calculated in the previous Problem to calculate the spin-orbit constant for the $2p$ state of hydrogen.

5. Modify the MATLAB program Hydrogen.m so that it solves the radial equation for hydrogen-like ion with nuclear charge $Z = 8$ and finds the energy eigenvalues for the $1s$, $2s$, $3s$, $2p$, $3p$, and $3d$ states. How do these eigenvalues differ from the eigenvalues found previously for hydrogen.

6. Write a MATLAB program that inputs the eigenfunctions produced by the modified program described in the previous Problem and plots the eigenfunction for the $3d$ state. How does this wave function differ from the $3d$ function found previously for hydrogen,

7. Write a MATLAB program that inputs the eigenfunctions produced by the program Hydrogen.m and calculates the radial part of transition integral for the hydrogen atom to make the transition $3d \rightarrow 2p$.

8. To which states can an electron in the $4p$ state of hydrogen decay?

9. What are the possible values of m_l and m_s for a $4f$ electron?

10. What are the possible values of the total angular momentum j for a single $4f$ electron? How much are these levels split by the spin-orbit interaction?

11. What are the possible values of the total angular momentum j for a single $2p$ electron? Taking $g_s = 2$, calculate the g_j value for each j-state.

12. Modify the MATLAB program Hydrogen.m so that it solves the radial equation for the He^+ ion with nuclear charge $Z = 2$ and finds the energy eigenvalues for the $1s$, $2s$, $3s$, $2p$, $3p$, and $3d$ states. How do these eigenvalues differ from the eigenvalues found previously for hydrogen?

13. Write a MATLAB program that inputs the $2p$ eigenfunction produced by program described in the previous Problem and calculates the average value of $1/r^3$ for the $2p$ electron of He^+.

14. Use the average value of $1/r^3$ for the $2p$ electron of He^+ calculated in the previous Problem to calculate the spin-orbit constant of He^+.

15. Using the results of Problems 14 and 4, calculate the ratio of the spin-orbit constant for the $2p$ state of He^+ to the spin-orbit constant for the $2p$ state of hydrogen.

Chapter 5

Many-electron atoms

Contents

5.1 The independent-particle model	117	Suggestions for further reading	138
5.2 Shell structure and the periodic table	120	Basic equations	138
5.3 The LS term energies	122	Summary	138
5.4 Configurations of two electrons	122	Questions	138
5.5 The Hartree-Fock method	126	Problems	139
5.6 Further developments in atomic theory	135		

We can obtain a good approximate description of a many-electron atom by supposing that each electron moves independently in an average potential due to the nucleus and the other electrons.

We begin this chapter by considering the helium atom which has two electrons. The problem of finding the wave functions and the energies of the two electrons can be simplified by assuming that each electron moves independently in an average field due to the nucleus and the electron. This approximation, which is known as the *independent-particle model*, provides a good, approximate description of the atom. If one also supposes that the average potential in which the electrons move is spherically symmetric, the single-electron wave functions can be written as a product of a radial function, a spherical harmonic, and a spin function as they are for hydrogen.

5.1 The independent-particle model

Each electron of a two-electron atom has a kinetic energy and a potential energy due to the nucleus and the other electron. As for the hydrogen atom, the energy may be represented by a Hamiltonian operator. The Hamiltonian for a two-electron atom may be written

$$H = -\frac{\hbar^2}{2m}\nabla_1^2 - \frac{1}{4\pi\epsilon_0}\frac{Ze^2}{r_1} - \frac{\hbar^2}{2m}\nabla_2^2 - \frac{1}{4\pi\epsilon_0}\frac{Ze^2}{r_2} + \frac{1}{4\pi\epsilon_0}\frac{e^2}{r_{12}}. \tag{5.1}$$

Here ∇_1^2 is the Laplacian operator for the first electron and r_1 is the distance of the first electron from the nucleus, while ∇_2^2 is the Laplacian operator for the second electron and r_2 is the distance of the second electron from the nucleus. The distance between the two electrons is denoted by r_{12}. The first term on the right-hand side of Eq. (5.1) represents the kinetic energy of the first electron and the second term on the right-hand side represents the potential energy of the first electron due to the nucleus with charge Z. Similarly, the third term corresponds to the kinetic energy of the second electron and the fourth term corresponds to the potential energy of the second electron due to the nucleus. The last term on the right-hand side of Eq. (5.1) is the potential energy of the two electrons due to their mutual Coulomb interaction. In addition to the contributions to the Hamiltonian H included in Eq. (5.1), the energy of a two-electron atom is also affected by the interaction of the spin of the electrons with the magnetic fields produced by their spin and their orbital motion (the so-called magnetic interactions).

Even if the magnetic interactions are neglected, the atomic Hamiltonian (5.1) is very complex, and the eigenfunctions of H can only be found within the framework of some approximation scheme. As a first approximation, we assume that each electron moves independently in an average potential due to the nucleus and the electrons. This is called the *independent-particle model*. We shall suppose that the potential for the two electrons is the same and denote the potential by $u(\mathbf{r})$. The independent-particle Hamiltonian can be obtained by replacing the Coulomb interaction between the electrons in Eq. (5.1) by the interaction of the electrons with the average field in which they move. This gives

$$H_0 = -\frac{\hbar^2}{2m}\nabla_1^2 - \frac{1}{4\pi\epsilon_0}\frac{Ze^2}{r_1} + u(\mathbf{r}_1) - \frac{\hbar^2}{2m}\nabla_2^2 - \frac{1}{4\pi\epsilon_0}\frac{Ze^2}{r_2} + u(\mathbf{r}_2).$$

This approximate Hamiltonian may be written

$$H_0 = h_0(1) + h_0(2),$$

where

$$h_0(i) = \frac{-\hbar^2}{2m}\nabla_i^2 - \frac{1}{4\pi\epsilon_0}\frac{Ze^2}{r_i} + u(\mathbf{r}_i), \quad i = 1, 2.$$

The single-particle Hamiltonian, $h_0(i)$, describes the motion of the ith electron in the average potential u.

5.1.1 Antisymmetric wave functions and the Pauli exclusion principle

The simplest wave function, which describes two electrons moving in an average potential, is the product function

$$\Phi = \phi_a(1)\phi_b(2), \tag{5.2}$$

where a stands for the quantum numbers of the first electron, and b stands for the quantum numbers of the second electron. If the single-electron wave functions satisfy the eigenvalue equations

$$h_0(1)\phi_a(1) = \epsilon_a\phi_a(1), \quad \text{and} \quad h_0(2)\phi_b(2) = \epsilon_b\phi_b(2),$$

then Φ will be an eigenfunction of the approximate Hamiltonian $H_0 = h_0(1) + h_0(2)$ corresponding to the sum of the single-electron eigenvalues $E_0 = \epsilon_a + \epsilon_b$. This is easy to prove. Operating on the wave function Φ with $H_0 = h_0(1) + h_0(2)$ and allowing each h_0 to act on the corresponding wave function ϕ, we get

$$H_0\Phi = [h_0(1) + h_0(2)]\phi_a(1)\phi_b(2) = [h_0(1)\phi_a(1)]\phi_b(2) + \phi_a(1)[h_0(2)\phi_b(2)] = [\epsilon_a + \epsilon_b]\phi_a(1)\phi_b(2). \tag{5.3}$$

Another eigenfunction of the approximate Hamiltonian H_0 can be obtained by interchanging the coordinates of the two electrons. Making the interchange $1 \leftrightarrow 2$ in the product function (5.2) gives

$$\Phi = \phi_a(2)\phi_b(1). \tag{5.4}$$

In the same way as before, this function can be shown to be an eigenfunction of H_0 corresponding to the eigenvalues $E_0 = \epsilon_a + \epsilon_b$.

The wave functions of an atom must have the property that no two electrons are in the same state. This property is called the *Pauli exclusion principle*. We can form a wave function satisfying the Pauli principle by taking the following antisymmetric combination of the wave functions (5.2) and (5.4)

$$\Psi = \frac{1}{\sqrt{2}}[\phi_a(1)\phi_b(2) - \phi_a(2)\phi_b(1)]. \tag{5.5}$$

The wave function (5.5) has the property that it changes sign if one interchanges the coordinates of the two electrons. (See Problem 2.) One can also confirm that the function Ψ will be equal to zero if a and b denote the same set of quantum numbers. The factor occurring before the square bracket in Eq. (5.5) ensures that the function Ψ is normalized. Since the wave function (5.5) is a sum of two terms, which are eigenfunctions of H_0 corresponding to the eigenvalue $E_0 = \epsilon_a + \epsilon_b$, Ψ must also have this property. The state (5.5) can be written in the form of a determinant

$$\Phi = \frac{1}{\sqrt{2}}\begin{vmatrix} \phi_a(1) & \phi_a(2) \\ \phi_b(1) & \phi_b(2) \end{vmatrix}.$$

These ideas can readily be generalized to an atom having N electrons. As for a two-electron atom, we assume that each electron moves independently of the other electrons in an average field due to the nucleus and the electrons. A system of N electrons moving in the attractive field of the nucleus and an average potential $u(\mathbf{r})$ due to the electrons can be described by the independent particle Hamiltonian

$$H_0 = h_0(1) + h_0(2) + \cdots + h_0(N),$$

where

$$h_0(i) = \frac{-\hbar^2}{2m}\nabla_i^2 - \frac{1}{4\pi\epsilon_0}\frac{Ze^2}{r_i} + u(\mathbf{r}_i), \quad i = 1, 2, \ldots, N.$$

As before, the single-particle Hamiltonians, $h_0(i)$, describe the motion of the ith electron in the average potential u.

The wave function of N electrons moving in an average field can be written as a linear combination of determinants

$$\Phi = \sqrt{(1/N!)}\begin{vmatrix} \phi_a(1) & \phi_a(2) & \ldots & \phi_a(N) \\ \phi_b(1) & \phi_b(2) & \ldots & \phi_b(N) \\ \ldots & \ldots & \ldots & \ldots \\ \phi_n(1) & \phi_n(2) & \ldots & \phi_n(N) \end{vmatrix}. \tag{5.6}$$

Here the symbol $N!$, which is called N factorial, is defined as $N! = 1 \cdot 2 \ldots N$. The wave function (5.6) is called a *Slater determinant*. An interchange of any two electrons amounts to interchanging two of the columns of the determinant and leads to a sign change. The wave function (5.6) is thus antisymmetric with respect to an interchange of two of the electrons. If any two of the single-electron wave functions are the same, two rows of the determinant will be identical, and the determinant must be equal to zero. This shows that the Pauli exclusion principle is satisfied. The states of a many-electron atom may thus be expressed in terms of Slater determinants.

5.1.2 The central-field approximation

The independent-particle model reduces the many-electron problem to independent single-electron problems. For atoms, we can generally make the additional assumption that the potential is spherically symmetric. This is called the *central-field approximation*. The h_0 operator for each electron then becomes

$$h_0 = \frac{-\hbar^2}{2m}\nabla^2 - \frac{1}{4\pi\epsilon_0}\frac{Ze^2}{r} + u(r), \tag{5.7}$$

where the potential $u(r)$ now depends only upon the distance of the electron from the nucleus.

As we have seen, the wave function for each electron satisfies the energy eigenvalue equation,

$$h_0\psi = \epsilon\psi. \tag{5.8}$$

The Schrödinger equation for each electron can be obtained by substituting (5.7) into Eq. (5.8) giving

$$\left[\frac{-\hbar^2}{2m}\nabla^2 - \frac{1}{4\pi\epsilon_0}\frac{Ze^2}{r} + u(r)\right]\phi = \epsilon\phi.$$

This equation differs from the Schrödinger equation for hydrogen by the term $u(r)$ which is the potential due to the electrons. As for the hydrogen atom, the wave function of an electron moving in a central-field can be written as the product of a radial function times a spherical harmonic and a spin function

$$\psi(r, \theta, \phi) = \frac{P_{nl}(r)}{r} Y_{lm_l}(\theta, \phi)\chi_{m_s}. \tag{5.9}$$

The function $P(r)$ satisfies the radial equation,

$$\left[-\frac{\hbar^2}{2m}\frac{d^2}{dr^2} + \frac{\hbar^2 l(l+1)}{2mr^2} - \frac{1}{4\pi\epsilon_0}\frac{Ze^2}{r} + u(r)\right]P_{nl}(r) = E\,P_{nl}(r), \tag{5.10}$$

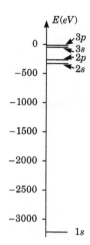

FIGURE 5.1 The energies of the occupied states of the argon atom.

which differs from the radial equation for hydrogen given in Chapter 4 only by the appearance of $u(r)$.

We would now like to discuss a feature of the energy levels of hydrogen which we have seen previously. Recall that the energy eigenvalues of hydrogen, which are illustrated in Fig. 4.5, depend upon the principal quantum number n but not on the orbital quantum number l. The $3s$ and $3p$ levels, for example, have the same energy. Other atoms do not have this property. The energies of the occupied orbitals of the argon atom obtained in an accurate calculation, to which we shall refer, are shown in Fig. 5.1. The energies of the $3s$ and $3p$ states of argon are -69.5 eV and -32.2 eV, respectively. The difference between the energies of electrons with the same value of n and different values of the quantum number of the orbital angular momentum l can be understood by considering the motion of the electrons in the charge distribution of the atom.

While the electron of a hydrogen atom always moves under the influence of one unit of charge, the effective nuclear charge for an electron in a many-electron atom increases as the electron approaches the nucleus passing through the charge cloud of other electrons which screen the nucleus. As illustrated by Fig. 4.3, electrons with a smaller value of the angular momentum come closer to the nucleus and see a larger effective nuclear charge. Since they move under the influence of a larger effective charge, the lower angular momentum states of a many-electron atom are more tightly bound.

5.2 Shell structure and the periodic table

As we have seen, we can suppose that the electrons in an atom move in a field which is spherically symmetric. While the energy of each electron in an atom depends upon the quantum numbers n and l, the energy of an electron moving in a spherically symmetric potential does not depend upon the azimuthal quantum numbers m_l and m_s, which define the z-component of the angular momentum vectors and thus determine the orientation of the angular momentum vectors in space. The energy of an electron moving in a spherically symmetric potential should not depend upon the orientation of the angular momentum vectors. The single-electron wave functions with the same value of n are said to form a *shell* and the single-electron wave functions with particular values of n and l are said to form a *subshell*. In the central-field approximation, the energy of an atom depends upon the n and l quantum numbers of electrons in each subshell. As for hydrogen, we shall denote the l-values $0, 1, 2, 3, 4, \ldots$ by the letters, $s, p, d, f, g, \ldots$. For instance, the state of an atom for which there are two electrons with the quantum numbers, $n = 1$ and $l = 0$, and two electrons with the quantum numbers $n = 2$ and $l = 0$, will be denoted simply $1s^2 2s^2$. The specification of the state of an atom by giving the number of electrons in each subshell is called the *electronic configuration*.

According to the Pauli exclusion principle, there can be only one electron in each single-electron state with the quantum numbers nlm_lm_s. Hence, a subshell with quantum numbers n and l can contain no more than $2(2l + 1)$ electrons. A subshell which has all orbitals occupied is said to be *filled* or *closed*, while a partially filled subshell is said to be *open*. The lowest energy for a particular atom can be obtained by successively filling the lowest-lying subshells. The electron configuration of the lowest state, which is called the *ground configuration*, would then consist of a number of closed subshells and at most one open subshell. This leads to the historically important *building-up principle* or *Aufbau principle* which was originally suggested by Bohr to explain the periodic table. Of course, as electrons are being added, the average central field changes. Also, the Coulomb interaction among the electrons causes departures from the central-field description. These factors lead

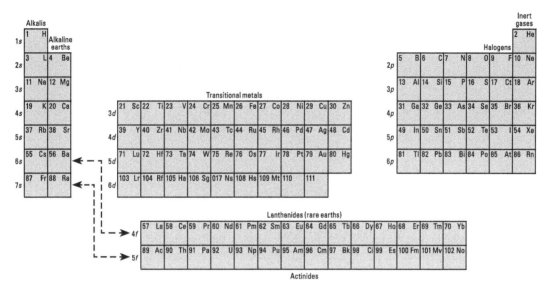

FIGURE 5.2 The periodic table of elements.

to some irregularities in the filling of the electron shells for heavier elements. The order and number of electrons in each subshell, however, is usually as follows

$$1s^2 2s^2 2p^6 3s^2 3p^6 4s^2 3d^{10}, \ldots .$$

The fact that the $4s$ subshell fills before the $3d$ can be understood by considering the values of the angular momentum of the two states. The $4s$ state has an angular momentum quantum number $l = 0$ and thus penetrates closer to the nucleus and is more tightly bound than the $3d$ state with angular momentum quantum number $l = 2$.

Using the above order of filling of the subshells of the atom, one can readily explain the appearance of the periodic table shown in Fig. 5.2. For the elements, H and He, the $1s$ subshell is being filled. The filling of the $2s$ subshell corresponds to the elements Li and Be, while the filling of the $2p$ subshell corresponds to the elements B through Ne. For the elements, Na and Mg, the $3s$ subshell is being filled, while the elements Al through Ar correspond to the filling of the $3p$ subshell. The transition elements for which the $3d$ subshell is being filled occur after the elements K and Ca for which the $4s$ subshell is filled. One may readily identify the ground configuration of light- and medium-weight elements using the order and the number of electrons for each shell given by the above sequence.

Example 5.1

Find the ground configuration of the titanium atom.

Solution

The neutral titanium atom has 22 electrons. These electrons would fill the $1s$, $2s$, $2p$, $3s$, $3p$, and $4s$ shells. The two remaining electrons can then be placed in the $3d$ shell. The ground configuration of titanium is thus

$$1s^2 2s^2 2p^6 3s^2 3p^6 4s^2 3d^2 .$$

In the periodic table shown in Fig. 5.2, the elements are arranged according to increasing nuclear charge which is shown in the upper left-hand corner of each square. The elements in the first column of the periodic table after hydrogen are called the *alkali* elements. These elements, which have a single electron outside filled shells, have the lowest ionization energies of the chemical elements. The nucleus of these atoms is effectively screened by the closed shells of electrons that surround it, and the outer electron is loosely bound by the atomic nucleus. An alkali atom easily contributes its outer, loosely-bound electron in chemical processes. As a result, these elements are very reactive, and they are also easily polarized by external fields. In contrast, the elements of the last column of the periodic table have closed shells. These elements, which are called the *rare* or *inert gases*, have very large ionization energies, and they are passive elements which do not take part readily in

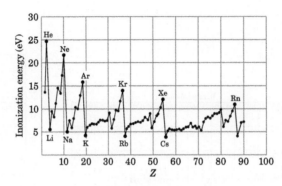

FIGURE 5.3 The ionization energy of neutral atoms. The nuclear charge is denoted by Z.

chemical processes. The elements in the second column of the periodic table are called the *alkali earth* elements, while the elements in the next to last column of the periodic table are called the *halogen* elements. The halogen elements, which need a single electron to complete their outer shell, are very reactive chemically. The ionization energy of the elements for Z up to 90 are illustrated in Fig. 5.3.

5.3 The LS term energies

The configurations of an atom can be described by specifying which single-electron states are occupied. As an example, we consider a np^2 configuration for which there are two electrons with the principle quantum number n and orbital angular momentum quantum number $l = 1$. Since each p-electron can have the m_l-values, $1, 0, -1$, and the m_s-values, $1/2, -1/2$, there are six ways of putting the first electron into the p-shell. The second p-electron cannot be put into the same state, and hence there are five ways of putting the second electron into the p-shell. Since the two electrons are identical, the order of our placing the electrons into the p-shell is unimportant, and, hence, there are a total of $(6 \cdot 5)/2 = 15$ distinct states of the configuration np^2. While these states all correspond to the same energy within the central-field approximation, the noncentral part of the Coulomb interaction between the electrons splits the np^2 configuration into a number of energy levels.

The noncentral part of the Coulomb interaction may be described by the operator V_{nc}, which is equal to the difference between the Coulomb interaction between the electrons and the interaction of the electrons with an average central-field. For a two-electron atom, the operator V_{nc}, which is responsible for the splitting of the energy levels of the configuration, is

$$V_{\mathrm{nc}} = \frac{1}{4\pi\epsilon_0} \frac{e^2}{r_{12}} - u(r_1) - u(r_2). \tag{5.11}$$

The potential $u(r)$, which is spherically symmetric, shifts the energy of all of the states having a definite value of n and l by the same amount. For this reason, the splitting of a configuration is due entirely to the Coulomb interaction between the electrons, which corresponds to the first term on the right-hand side of Eq. (5.11).

We consider now the effect of the Coulomb interaction upon a single configuration. The energy levels and the wave functions of a configuration can be described using quantum numbers of the total angular momenta obtained by adding the angular momenta of the individual electrons. The orbital angular momenta of the electrons can be combined to form a total orbital quantum number L, and the spin of the electrons can be combined to form a total spin angular momentum S. The Coulomb interaction among the electrons splits a configuration into a number of distinct energy levels. These levels having definite values of L and S are called *LS terms*.

5.4 Configurations of two electrons

In the following, we shall distinguish between configurations of electrons having the same quantum numbers n and l and those for which the quantum numbers n and l are different. Two electrons having the same principle quantum number n and the same orbital angular momentum l are said to be *equivalent*, while electrons having different values of n or different values of l are said to be *nonequivalent*. We consider first the case of two equivalent electrons.

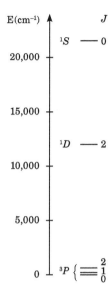

FIGURE 5.4 The energy levels of the $1s^2 2s^2 2p^2$ configuration of the carbon atom.

5.4.1   Configurations of equivalent electrons

The possible values of the total orbital angular momentum L and the total spin angular momentum S may be obtained by using the rule for the addition of angular momenta given in Section 4.3.4. For a configuration of two equivalent electrons, though, the possible values of L and S are limited by the following rule.

Condition upon L and S for a configuration of equivalent electrons

$S + L$ must be even for a configuration of two equivalent electrons

---

It is possible to show that only the states for which $S + L$ is even satisfy the Pauli exclusion principle.

As an example of the energy levels of a configuration of equivalent electrons, we consider an np^2 configuration for which there are two electrons with the principle quantum number n and the orbital quantum number $l = 1$.

Example 5.2

Find the possible S and L values for the np^2 configuration.

Solution

The argument used in Example 4.3 applies to any configuration of two electrons. For two electrons, the quantum number of the total spin S can have the values 0 or 1. The quantum number of the total orbital angular momentum L may be obtained using the rule for the addition of angular momenta with $j_1 = 1$ and $j_2 = 1$. The maximum value of L is equal to $j_1 + j_2$ and is hence equal to two, while the minimum value of L is $|j_1 - j_2|$ and is equal to zero. According to the rule for the addition of angular momenta, the value one which falls between the maximum and minimum values of L is also an acceptable value of L. We thus have $L = 2, 1, 0$. Since this configuration consists of equivalent electrons $S + L$ must be even. The $2p^2$ configuration thus has the following LS terms: (S=1,L=1), (S=0,L=2), and (S=0,L=0).

The LS terms of a configuration are usually given in the spectroscopic notation in which one uses a capital letter to denote the total orbital angular momentum and gives the value of $2S + 1$ as a raised prefix as follows: ^{2S+1}L. The LS terms of the np^2 configuration are thus denoted ^3P, ^1D, and ^1S, respectively. We leave as an exercise the problem of determining the LS terms of the ns^2, nd^2, and nf^2 configurations. (See Problem 8.)

The lowest energy levels of the carbon atom are shown in Fig. 5.4. These energy levels all belong to the $1s^2 2s^2 2p^2$ configuration. The filled $1s$ and $2s$ shells are spherically symmetric and do not affect the relative position of the energy levels.

The three lowest levels are associated with the ^3P state, and the next higher levels correspond to the ^1D and ^1S states. The separation between the three LS terms of the configuration is due to the Coulomb interaction between the two $2p$ electrons, while the fine-structure splitting of the ^3P term into sublevels with distinct values of J is mainly due to the spin-orbit interaction.

The vertical scale in Fig. 5.4 is given in units of cm^{-1}. To see how this way of labeling energy levels arises, we recall that the energy of a photon is related to the wavelength of light by Eq. (I.31) of the introduction. Solving this equation for $1/\lambda$, which is the wave number of the light, we have

$$\frac{1}{\lambda} = \frac{E}{hc} \, .$$

Books on lasers and spectroscopy generally give the values of $1/\lambda$ corresponding to energy levels in units of cm^{-1}. The value of $1/\lambda$ which corresponds to an energy of one electron volt may be obtained by substituting $E = 1$ eV and the value of hc given in Appendix A into the above equation giving

$$\frac{1}{\lambda} = \frac{1 \, \text{eV}}{1239.8424 \, \text{eV} \cdot \text{nm}} = 8.06554 \times 10^{-4} \, \text{nm}^{-1} = 8065.54 \, \text{cm}^{-1} \, . \tag{5.12}$$

A photon energy of one eV thus corresponds to light with a wave number of $8065.54 \, cm^{-1}$.

In addition to the electrostatic splitting of the LS terms of a configuration, the spin-orbit interaction considered in Chapter 4 can split the LS terms of a configuration into sublevels with different values of J. For example, the ^3P term of the np^2 configuration, which we have just considered, has $S = 1$ and $L = 1$. Using the rule for the addition of angular momentum given in Section 4.3.4, one may readily show that S and L may couple together to form a total angular momentum with quantum number J equal to 2, 1, or 0. As shown in Fig. 5.4, the spin-orbit interaction splits the ^3P term into three sublevels $J = 2$, $J = 1$, and $J = 0$. The ^1D terms of the np^2 configuration has $S = 0$ and $L = 2$. Using the rule for adding angular momenta, we find that the total angular momentum J of this term must be equal to 2. Similarly, the ^1S term, for which $S = 0$ and $L = 0$, can only have J equal to zero. The ^1D and ^1S terms, each of which correspond to a single value of J, are not split by the spin-orbit interaction. From the splitting of the ^3P term shown in Fig. 5.4, one can see that the $J = 0$ sublevel lies lowest. We also note that for the np^2 configuration the $2S + 1$ value for each LS term gives the number of sublevels into which the term is split by the spin-orbit interaction. Since this is usually the case, the $2S + 1$ value for a particular term is referred to as the *multiplicity* of the term.

The lowest LS term of a configuration of equivalent electrons in LS coupling is given by *Hund's rule*. According to this rule, the lowest lying LS term has the maximum value of S and hence the largest multiplicity $2S + 1$. If there are several terms with the maximum value of S, the state with the maximum value of L lies lowest. Since the ^3P term is the only LS term of the $2p^2$ configuration for which $S = 1$, this term lies lowest. Among the singlet states, the term ^1D with larger L lies below ^1S, but Hund's rule can not always be used in this way to determine the ordering of the LS terms. For example, the lowest lying configuration of the zirconium atom with Z equal to 40 is the $4d^2 5s^2$ configuration for which the partially filled $4d$ shell lies inside an outer closed $5s$ shell. The allowed terms of the $4d^2$ configuration are ^1S, ^1D, ^1G, ^3P, and ^3F. According to Hund's rule, the ground term is a triplet for which L is larger. Of the two triplets the ^3F term has the larger L-value and has been found experimentally to lie lowest. Among the singlet terms, though, the ^1D lies below the ^1G term.

For a configuration of equivalent electrons, the ordering of the different J sublevels depends upon whether the shell is less or more than half filled. The sublevel with the minimum value of J lies lowest if the shell is less than half filled, while the maximum value of J lies lowest if the shell is more than half filled. Since six electrons are needed to fill a p-shell, the $2p^2$ configuration has a p-shell which is less than half filled. As shown in Fig. 5.4, the $J = 0$ sublevel of the ^3P term in the $2p^2$ configuration of carbon lies below the $J = 1$ and $J = 2$ sublevels. The $2p^4$ configuration of oxygen, which has two vacancies in the $2p$-shell, has the same LS terms as a $2p^2$ configuration, and in this case the ^3P term also lies lowest. For the $2p^4$ configuration, however, the $J = 2$ sublevel of the ^3P term lies below the $J = 1$ and $J = 0$ sublevels. The ordering of the J sublevels is reversed.

The spin-orbit interaction causes the splitting of the J-levels of the ^3P term of carbon and mixes the $J = 2$ sublevels of the ^3P and ^1D terms. While the mixing of different LS terms is very small for carbon, this effect increases in importance as the nuclear charge and the spin-orbit interaction increase. The states of heavier atoms generally consist of components with different values of L and S.

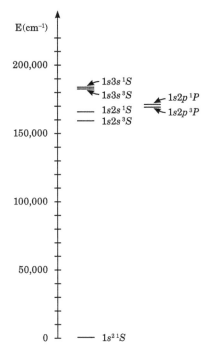

FIGURE 5.5 The lowest energy levels of the helium atom.

5.4.2 Configurations of two nonequivalent electrons

Two nonequivalent electrons have different values of the principle quantum number n or the orbital angular momentum quantum number l. The possible values of L and S for a configuration of nonequivalent electrons can be obtained by using the rule for the addition of angular momentum. There is no additional condition in this case which limits the number of possible LS terms. As an example of a configuration with nonequivalent electrons, we consider a configuration having one electron with principle quantum number n and orbital angular momentum quantum number $l = 0$ and an additional electron with principle quantum number n' and orbital angular momentum quantum number l. Configurations of this kind may be denoted generally by $ns\, n'l$.

Example 5.3

Find the possible S and L values for a $ns\, n'l$ configuration.

Solution

As for configurations of two equivalent electrons considered previously, the spins of the two electrons of a $ns\, n'l$ configuration may be coupled together to produce states with $S = 0, 1$. The total orbital angular momentum quantum number L may be obtained using the rule for the addition of angular momenta with $j_1 = 0$ and $j_2 = l$ giving $L = l$. The sum of the total spin quantum number S and the total orbital quantum number L need not be even for nonequivalent electrons. Hence, a $ns\, n'l$ configuration has two LS terms, one for which $S = 1$ and $L = l$ and another for which $S = 0$ and $L = l$.

The lowest energy levels of the helium atom are shown in Fig. 5.5. All of the lowest levels of helium have one electron in the $1s$ level and a single additional electron. The ground configuration is $1s^2$, while the next low-lying configuration is $1s2s$. As shown in Example 5.3, the $1s2s$ configuration has two LS terms, the ^3S and ^1S terms. The next two configurations are $1s2p$ with the terms, ^3P and ^1P, and the $1s3s$ configuration with the terms, ^3S and ^1S. As for the energy levels of carbon shown in Fig. 5.4, the vertical scale in Fig. 5.5 is given in cm^{-1}.

The description of LS terms given in this section will be used in the next chapter in describing laser transitions and in describing laser cooling and trapping of atoms in magnetic fields. Even for medium- and heavy-weight atoms for which the spin-orbit interaction significantly mixes states having different values of S and L, the state of the atom is still often specified by giving the LS symmetry of the dominant component of the state.

5.5 The Hartree-Fock method

There are many similarities between the central-field description of many-electron atoms and the description of the hydrogen atom given in the previous chapter. In each case, the wave function of the electrons is of the form (5.9) having a radial part which we have written $P_{nl}(r)/r$, an angular part equal to a spherical harmonic $Y_{lm_l}(\theta, \phi)$, and a spin function. For both the central-field model and the hydrogen atom, the energy of an electron is independent of the azimuthal quantum number m_l and m_s which determine the orientation of the orbital and spin angular momentum of the electron. The spherical harmonics $Y_{lm_l}(\theta, \phi)$ are well-defined functions of the angular variables which are the same for any atom. In contrast, the radial functions $P_{nl}(r)$, which vary depending upon the nuclear charge and the number of electrons, have all of the special features that we would associate with a particular atom.

Equations for the radial functions of a many-electron atom may be obtained by requiring that the energy of the atom be stationary with respect to small variations in the form of the wave functions. This condition called the variational principle leads to a set of equations known as the *Hartree-Fock equations* that can be solved for the radial wave functions $P_{nl}(r)$.

In atomic units, the Hartree-Fock equations for the $1s^2 2s^2 \, ^1S$ state of beryllium are

$$\left[-\frac{1}{2}\frac{d^2}{dr^2} - \frac{Z}{r} + J_{1s} + (2J_{2s} - K_{2s}) \right] P_{1s} = \epsilon_{1s} P_{1s} \tag{5.13}$$

$$\left[-\frac{1}{2}\frac{d^2}{dr^2} - \frac{Z}{r} + (2J_{1s} - K_{1s}) + J_{2s} \right] P_{2s} = \epsilon_{2s} P_{2s} \tag{5.14}$$

Each term in the Hartree-Fock equations correspond to a particular physical interaction. The first term in Eq. (5.13) corresponds to the kinetic energy of the $1s$ electron and the second term corresponds to the potential energy of the $1s$ electron due to the nucleus. The third term corresponds to the potential energy of the $1s$ electron due to the charge distribution of the other $1s$ electron and the last two terms correspond to the electrostatic potential energy of the $1s$ electron due to the charge distribution of the $2s$ electrons. The product of the operators, J_{1s} and K_{1s} times a radial function P_{nl} are defined by the equations

$$J_{1s} P_{nl}(r_1) = \int P_{1s}(r_2)(1/r_{12}) P_{1s}(r_2) dr_2 \times P_{nl}(r_1) \tag{5.15}$$

$$K_{1s} P_{nl}(r_1) = \int P_{1s}(r_2)(1/r_{12}) P_{nl}(r_2) dr_2 \times P_{1s}(r_1) \tag{5.16}$$

For Eq. (5.15), the potential energy of the $1s$ wave function is multiplied times the nl wave function. J_{1s} is called the *direct term* or the *direct energy*. In the second of these two equations, the P_{nl} is drawn inside the integral and a P_{1s} function comes out. The second term is called the *exchange term* or the *exchange energy*.

Because the Coulomb interaction is independent of spin, the exchange interaction can only effects electrons with the same spin orientation. With this in mind, we can easily understand the structure of the Hartree-Fock equations (5.13) and (5.14). If we suppose that one electron in each shell has spin up and the other electron has spin down, the third term in Eq. (5.13) is the direct interaction of the $1s$ electron with the other $1s$ electron. The fourth term in the equation gives the direct interaction between the $1s$ electron and the two $2s$ electrons while the fifth tern gives the exchange interaction between the $1s$ electron and the $2s$ electron with the same spin orientation.

The Hartree-Fock equations for an atom have the same form as the radial equation (5.10) with a specific form for the potential energy function $u(r)$ which depends upon the wave functions of the occupied single-electron states. Since the equations for the radial wave functions provided by the Hartree-Fock theory depend upon the wave functions themselves, these equations can only be solved using an iterative approach. One first makes initial estimates of the radial wave functions, then finds the potential energy of each electron moving in the field of the nucleus and the other electrons. The program then obtains new radial wave functions by solving the Hartree-Fock equations and uses these wave functions to reevaluate the potential energy terms in the equations. The iteration process is continued until self consistency is achieved with the most recent solutions of the Hartree-Fock equations being identical to the previous solutions.

The wave functions we shall use for the examples in this chapter were obtained using an interactive Hartree-Fock program developed by Charlotte Froese Fischer, who early in her career worked as a research assistant of Douglas Hartree at Cambridge University in England. This program is described in a recent book by Fischer, Brage, and Johnsson and is available at the web site *http://nlte.nist.gov/MCHF/hf.html*. In the following section, we shall describe a computer applet, which serves as an interface to Charlotte Fisher's program. Using this applet, one can view the wave functions of many-electron atoms and discover for oneself many striking and interesting features of the periodic table.

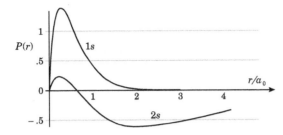

FIGURE 5.6 The Hartree-Fock $1s$ and $2s$ radial wave function $P(r)$ of beryllium.

Before discussing the properties of radial functions $P_{nl}(r)$ for different atoms, we first recall some of the general properties of the radial wave functions of the hydrogen atom. Near the nucleus, the functions $P_{nl}(r)$ of hydrogen all have the general form

$$P_{nl}(r) = Ar^{l+1}$$

with the constant A and the orbital angular momentum quantum number l determining the behavior of the radial function. As a consequence, wave functions with larger values of l rise more slowly near the origin and are localized farther from the nucleus than wave functions for lower values of l. One can get some insight into how functions of the form r^{l+1} vary near the origin by comparing the functions, r and r^2. The function r rises linearly near the origin, while the function r^2 having a parabolic form rises more slowly.

The radial wave functions of many-electron atoms all have the same dependence upon r near the nucleus as do the hydrogen wave functions, and they all fall off exponentially for large values of r as do the hydrogen wave functions. Another common feature of the radial functions of hydrogen and of many-electron atoms is that the region of space where the wave function of an electron has a significant value depends upon the energy of the electron with the wave functions of more tightly bound electrons being localized closer to the nucleus and the wave functions of less tightly bound electrons having a greater spatial extension.

The Hartree-Fock radial wave functions for beryllium are shown in Fig. 5.6. These functions have the same general appearance as the $1s$ and $2s$ radial functions for hydrogen shown in Fig. 4.8. However, the wave functions for beryllium do not have the sharp maxima and minima of their hydrogenic counterparts. The difference in the appearance of the wave functions of beryllium can be attributed to the screening of the beryllium nucleus by the electrons. While the electron of a hydrogen-like ion always moves under the influence of Z units of charge, the effective nuclear charge for an electron of the beryllium atom depends upon the distance of the electron from the nucleus. As an electron in beryllium approaches the nucleus, it passes through the charge cloud of other electrons which screen the nucleus and the effective nuclear charge for the electron increases. Hence, the wave functions of beryllium have a more diffuse appearance than the wave functions of hydrogen.

The screening of the nucleus by the electrons is an effect that occurs generally for atoms. Fig. 5.7 shows the Hartree-Fock $3p$ radial function for the ground configuration of the argon atom. In Fig. 5.7(A), the Hartree-Fock wave function is compared with the $3p$ function of a hydrogen-like ion with a value of Z chosen so that the function matches the Hartree-Fock function in the tail region. This corresponds to a nuclear charge of 6. Fig. 5.7(B) compares the same Hartree-Fock function with another hydrogenic function for which the nodal points match. The outer portion of the second hydrogenic function, which corresponds to a nuclear charge of 12, completely fails to represent the Hartree-Fock function in the tail region. By comparing these two figures, one can see that the effective nuclear charge of argon increases as an electron approaches the nucleus.

5.5.1 The Hartree-Fock applet

Our Hartree-Fock applet, which is shown in Fig. 5.8(A), can be found at the web site for this book. The Hartree-Fock applet was developed by Charlotte Fischer and Misha Saparov in 1997 and modified for this textbook by Simon Rochester and myself. The applet has the Hartree-Fock program developed by Charlotte Fischers and her collaborators working in the background and a Java interface. The Hartree-Fock equations solved with directives coming from the applet are obtained by applying the variational principle to an expression for the energy of the atom. As can be seen in Fig. 5.8(A), the applet comes up showing a periodic table with the atom lithium highlighted. In the lower left-hand corner of the web page, the atomic number of lithium ($Z = 3$) and the ground state of lithium ($1s^2 2s^1{}^2S$) are shown. The $1s$ shell of lithium is filled,

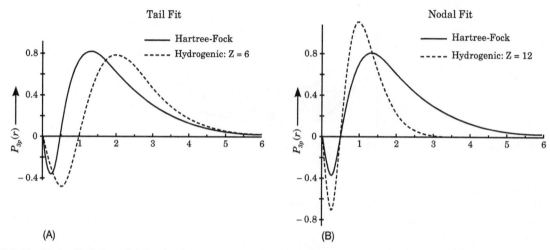

FIGURE 5.7 The Hartree-Fock $3p$ radial function for the ground state of the argon atom. (A) The Hartree-Fock wave function is compared with a hydrogenic $3p$ function with a value of Z chosen so that the function matches the Hartree-Fock function in the tail region. (B) The same Hartree-Fock function is compared with another hydrogenic function for which the nodal points match.

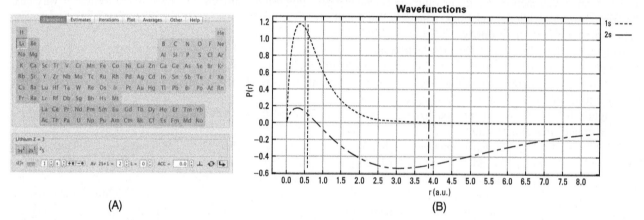

FIGURE 5.8 (A) The Hartree-Fock applet with the Li atom highlighted. The nuclear charge of Lithium, $Z = 3$, and the ground state of lithium, $1s^2 2s^1 S$, are shown in the lower left-hand corner. (B) The radial wave functions $P(r)$ for the $1s$ and $2s$ electrons of lithium. The average distances of the electrons from the nucleus are represented in this figure by vertical lines of the same kind as the corresponding wave functions.

and there is a single electron in the $2s$ shell. The total spin and orbital angular momentum of the filled $1s$ shell are equal to zero. Hence, the spin and orbital angular momentum of the ground state of lithium is entirely due to the $2s$ electron, which has a spin of one-half and an orbital angular momentum equal to zero. This LS term, which is designated ^2S in spectroscopic notation, is shown in them lower left-hand corner of the applet.

Hartree-Fock calculations can be performed for a particular LS term or for the average energy of configurations. For a configuration having several LS terms, one can perform a Hartree-Fock calculation for the average energy by clicking on the "Av" tab in the bottom of the page. Since the lowest configuration of lithium has a single LS term, the average energy of the configuration is the same as the energy of the single LS term. The program does not then allow one to perform a Hartree-Fock calculation for the average energy. If one were to click the "Av" tab in this case, the red arrow in the lower right-hand corner of the screen becomes dark – implying that a Hartree-Fock calculation can not be performed. One can get out of this mode by simply clicking on the "Av" tab again with the arrow in the lower right corner then becoming red again. When one clicks on the red arrow for lithium, the wave functions shown in Fig. 5.8(B) appears on the screen. The lines corresponding to the $P(r)$ functions of the $1s$ and $2s$ electrons each have distinctive colors on the applet. The wave functions of the $1s$ and $2s$ subshells are distinguished in the black-white image shown in Fig. 5.8(B) by using different kinds of lines to correspond to the $1s$ and $2s$ wave functions. The curve for the $1s$ function is dashed while the curve for the $2s$ function has dots and dashes. Vertical lines of the same kinds as the corresponding wave functions indicate the average value of r for each orbital. One thus immediately sees that the inner $1s$ orbital has an average value of r equal to about 0.6 times the Bohr radius (one atomic unit) while the $2s$ wave function has an average value of r equal to about 3.8 times

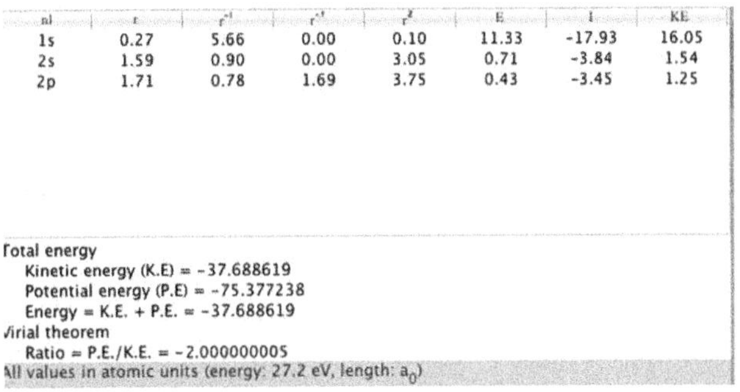

nl	r	r^{-1}	r^{-3}	r^2	E	i	KE
1s	0.27	5.66	0.00	0.10	11.33	-17.93	16.05
2s	1.59	0.90	0.00	3.05	0.71	-3.84	1.54
2p	1.71	0.78	1.69	3.75	0.43	-3.45	1.25

Total energy
   Kinetic energy (K.E) = -37.688619
   Potential energy (P.E) = -75.377238
   Energy = K.E. + P.E. = -37.688619
Virial theorem
   Ratio = P.E./K.E. = -2.000000005
All values in atomic units (energy: 27.2 eV, length: a_0)

FIGURE 5.9 Web page with the average value of the powers of r and the kinetic and potential energies for the carbon atom.

the Bohr radius. More accurate values of the average value of r for the 1s and 2s orbitals can be obtained by clicking the "Averages" tab on the top of the web pages. The average value of r for the 1s wave function is 0.573 atomic units, while the average value for the 2s wave function is 3.873 atomic units.

To provide an illustration of the kind of information that can be obtained using the Hartree-Fock applet, we shall now show in detail how to use the applet to get information about the carbon atom. On the web page showing a periodic table, one chooses carbon by clicking on the letter "C" in the second row of the periodic table. We would like to perform Hartree-Fock calculations for carbon to study the electrostatic and magnetic structure of the energy levels of the atom. Since the spin-orbit interaction only causes a splitting of the lowest ^3P term of carbon, we would like to run the applet for this particular LS state. This can be done by clicking on the red arrow in the lower right-hand corner of the web page with the periodic table. The wave functions of carbon then come up on the screen with the 1s being red, the 2s being blue, and the 2p being green. There are vertical lines with the colors of the three electrons indicating their average value of r. Of course, the 1s electron is closer to the nucleus than the other two electrons. One can then click on the *Averages* tab on the top of the web page for carbon to obtain the web page shown in Fig. 5.9. The average distance r from the nucleus for each of the shells is given in the second column of the table in units of the Bohr radius a_0, which is the atomic unit of distance. Notice that the average distance of a 1s electron from the nucleus is only $0.27\,a_0$, while the distances of the 2s and 2p electrons from the nucleus are $1.59\,a_0$ and $1.71\,a_0$, respectively.

We now consider other properties of the electrons in carbon. Column six of the table shown in Fig. 5.9 gives the energy of the electrons in atomic units. We recall that the atomic unit of energy is the absolute value of the potential energy of two electrons or of an electron and a proton separated by a distance equal to the Bohr radius. This unit of energy, which is called the *hartree*, is equal to 27.2 electron volts. The single-electron energy of the 1s electron of carbon is equal to 11.33 atomic units (a.u.) or 308 eV, while the single-electron energies of the 2s and 2p electrons are 0.71 a.u. and 0.43 a.u., respectively. The 1s electrons are much more tightly bound than the 2s and 2p electrons. The column at the far right of the table shown in Fig. 5.9 gives the average kinetic energies of the electrons in the carbon atom, while the next column to the left gives the average value of the kinetic energy plus the potential energy due to the nucleus – the so-called single electron energy.

In the lower portion of Fig. 5.9, the total energy of the electrons and their total kinetic and potential energies are shown. At the very bottom of the table, the ratio of the potential and the kinetic energy are also shown. According to the *virial theorem*, which applies generally to particles moving in a potential field, the ratio of the potential and the kinetic energies of the particles should be equal to minus two. The fact that this ratio is so close to minus two for carbon is an indication that the Hartree-Fock calculation converged properly.

One can obtain more information about the properties of the carbon atom by clicking the *Other* tab at the top of the web page. The web page which then comes up is shown in Fig. 5.10. The first set of entries enables one to calculate the average value of r^k for any one of the orbitals occupied in the $1s^2 2s^2 2p^2$ configuration of carbon. We have chosen the orbital to be the 2p and the power $k = 1$ to calculate the average value of r for the 2p electron. Clicking on the arrow to the right we find that the average value of r for the 2p is equal to 1.714 a.u., which is consistent with the average value of r given in Fig. 5.9. Similarly, by setting $k = -3$ and clicking on the arrow to the right, we find that the average value of $1/r^3$ for the 2p electron is equal to 1.692.

The second set of places to enter data in Fig. 5.10 enable one to calculate the values of the two-electron *Slater integrals* which determine the strength of the Coulomb interaction between the electrons of an atom. For a configuration of equivalent electrons, the Slater integrals are defined by the equation

$$F^k(a, b) = \int \int P_a(r_1) P_b(r_2) \frac{r_<^k}{r_>^{k+1}} P_a(r_1) P_b(r_2)\, dr_1 dr_2, \tag{5.17}$$

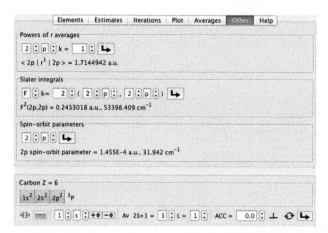

FIGURE 5.10 The average value of powers of r and the two-electrons Slater integrals, $F^k(2p, 2p)$ and spin-orbit constant ζ for the carbon atom.

where $r_>$ denotes the greater of the two radial distances, r_1 and r_2, and $r_<$ denoted the lesser of the two radial distances.

The three LS levels of the $1s^2 2s^2 2p^2$ configuration of carbon are shown in Fig. 5.4. Using the formulas given in Appendix F for a configuration of two equivalent p electrons, the contribution of the Coulomb interactions to the energy of these three levels is given by the following three equations

$$E(^1S) = F^0(2p, 2p) + \frac{10}{15} F^2(2p, 2p)$$

$$E(^1D) = F^0(2p, 2p) + \frac{1}{25} F^2(2p, 2p) \qquad (5.18)$$

$$E(^3P) = F^0(2p, 2p) - \frac{5}{25} F^2(2p, 2p)$$

We note that the Slater integer $F^0(2p, 2p)$ shifts all of the LS terms by the same amount, while the Slater integral $F^2(2p, 2p)$ is responsible for the splitting of the energy levels. In addition to the Coulomb interaction between the $2p$ electrons, the atomic Hamiltonian contains other terms corresponding to the kinetic and potential energy of the electrons and the interact of the $2p$ electrons with electrons in closed shells. Since these additional effects all shift the energy of the LS terms by the same amount, they can be included in the expressions for the LS energies by replacing the Slater integral $F^0(2p, 2p)$ in Eq. (5.18) by another constant E^0.

In atomic structure calculations, it is often convenient to use the average energy rather than the energies of the particular LS terms. The average energy for the $2p^2$ configuration can be obtained by taking a weighted average of the three LS terms. We have

$$E_{\text{Av}} = \left[9 \times E(^3P) + 5 \times E(^1D) + 1 \times E(^1S) \right] / 15 = E^0 - \frac{2}{25} F^2(2p, 2p). \qquad (5.19)$$

The equations for the term energies of the $1s^2 2s^2 2p^2$ configuration and the equation for the average energy can now be summarized in the following way:

$$E(^1D) = E^0 + \frac{a}{25} F^2(2p, 2p), \qquad (5.20)$$

where $a = 10, 1, -5$ for the ^1S, ^1D, and ^3P states, and $a = -2$ for the average energy of the configuration.

To determine the Slater integral $F^2(2p, 2p)$, which is responsible for the splitting of the $2p^2$ configurations, we have chosen $k = 2$ in the second set of entries on the web page shown in Fig. 5.10, and we have chosen the orbital to be the $2p$. By clicking on the arrow to the right of the second set of entries, we find that the Slater integral $F^2(2p, 2p)$ is equal to $53,398.409$ cm^{-1}. Similarly, to obtain the value of the spin-orbit constant for the $2p^2$ configuration, we have again chosen the orbital to be the $2p$ in the third row of entries on the web page shown in Fig. 5.10 and clicked on the arrow on the right to obtain the spin-orbit constant $\zeta = 31.94$ cm^{-1}.

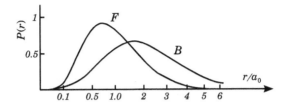

FIGURE 5.11 The Hartree-Fock $2p$ wave functions for the boron and florine atoms.

5.5.2 The size of atoms and the strength of their interactions

Using the Hartree-Fock applet, we have performed similar calculations for the other atoms in the second row to the periodic table obtaining the values given in Table 5.1.

TABLE 5.1 The average values of r and $1/r^3$, the Slater integral $F^2(2p, 2p)$, and the spin-orbit constant $\zeta(2p)$ for the elements boron through fluorine. The values of r and $1/r^3$ are given in atomic units, while the values of $F^2(2p, 2p)$ and ζ are given in cm^{-1}.

	$B(5)\,2p$	$C(6)\,2p^2$	$N(7)\,2p^3$	$O(8)\,2p^4$	$F(9)\,2p^5$
$<r>$	2.205	1.714	1.410	1.232	1.085
$<1/r^3>$	0.776	1.692	3.100	4.974	7.545
$F^2(2p, 2p)$	...	53,398	64,837	73,756	83,588
$\zeta(2p)$	10.12	31.94	75.58	148.52	266.50

The average value $<r>$ shown in Table 5.1 gives the average distance of the $2p$ electron from the nucleus in units of the Bohr radius a_0. As one goes from one atom to the next in this row of the periodic table, the nuclear charge increases and the $2p$ electrons are drawn in toward the nucleus. The average value of r decreases as one moves along the row and the distance between the two electrons also decreases. As a result, the average value $1/r^3$ and the magnitude of the Coulomb interactions between the $2p$ electrons increase. The average value of $1/r^3$ increases by about a factor of 10 for the elements of this series, while the spin-orbit parameter ζ increases by a factor of 26. The large increase in the spin-orbit parameter is due to the increase in the average value of $1/r^3$ and the decrease of the screening of the other electrons as the $2p$ electrons are drawn in toward the nucleus.

Fig. 5.11 shows the Hartree-Fock radial $2p$ functions for the boron and fluorine atoms. The average value of r is $2.2\,a_0$ for boron and $1.1\,a_0$ for fluorine, where as before a_0 denotes the Bohr radius. The fluorine atom, which has 9 electrons, is much smaller than the boron atom, which has only five electrons. The same effect occurs for the other rows of the periodic table. The first atom of each row of the periodic table has an electron in a shell which is not occupied for lighter elements. The first atom of each row is thus a good deal larger than the last atom of the previous row. However, the size of the atoms decrease as one goes from one atom to the next along any row of the periodic table. The average value r for the single $3p$ electron of aluminum is equal $3.4\,a_0$, while for chlorine, which has five $3p$ electrons, the average value of r for the $3p$ is $1.8\,a_0$.

The increase in the Coulomb interaction between the $2p$ electrons as one goes along this row of the periodic table causes the LS term structure of the atoms to expand. Fig. 5.12 shows the experimentally determined positions of the LS terms of the $2p^2$ configuration of carbon, the $2p^3$ configuration of nitrogen and the $2p^4$ configuration of oxygen. We note that the $2p^2$ configuration of carbon extends over a range of 21,600 cm^{-1}, while the $2p^3$ configuration of nitrogen extends over 28,800 cm^{-1}, and the $2p^4$ configuration of oxygen extends over 33,700 cm^{-1}. The small splitting of the LS terms due to the spin-orbit interaction is neglected in Fig. 5.12. In keeping with the ordinary spectroscopic notation, the ^4S term of nitrogen has $S = 3/2$ and $L = 0$, while the ^2D state has $S = 1/2$ and $L = 2$, and the ^2P state has $S = 1/2$ and $L = 1$.

As a second example of how calculations can be carried using the Hartree-Fock applet we consider the $1s2p$ configuration of helium. The experimentally observed splitting of the LS terms of this configuration is illustrated in Fig. 5.5. There are two LS terms – ^1P and ^3P – with the triplet state lying lowest. Using the formulas given in Appendix F for a $nsn'l$ configuration, the contributions of the Coulomb interactions to the energy of these two states are given by the

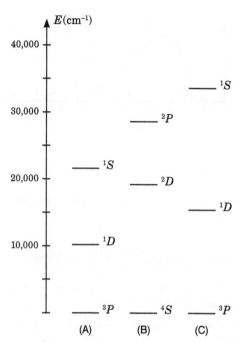

FIGURE 5.12 The lowest lying LS terms of (A) the $2p^2$ configuration of carbon, (B) the $2p^3$ configuration of nitrogen, and (C) the $2p^4$ configuration of oxygen.

following equations

$$E(^1P) = F^0(1s, 2p) + \frac{1}{3}G^1(1s, 2p) \tag{5.21}$$

$$E(^3P) = F^0(1s, 2p) - \frac{1}{3}G^1(1s, 2p), \tag{5.22}$$

where the exchange integral $G^1(1s, 2p)$ is defined

$$G^k(1s, 2p) = \int \int P_{is}(r_1) P_{2p}(r_2) \frac{r_<^k}{r_>^{k+1}} P_{2p}(r_1) P_{is}((r_2) \, dr_1 dr_2. \tag{5.23}$$

Unlike the Slater integral $F^k(a, b)$ defined by Eq. (5.17), the order of the two radial functions are reversed on the right. The Slater integral F^k introduced before is called a *direct* integral while the integral G^k introduced here is called an *exchange* integral. The splitting of the ^1P and ^3P terms is due to the single exchange integral $G^1(1s, 2p)$.

As before, we can define an average energy by taking a weighted average of the energies of the two states. We have

$$E_{Av} = \left[3 \times E(^3P) + 1 \times E(^1P) \right] / 4 = E^0 - \frac{1}{6}G^1(1s, 2p), \tag{5.24}$$

where as before the constant E^0 includes the effect of the Slater integral $F^0(1s, 2p)$ and of the single-particle energies of the $1s$ and $2p$ which shift the two levels of the $1s2p$ configuration by the same amount.

To give a better understanding of the splitting of the ^3P and ^1P levels of helium, we have done Hartree-Fock calculations for the lowest ^3P term, for the ^1P term, and for the configuration average. If one clicks on the He atom in the periodic table, the web page comes up for helium with the ground $1s^2\,^1S$ state of helium showing in the lower left-hand corner. To do the calculation for the $1s2p^3P$ state, one uses the "-e" button to remove one $1s$ electron from the configuration and then uses the up arrows in the lower left to change the orbital from $1s$ to $2p$ and then click the $+e$ button to add a $2p$ orbital. The state then shows as $1s^12p^{1\,1}S$. One can then use the up arrows farther along to the right to change $2S + 1$ to 3 and L to 1. The Hartree-Fock applet for the $1s2p^3P$ state can then be run by clicking the red arrow on the lower right with the wave functions of the $1s$ and $2p$ coming up on the screen. From the vertical lines on the screen, one can see that the average value of r for the $1s$ wave function is about 0.75 atomic units and the average value of r for the $2p$ wave function is about

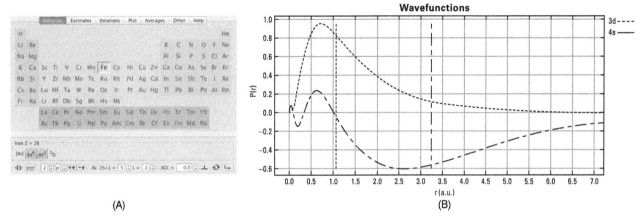

(A)

(B)

FIGURE 5.13 The web page for iron Fe. As for lithium, the nuclear charge and ground state are shown in the lower left-hand corner. For the ground state, six $3d$ electrons and two $4s$ electrons are outside an $[Ar]$ core. (B) The radial wave functions $P(r)$ for the $3d$ and $4s$ electrons of iron. As for lithium, vertical lines of the same kinds as the corresponding wave functions give the average value of r for each wave function.

4.65 atomic units. To get more accurate values of the average values of r for the electrons, one can click on the "averages" tab or on the "other" tab. Since the results on the "other" screen are more accurate and one can also find the value of the exchange integral on the same page, it is just as well to click on "other" tab. Then by changing $1s$ to $2p$ and $k = 0$ to $k = 1$ in the first row of entries on the screen, and clicking on the arrow on the far right, one finds that the average value of r for the $1s$ is 0.756. One can also find that the average value of r for the $2p$ is 4.656. The value of the exchange integral $G^1(1s, 2p)$ can be found using the second row of the web page shown by clicking on the first up arrow to convert the integral from an "F" to a "G", using the second up arrow to change k to 1, and using the final set of arrows to calculate G for the $1s$ and $2p$. By clicking on the arrow to the far right of the second row, the exchange integral $G^1(1s, 2p)$ is found to be 4229.5 cm^{-1}. The same process can be followed to find the average values of r and the exchange integral for the $1s2p\,^1P$ state and for the configuration average. The calculation for the configuration average is easier because one only has to get the configuration right and then hit the "Av" tab. The values obtained by these Hartree-Fock calculations are summarized in Table 5.2.

TABLE 5.2 The values of $< r >_{1s}$, $< r >_{2p}$, and $G^1(1s, 2p)$ for the $1s2p\,^3P$ state, the configuration average, and the $1s2p\,^1P$ state of helium. The values of r are given in atomic units, while the value of $G^1(1s, 2p)$ is given in cm^{-1}.

	^3P	E_{Av}	^1P
$< r >_{1s}$	0.756	0.753	0.748
$< r >_{2p}$	4.656	4.806	5.141
$G^1(1s, 2p)$	4229.5	3449.2	2057.6

The columns in Table 5.2 are arranged according to increasing energy. The ^3P state lies lowest with the average energy lying a little higher. For the highest ^1P state, the $2p$ wave function is more diffuse with the average value of r being 5.141 atomic units. The basic premise of the central field approximation that one has a single set of atomic orbitals for all of the states of a configuration has become somewhat questionable. The $2p$ wave function corresponding to the higher-lying ^1P state is significantly more diffuse than the $2p$ for the low-lying ^3P state. With the $2p$ becoming more diffuse, the screening of the $1s$ is reduced and the $1s$ is drawn in slightly toward the nucleus becoming more compact for the ^1P state. To calculate the Coulomb splitting of the configuration using Eqs. (5.21) and (5.22), we use the value of the $G^1(1s, 2p)$ corresponding to the configuration average giving the splitting

$$\Delta E = E(^1P) - E(^3P) = \frac{2}{3}G^1(1s, 2p) = 2299.4 \text{ cm}^{-1}$$

We see this splitting of the $1s2p$ configuration is similar to the splitting of the configurations shown in Fig. 5.5.

Our applet can be used to obtain wave functions for all the atoms in the periodic table. Clicking on the iron atom (Fe), for example, we obtain the page shown in Fig. 5.13(A). Looking at the lower left-hand corner of this page, one sees that the

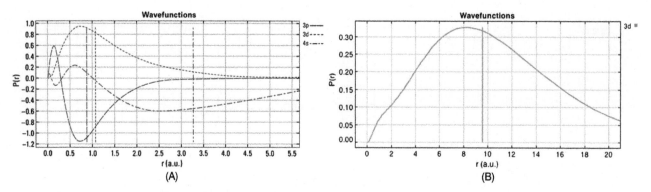

FIGURE 5.14 (A) The radial wave functions for the 3p, 3d, and 4s electrons of iron. (B) The radial wave functions for the 3d electron of potassium.

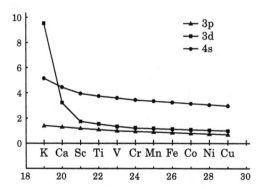

FIGURE 5.15 The average values of r of the 3p, 3d, and 4s electrons for the elements potassium through copper.

iron atom with nuclear charge $Z = 26$ has a ground state $[Ar]3d^6 4s^2\,^5D$ with six 3d electrons and two 4s electrons outside an argon core. The designation ^5D of the ground state means that the total spin of the state is $S = 2$ and the total orbital angular momentum is $L = 2$. In spectroscopic notation, the total spin S of a state is indicated by giving the value of the multiplicity of the state $(2S + 1)$ as a superscript before the letter corresponding to the total orbital angular momentum L. Clicking on the red arrow in the lower right-hand corner, we obtain the wave functions shown in Fig. 5.13(B). The curve for the 3d function is dashed while the curve for the 4s function has dots and dashes. From the vertical lines, we see that the 3d has an average value of r equal to about 1.1 atomic units, while the 4s electron has an average value of r equal to about 3.3 atomic units. The 3d orbitals of iron lie inside the 4s shell.

The wave functions for electrons within the argon core can also be plotted using the Hartree-Fock applet. The details of how to use the applet to obtain the wave functions of electrons within the argon core and to carry out other Hartree-Fock calculations can be found in Appendix F. The 3p, 3d, and 4s radial wave functions of iron are shown in Fig. 5.14(A). The wave function of the 3p has an average value of r equal to about 0.85 atomic units and thus lies slightly inside the 3d and far inside the outer 4s. The 3d radial function of iron is very different from the 3d functions of lighter elements. Fig. 5.14(B) show the 3d wave function of an excited state of potassium which has a single 3d electron outside an argon core. The average value of r for the 3d electron of potassium is about 9.5 atomic units, while the average value of r for the 3d electron of iron is about 1.1 atomic units.

For light atoms, the 3d orbital is quite diffuse and hydrogenic; however, as the nuclear attraction becomes strong in comparison to the centrifugal repulsion, the d orbital is drawn into the core of the atom. In the region of the periodic table around potassium, the 3d orbital begins to collapse becoming quite similar in spatial extent to the already filled 3s and 3p orbitals. The average values of r of the 3p, 3d, and 4s electrons for the elements potassium through copper are shown in Fig. 5.15. The collapse of the 3d shell is apparent.

In a similar manner, the 4d shell begins to collapse around rubidium $(Z = 37)$, producing the second transition-metal row of the periodic table. A still more dramatic shell collapse occurs shortly beyond cesium, when the much larger centrifugal repulsion for the $l = 3$ electrons is finally overcome by the attractive nuclear field, and the 4f orbital is drawn in toward the nucleus. The collapse of the 4f shell affects the order of the electronic configurations of these elements. For the singly ionized lanthanum ion, the $5d^2$ configuration lies below the $4f^2$, while for the doubly ionized cesium ion, the $4f^2$ configuration lies below the $5d^2$. The collapse of the 4f wave function places the 4f well inside the filled 5s and 5p shells.

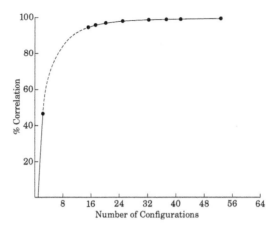

FIGURE 5.16 Percentage of Correlation Energy vs the number of configurations for the ground state of beryllium.

Rare-earth elements, which differ only by the number of electrons in the inner $4f$ shell, are well known for their nearly identical chemical properties.

While the single-configuration Hartree-Fock method give a good qualitative description of most atoms, the results it attains often leaves room for improvement. To understand when the Hartree-Fock method is likely to do well and when it is likely to fall short, it is good to remember that the method is based on a variational principle. The wave functions that are most tightly bound such as the $1s$ make a very large contribution to the total energy of the atom, and these functions are very well determined by the Hartree-Fock method. However, the weakly bound outer electrons, which are mainly responsible for the chemical and optical properties of atoms, do not contribute very much to the total energy of the atom, and they are not well-determined by the Hartree-Fock method.

Over the years, chemists and physicists have used various methods to improve upon the single-configuration Hartree Fock description of atoms and molecules. Theoretical chemists have mainly relied upon basis set methods in which atomic and molecular wave functions are expanded in terms of sets of analytical functions with adjustable parameters. These parameters are determined by requiring that the total energy of an atom or molecule be stationary with respect to variations of these parameters. Very accurate calculations have been reported using basis set methods. Here I only refer to the accurate calculation carried out by Carlos Bunge at the Autonomous University of Mexico. (C. Bunge, Phys. Rev A, 1976, Volume 14, page 1965)

In keeping with the computational nature of this book, we now consider the numerical methods that have been developed to improve upon the single-configuration Hartree-Fock model.

5.6 Further developments in atomic theory

One way of going beyond the single-configuration Hartree-Fock theory is to include more than one configuration in the wave function. For example, rather than describe the ground state of beryllium by the wave function $\Phi = 1s^2 2s^2\ ^1S$ of a single configuration, we can describe the lowest state of beryllium by the wave function $\Phi = a\phi(1s^2 2s^2\ ^1S) + b\phi(1s^2 2p^2\ ^1S)$ which includes the $1s^2 2s^2$ and $1s^2 2p^2$ configurations. For particular values of the a and b, the expectation value the energy of the multi-configurational state is minimized by varying the form of the single-electron wave functions, and then the two-by-two matrix is diagonalized to get new values of the energy. For the two-configuration model space spanned by the $1s^2 2s^2\ ^1S$ and $1s^2 2p^2\ ^1S$ configurations, the ratio of the b and a coefficients is found to be about 0.3. Of course, this approach can be extended to include more than two configurations.

The energy of a many-electron state is equal to the work required to pull the electrons entirely apart. For nonrelativistic calculations of the kind we have discussed thus far, one must also remove the relativistic contribution to the energy. The *correlation energy* of the atom is defined to be the difference between the total nonrelativistic energy of an atom and the energy produced by a Hartree-Fock calculation. The correlation energy is always positive because electrons moving in an average potential come closer to each other than they would in an atom in which the electrons who repel each other would avoid close contact. The Hartree-Fock theory does not include the effect of such correlated motions.

Fig. 5.16 shows the percentage of the correlation energy of beryllium obtained by including different numbers of configurations. One can see that one picks up about half of the correlation energy with a single configuration ($1s^2 2p^2$) and

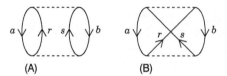

FIGURE 5.17 Second-order Goldstone diagram.

more than 95% of the correlation energy with 20 or 30 configurations, but the curve then flattens out so that one obtains only a very small contribution to the correlation energy by adding more configurations.

Another general way to improve upon the single configuration Hartree-Fock model is to use many-body perturbation theory. In a perturbation calculation, the Hamiltonian is divided into a term H_0 representing the energy of electrons moving independently in an average potential and a perturbation V which is the difference of the Coulomb interaction and the interaction between the electrons with the average potential.

$$H = H_0 + V. \tag{5.25}$$

H_0 is given by the equation

$$H_0 = \sum_i N \left[-\frac{1}{2}\nabla_i^2 - \frac{Z}{r_i} + \sum_i u_i \right] \tag{5.26}$$

where r_i denotes the distances of the electron i from the nucleus and where we have again used atomic units with the charge and mass of the electron, $\hbar$, and $4\pi\epsilon_0$ all equal to one. V is given by the equation

$$V = \sum_{i<j} \frac{1}{r_{ij}} - \sum_i u_i \tag{5.27}$$

We can take the Hartree-Fock potential to be the potential in the independent-particle theory so that V is the difference between the full Hamiltonian and the Hartree-Fock potential energy.

Atomic many-body calculations were first performed by Hugh Kelly at University of Virginia who successfully calculated the correlation energy of a number of second- and third-row elements (H.P. Kelly (1963) Phys. Rev. Volume 131, page 684 and Phys. Rev. (1968) Volume 173, page 142). Some years later, Katherine Rajnak and I at the Argonne National Laboratory preformed accurate many-body calculations of the effective interaction between the $4f$ electrons of rare earth ions using an effective operator form of many-body theory. (J. Morrison and K. Rajnak, Phys. Rev. A (1971) Volume 4, page 536, and J. Morrison J. Phys. B (1973) Volume 6, page 2205)

Atomic many-body calculations and the calculations of particle physics to be described in later chapters are performed using diagrams. While the diagrams used in particle physics are called Feynman diagrams, the diagrams used in atomic and molecular structure calculations, which depend upon a time-independent form of perturbation theory, are called *Goldstone diagrams*. Such diagrams were first introduced by Jeffrey Goldstone who was a nuclear theorist working under the guidance of Hans Bethe.

The two second-order Goldstone diagrams that occur in a perturbation calculation of the correlation energy of beryllium are show in Fig. 5.17. In these diagrams, a and b correspond to orbitals that are occupied in the ground state and r and s denote excited electrons. The diagram shown in Fig. 5.17(A), for which the excited orbitals, r and s, are in the same order on the bottom and the top of the diagram, is called a *direct* diagram, while the second diagram for which order of the orbitals r and s are reversed is called an *exchange* diagram.

The Goldstone diagrams shown in Fig. 5.17(A) corresponds to the mathematical expression

$$\sum_{rs} \frac{\int \psi_a \psi_b (1/r_{12}) \psi_r \psi_s dv_1 dv_2 \int \psi_r \psi_s (1/r_{12}) \psi_a \psi_b dv_1 dv_2}{\epsilon_a + \epsilon_b - \epsilon_r - \epsilon_s}. \tag{5.28}$$

Beginning at the bottom of each diagram and moving upward, there is a Coulomb matrix element for each dotted line corresponding to a Coulomb interaction except the last and there is an energy denominator that can be obtained by drawing a horizontal line above the dotted line and including a positive energy for each line directed downward and a negative energy for every line moving upward.

A general feature to notice in Eq. (5.28) is that the integrals that depend upon the overlap of the wave functions appear twice while the energy denominator appears once. This explains the importance of the $2s^2 2p^2$ configuration of beryllium that lies far above the ionization energy of a single $2s$ electron.

The angular part of the integrals appear in Eq. (5.28) involve the spherical harmonics and are the same for all excitations into particular angular momentum states. Denoting the radial function with quantum numbers n_a and l_a simply by P_a and denoting the other radial functions in a similar way, the radial part of the direct diagram is

$$\sum_{n_r n_s} \frac{\int P_a(1) P_b(2)(1/r_{12}) P_r(1) P_s(2) dr_1 dr_2 \int P_r(1) P_s(2)(1/r_{12}) P_a(1) P_b(2) dr_1 dr s_2}{\epsilon_a + \epsilon_b - \epsilon_r - \epsilon_s}. \tag{5.29}$$

A feature of many-body calculations which make them technically difficult is that the sums over excited states includes large contributions from wave functions of free-electron states having large values of the kinetic energy. Consider, for example, the excitation of a $2s$ or $2p$ electron into d-states. The lowest lying d state is the $3d$ state whose wave function lies far outside the $2s$ and $2p$ wave functions. None of the bound d-functions or the free d-states with small values of the kinetic energy have a significant overlap with the $2s$ and $2p$ wave function in the initial state of the atom; however, as the kinetic energy of d electrons increases, the wave length of d electrons becomes shorter and the first maxima the free d wave functions moves back into the same region of space as the $2s$ and $2p$. The first oscillation of these energetic d wave functions make a very large contribution to the direct and exchange diagrams while the effect of the rapidly oscillating functions on the slowing declining tail of the bound functions cancel out leaving a large contribution to the Goldstone diagrams. H. Kelly and I found it necessary to extend our calculations to very high energy and use grids that would faithfully represent rapidly oscillating functions. In the end, of course, one integrates over the continuum, but highly excited states make very large contributions to the final result.

The possibility of calculating the stationary properties of atoms in a less time-consuming fashion became clear in 1970 when Nichols Winter and Vincent McCoy at Cal Tech solved the two-electron pair equation for excited states of helium. To understand how their work can be applied to many-body calculations of atomic physics, we suppose that the radial single-electron functions satisfy the eigenvalue equation

$$h_0 P_{nl} = \epsilon_{nl} P_{nl} \tag{5.30}$$

and consider the product of the operator $[\epsilon_a + \epsilon_b - h_0(1) - h_0(2)]$ times the following wave packet of excited states

$$\rho(ab \to rs, r_1, r_2) = \sum_{n_r, n_s} \frac{P_r(1) P_s(2) \int P_r(1) P_s(2)(1/r_{12}) P_a(1) P_b(2) dr_1 dr s_2}{\epsilon_a + \epsilon_b - \epsilon_r - \epsilon_s}. \tag{5.31}$$

Allowing the operators $h_0(1)$ and $h_0(2)$ to act on the radial functions $P_r(1)$ and $P_s(2)$ and using Eq. (5.30), we see that the term $[\epsilon_a + \epsilon_b - h_0(1) - h_0(2)]$ cancels the energy denominator and we obtain the equation

$$[\epsilon_a + \epsilon_b - h_0(1) - h_0(2)]\rho(ab \to rs, r_1, r_2) = \sum_{n_r, n_s} P_r(1) P_s(2) \int P_r(1) P_s(2)(1/r_{12}) P_a(1) P_b(2) \tag{5.32}$$

The sums over the excited orbitals, P_r and P_s in this last equation can be shown to force the orthogonality of the right-hand side with the occupied orbitals, P_a and P_b. The pair equation can thus be written

$$[\epsilon_a + \epsilon_b - h_0(1) - h_0(2)]\rho(ab \to rs, r_1, r_2) = 1/r_{12} P_a(1) P_b(2) - ortho, \tag{5.33}$$

where "*ortho*" refers to the terms that must be added to the right-hand side of the equation to make the right-hand side orthogonal to the functions $P_a(r_1) P_b(r_2)$. Using Eq. (5.31), the radial part of second-order direct diagram given by Eq. (5.29) can thus be written simply

$$\int P_a(r_1) P_b(r_2) \rho(ab \to rs, r_1, r_2) dr_1 dr_2.$$

The direct diagram can thus be evaluated by evaluating a single integral involving the occupied orbitals and the pair function. The exchange diagram and higher-order diagrams can be evaluated in a similar way. One thus avoids the need of constructing the excited states and integrating over the continuum by solving two-electron (pair) equations to find the linear combinations of excited states that contribute to the Goldstone diagrams. This way of carrying out atomic many-body calculations is described in the book by Ingvar Lindgren and John Morrison that is cited at the end of this chapter.

In comparing the multi-configuration Hartree-Fock method with many-body perturbation theory, one sees that each of these methods does very well to solve one part of the correlation problems. The multi-configuration Hartree-Fock theory does very well to include the effect of strongly interacting configurations which contribute significantly to the ground state energy of the atom, while perturbation theory does very well in including the effect of the large number of weakly interacting configurations that do not significantly affect the form of the occupied orbitals. My own preference is to use the multi-configuration Hartree-Fock method for the strongly interacting configurations and use perturbation theory only for effects which are very small. (J. Morrison and C. Fischer (1987) Phys. Rev. A Volume 35, page 2429)

Suggestions for further reading

Charlotte Fischer, T. Brage, and P. Johnsson, *Computational Atomic Structure: An MCHF Approach* (CRC Press, January 1997).
Ingvar Lindgren and John Morrison, *Atomic Many-Body Theory* (Springer, 2013).

Basic equations

Definition of atomic units

$$e = m = \hbar = 4\pi\epsilon_0 = 1$$

Atomic unit of distance

$$1 \text{ atomic unit} = a_0 \text{ (Bohr radius)}$$

Atomic unit of energy

$$1 \text{ atomic unit} = 27.2 \text{ eV}$$

This unit of energy called the hartree is equal to the absolute value of the potential energy of an electron and a proton separated by a distance of a_0.

Summary

The electrons in many-electron atoms move in a potential field due to the nucleus and the other electrons. The wave functions of the electrons in this complex environment can only be obtained within the framework of some approximation scheme. One approximation, which has proved to be very useful, is to assume that the electrons move in an electric potential that is spherically symmetric. The central-field model provides a theoretical basis for the atomic shell model and for the regularities observed in the chemical elements.

The wave function of an electron moving in a central-field is of the general form

$$\psi(r, \theta, \phi) = \frac{P_{nl}(r)}{r} Y_{lm_l}(\theta, \phi)\chi_{m_s} \, ,$$

where the radial part of the wave function is expressed as a function $P(r)$ divided by r. The angular part of the wave function, $Y_{lm_l}(\theta, \phi)$, is called a spherical harmonic. The factor χ_{m_s} represents the spin part of the wave function.

The radial functions of many-electrons atoms can be obtained using the Hartree-Fock theory described in this chapter. A Hartree-Fock applet available at the web site of this book enables one to plot atomic wave functions and calculate the size of the atom and the strength of the interactions between the electrons.

Questions

1. Upon which assumption is the independent particle based?
2. What is the Pauli exclusion principle?
3. An electron moving in a central field can be described by the quantum numbers, n, l, m_s, and m_l. Upon which of these quantum numbers does the energy of the electron generally depend?

4. Give the principal quantum number n and the orbital angular momentum number l of a $3d$ electron.
5. Give the possible values of m_s and m_l for a $3d$ electron.
6. How many electrons does it take to fill the $1s$, $2p$, and $3d$ shells?
7. Give the ground configuration of (a) N ($Z = 7$), (b) Si ($Z = 14$), (c) Ca ($Z = 20$).
8. What interaction is responsible for the separation of the ^1S, ^1D, and ^3P terms of carbon?
9. What interaction is responsible for the fine splitting of the ground ^5D state of Fe?
10. The lithium atom has a nuclear charge $Z = 3$. What are the three lowest configurations of lithium?
11. The magnesium atom has a nuclear charge $Z = 12$. What are the three lowest configurations of magnesium?
12. Does the Coulomb interaction between the outer $2p$ electrons increase or decrease as one moves across the periodic table from boron to fluorine? Explain your answer.
13. Does the spin-orbit interaction of the $2p$ electrons increase or decrease as one moves across the periodic table from boron to fluorine? Explain your answer.
14. How would the size of a carbon atom be affected if one were to remove a single $2p$ electron?
15. Suppose that one were to remove the two electrons from the helium atom. Would it take more or less energy to remove the second electron once the first electron had been removed? Explain your answer.
16. Describe how you would use the Hartree-Fock applet to find the average values of r of the electrons in an atom.
17. Describe how you would use the Hartree-Fock applet to find the ionization energy of the lithium atom.
18. What is the virial theorem?

Problems

1. What are the possible values of m_l and m_s for a single $4f$ electron. How many distinct states does an atom with two $4f$ electrons have?
2. Show that the wave function given by Eq. (5.5) changes sign if the coordinates of the two electrons are interchanged.
3. Suppose that a lithium atom has one electron in the $1s$ shell with its spin up ($m_s = +1/2$), another electron in the $1s$ shell with its spin down ($m_s = -1/2$) and a third electron in the $2s$ shell with spin up ($m_s = +1/2$). Denoting the wave functions for these three states by $\psi_{10}\alpha$, $\psi_{10}\beta$, and $\psi_{20}\alpha$, construct a Slater determinant representing the state of the lithium atom.
4. Using the periodic table shown in Fig. 5.2 determine the ground configuration of the following elements: fluorine (F), magnesium (Mg), silicon (Si), potassium (K), and cobalt (Co).
5. (a) Give all elements with a p^4 ground configuration. (b) Give all elements with a d^5 ground configuration.
6. What are the three lowest configurations of carbon?
7. What are the three lowest configurations of nitrogen?
8. Determine the LS terms of the ns^2, nd^2, and $4f^2$ configurations. Using Hund's rules, determine the lowest LS term of the $4f^2$ configuration.
9. Using the LS terms of a nf^2 configuration found in the previous problem and taking into account the possible values of M_S and M_L for each LS term, find the total number of states of a $4f^2$ configuration. Compare your result with the result of Problem 1.
10. Give the possible values of the total spin (S) and total orbital angular momentum (L) of the $1s^2 2s^2 2p3d$ configuration of carbon in spectroscopic notation. In obtaining this result, the effect of filled shells of electrons may be neglected.
11. Give the possible values of the total spin (S) and total orbital angular momentum (L) for the $[Xe]4f\,5d$ and $[Xe]4f^2$ configurations of Ce^{2+}.
12. Use the rule for the addition of angular momentum given in the preceding chapter to determine the possible values of the total angular momentum J for the LS terms of the $4f^2$ configuration of Ce^{2+}. Use Hund's rule to determine the values of S, L, and J for the lowest state of the $4f^2$ configuration.
13. For which of the following elements would you expect the magnitude of the Coulomb interaction between the $3p$ electrons to be largest: Si, P, S?
14. Using the Hartree-Fock applet obtain the values of the integral, $F^2(3p, 3p)$, for the elements, Si, P, and S, and show that your results are consistent with the result of Problem 13.
15. For which of the following elements would you expect the magnitude of the spin-orbit interaction of the $3p$ electrons to be largest: Si, P, S?
16. Using the Hartree-Fock applet obtain the average values of $1/r^3$ and the values of the spin-orbit constant ζ for $3p$ electrons of the elements, Si, P, and S. Explain the variation of ζ for these elements.
17. For which of the following elements do the states most nearly correspond to pure LS terms: O, S, Se?

18. Using the Hartree-Fock applet to obtain the average value of r for the $4f$ and $5d$ electrons of La^+, Ce^{2+}, and Pr^{3+}. Are your results consistent with the idea that the $4f$ shell collapses in this region of the periodic table?

19. Using the Hartree-Fock applet, obtain the total energy of the neutral neon atom and the neon ion for which a single $2p$ electron has been removed. Use these results to find the binding energy of the $2p$ electrons of neon. How does your result compare with the single-electron energy for the p electron?

20. The energy of ^1S and ^3S states of the excited $1s2s$ configuration of helium are given by the following formulas

$$E(^1S) = E^0 + G^0(1s, 2s)$$
$$E(^3S) = E^0 - G^0(1s, 2s)$$

By performing a Hartree-Fock calculation for the $1s2s$ configuration of helium using the average energy of the configuration, evaluate the exchange integral $G^0(1s, 2s)$ and calculate the splitting to the singlet and triplet levels of this configuration.

21. Using the special techniques described in Appendix F, perform Hartree-Fock calculations for the ^3S and ^1S terms of the excited $1s2s$ configuration of helium and calculate the average value of r for the $1s$ and $2s$ wave functions for Hartree-Fock calculated with these two LS terms.

22. The selection rules obtained in the preceding chapter for hydrogen apply to many-electron atoms as well. For many-electron atoms, additional selection rules give the possible changes of the total angular momentum. The selection rule for the total spin (S) is $\Delta S = 0$, while the selection rule for the total orbital angular momentum is $\Delta L = 0$ or ± 1. To this last rule must be added the requirement that a transition from a state for which $L = 0$ to another state for which $L = 0$ is forbidden. Using these rules determine the possible transitions between the LS terms of the $[Xe]4f\,5d$ and $[Xe]4f^2$ configurations of Ce^{2+} considered in Problem 11.

Chapter 6

The emergence of masers and lasers

Contents

6.1 Radiative transitions	141	Basic equations	150
6.2 Laser amplification	142	Summary	151
6.3 Laser cooling	146	Questions	151
6.4 * Magneto-optical traps	147	Problems	151
Suggestions for further reading	150		

Physicists now routinely use laser cooling and evaporative cooling to reduce the temperature of clouds of atoms to a few hundred nanoKelvin.

We will now discuss the ideas and experiments which led to the development of modern lasers. The word laser is an acronym for "light amplification by the stimulated emission of radiation". We have seen that radiation with a photon energy equal to the difference between two energy levels can be absorbed by an atom allowing the atom to make a transition to a higher energy level and radiation can also stimulate an atom to fall down to a lower level and emit radiation. The idea of stimulated emission was originally due to Einstein. Prior to Einstein's study of radiation processes in 1916, physicists had supposed that the only way light could interact with an atom was through absorption or spontaneous emission.

6.1 Radiative transitions

Many of the ideas and formulas used in Chapter 4 to describe radiative transition for the hydrogen atom can be applied to many-electron atoms as well. We begin our treatment of atomic transitions here by considering again transitions between the two energy levels shown in Fig. 4.11. The energy of the upper level in this figure is denoted by E_2 and the energy of the lower level is denoted by E_1. We suppose that there are N_2 atoms in the upper level and the degeneracy of the level is g_2, while there are N_1 atoms in the lower level and the degeneracy is g_1. The rate with which atoms absorb light and make a transition from the lower level to the upper level is equal to the product of the Einstein coefficient B_{12}, the density of the radiation field $\rho(f)df$, and the number of atoms in the lower level N_1

$$R_\uparrow = B_{12}\rho(f)df N_1 \,.$$

The rate with which atoms emit light and make a spontaneous transition from the upper level to the lower level is equal to the product of the Einstein coefficient A_{21} and the number of atoms in the upper level N_2, while the rate with which atoms emit light and make a stimulated transition from the upper to the lower level is equal to the product of the Einstein coefficient B_{21}, the density of the radiation field $\rho(f)df$, and the number of atoms in the upper level N_2

$$R_\downarrow = A_{21}N_2 + B_{21}\rho(f)df N_2 \,. \tag{6.1}$$

When atoms make spontaneous transitions, each atom independently emits a single photon. The radiation emitted spontaneously from a collection of atoms is random in nature. In contrast, the process by which one photon stimulates an atom to emit a second photon involves the correlated motion of two photons. Unlike electrons, which satisfy the Pauli exclusion principle, photons can be in the same state and are susceptible to collective motion. The process of stimulated emission is not random, but is driven by and coherent with the applied signal. For laser transitions, stimulated emission plays a central role, while the process of spontaneous emission is extraneous. Discarding the term with A_{21} from Eq. (6.1), the ratio of the transition rate for emission to the transition rate for absorption then becomes

$$R = \frac{R_\downarrow}{R_\uparrow} = \frac{B_{21}\rho(f)df N_2}{B_{12}\rho(f)df N_1} = \frac{N_2}{N_1} \cdot \frac{B_{21}}{B_{12}} \,.$$

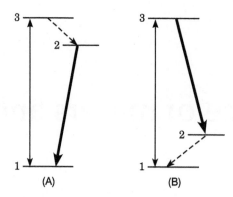

FIGURE 6.1 Two three-level schemes.

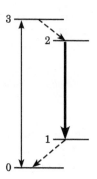

FIGURE 6.2 A four-level schemes.

Using Eq. (4.21), this equation can be written

$$R = \frac{N_2}{N_1} \cdot \frac{g_1}{g_2} = \frac{(N_2/g_2)}{(N_1/g_1)} \, . \tag{6.2}$$

The ratio of N_2/g_2 to N_1/g_1, which appears here, describes the relative populations of the upper and lower energy levels. As we shall see in the next chapter, atoms and molecules in thermal equilibrium tend to drop down to the lowest energy level available. For two independent levels in thermal equilibrium, the ratio $(N_2/g_2)/(N_1/g_1)$ is always less than one. According to Eq. (6.2), laser amplification can never occur for a two-level system.

6.2   Laser amplification

Two three-level schemes which can lead to laser transitions are shown in Fig. 6.1. In these schemes, atoms are pumped up from level 1 to level 3 by a source of electromagnetic radiation. Atoms in Fig. 6.1(A) decay from level 3 to level 2 and then make a laser transition from level 2 to level 1, while atoms in Fig. 6.1(B) make a laser transition from level 3 to level 2 and then decay down to level 1. Each of these schemes offer certain difficulties. The possibility of laser amplification in the first scheme depends upon the population in level 2 being greater than the population in level 1. Since any atom in level 3 can return to level 1, the relaxation from level 3 to level 2 must occur very rapidly. A disadvantage of this laser scheme is that atoms which have decayed by the laser transition to level 1 can absorb energy from the laser beam and return to level 2. The possibility of laser amplification in the second scheme depends upon the population in level 3 being greater than the population in level 2. The difficulty with this scheme is that the atoms in level 3 can be stimulated by the pumping signal to return to level 1.

These difficulties are overcome by the four-level system shown in Fig. 6.2. In this scheme, atoms are pumped up from the ground level, which we have denoted by 0, to level 3. The atoms then decay rapidly from level 3 to level 2, and the laser transition occurs from level 2 to level 1. Following the laser transition, atoms in level 1 decay rapidly down to the ground level, thus removing the possibility that an atom which has made the laser transition can absorb light removing energy from the beam. The lasers, which we shall consider in this book, provide a number of different variations of these three- and four-level laser schemes.

FIGURE 6.3 The ammonium molecule NH_3. The nitrogen atom can vibrate about equilibrium positions on either side of the plane defined by the three hydrogen atoms.

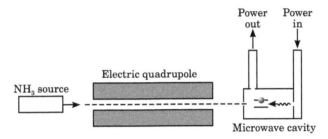

FIGURE 6.4 The ammonium maser.

The first research which confirmed the occurrence of stimulated emission was carried out by A. Ladenburg and his collaborators in Berlin between 1926 and 1930 — more than a decade after Einstein's prediction. Twenty more years passed before stimulated emission was used in practical devices to amplify beams of electromagnetic radiation. The first devices using stimulated emission operated in the microwave region of the spectrum and were called *masers* which stands for the microwave amplification by the stimulated emission of radiation. In the microwave region, the probability of spontaneous emission, which is inversely proportional to the cube of the wavelength, is very small. The idea of the maser developed almost simultaneously in the United States and the Soviet Union with the first operating maser being built by a research group at Columbia University under the direction of Charles Townes.

In Townes' own mind, the development of the maser was connected to the development of microwave technology during the Second World War. Townes worked during the war at the Bell Telephone Laboratories where he was involved in a project to push the operational frequency of radar higher. In 1947, he accepted an invitation from I. Rabi to leave Bell Laboratories and join the physics department of Columbia University where he became an authority on microwave spectroscopy. In 1950, Townes was asked to be the chair of a study commission on millimeter waves. He worked on the committee for nearly two years and was dissatisfied with its progress. One day in 1952, Townes was sitting on a park bench in Washington DC before a committee meeting. He tried to formulate in his own mind why he had failed to produce a useful millimeter wave generator. It was clear to him they needed a small resonator coupled to an electromagnetic field. He then realized that the resonator he was imagining had the properties of a molecule. Returning to Columbia University he began a project which led to the construction of the ammonia maser.

The ammonia molecule NH_3, which is illustrated in Fig. 6.3, is like a pyramid with three hydrogen atoms at the vertices of the base of the pyramid and the nitrogen atom at the apex. The nitrogen atom can vibrate about equilibrium positions on either side of the plane defined by the hydrogen atoms. As for the harmonic oscillator discussed in Chapter 2, the energy levels associated with the vibration of the nitrogen atom about each of these equilibrium positions are equally spaced. Since the nitrogen atom can go through an inversion in which it passes between these two equivalent vibrational positions, the energy levels corresponding to these two sets of vibrational states are coupled together. A quantum mechanical treatment of the molecule shows that the vibrational energy levels of the molecule are split into doublets with the energy separation between the two members of each doublet corresponding to the small inversion frequency of the molecule. The wave function for the lower member of each doublet is symmetric with respect to an inversion through the plane of the hydrogen atoms, while the wave function for the upper level is antisymmetric with respect to inversion.

The small research group which Townes formed for this project included a post-doctoral associate, Herbert Zeiger, and a doctoral student, James Gordon. Working together they constructed the apparatus illustrated in Fig. 6.4. A beam of ammonia molecules was emitted by the source and passed through an electrostatic field in the focusing region. The upper inversion levels of molecules in the focusing region experienced a force directed toward the axis of the beam, while the

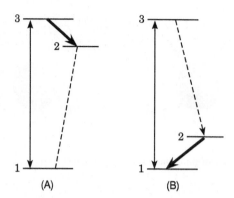

FIGURE 6.5 The three-level schemes proposed by Prokhorov and Basov.

lower inversion levels experienced a force directed away from the axis. As a result, virtually all of the molecules arriving in the microwave cavity were in the upper states. The group was thus able to obtain a stable population inversion and an amplification of microwave radiation at the inversion frequency.

Other methods of obtaining population inversions in molecular beams were proposed by Alexander Prokhorov and Nikolai Basov who were trying to develop accurate frequency and time standards at the Lebedev Institute of Physics in Moscow. In order to create a population inversion in a beam of molecules and thus increase the intensity of the light emitted by the molecules, Prokhorov and Basov proposed the three-level schemes shown in Fig. 6.5. They suggested that a population inversion could be created between level 3 and level 2 shown in Fig. 6.5(A) or between levels 2 and 1 shown in Fig. 6.5(B) by optically pumping molecules from level 1 to level 3. Although these three-level schemes did not lead to a successful molecular maser, Nicolas Bloembergen proposed a similar three-level scheme shortly afterwards that led to tunable masers using paramagnetic crystals.

The ammonia maser was reported by Gordon, Zeiger, and Townes in 1954, and the proposal of Bloembergen was published in 1956. Since that time many solid state masers have been built for applications in radio astronomy or as components in radar receivers. With the successful development of masers, many physicists began looking into the possibility of applying the maser idea to infrared and optical wavelengths. The physical conditions to produce stimulated emission at optical wavelengths is different than it is for microwaves. The wavelength of visible light is much shorter than the length of any practical cavity, and hence the resonance cavity of a laser must support a large number of modes within the frequency range of interest. Charles Townes and Arthur Schawlow studied the theoretical aspects of light amplification in the visible region outlining the design consideration that should be taken into account to construct lasers. Their results were published in The Physical Review in 1958. Our understanding of how to amplify visible light has developed over the years and many different lasers have appeared on the market.

Any functioning laser has three key components: an optical medium which generates the laser light, a source of power which excites the medium and creates a population inversion and an optical cavity which concentrates the laser light. The population inversion that occurs leads to an amplification of the signal or a laser gain which is equal to the difference between stimulated emission and absorption at that wavelength. In order to limit the number of vibration modes, laser cavities are open with only a pair of small mirrors at opposite ends. One of the mirrors is generally only partially reflecting and allows light to leak out of the cavity and form the laser beam. The fraction of light that is allowed out of the cavity depends on the laser gain. For the laser process to sustain itself, the total gain must equal the sum of the cavity loses and the fraction of the energy allowed to leak out of the cavity. Schawlow advocated having plane mirrors at the ends the laser cavity. The simplicity of the plane mirror resonator is offset by the practical difficulty of aligning the mirrors precisely enough for stable laser operation. Some lasers have spherical mirrors at the ends of the cavity to focus the laser beam, while others with parallel plane mirrors have something else within the laser cavity to provide the needed focusing. The output beam of a laser is well-direct and consists of waves which are in phase with each other.

The first laser was developed by Theodore Maiman at the Hughes Research Laboratories in Southern California in July of 1960. The optical medium of this laser consisted of a synthetic ruby crystal 1 cm long coated on either side by silver. When the crystal was irradiated by a pulse of light from a xenon flash-lamp, a population inversion was achieved and a pulse of laser light was emitted in the red region of the spectra at 694.3 nm. The design of the laser is illustrated in Fig. 6.6. The ruby crystal consists of a lattice of sapphire (Al_2O_3) in which a small amount of the aluminum is replaced by chromium by adding Cr_2O_3 to the melt in the growth process. The electronic energy levels of the ruby crystal, which are shown in Fig. 6.7, are due to the optically active $3d$ electrons of the chromium Cr^{3+} ion in the host lattice. The blue and

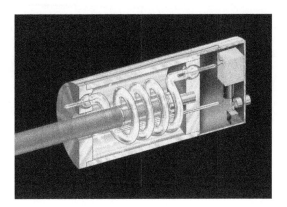

FIGURE 6.6 The Ruby laser.

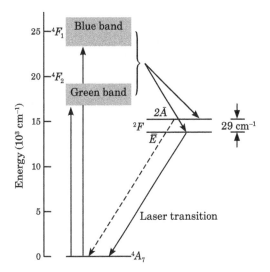

FIGURE 6.7 The electronic energy levels of the ruby crystal.

green wavelengths of the flash-lamp excite the chromium ion into the broad $^4F_2$ and $^4F_1$ bands of the excited levels shown in the figure. These levels then decay rapidly into the two metastable levels denoted by $2\bar{A}$ and $\bar{E}$. The 694.3 nm light is produced in transitions from the $\bar{E}$ level down to the ground state.

A few months after Maiman demonstrated the ruby laser, P.P. Sorokin and M.J. Stevenson reported a four-level laser using a CaF_2 crystal in which a small amount of calcium is replaced by uranium. Although this laser was not found to be very useful, Sorokin and Stevenson and a number of others soon developed workable lasers in which rare earth elements are doped into divalent and trivalent crystals. The most important of this group of lasers has proved to be the neodymium YAG laser for which neodymium is doped into a yttrium aluminum garnet. The electronic energy levels for this neodymium laser correspond to the states of the $4f$ electrons of the neodymium Nd^{3+} ion which replaces the Y^{3+} ions in the host lattice.

A helium-neon laser, which achieved a population inversion in a gas discharge tube, was demonstrated by A. Javan, W. Bennett Jr, and D. Herriott in January 1961. This first helium-neon laser operated at 1.15 micrometers in the near infrared. Later other researchers developed a version of the helium-neon laser — now widely used — which operates at 632.8 nanometers. The laser medium of the He-Ne laser consists of helium gas with about one-tenth as much neon inside a quartz plasma discharge tube as shown in Fig. 6.8. The pumping mechanism of the He-Ne laser is slightly more complex than the pumping mechanisms considered so far. Free electrons in the gas discharge collide with helium atoms exciting them to the metastable $1s2s\,^3S$ and $1s2s\,^1S$ states. Collisions between the helium and neon atoms then populate the excited $2p^54s$ and $2p^55s$ levels of neon which happen to have about the same energies as the $1s2s\,^3S$ and $1s2s\,^1S$ levels of helium. The lowest lying energy levels of helium and neon are shown in Fig. 6.9. The 1.15 μm laser line corresponds to a transition between the $2p^54s$ and the $2p^53p$ configurations of neon, while the 632.8 nm line corresponds to a transition between the $2p^55s$ and the $2p^53p$ levels. Another laser transition with a wavelength of 3.39 μm corresponds to the transition between $2p^55s$ and the $2p^54p$ levels. The energy levels in Fig. 6.9 are designated using the unfortunate notation of Paschen. The

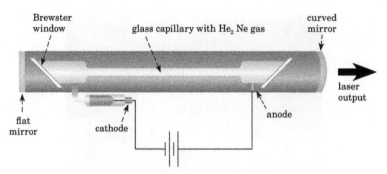

FIGURE 6.8 The $He - Ne$ laser.

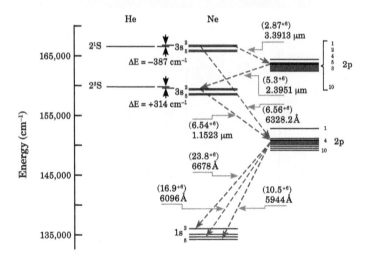

FIGURE 6.9 The lowest lying levels of helium and neon.

intention of this widely used notation was to fit the spectra of other atoms to a hydrogen-like theory. Thus, for example, the levels of the $2p^5 3s$ configuration, which are the lowest levels of neon for which there is an unpaired s-electron are denoted by $1s$, while the levels of the higher lying $2p^5 4s$ and $2p^5 5s$ configurations are denoted $2s$ and $3s$, respectively. The lowest lying levels of neon having a p-electron in an unfilled shell correspond to the $2p^5 3p$ and $2p^5 4p$ configurations, which are denoted $2p$ and $3p$, respectively.

6.3 Laser cooling

The idea that free atoms could be cooled by the scattering of laser light was first suggested by D. Wineland and H. Dehmelt and by T. Hänsch and A. Schawlow in 1975. Steven Chu, Claude Cohen-Tannoudji, and William Phillips received the Nobel prize in physics in 1997 for developing methods to cool and trap atoms with laser light. More recently, Eric Cornell, Wolfgang Ketterlie, and Carl Wieman received the Nobel Prize in 2001 for cooling a gas consisting of alkaline atoms to such a low temperature that the great majority of atoms fell into their lowest quantum state. This effect, which is called *Bose-Einstein condensation*, will be treated in the next chapter.

The possibility of laser cooling depends upon the fact that atomic absorption near the maximum of an absorption curve is strongly frequency dependent and that the frequency of light absorbed by a moving atom is shifted by the Doppler effect. To illustrate what is involved, a typical absorption curve is depicted in Fig. 6.10. In this figure, the intensity of absorption is plotted versus frequency and a single laser frequency (f) below the absorption maximum is shown. Frequencies below the absorption maximum are said to be *red detuned* and frequencies above the absorption maximum are said to be *blue detuned*. As atoms move toward the source of laser light, the Doppler effect shifts the frequency of the absorbed light in the direction of the absorption maximum and the amount of light absorbed by the atom increases. The momentum of the photons absorbed by the atom slow its motion. In contrast, when atoms move away from the laser source, the Doppler effect shifts the frequency of the laser light away from the absorption maximum and the amount of the light absorbed by the atoms decreases. Any light absorbed by atoms as they move away from the laser would increase their velocity. Thus, an

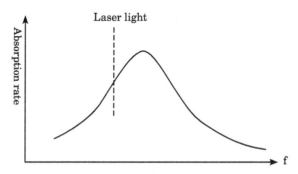

FIGURE 6.10 Absorption rate versus frequency for an atomic transition. A single frequency of laser light is shown below the absorption maximum.

atom in the beam of a continuous laser tuned to a frequency slightly below its absorption maximum finds itself in an unequal situation in which it absorbs many more photons when it moves toward the laser than when it moves away from the laser. The impact of the photons absorbed as an atom moves toward a laser slows the motion of the atom. Modern experiments in which clouds of alkali atoms are cooled by laser beams incident from three different directions now routinely achieve temperatures in the milli-Kelvin range. This is not cold enough to produce Bose-Einstein condensation, but it is cold enough for the atoms to be trapped in a magnetic field where they can be further cooled by evaporative cooling.

6.4   * Magneto-optical traps

Magnetic trapping of atoms is due to the Zeeman effect, which was discussed in Chapter 4. In an inhomogeneous magnetic field, each atom has a potential energy which depends with its spatial coordinates. We shall suppose that the magnetic field is weak and that the energy of the atom is linear in the magnetic field. Denoting the magnetic moment of the atom by μ and the magnetic field by B, the energy of the atom due to the field can be written

$$E_{mag} = -\mu \cdot \mathbf{B}. \tag{6.3}$$

In the magnetic field, an atom experiences a force in the direction in which its potential energy decreases. If the magnetic moment of the atom is positive, the force on the atom will be directed toward regions of higher field, while if the magnetic moment is negative, the force on the atom will be directed toward regions of lower field. For this reason, states with positive magnetic moment are referred to as *high-field seekers* and states with negative magnetic moment are referred to as *low-field seekers*.

The task of constructing a magnetic trap for neutral atoms is to produce a magnetic field with either a local maximum or a local minimum. It can be shown generally that a magnetic field $\mathcal{B}$ cannot have a local maximum in regions where there is no electrical currents. So, only local minima are possible and the only atomic states that can be trapped are the low-field seekers. The simplest kind of magnetic trap is the so-called *quadrupole trap* for which the magnetic field varies linearly with coordinates for small displacements from a local minimum. Such a magnetic field may be produced by a pair of opposed Helmholtz coils.

A magnetic field always has the property that the sum of the changes of the field in three orthogonal directions is equal to zero. Denoting the rate of changes of the magnetic field in the x- and y-directions by B', the magnetic field in the vicinity of the origin can be written

$$\mathbf{B} = B'(x, y, -2z).$$

The magnitude of the magnetic field, which is

$$B = B'(x^2 + y^2 + 4z^2)^{1/2},$$

has a minimum at the origin and has a dependence upon the spatial coordinates that is symmetric about the z-axis.

The quadrupole trap has one important disadvantage associated with the fact that the magnetic field is zero at the origin. We shall find shortly that when the magnetic field is zero, atomic states that have negative magnetic moments and are trapped by the field are degenerate with other states having positive magnetic moments. The motion of atoms in low-field seeking states can cause them to make transitions to high-field seeking states in which they are ejected from the trap. The quadrupole trap has a *hole* near the node of the field, and this limits the time during which atoms can be held in the trap.

In their efforts to produce Bose-Einstein condensation in dilute gases, E. Cornell and C. Wieman of the University of Colorado at Bolder used a modified quadrupole trap known as a time-averaged orbiting potential (TOP) trap. In this trap, a rotating magnetic field is superimposed on the quadrupole field. Suppose that the component of the oscillating magnetic field in the x-direction is $B_0 \cos(\omega t)$ and the component of the magnetic field in the y-direction is $B_0 \sin(\omega t)$. At a particular time t, the total magnetic field will then be

$$\mathbf{B} = (B'x + B_0 \cos(\omega t), B'y + B_0 \sin(\omega t), -2B'z).$$

The effect of the oscillating field is to move the instantaneous position of the node of the magnetic field. Cornell and Wieman chose the frequency of the oscillating magnetic field to be much faster than the rate of collisions between the slowly moving rubidium atoms, but slower than the transition frequencies between the different energy levels of rubidium atoms in the magnetic field. Under these circumstances, the rubidium atoms remain in a single electronic state while moving in an effective potential that is the time-average of the instantaneous potential over one rotational period of the field. Shortly after Cornell and Wieman reported their results, W. Ketterlie achieved Bose-Einstein condensation at the Massachusetts Institute of Technology in a dilute gas of sodium atoms. Ketterlie used the oscillating electric field of a blue-detuned laser to repel sodium atoms from the hole in the center of his quadrupole trap. The repulsion of atoms by a blue-detuned laser is similar to the process of laser cooling considered previously. As atoms move away from a source of laser light tuned to a frequency above resonance, the Doppler effect shifts the frequency of absorbed light in the direction of the absorption maximum and the absorption of light increases the velocity of the atoms. Ketterlie used a laser beam at a frequency above resonance (blue detuned) to repel atoms from the hole in the center of his quadrupole trap.

The experiment of Ketterlie is an early example of the use of optical lasers to confine gas particles. C.S. Adams and his coworkers at Stanford University confined sodium atoms in a dipole trap consisting of two crossed red-detuned lasers. Blue-detuned traps have the advantage over red-detuned traps in that atoms are repelled by blue-detuned laser beams and can be confined in the dark by surrounding them with a repulsive wall of blue-detuned light. Atoms in the dark are not then heated by absorbing laser light. In a recent article cited at the end of next chapter, W. Phillips and his collaborators at the National Institute of Standards and Technology report on an experiment in which they have trapped atoms in an optical lattice created by the interference of two laser beams. Cooled atoms trapped in an interference pattern of this kind provides an unprecedented opportunity for exploring the properties of atoms in a lattice with a high degree of control over the placing of the atoms.

In order to understand the experiments that have succeeded in producing Bose-Einstein condensation and, in particular, to understand how atoms with larger kinetic energies can be selectively ejected from magnetic traps during evaporative cooling, we now study the energy levels of alkali atoms in a magnetic field. Bose-Einstein condensation has been attained in alkali gases at temperatures of a hundred nanoKelvin. At such low temperatures, alkali atoms are in their lowest s-state with only the lowest hyperfine levels of the atom being populated.

For an alkali atom in an s-state, the hyperfine interaction is the interaction of the spin of the outer electron with the magnetic field due to the atomic nucleus. Just as the spin-orbit interaction considered in Chapter 4 can be described by an effective Hamiltonian depending upon the spin and orbital angular momenta, the hyperfine interaction of the s-states of alkaline atoms can be described by an effective Hamiltonian depending upon the spin of the outer electron ($\mathbf{S}$) and the spin of the nucleus ($\mathbf{I}$)

$$H_{\text{hyp}} = A\mathbf{S} \cdot \mathbf{I}. \tag{6.4}$$

The constant A in this last equation plays the same role as the spin-orbit constant ζ in Eq. (4.53).

The energy levels of the combined electron-nuclear system can be described in a natural way by introducing the total angular momentum $\mathbf{F}$ which is the sum of $\mathbf{S}$ and $\mathbf{I}$

$$\mathbf{F} = \mathbf{S} + \mathbf{I}.$$

The spin quantum number of the outer electron is generally denoted by S, while the quantum number associated with the spin of the nucleus is denoted by I and the quantum number of the total angular momentum is denoted by F. As the spin-orbit interaction described in Chapter 4, the hyperfine interaction causes a splitting between the states having different values of the total angular momentum. For the s-states of the alkali atoms, the splitting between the levels for which $F = I \pm 1/2$ is equal to the product of the constant A in Eq. (6.4) and $I + 1/2$. This result is entirely analogous to the splitting of the energy levels of the hydrogen atom due to the spin-orbit interaction. The spin-orbit interaction causes a splitting of the $j = l \pm 1/2$-levels of a single electron equal to the product of the spin-orbit constant ζ and $l + 1/2$.

The z-component of the magnetic moment of the atom can be written

$$\mu_z = -g_s \left(\frac{\mu_B}{\hbar} \right) s_z + g_N \left(\frac{\mu_N}{\hbar} \right) I_z .$$

Here the Bohr magneton μ_B has the value $e\hbar/2m$, where m is the mass of the electron, and the nuclear magneton μ_N has the value $e\hbar/2m_p$, where m_p is the mass of the proton. The constant g_s is the g-value of the spin of the electron and g_N is the g-value of the nucleus. Nuclear g-values are determined very accurately by magnetic resonance experiments.

As in Chapter 4, we can argue that in a weak magnetic field the magnetic moments associated with the spin of the electron and the nucleus precess much more rapidly about each other than they precess about the external magnetic field, and we may replace the angular momentum vectors with their projection along the total angular momentum **F**. The **F** vector itself precesses about the magnetic field vector **B** and has a fixed projection upon **B**. As indicated in Problem 6, the component of the magnetic moment of the atom along the magnetic field can be written

$$\mu_z = -g_F \mu_B M_F, \tag{6.5}$$

where

$$g_F = g_s \left[\frac{F(F+1) - I(I+1) + S(S+1)}{2F(F+1)} \right] - g_N \left(\frac{\mu_N}{\mu_B} \right) \left[\frac{F(F+1) + I(I+1) - S(S+1)}{2F(F+1)} \right] . \tag{6.6}$$

Because the mass of the proton is about 1836 times the mass of the electron, the nuclear magneton is much smaller than the Bohr magneton, and the nuclear contribution to g_F represented by the second term in Eq. (6.6) is very small. Neglecting the nuclear contribution to g_F and using the fact that g_s is approximately equal to two, the value of g_F becomes

$$g_F = \frac{F(F+1) - I(I+1) + S(S+1)}{F(F+1)} . \tag{6.7}$$

The energy of the atom due the magnetic field B can be obtained by substituting Eq. (6.5) into Eq. (6.3) to obtain

$$E_{\text{mag}} = g_F \mu_B B M_F .$$

As an example, we consider the s-states of the alkali atoms, ^{87}Rb, ^{23}Na, and ^{7}Li, for which $S = 1/2$ and $I = 3/2$. Applying the rule given in Chapter 4 for adding angular momenta, the possible values of the total angular momentum of the atom can have the values $F = 2$ and $F = 1$. In the absence of a magnetic field, the two lowest states of these alkali atoms correspond to these two hyperfine levels. The $F = 2$ level corresponds to five states for which $M_F = 2, 1, 0, -1, -2$, and the $F = 1$ level corresponds to three states for which $M_F = 1, 0, -1$. Using Eq. (6.7), one can readily show that for the $F = 2$ states, $g_F = 1/2$, while for the $F = 1$ states, $g_F = -1/2$. It then follows immediately from Eq. (6.5) that the $F = 2$ states with positive values of M_F will have negative magnetic moments and be drawn to regions of low magnetic field, while states with negative values of M_F will have positive magnetic moments and not be trapped by the magnetic field. In contrast, for $F = 1$, the sublevel with $M_F = -1$ will be trapped by the magnetic field, while the $M_F = 1$ state will not be trapped.

An external magnetic field splits the $F = 2$ level into five sublevels corresponding to the different values of M_F and splits the $F = 1$ level into three sublevels. The splitting of the levels due to the magnetic field is shown in Fig. 6.11. In this figure, the vertical axis gives the energy in units of the constant A in Eq. (6.4) and the horizontal axis gives $g_F \mu_B B$ in units of A. An axially symmetric magnetic field splits the two hyperfine levels and mixes states having the same values of the quantum number M_F. In Fig. 6.11, we note that the curves corresponding to the energy levels $M_F = \pm 2$ are straight lines. This is due to the fact that each of these levels, which are the only states with $M_F = 2$ and $M_F = -2$, are unaffected by the other states. In contrast, the energy curves for the $M_F = 1, 0, -1$ levels bend as the strength of the magnetic field increases. The state with $F = 2, M_F = 1$ for example, mixes with the state with $F = 1, M_F = 1$.

In all the earliest experiments producing Bose-Einstein condensation, evaporative cooling was forced by using radio frequency signals with frequencies corresponding to transition between energy levels in large magnetic fields. Only atoms with large kinetic energies could penetrate into the high-field regions at the edges of the magnetic trap. The rf signal there caused atoms to make transitions to other states for which the sign of μ_z changed, and the atoms were no longer held by the magnetic trap. Since atoms with large kinetic energy were lost, the average kinetic energy and temperature of the remaining atoms decreased. Bose-Einstein condensation has been observed for hydrogen with $S = 1/2$ and $I = 1/2$, for ^{85}Rb with $S = 1/2$ and $I = 5/2$, and for metastable states of ^{4}He with $S = 1$ and $I = 0$.

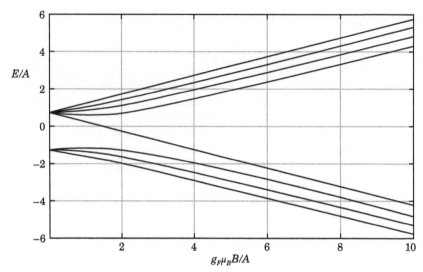

FIGURE 6.11 The splitting of $F = 1$ and $F = 2$ hyperfine levels in a magnetic field. The vertical axis gives the energy in units of the constant A and the horizontal axis gives $g_F \mu_B B$ in units of A.

The use of lasers and carefully designed magnetic traps to produce Bose-Einstein condensation and to produce lattices of atoms with controlled inter-atomic separations shows the power of modern laser technology to study and to design microscopic systems. The possibilities open to contemporary physics reaches far beyond the world imagined by Albert Einstein and, yet, shows the far-reaching consequences of his ideas.

Suggestions for further reading

A. Siegman, *Lasers* (Mill Valley, California: University Science Books, 1986).
J. Verdeyen, *Laser Electronics*, Third Edition (Englewood Cliffs, New Jersey: Prentice Hall, 1995).
M. Bertolotti, *Masers and Lasers: An Historical Approach* (Bristol: Adam Hilger, Techno House, 1983).
H.J. Metcalf and P. van der Straten, *Laser Cooling and Trapping* (New York: Springer-Verlag, 1999).
C.J. Pethick and H. Smith, *Bose-Einstein Condensation in Dilute Gases* (Cambridge: Cambridge University Press, 2002).

Basic equations

Hamiltonian of outer electron in the magnetic field of nucleus

$$H_{\text{hyp}} = A\mathbf{S} \cdot \mathbf{I}$$

Total angular momentum of electron and nucleus

$$\mathbf{F} = \mathbf{S} + \mathbf{I}$$

in each of these last two equations, $\mathbf{S}$ and $\mathbf{I}$ are the spin angular momenta of the electron and the nucleus, respectively.

The z-component of magnetic moment of outer electron

$$\mu_z = -g_F \mu_B M_F,$$

where

$$g_F = \frac{F(F+1) - I(I+I) + S(S+1)}{F(F+1)}$$

and M_F is the z-component of the total angular momentum.

The energy of the outer electron due the magnetic field B

$$E_{\text{mag}} = g_F \mu_B B M_F$$

Summary

The idea that a microscopic system can be stimulated to emit radiation was originally due to Einstein. After Einstein's prediction, more than a decade passed before the occurrence of stimulated emission was confirmed by the research of A. Ladenburg and his collaborators in Berlin. Twenty more years passed before stimulated emission was used in practical devices to amplify beams of electromagnetic radiation. The first device using stimulated emission, which operated in the microwave region of the spectrum, was built by a research group at Columbia University under the direction of Charles Townes.

The use of stimulated emission to amplify a beam of radiation depends upon creating a population inversion in which more atoms are in a higher-lying level than in a lower level. Population inversions have been created in a number of innovative ways. In the first laser developed by Theodore Maiman, the population inversion was created by pulses of light from a xenon flashlamp. The population inversion in the popular helium-neon laser is created by collisions between neon and helium atoms.

Modern experiments in which clouds of alkali atoms are cooled by laser beams with a frequency slightly below an absorption maximum now routinely achieve temperatures in the milli-Kelvin range. At these temperatures, the atoms can be trapped in a magnetic or optical field where they can be further cooled by evaporative cooling.

Questions

1. What is meant by stimulated emission?
2. Sketch a three-level laser scheme and describe the difficulties offered by this scheme.
3. The first working maser amplified radiation with a frequency corresponding to a transition of the ammonia molecule NH_3. Describe this transition of ammonia.
4. Sketch the energy levels of the ruby crystal showing the laser transition.
5. What physical process is responsible for populating the levels of neon that are responsible for the laser transition in the helium-neon laser?
6. How should the frequency of laser light used to cool a cloud of atoms be related to the absorption frequency of the atoms?
7. What happens when an atom being cooled by a laser beam approaches the laser?
8. How is a cloud of atoms cooled by evaporative cooling? What is the basic idea of this cooling process?
9. Write down an equation describing the energy of a magnetic moment μ in a magnetic field $\mathbf{B}$.
10. What effect does a magnetic field have upon atoms with a positive magnetic moment?
11. How does the fact that a quadrupole trap has a zero field at its center adversely effect the functioning of the trap?
12. How did Eric Cornell and Carl Wieman overcome the difficulty mentioned in the preceding question?
13. Write down an expression for the Hamiltonian of the spin $\mathbf{S}$ of the outer electron of an alkali atom interacting with the magnetic field produced by the spin $\mathbf{I}$ of the nucleus.
14. Write down the equation defining the total angular momentum $\mathbf{F}$ of an alkali atom.
15. Write down an equation describing how the z-component of the magnetic moment μ depends upon g_F and M_F.
16. What role do the ideas of Einstein play in the recent experiments of laser and evaporative cooling?

Problems

1. Using the energy levels of the ruby crystal shown in Fig. 6.7, estimate the ranges of frequencies necessary to excite an electron in the ground state up to the $^4F_1$ and $^4F_2$ bands.
2. Calculate the wavelength of light corresponding to the transition between the $2p_3$ and the $1s_3$ levels of neon.
3. For the levels of the neon atom having an excited p-electron, the orbital angular momentum of the excited electron can be coupled to the angular momentum J of the $2p$ core to form an angular momentum K and the spin of the excited electron can then be coupled to K to form a total angular momentum. The energy levels in this coupling scheme – called *JK coupling* – typically have a doublet structure with the two levels corresponding to the possible values of the total angular momentum. What are the possible values of K and the total angular momentum for the $2p^5(^2P_{3/2})3p$

levels of neon? The JK coupling scheme often gives a fairly good description of configurations for which the wave function of a highly excited electron has a small spatial overlap with the wave functions of an inner core of electrons.

4. The most common isotope of hydrogen has a nuclear spin $I = 1/2$. Using this fact, find what the possible values of F and M_F for hydrogen. Draw a figure showing the splitting of the hyperfine levels of hydrogen in a magnetic field.

5. For the hyperfine levels considered in the previous problems, calculate the values of g_F for each state and describe the transitions that would be used for evaporative cooling of a cloud of hydrogen atoms.

6. Suppose that a nucleus with a spin angular momentum $\mathbf{I}$ and an electron having a spin that we denote by $\mathbf{S}$ are in a weak magnetic field. The magnetic moment of the atom can be written

$$\mu_z = -g_s \mu_B S_z + g_N \mu_N I_z .$$

If the spin of the electron precesses much more rapidly about the total angular momentum $\mathbf{F}$ than it processes about the magnetic field, S_z may be replaced by its projection along $\mathbf{F}$, and we have

$$S_z \rightarrow \frac{(\mathbf{S} \cdot \mathbf{F}) F_z}{F(F+1)} .$$

A similar replacement may be made for I_z. Using arguments of this kind, show that the component of the magnetic moment of the atom along the magnetic field can be written

$$\mu_z = -g_F \mu_B M_F,$$

where g_F is given by Eq. (6.6).

7. The ground state of ^{85}Rb has $S = 1/2$ and $I = 5/2$. What are the possible values of the total angular momentum F for this isotope? Sketch the splitting of the hyperfine levels in a magnetic field indicating which levels are pure and correspond to straight lines and which levels are mixed and correspond to curves.

Chapter 7

Diatomic molecules

Contents

7.1 The hydrogen molecular ion   153   References   166
7.2 The Hartree-Fock method   163   Summary   167
7.3 Exoplanets   164   Questions   168

Diatomic molecules make up about 98% of the atmosphere of the Earth by weight and they could well make up a substantial part of the atmospheres of the exoplanets that support life.

7.1 The hydrogen molecular ion

We begin this chapter by considering the hydrogen-like molecular ion illustrated in Fig. 7.1 with a charge of $Z_a e$ at one nuclear site, a charge of $Z_b e$ at a second nuclear site, and a single electron. With $Z_a = 1$ and $Z_b = 1$, the hydrogen-like ion is the hydrogen molecular ion H_2^+ which is the simplest of diatomic molecule, while for $Z_a = 2$ and $Z_b = 0$ the hydrogen-like molecular ion is just the helium ion He^{2+} which we shall find provides a simple test case for our theories.

In atomic units with $\hbar$, $4\pi\epsilon$ and the charge and mass of the electron all equal to one, the Schrödinger equation for the hydrogen-like ion is

$$\left[\frac{-1}{2}\nabla^2 - \frac{Z_a}{r_a} - \frac{Z_b}{r_b} \right]\psi(\mathbf{r}) = E\psi(\mathbf{r}), \tag{7.1}$$

where r_a and r_b denote the distances of the electron from the two nuclei separated by R atomic units, and the charges of the two nuclei in units of the electron charge e are denoted by Z_a and Z_b.

Diatomic molecules are usually described in *spheroidal coordinates*

$$\begin{aligned}
\xi &= (r_a + r_b)/R, &\quad \text{where } 1 \le \xi < \infty, \\
\eta &= (r_a - r_b)/R, &\quad \text{where } -1 \le \eta \le 1, \\
\phi &= \phi, &\quad \text{where } 0 \le \phi \le 2\pi,
\end{aligned} \tag{7.2}$$

in which r_a and r_b are the distances from the two nuclei and ϕ measures rotations about the molecular axis.

With these coordinates, the Laplacian operator assumes the form

$$\nabla^2 = \frac{4}{R^2(\xi^2 - \eta^2)}\left[\frac{\partial}{\partial\xi}(\xi^2 - 1)\frac{\partial}{\partial\xi} + \frac{\partial}{\partial\eta}(1 - \eta^2)\frac{\partial}{\partial\eta} + \left(\frac{1}{\xi^2 - 1} + \frac{1}{1 - \eta^2} \right)\frac{\partial}{\partial\phi^2} \right] \tag{7.3}$$

and the potential energy of the electron in the field of the two nuclei is

$$\frac{Z_a}{r_a} + \frac{Z_b}{r_b} = \frac{2}{R^2(\xi^2 - \eta^2)}\left[R(Z_a + Z_b)\xi - R(Z_a - Z_b)\eta \right] \tag{7.4}$$

This last equation may be written more compactly if one uses the special symbols, Z and Δ, to represent the sum and difference of the two nuclear charges. The potential energy due to the two nuclei then becomes

$$\frac{Z_a}{r_a} + \frac{Z_b}{r_b} = \frac{2}{R^2(\xi^2 - \eta^2)}(RZ\xi - R\Delta\eta)$$

Modern Physics with Modern Computational Methods. https://doi.org/10.1016/B978-0-12-817790-7.00014-7

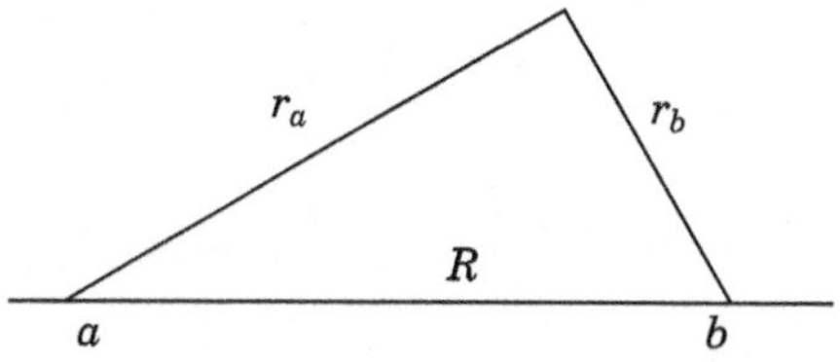

FIGURE 7.1 Positions of the nuclei and the electron in the hydrogen molecular ion.

where $Z = Z_a + Z_b$ and $\Delta = Z_a - Z_b$. If one supposes that the single-electron wave function of the general form

$$\psi(\xi, \eta, \phi) = \frac{e^{im\phi}}{\sqrt{2\pi}} u(\xi, \eta), \tag{7.5}$$

the Schrödinger equation for a hydrogen-like molecular ion can then be written

$$-\frac{\partial}{\partial \xi}(\xi^2 - 1)\frac{\partial u}{\partial \xi} + \frac{\partial u}{\partial \eta}(1 - \eta^2)\frac{\partial u}{\partial \eta} + \left(\frac{m}{\xi^2 - 1} + \frac{m}{1 - \eta^2}\right)u + (R\Delta\eta - RZ\xi)u = \frac{R^2}{2}(\xi^2 - \eta^2)\epsilon u, \tag{7.6}$$

which is in the form of a generalized eigenvalue problem.

In contrast to the eigenvalue problems considered in the early chapters of this book which depended upon differential operators with a single independent variable, Eq. (7.6) depends upon two independent variables, ξ and η. It follows immediately from Eq. (7.2) that the ξ coordinate of a point on the molecular axis between the nucleus a and the nucleus b has the value one and is larger than one for every other point. The η coordinate will have the value of minus one for the a nucleus and the value one for the b-nucleus. The physical region is thus a rectangle with $\xi = 1$ on the lower boundary and with η equal to -1 and $+1$ on the left and right boundaries.

In Chapters 4 and 5, we found that atomic orbitals oscillated very rapidly near the atomic nucleus and then gradually declined to zero for large distances. The wave functions of a diatomic molecule can be expected to have similar properties varying rapidly near the nuclei and falling off gradually for large distances.

One can ensure that the grid points are clustered near the nuclei where the wave functions oscillate most rapidly by making the variable transformations

$$\eta = \cos(v) \quad \text{where } 0 \le v \le \pi, \tag{7.7}$$

$$\xi = \cosh(\mu) \quad \text{where } 0 \le \mu \le \infty. \tag{7.8}$$

The physical region then extends from zero on the left to π on the right and from zero to a large enough value of μ that the molecular wave functions can be expected to be approximately equal to zero.

The Schrödinger equation then becomes

$$-\frac{\partial^2 u}{\partial v^2} - \left(\frac{\cos v}{\sin \mu}\right)\frac{\partial u}{\partial v} - \frac{\partial^2 u}{\partial \mu^2} - \left(\frac{\cosh \mu}{\sinh \mu}\right)\frac{\partial u}{\partial \mu} + \left(\frac{m}{\xi^2 - 1} + \frac{m}{1 - \eta^2}\right)u + (RZ\eta - R\Delta\xi)u = \frac{R^2}{2}(\xi^2 - \eta^2)\epsilon u. \tag{7.9}$$

For the solutions of a partial differential equation to be well-defined, the wave functions must satisfy boundary conditions on the exterior boundaries. Eq. (7.7) ensures that the derivative of the wave function is equal to zero at the left and right boundaries. We have

$$\frac{\partial u}{\partial v} = \frac{\partial u}{\partial \eta}\frac{\partial \eta}{\partial v} = -\sin v\frac{\partial u}{\partial \eta} = 0 \quad \text{for } v = 0 \quad \text{or} \quad v = \pi$$

Similarly, using Eq. (7.8) the value of the derivative of the wave function at lower boundary may be shown to be

$$\frac{\partial u}{\partial \mu} = \frac{\partial u}{\partial \xi}\frac{\partial \xi}{\partial \mu} = \sinh \mu \frac{\partial u}{\partial \xi} = 0 \quad \text{for } \mu = 0 \,,$$

while the wave function u goes to zero for large values of ξ. The derivative of the wave function thus goes to zero on the left and right boundaries and on the lower boundary, and the function is zero on the upper boundary.

In the following, we shall consider the solutions of the equation with $m = 0$ and will denote the ν variable by x and the μ variable by y. The Schrödinger equation can then be written

$$-ca(x)\frac{d^2u}{dx^2} - cad(x)\frac{du}{dx} - cb(y)\frac{d^2u}{dy^2} - cbd(y)\frac{du}{dy} + cc(x,y) = \frac{R^2}{2}(\xi^2 - \eta^2)\epsilon u' \tag{7.10}$$

where $ca(x) = 1$, $cad(x) = \cos x / \sin x$, $cb(y) = 1$, $cbd(y) = \cosh y / \sinh y$, and

$$cc(x,y) = R\Delta\cos(x) - RZ\cosh(y).$$

In Eq. (7.10), $ca(x)$ and $cad(x)$ are the coefficients of the second and first derivative with respect to x, $cb(y)$, and $cbd(y)$ are the coefficients of the second and first derivative with respect to y, and $cc(x, y)$ is the coefficient of the function $u(x, y)$. The advantage of expressing partial differential and eigenvalue problems in such a generic form is that a program for solving a particular partial differential equation or eigenvalue problem can be converted into a program for solving another problem by changing the definition of the functions, ca, cad, cb, cbd, and cc without otherwise changing the program.

In this section we will solve the eigenvalue equation (7.10) using the spline collocation method used in Chapter 3 to solve the eigenvalue problems of an electron in one dimensional well and of a simple harmonic oscillator. We recall that an approximate solution of an ordinary differential equation can be expressed as a linear combination of the Hermite splines illustrated in Fig. 3.3 of Chapter 3. We have

$$u(x) = \sum_{i=0}^{N}[\alpha_i v_i(x) + \beta_i s_i(x)]. \tag{7.11}$$

There are N intervals over the entire range of the variable x and within each intervals Gauss quadrature points are defined by the equations

$$\xi_{i1} = x_{i-1} + \frac{3 - \sqrt{3}}{6}h, \quad \xi_{i2} = x_{i-1} + \frac{3 + \sqrt{3}}{6}h.$$

For the Gauss quadrature points, ξ_{i1} and ξ_{i2}, within the i-th interval, four functions of the spline basis, v_{i-1}, s_{i-1}, v_i, and s_i, have nonzero values. These functions are illustrated in Fig. 3.4 of Chapter 3. The values of the solution at the two Gauss points within the ith interval are

$$u(\xi_{i1}) = \alpha_{i-1}v_{i-1}(\xi_{i1}) + \beta_{i-1}s_{i-1}(\xi_{i1}) + \alpha_i v_i(\xi_{i1}) + \beta_{i-1}s(\xi_{i1}) \tag{7.12}$$

$$u(\xi_{i2}) = \alpha_{i-1}v_{i-1}(\xi_{i2}) + \beta_{i-1}s_{i-1}(\xi_{i2}) + \alpha_i v_i(\xi_{i2}) + \beta_{i-1}s(\xi_{i2}) \tag{7.13}$$

Denoting the values of the spline functions at the first Gauss point (ξ_{i1}) by b_{11}, b_{12}, b_{13}, and b_{14} and the values of the spline functions at the second Gauss point (ξ_{i2}) by b_{21}, b_{22}, b_{23}, and b_{24}, these last two equation can be written simply

$$u(\xi_{i1}) = b_{11}\alpha_{i-1} + b_{12}\beta_{i-1} + b_{13}\alpha_i + b_{14}\beta_i \tag{7.14}$$

$$u(\xi_{i2}) = b_{21}\alpha_{i-1} + b_{22}\beta_{i-1} + b_{23}\alpha_i + b_{24}\beta_i, \tag{7.15}$$

or in matrix form

$$\begin{bmatrix} u(\xi_{i1}) \\ u(\xi_{i2}) \end{bmatrix} = [\mathbf{B}_i] \times \begin{bmatrix} \alpha_{i-1} \\ \beta_{i-1} \\ \alpha_i \\ \beta_i \end{bmatrix} \tag{7.16}$$

where

$$[\mathbf{B}_i] = \begin{bmatrix} b_{11} & b_{12} & b_{13} & b_{14} \\ b_{21} & b_{22} & b_{23} & b_{24} \end{bmatrix} \tag{7.17}$$

Notice that each column of the matrix B_i corresponds to a spline coefficient and each row corresponds to a collocation point.

Treating each of the subintervals in a similar manner, the vector $\mathbf{u_G}$ consisting of the values of the approximate solution at the Gauss points can be written as the product of a matrix times a vector

$$\mathbf{u_G} = \mathbf{Bu},$$

where the vector

$$\mathbf{u} = [\alpha_0, \beta_0, \alpha_1, \beta_1, ..., \alpha_N, \beta_N]^T$$

consists of the spline coefficients and the vector

$$\mathbf{u_G} = [u(\xi_{11}), u(\xi_{12}), u(\xi_{21}), u(\xi_{22}), ..., u(\xi_{N1}), u(\xi_{N2})]^T,$$

consists of the values of the solution at the Gauss quadrature points. There are $N + 1$ nodes, $x_0, x_1 ..., x_N$, with each node having two splines and N subintervals with each interval having two Gauss quadrature points. Because there are $2N + 2$ splines and $2N$ collocation points, $\mathbf{B}$ is a rectangular matrix having $2N + 2$ columns and $2N$ rows with the structure

$$\mathbf{B} = \begin{bmatrix} \boxed{\mathbf{B}_1} & & & & \\ & \boxed{\mathbf{B}_2} & & & \\ & & \ddots & & \\ & & & \boxed{\mathbf{B}_{N-1}} & \\ & & & & \boxed{\mathbf{B}_N} \end{bmatrix}. \tag{7.18}$$

Two adjacent blocks $\mathbf{B}_i$ and $\mathbf{B}_{i+1}$ overlap in two columns.

The first and second derivatives of the approximate solution can also be represented by matrices with the same block structure. We express the vector $\mathbf{u'_G}$ consisting of the values of the first derivative of the solution at the Gauss points as

$$\mathbf{u'_G} = \mathbf{Cu}, \tag{7.19}$$

where

$$\mathbf{u} = [\alpha_0, \beta_0, \alpha_1, \beta_1, ..., \alpha_N, \beta_N]^T \tag{7.20}$$

$$\mathbf{u'_G} = [u'(\xi_{11}), u'(\xi_{12}), u'(\xi_{21}), u'(\xi_{22}), ..., u'(\xi_{N1}), u'(\xi_{N2})]^T. \tag{7.21}$$

The negative of the second derivative of the function at the Gauss points can be written

$$-\mathbf{u''_G} = \mathbf{Au}, \tag{7.22}$$

where

$$\mathbf{u} = [\alpha_0, \beta_0, \alpha_1, \beta_1, ..., \alpha_N, \beta_N]^T \tag{7.23}$$

$$\mathbf{u''_G} = [u''(\xi_{11}), u''(\xi_{12}), u''(\xi_{21}), u''(\xi_{22}), ..., u''(\xi_{N1}), u''(\xi_{N2})]^T. \tag{7.24}$$

We shall now use the spline collocation method to find the energy eigenvalues of the hydrogen molecular ion. For this purpose, we define a partitioning of the two coordinate axes which we denote simply by x (in place of ν) and y (in place of μ) to be

$$a = x_0 < x_1 < x_2 \cdots < x_M = b,$$

and

$$c = y_0 < y_1 < y_2 \cdots < y_N = d.$$

Functions of x and y may be approximated using a basis which consists of products of the members of the Hermite basis considered previously. This approximation has the form

$$u(x, y) = \sum_{i=0}^{N} \sum_{j=0}^{N} [\alpha_{ii} v_i(x) v_j(y) + \beta_{ii} s_i(x) v_j(y) + \gamma_{ii} v_i(x) s_j(y) + \delta_{ii} s_i(x) s_j(y)].$$

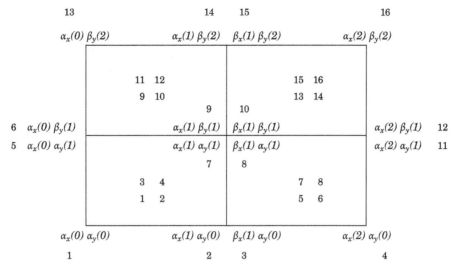

FIGURE 7.2 The four by four grid used for the MATLAB program $H2p.m$ described in this Chapter.

The values of a function and its partial derivatives at the collocation points may be expressed using tensor products of the matrices $\mathbf{A}_x$, $\mathbf{B}_x$, and $\mathbf{C}_x$ for the operators corresponding to the x coordinates and matrices $\mathbf{A}_y$, $\mathbf{B}_y$, and $\mathbf{C}_y$ for operators corresponding to the y coordinates. Using spline collocation, the discrete representation of the Schrödinger equation (7.10) can be written

$$\mathbf{C}_{mat}\mathbf{u} = \mathbf{D}(\frac{R^2}{2}(\xi^2 - \eta^2))\epsilon\mathbf{B}_x \otimes \mathbf{B}_y\mathbf{u}, \tag{7.25}$$

where the collocation matrix $\mathbf{C}_{mat}$ may be written

$$\mathbf{C}_{mat} = \mathbf{A}_x \otimes \mathbf{B}_y - \mathbf{D}(cad(x))\mathbf{C}_x \otimes \mathbf{B}_y + \mathbf{B}_x \otimes \mathbf{A}_y - \mathbf{D}(cbd(y))\mathbf{B}_x \otimes \mathbf{C}_y + \mathbf{D}(cc(x,y))\mathbf{B}_x \otimes \mathbf{B}_y. \tag{7.26}$$

The diagonal matrices, $\mathbf{D}(cad(x))$, $\mathbf{D}(cbd(y))$, $\mathbf{D}(cc(x,y))$, and $\mathbf{D}(\frac{R^2}{2}(\xi^2 - \eta^2))$ have the values of the functions at the Gauss points appearing along the diagonal. The functions for the terms on the left hand side of Eq. (7.26) for the collocation matrix are

$$cad(x) = \cos(x)/\sin(x)$$
$$cbd(y) = \cosh(y))/\sinh(y)$$
$$cc(x,y) = R(Z_a - Z_b) * \cos(x) - R(Z_a + Z_b) * \cosh(y)$$

The discrete form of the Schrödinger equation has the form of the generalized eigenvalue problem for the finite well considered in Chapter 3.

As for the MATLAB® programs described in earlier chapters, we will first consider a MATLAB program using a very sparse grid with only two intervals or four grid points in each direction. The $\mathbf{A}_x$, $\mathbf{B}_x$, and $\mathbf{C}_x$ matrices and the $\mathbf{A}_y$, $\mathbf{B}_y$, and $\mathbf{C}_y$ matrices will then each have four rows and six columns and the tensor product matrices appearing in Eq. (7.25) will each have sixteen rows and thirty six columns. In the third and fourth chapters of this book, we found that boundary conditions can be imposed by deleting the columns of the matrices corresponding to the splines with zero coefficients. Fig. 7.2 shows our four by four grid. Notice that the four collocation points are numbered within each rectangle and the nonzero spline indices are shown at each vertex of the grid. Recall that boundary conditions requiring that the wave function be zero on the boundary are called Dirichlet boundary conditions and boundary conditions requiring that the derivative of the wave function be zero on the boundary are called Neumann boundary conditions. Only the $\alpha_x(0)\alpha_y(0)$ pair appears at the lower left-hand corner of Fig. 7.2. The $\beta_x(0)$ spline does not appear because of the Neumann boundary condition on the left boundary and the $\beta_y(0)$ does not occur because of the Neumann boundary condition on the lower boundary. More generally the $\beta_x(0)$, $\beta_x(2)$, and $\beta_y(0)$ do not appear in Fig. 7.2 because of the Neumann boundary conditions on the two vertical boundaries and the Neumann boundary condition on lower boundary. The $\alpha_y(2)$ spline does not contribute because of the Dirichlet boundary condition on the upper boundary. There are a total of sixteen spline pairs that contribute and sixteen collocation points. With this grid, the collocation matrix has sixteen rows and sixteen columns.

The products of splines used in such a calculation are given in Table 7.1.

TABLE 7.1 Pairs of splines for a spline collocation calculation of the lowest eigenvalues for the hydrogen molecular ion.

Number	Splines	isp	Number	Splines	isp
1	$\alpha_x(0)\alpha_y(0)$	1	19	$\alpha_x(0)\beta_y(1)$	6
2	$\beta_x(0)\alpha_y(0)$		20	$\beta_x(0)\beta_y(1)$	
3	$\alpha_x(1)\alpha_y(0)$	2	21	$\alpha_x(1)\beta_y(1)$	9
4	$\beta_x(1)\alpha_y(0)$	3	22	$\beta_x(1)\beta_y(1)$	10
5	$\alpha_x(2)\alpha_y(0)$	4	23	$\alpha_x(2)\beta_y(1)$	12
6	$\beta_x(2)\alpha_y(0)$		24	$\beta_x(2)\beta_y(1)$	
7	$\alpha_x(0)\beta_y(0)$		25	$\alpha_x(0)\alpha_y(2)$	
8	$\beta_x(0)\beta_y(0)$		26	$\beta_x(0)\alpha_y(2)$	
9	$\alpha_x(1)\beta_y(0)$		27	$\alpha_x(1)\alpha_y(2)$	
10	$\beta_x(1)\beta_y(0)$		28	$\beta_x(1)\alpha_y(2)$	
11	$\alpha_x(2)\beta_y(0)$		29	$\alpha_x(2)\alpha_y(2)$	
12	$\beta_x(2)\beta_y(0)$		30	$\beta_x(2)\alpha_y(2)$	
13	$\alpha_x(0)\alpha_y(1)$	5	31	$\alpha_x(0)\beta_y(2)$	13
14	$\beta_x(0)\alpha_y(1)$		32	$\beta_x(0)\beta_y(2)$	
15	$\alpha_x(1)\alpha_y(1)$	7	33	$\alpha_x(1)\beta_y(2)$	14
16	$\beta_x(1)\alpha_y(1)$	8	34	$\beta_x(1)\beta_y(2)$	15
17	$\alpha_x(2)\alpha_y(1)$	11	35	$\alpha_x(2)\beta_y(2)$	16
18	$\beta_x(2)\alpha_y(1)$		36	$\beta_x(2)\beta_y(2)$	

Before discussing our MATLAB program for finding the energy eigenvalues and the wave function, we should first mention that we have developed Fortran programs that use spline collocation to solve the partial differential equations and eigenvalue equations that arise in the theory of diatomic molecules and our MATLAB programs were first developed to check our Fortran program for solving the two-electron pair equation. Our Fortran programs and our MATLAB programs work in unison with each other. The collocation points shown in Fig. 7.2 are labeled as they are in our Fortran programs. In the square in the lower left-hand corner the collocation points are labeled 1, 2, 3, and 4, and they are labeled for the rectangle in the lower right as 5, 6, 7, and 8. Reading from bottom to top and from left to right in Fig. 7.2, the indices of the collocation points are defined by the array *icol* given below

$$icol = 1\ 2\ 5\ 6\ 3\ 4\ 7\ 8\ 9\ 10\ 13\ 14\ 11\ 12\ 15\ 16 \tag{7.27}$$

Similarly, the array *isp* defined below shows how the spline variables are defined

$$isp = 1\ 3\ 4\ 5\ 13\ 19\ 15\ 16\ 21\ 22\ 17\ 23\ 31\ 33\ 34\ 35 \tag{7.28}$$

Reading down the third and sixth columns of Table 7.1, one can see that the first spline with a nonzero coefficient is spline number 1, the second spline with nonzero coefficient is spline number 3, and for instance the fifth and sixth splines with nonzero coefficients are splines number thirteen and nineteen. Since the rows of the colocation matrix correspond to the collocation points and the columns of the collocation matrix correspond to the spline variables, the arrays *icol* and *isp* determine how the rows and the columns of the collocation matrix produced by MATLAB must be rearranged to bring the collocation matrix produced by the MATLAB program into agreement with the collocation matrix produce by our fundamental Fortran programs. MATLAB functions for producing the *icol* and *isp* arrays for larger numbers of grids are given in Appendix CC.

MATLAB Program H2p.m

A MATLAB program for the hydrogen molecular ion using the spline collocation method.

```
r=2.0;
za=1.0
```

```
zb=1.0
xmax=pi;
nx=2
deltax=xmax/nx;
deltaxs=deltax*deltax;
ny=2;
ymax=3.4;
deltay=ymax/ny;
deltays=deltay*deltay;

% Gauss Points
xi1=(3-sqrt(3))/6
xi2=(3+sqrt(3))/6

% Coefficients A, B, and C matrices
p1=(9 -4*sqrt(3))/18;
p2=(9+4*sqrt(3))/18;
p3=(3-sqrt(3))/36;
p4=(3+sqrt(3))/36;
p5=2*sqrt(3);
p6=sqrt(3)-1;
p7=sqrt(3)+1;

% Construct 2 x 4 A, B,and C matices
Ax=[p5 p7 -p5 p6;-p5 -p6 p5 -p7]/deltaxs;
Ay=[p5 p7 -p5 p6;-p5 -p6 p5 -p7]/deltays;

Bx=[p2 p4 p1 -p3; p1 p3 p2 -p4]
By=[p2 p4 p1 -p3; p1 p3 p2 -p4]

Cx=[1.0 -1.0/p5 -1.0 1.0/p5;1.0 1.0/p5 -1.0 -1.0/p5]/deltax;
Cy=[1.0 -1.0/p5 -1.0 1.0/p5;1.0 1.0/p5 -1.0 -1.0/p5]/deltay;

% Construct Full A, B, and C matrices
Axmat=zeros(2*nx, 2*nx+2);
for row=1:2:2*nx;
  Axmat(row:row+1, row:row+3)=Ax;
end

Aymat=zeros(2*ny, 2*ny+2);
for row=1:2:2*ny;
  Aymat(row:row+1, row:row+3)=Ay;
end

Bxmat=zeros(2*nx, 2*nx+2);
for row=1:2:2*nx
  Bxmat(row:row+1, row:row+3)=Bx;
end

Bymat=zeros(2*ny, 2*ny+2);
for row=1:2:2*ny
  Bymat(row:row+1, row:row+3)=By;
end
Bymat;
```

```
Cxmat=zeros(2*nx, 2*nx+2);
for row=1:2:2*nx   Cxmat(row:row+1, row:row+3)=Cx;
end


Cymat=zeros(2*ny, 2*ny+2);
for row=1:2:2*ny
  Cymat(row:row+1, row:row+3)=Cy;
end

% Construct Diagonal Dcad matrix
cad = zeros(1,4*nx*ny);


jj = 0;
for j = 1:ny
   x = 0.0;
   for i = 1:nx
      x1 = x + xi1*deltax;
      x2 = x + xi2*deltax;
      cad(jj+1) = cos(x1)/sin(x1);
      cad(jj+2) = cos(x2)/sin(x2);
      cad(jj+3) = cos(x1)/sin(x1);
      cad(jj+4) = cos(x2)/sin(x2)
      jj = jj+4;
      x = x + deltax;
   end
end


Dcad = zeros(4*nx*ny,4*nx*ny);
Dcad = diag(cad);

% Construct Diagonal Dcbd matrix
cbd = zeros(1,4*nx*ny);


jj = 0;
y = 0.0;
for j = 1:ny
   y1 = y + xi1*deltay;
   y2 = y + xi2*deltay;
   for i = 1:nx
      cbd(jj+1) = cosh(y1)/sinh(y1);
      cbd(jj+2) = cosh(y1)/sinh(y1);
      cbd(jj+3) = cosh(y2)/sinh(y2);
      cbd(jj+4) = cosh(y2)/sinh(y2);
      jj = jj+4;
   end
   y = y + deltay;
end
Dcbd = zeros(4*nx*ny,4*nx*ny);
Dcbd = diag(cbd);

% Construct Diagonal Dcc matrix
jj = 0;
y = 0.0;
```

```
for j = 1:ny
   y1 = y + xi1*deltay;
   y2 = y + xi2*deltay;
   x = 0.0;
   for i = 1:nx
      x1 = x + xi1*deltax;
      x2 = x + xi2*deltax;
      cc(jj+1) = r*(za+zb)*cosh(y1)-r*(za-zb)*cos(x1);
      cc(jj+2) = r*(za+zb)*cosh(y1)-r*(za-zb)*cos(x2);
      cc(jj+3) = r*(za+zb)*cosh(y2)-r*(za-zb)*cos(x1);
      cc(jj+4) = r*(za+zb)*cosh(y2)-r*(za-zb)*cos(x2);
      jj = jj+4
      x = x + deltax;
   end
   y = y + deltay;
end
Dcc = zeros(4*nx*ny,4*nx*ny);
Dcc = diag(cc)

% The icol array
icol(1)=1;
icol(2)=2;
icol(3)=5;
icol(4)=6;
icol(5)=3;
icol(6)=4;
icol(7)=7;
icol(8)=8;
icol(9)=9;
icol(10)=10;
icol(11)=13;
icol(12)=14;
icol(13)=11;
icol(14)=12;
icol(15)=15;
icol(16)=16;
icol'

% The isp Array
isp = zeros(1,(2*nx+2)*(2*ny+2))
isp(1)=1;
isp(2)=3;
isp(3)=4;
isp(4)=5;
isp(5)=13;
isp(6)=19;
isp(7)=15;
isp(8)=16;
isp(9)=21;
isp(10)=22;
isp(11)=17;
isp(12)=23;
isp(13)=31;
isp(14)=33;
```

```
isp(15)=34;
isp(16)=35;
isp'

BxBy = kron(Bxmat,Bymat)
ColBxBy=zeros(4*ny*ny,4*nx*ny);

for j = 1:4*nx*ny
   for i = 1:4*nx*ny
      ColBxBy(i,j) = BxBy(icol(i),isp(j));
   end
end

AxBy = kron(Axmat,Bymat)
ColAxBy=zeros(4*nx*ny,4*nx*ny);

for j = 1:4*nx*ny
   for i = 1:4*nx*ny
      ColAxBy(i,j) = AxBy(icol(i),isp(j));
   end
end

CxBy = kron(Cxmat,Bymat)
ColCxBy=zeros(4*nx*ny,4*nx*ny);

for j = 1:4*nx*ny
   for i = 1:4*nx*ny
      ColCxBy(i,j) = CxBy(icol(i),isp(j));
   end
end

BxAy = kron(Bxmat,Aymat)
ColBxAy=zeros(4*nx*ny,4*nx*ny);

for j = 1:4*nx*ny
   for i = 1:4*nx*ny
      ColBxAy(i,j) = BxAy(icol(i),isp(j));
   end
end

BxCy = kron(Bxmat,Cymat)

ColBxCy=zeros(4*nx*ny,4*nx*ny);

for j = 1:4*nx*ny
   for i = 1:4*nx*ny
      ColBxCy(i,j) = BxCy(icol(i),isp(j));
   end
end

Colmat = ColAxBy + Dcad*ColCxBy + ColBxAy + Dcbd*ColBxCy + Dcc*ColBxBy

Colmat
```

The first lines of this MATLAB program give the charges (z_a and z_b) of the two nuclear centers and defines the grid. The program then gives the coefficients ($xi1$ and $xi2$) necessary to define the Gauss quadrature points and the p coefficients necessary to form the A, B, and C matrices. After forming the 2×4 A, B, and C matrices, the program then forms the full Ax, Bx, Cx and Ay, By, Cy matrices, each of which has $2nx$ rows and $2ny + 2$ columns.

The functions $cad(x)$, $cbd(y)$, and $cc(x, y)$ that appear in Eq. (7.25) are represented in the theory by diagonal matrices, $Dcad$, $Dcbd$, and Dcc with the values of the functions at the collocation points along the diagonal of the matrices. After forming the D matrices and defining the $icol$ array to label the collocation points and the isp array to label the splines with nonzero coefficients, the program then forms the tensor products of the Ax, Bx, Cx and the Ay, By, and Cy matrices. For the case $nx = 2$ and $ny = 2$, each of the tensor products have 16 rows and 36 columns. With the formation of each tensor product, the program then forms the corresponding matrix with 16 rows and 16 columns. These matrices are then used to form the collocation matrix $Colmat$. MATLAB functions, icol.m and isp.m, which may be used for larger values of nx and ny, are given in Appendix CC.

As noted earlier, Eq. (7.25) has the same form as the generalized eigenvalue equation for a particle moving in a finite well. As we found in Chapter 3, the eigenvalues for a generalized eigenvalue problem of this kind can be found using the MATLAB function eig. They can also be found using routines from the Fortran package LAPACK. The lowest eigenvalues of the hydrogen-like ion obtained using a grid with nx and ny equal to 96 are given in Table 7.2 together with the values obtained by my collaborator Jacek Kobus using an accurate finite difference formula for the derivatives and the values obtained by J.D. Power who achieved a power series solution of the Schrödinger equation.

TABLE 7.2 The lowest three eigenvalues of the hydrogen molecular ion using spine collocation are compared with the eigenvalues of Jacek Kobus using his Hartree Fock program [1] and the values of J.D. Powers [2] obtained by a power series.

nlm	Spline Collocation	Hartree-Fock	Power Series
200	−0.3608649	−0.3608649	−0.3608649
210	−0.6675344	−0.6675344	− 0.6675344
100	−1.1026342	−1.1026342	−1.1026342

For a grid with 192 collocation points in each direction, the spine collocation method agrees with Jacek Kobus's Hartree-Fock calculation and Power's power series values to seven significant figures.

7.2 The Hartree-Fock method

The Hartree-Fock method provides a good approximate description of diatomic molecules near their equilibrium positions and serves as the starting point of calculations using many-body perturbation theory. Numerical solutions of the Hartree-Fock equations for diatomic molecules were first reported by E.A. McCullough Jr [3] and were reported by Laaksonen and his collaborators [4] and by J. Kobus [1]. More recently, we have reported numerical solutions using the spline collocation method [5,6].

The Hartree-Fock equations for diatomic molecules include all of the terms for the hydrogen molecular ion and additional terms corresponding to the direct and exchange Coulomb interaction of an electron with the other electrons in the molecule. The Hartree-Fock equations for a diatomic molecule can be written

$$-\frac{\partial^2 u}{\partial v} - \left(\frac{\cos v}{\sin \mu}\right)\frac{\partial u}{\partial v} - \frac{\partial^2 u}{\partial \mu} - \left(\frac{\cosh \mu}{\sinh \mu}\right)\frac{\partial u}{\partial \mu} + \left(\frac{m}{\xi^2 - 1} + \frac{m}{1 - \eta^2}\right)u$$
$$+ (RZ\eta - R\Delta\xi)u + \frac{R^2}{2}(\xi^2 - \eta^2)\sum_b(2J_b - K_b)u = \frac{R^2}{2}(\xi^2 - \eta^2)\epsilon u, \quad (7.29)$$

where J_b represents the direct Coulomb interaction of the electron with another electron in the b shell and K_b represents the corresponding exchange interaction. Fig. 7.3 gives the total energy of the OH molecule as a function of the separation of the O and H atoms for a Hartree-Fock calculation of OH.

In this chapter, we what role accurate numerical calculations of the properties of diatomic molecules might play in helping to find life on other planets. I have decided not to focus on the O_2 molecule, which being homo-nuclear has no dipole moment. The single absorption lines of O_2 observed in the spectra of exoplanets can not be due to electric dipole

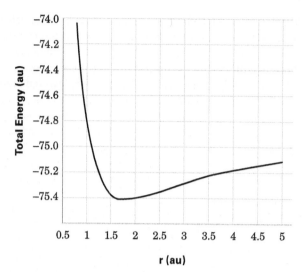

FIGURE 7.3 Total energy of the OH molecule as a function of the separation of the O and H atoms.

transitions but must be due to higher order – highly forbidden processes. Not wanting to carry out magnetic octopole calculations myself, I have decided to consider instead light emitted and absorbed by the OH molecule which is the cause of the intense airglow spectrum in the far-red and near-infrared parts of the spectrum. The OH molecule is produced in highly excited vibrational state by the reaction

$$O_3 + H \rightarrow OH^* + O_2$$

with the production rate or OH being highest in Earth's atmosphere at altitudes of 70–90 km. At such altitudes, a little atomic hydrogen exists as the result of sunlight photodissociation of H_2O and there is enough ozone present. The airglow spectrum, which is known as the Meindel vibrational spectrum, cannot be observed from ground-based observatories very effectively because the terrestrial airglow is blindingly bright in the near-infrared, but it can be an important tracer of oxygen bearing exoplanets observed with telescope in space.

As we have seen in Chapter 5, the Goldstone diagrams that arise in many-body perturbation theory can be evaluated by solving two-electron pair equations for the linear combination of excited orbitals that contribute to the Goldstone diagrams. Because the spherical harmonics for each electron of an atom can be factored out of the two-electron wave function, the first-order pair equation for atoms only depends upon the two radial coordinates, r_1 and r_2. In contrast to this simple two-variable partial differential equation, the pair functions for diatomic molecules depend upon five independent variable: two variable to locate each electron in the plane of the molecule and an angular variable that determines how much a vector pointing to the first electron must be rotated about the molecular axis so that it lies in the plane of the second electron. With one hundred grid points for each variable, the collocation matrix has $100^5 = 10^{10}$ rows and columns. As shown in a recent article [7], the huge matrices that arise in molecular many-body calculations can be reduced to a number of smaller problems by using the technique of domain decomposition to divide the entire region into small subregions in which equations are solved independently. A number of these subregions can be assigned to each processor of a parallel computer.

7.3 Exoplanets

Thirty years have now passed since the identification of the first exoplanet orbiting a star other than our Sun, and we have now identified thousands of exoplanets orbiting distant stars. A fine review of the progress that has been made to identified habitable worlds where life as we know it may be found has been written by Lisa Kaltenegger at Cornell University [8].

The size of the planets can be determined by the length of time the planet obscures light coming from its host star when it passes in front of the star and the radial velocity of the planet may be used to determine its approximate mass. If both mass and radius are known, the mean density of the planet can be used to derive its composition and compare it to planets in our own Solar System. Exoplanets with a mass of less than 10 Earth masses are commonly considered rocky planets and exoplanets with a mass known to be more than 10 Earth masses are considered gas planets. In our own solar system, Mercury, Venus, Earth, and Mars are rocky planets, while Jupiter and Saturn are gas planets.

The surface temperature of a planet depends on the solar flux it receives from its star but also depends upon its atmosphere. Present-day Venus, for example, receives 1.9 times the solar flux at Earth's orbit, and present-day Mars receives 0.4 time the Earth's flux. Any rocky planet that receives more flux than present-day Venus is judged too hot to be habitable. The albedo of a planet is related to the fraction of the incident light which is reflected. The temperature of planets with zero albedo that absorbs all light incident upon it can be calculated assuming the planet is a black body.

Only if life modifies the atmosphere or surface of a planet can it be detectable from space. This practical reality limits our search for life to rocky planets with water on its surface. While a bacteria living in an underground cave might not be identifiable from space, bacteria living on a mountainside or a body of water probably would be. We are also inclined to assume that extraterrestrial life shares fundamental characteristics with life on Earth in that it requires liquid water as a solvent and has a carbon-based chemistry. It would be difficult for us to evaluate signs of life that is different from our own.

One systematic way to review the possible candidates for habitable planets is to determine the stellar flux falling on the planets orbiting nearby stars. Around each star two concentric spheres can be drawn and between these two spheres is a habitable zone where water can be found on the surface of a rocky planet. Very many planets lying within habitable zones have been located in this way.

Another systematic way of looking for habitable planets is to return to our basic chemistry. Life in its familiar forms requires oxygen (O_2) and water (H_2O). Add to this picture radiation coming from a nearby star and we immediately also have atomic oxygen produced by the photo-disassociation of oxygen molecules (O_2) and ozone (O_3) producer when free oxygen atoms recombine with O_2. The photo-disassociation of water molecules also gives atomic hydrogen, and as we have seen before, atomic hydrogen can combine with ozone to produce OH by the reaction

$$O_3 + H \rightarrow OH^* + O_2.$$

The OH molecules produced in this way are typically in a highly excited vibrational states from which it emits a recognizable spectrum.

Over the years, quantum chemists have performed ten of thousands of basis set calculations that use the variational principle to converge on the ground states of molecules. There is a wealth of theoretical formation about the ground state properties of molecules, but very little theoretical information about excited states. The excited spectra of molecules is a vast, unexplored territory. However, the recent development of accurate numerical methods for performing many-body calculations on diatomic molecules makes it possible to identify the spectral lines and calculate transition probabilities of the electronic transitions of OH and other diatomic molecules in habitable worlds.

Now if we add carbon and nitrogen to our pictures, hydrogen and oxygen combine with these elements to form methane (CH_4) and nitrogen dioxide (NO_2). Methane and nitrogen dioxide are mainly produced on Earth by biological activity. Several researchers have argued that to verify life is present on an exoplanet one should monitor continuously O_2 and CH_4 or O_2 and NO_2. Both of these kinds of molecules are unstable reacting with each other. Only if the O_2 and CH_4 or O_2 and NO_2 are being continuously produced will the populations of these compounds remain constant.

Whatever theoretical approach one adopts, a great deal of experimental evidence has accumulated about the spectra of the molecules in planetary atmospheres. Fig. 7.4 shows an absorption spectra of Venus, the Earth, and Mars in the visible and infrared regions of the spectra. For Mars and Venus, only CO_2 features are observable at this resolution, whereas Earth shows absorption features of O_2 and H_2O in the visible and CO_2, N_2O, and H_2O in the infra red.

Beyond our solar system is a wide variety of stars with different colors, masses, and luminosities. After many years of stellar spectra being classified with capital letter, Annie Cannon at the Harvard Observatory, reordered the sequence of stars according to their surface temperature. In order of declining surface temperature, stars are denoted by the letters O, B, F, G, K, and M. (This sequence of letters can be remembered by the phrase "Oh, be a fine girl kiss me.") The letters are broad temperature classes, with the numerals denoting subclasses (0 through 9 from hot to cool within each class) and the luminosity of a star being denoted by the Roman numerals I to V with I corresponding to supergiants and V being a main sequence star. Cool M stars are the most common type of star in our galaxy and make up 75% of the stars in the solar neighborhood. They are also excellent candidates for terrestrial planet in the habitable zone. Fig. 7.5 shows the visible and infrared spectra of two cooler main sequence stars (K7V and M9V) with absorption lines for O_2, H_2O, CH_4, and N_2O.

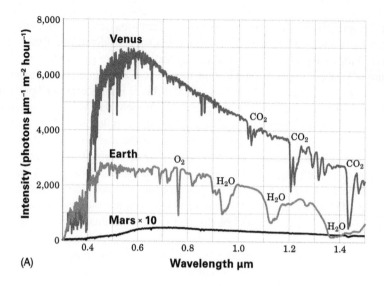

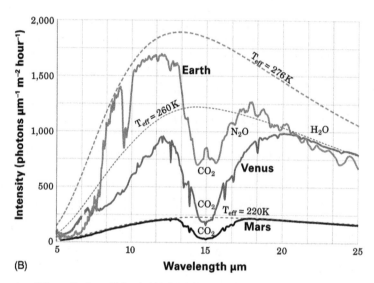

FIGURE 7.4 The absorption spectra of Venus, Earth, and Mars in (A) the visible and (B) in the infrared.

To all of this we should add that the Earth has a long history in which it has passed through ages in which its surface features and the composition of its atmosphere was quite different. The same could well be true of other planets in habitable zones. Our nearest star – Proxima Centauri – is a red dwarf that is two hundred million years older than our Sun with a planet in its habitable zone. Who would be prepared to suggest what kind of civilization lives on this ancient planet?

References

1. J. Kobus, Comp. Phys. Commun. **184**, 799 (2013).
2. J.D. Powers, Philos. Trans. R. Soc. London Ser. A **274**, 663 (1973).
3. E.A. McCullough, Jr., Comput. Phys. Rep. 4, 265 (1986).
4. L. Laaksonen, P. Pykkö, and D. Sundholm, Comput. Phys. Rep. **4**, 313 (1986).
5. J. Morrison et al., Commun. Comp. Phys. **5**, 959 (2009).
6. J. Morrison et al., Comunm. Comp. Phys. **19**, 632 (2016).
7. J. Morrison and J. Kobus, Adv. Quant. Chem. **76**, 104 (2118).
8. L. Kaltenegger, Annu. Rev. Astron. Astrophys. **55**, 433 (2017).

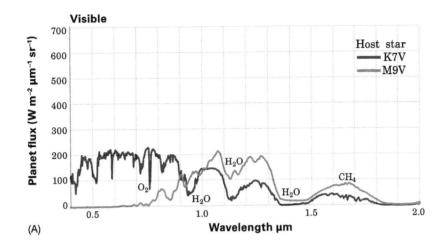

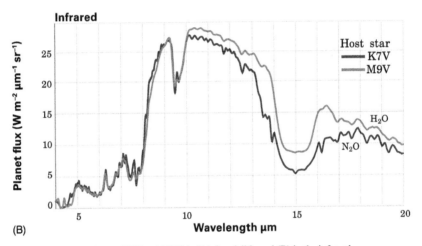

FIGURE 7.5 The absorption spectra of the cool stars K7V and M9V in (A) the visible and (B) in the infrared.

Summary

This chapter begins by considering the hydrogen-like molecular ion with a single electron and two nuclei with changes Z_a and Z_b. This simple molecule is described using spheroidal coordinates

$$
\begin{aligned}
\xi &= (r_a + r_b)/R, && \text{where } 1 \leq \xi < \infty, \\
\eta &= (r_a - r_b)/R, && \text{where } -1 \leq \eta \leq 1, \\
\phi &= \phi, && \text{where } 0 \leq \phi \leq 2\pi,
\end{aligned}
\tag{7.30}
$$

in which r_a and r_b are the distances of the electron from the two nuclei, R is the distance between the two nuclei, and ϕ measures rotations about the molecular axis.

The Schrödinger equation, which is expressed in spheroidal coordinates, is an eigenvalue equation with two independent variables. This equation is similar to the eigenvalue equation for a particle in a finite well which was solved in Chapter 3 using a finite difference approximation of the derivatives and also using the spline collocation method. A MATLAB program is given for solving the eigenvalue problem for a hydrogen molecular ion using spline collocation. Our MATLAB program is described in detail for a smaller grid but it is found that the spline collocation method produces results accurate to seven significant figures with a larger grid having 192 grid points for each variable.

The Hartree-Fock equations are then described. In addition to the terms in the eigenvalue equation for a hydrogen-like ion, the Hartree-Fock equations have terms corresponding to the direct and exchange interactions of an electron with the other electrons in the diatomic molecule. Numerical results are given for the OH molecule which is responsible for the powerful airglow spectra produced in the higher levels of our atmosphere and may be observed in emission and absorption spectra of exoplanets orbiting distant stars.

We then consider the possibility of detecting life on exoplanets. For life to be remotely visible from space, the habitable zone around distant stars is restricted to the zone where rocky planets can be found with water on their surfaces. Very many planets that satisfy these criteria have already been observed, and efforts continue to be made to find definitive signs of life. One strategy that has been suggested to verify life is present on an exoplanet is to monitor continuously O_2 and CH_4 or O_2 and NO_2. Both of these kinds of molecules are unstable reacting with each other. Only if the O_2 and CH_4 or O_2 and NO_2 are being continuously produced, will the populations of these compounds remain constant. On Earth, the CH_4 and NO_2 molecule are mainly produced by life.

The Earth has a long history in which it has passed through ages in which its surface features and the composition of its atmosphere was quite different. The same could well be true of other habitable planets. Our nearest star – Proxima Centauri – is a red dwarf that is two hundred million years older than our Sun with a planet in its habitable zone.

Questions

1. How are the spheroidal coordinates ξ and η defined?
2. What will be the value of ξ and η for a point on the molecular axis at the midpoint of the two atoms?
3. Give the values of ξ and η for a point on the charge Z_a.
4. Suppose one were to perform a spine collocation calculation on a grid with $nx = 4$ and $ny = 4$. How large would the full A_x, B_x, C_x and the A_y, B_y, and C_y matrices be?
5. For the grid with $nx = 4$ and $ny = 4$ grid in the previous problem, how many rows and columns would the tensor product of the Bxmat and Bymat matrices have?
6. For the grid with $nx = 4$ and $ny = 4$ grid in the previous problem, which of the following pairs of splines would have nonzero coefficients $\alpha_x(0)\alpha_y(0)$, $\alpha_x(0)\beta_y(0)$, $\beta_x(0)\alpha_y(1)$, $\alpha_x(1)\beta_y(1)$, $\alpha_x(1)\alpha_y(4)$, $\alpha_x(1)\beta_y(4)$?
7. Which terms in the Hartree-Fock equation describe the direct and exchange interaction of the electron with the other electrons in the molecule?
8. Make a sketch of the energy of the OH molecule as a function of the interatomic separation.
9. How can the vibrational frequencies of the OH molecule be calculated?
10. Give the reaction formula describing the molecular process in which atomic hydrogen combines with ozone to produce OH.
11. In basis set calculations of molecular properties, the wave function of a molecule is described as a linear combination of atomic orbitals or Gaussian functions with adjustable parameters. How are these parameters determined?
12. What properties must a planet have for life on the planet to be observable from space?
13. Upon which factors does the surface temperature of a planet depend?
14. How is the albedo of a planet defined?
15. When an exoplanet passes in front of its host star, it absorbs light emitted by the star. What properties of the atmosphere of the exoplanet can be deduced from the absorption spectra?
16. To which spectral classes do stars with the hottest and coolest surface temperatures correspond?
17. Have the surface features and atmosphere of the Earth always been the same as they are today?

Chapter 8

Statistical physics

Contents

8.1 The nature of statistical laws	169	8.7 Free-electron theory of metals	194
8.2 An ideal gas	172	Suggestions for further reading	199
8.3 Applications of Maxwell-Boltzmann statistics	174	Basic equations	199
8.4 Entropy and the laws of thermodynamics	185	Summary	201
8.5 A perfect quantum gas	189	Questions	201
8.6 Bose-Einstein condensation	193	Problems	202

Since each assignment of particles to the available states is equally likely to occur, the most probable distribution is the one that can occur in the most possible ways.

Solids and liquids contain an enormous number of individual atoms. An amount of material consisting of a few grams typically contains 10^{22} to 10^{23} atoms. For systems containing so many atoms, the questions we ask are necessarily of a more general character. We may want to know, for example, which fraction of the particles of a gas have kinetic energies in a particular range. The laws of statistical physics give us general ways of answering questions of this kind.

8.1 The nature of statistical laws

One of the most far-reaching ideas of statistical physics is that the probability of a particular outcome depends upon the number of possible ways it can occur. Suppose, for instance, that we intended to flip a coin four times, and we wanted to know the probability of getting two heads and two tails. We can answer this question by listing all of the possible sequences of heads and tails and seeing which fraction of the sequences gives two heads and two tails. The possible outcomes of flipping a coin four times are given in Table 8.1. In this table, the occurrence of heads is denoted by H and the occurrence of tails is denoted by T.

TABLE 8.1 Possible outcomes of flipping a coin four times.

HHHH	Four Heads
HHHT, HHTH, HTHH, THHH	Three Heads and One Tail
HHTT, HTHT, THHT, HTTH, THTH, TTHH	Two Heads and Two Tails
TTTH, TTHT, THTT, HTTT	One Head and Three Tails
TTTT	Four Tails

Of the 16 possible sequences of heads and tails, 6 sequences correspond to two heads and two tails. Since each sequence is equally likely to occur, the probability of getting two heads and two tails is $6/16 = 3/8$. Similarly, the probability of getting three heads and one tail or one head and three tails is $4/16 = 1/4$, while the probability of getting all heads or all tails is $1/16$. All of these probabilities must add up to one. We have

$$\frac{3}{8} + \frac{2}{4} + \frac{2}{16} = 1.$$

The basic principle we have used to calculate probabilities is that since each sequence of heads and tails is equally likely to occur, the probability of a particular distribution of heads and tails depends upon the number of ways the distribution can occur.

Modern Physics with Modern Computational Methods. https://doi.org/10.1016/B978-0-12-817790-7.00015-9

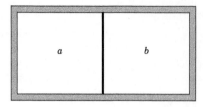

FIGURE 8.1 A container which has been divided into two compartments by inserting a wall.

The concepts of *sequence* and *distribution* played an important role in the above discussion. A sequence refers to a particular way of selecting outcomes in a particular order, while a distribution does not imply a particular order. Each distribution corresponds to one or more sequences of outcomes. For example, the distribution of one head and three tails can occur in four possible ways, $HTTT, THTT, TTHT, TTTH$ and thus the distribution of one head and three tails corresponds to four sequences.

As a further application of the principle of statistics, we consider the number of ways six gas molecules can be put in a container which has been divided into two compartments by inserting a wall. As illustrated in Fig. 8.1, we shall denote the two compartments by a and b. The number of molecules in the two compartments will be denoted by n_a and n_b. We shall suppose that the gas molecules are distinguishable and use the word *allotment* to denote a particular assignment of the distinguishable particles to the two compartments. In one allotment of the particles, for example, the first and the sixth molecule might be assigned to compartment a and the other molecules assigned to compartment b. This allotment of the molecules could be obtained by first choosing the first molecule and then choosing the sixth molecule, or one could choose the two molecules for compartment a in the opposite order. The two ways of selecting the first and sixth molecules for compartment a will be regarded to be equivalent. No physical effect can depend upon the order one selects the molecules for a particular allotment.

Notice that the term *allotment* includes information about which molecules are assigned to each compartment, while the term *distribution* gives only the number of molecules in each compartment. The number of possible ways the six gas molecules can be assigned to the two compartments in the container are given in Table 8.2.

TABLE 8.2 Possible distributions of six molecules between two compartments.

Distributions		Weights	Probabilities
n_a	n_b	W	P
0	6	1	1/64
1	5	6	6/64
2	4	15	15/64
3	3	20	20/64
4	2	15	15/64
5	1	6	6/64
6	0	1	1/64
		64	1

Each of the rows of Table 8.2 corresponds to a possible distribution of the six molecules between the two compartments. In the first distribution, n_a is equal to zero and n_b is equal to six. This means that all six molecules are placed in compartment b. In the second distribution, one molecule is placed in compartment a and the other five molecules are placed in compartment b. The other rows of the table correspond to other ways of distributing the six molecules between the two compartments.

Each of the distributions shown in Table 8.2 corresponds to one or more allotment of the molecules to the two containers. For instance, the second distribution shown in Table 8.2 can be obtained by assigning any one of the six molecules to compartment a and assigning all of the other molecules to compartment b. For this distribution, there are six possible allotments of the molecules to the two compartments corresponding to the six distinct ways of choosing the molecule for compartment a. The number of allotments of the molecules to the two compartments according to a particular distribution is called the *statistical weight* of the distribution and is denoted by W. The weights for the distributions of the six molecules are given in the third column of Table 8.2. The first distribution corresponds to one allotment, while the second distribution corresponds to six allotments. To obtain the third distribution shown in Table 8.2, one can select the first molecule for compartment a in six possible ways and chose the second molecule in five different ways. However, as we have seen, the order of selecting the two molecules for compartment a is unimportant. Enumerating the ways the molecules are selected has the effect of counting each allotment of the molecules twice. The third distribution thus corresponds to $(6 \cdot 5)/2 = 15$ allotments of the molecules. The distribution for which three molecules have been assigned to each compartment has the largest statistical weight, while the distributions for which two molecules are in one compartment and four molecules are in the other compartment have the next largest weights. If the wall separating the two compartments were a permeable membrane allowing molecules to pass freely from one compartment to another, the number of molecules in each compartment would vary with time. If we suppose, however, that each allotment of the molecules between the two compartments is equally likely to occur and that the volume of the two compartments is the same, the distributions for which the molecules are fairly evenly divided between the two compartments would occur most often. As for the example of flipping a coin, the probability of each distribution is equal to the number of distinct ways the distribution can occur divided by the total number of assignments of the molecules to the two compartments. The probabilities P for the distributions of the six molecules are given in the fourth column of Table 8.2.

A general formula for the weights corresponding to the distribution of the six molecules between the two compartments can be given using the factorial notation. For any positive integer n, n *factorial*, which is written $n!$, is equal to $1 \cdot 2 \ldots (n - 1) \cdot n$. (The special symbol $0!$ is given the value 1.) The number $n!$ is equal to the total number of permutations of the order of n objects. Using the factorial notation, the weights appearing in Table 8.2 can be given by the following formula

$$W = \frac{6!}{n_a! n_b!} . \tag{8.1}$$

The factor $6!$ which occurs in Eq. (8.1) is the total number of permutations of the six molecules, while $n_a!$ and $n_b!$ are the total number of permutations of the molecules that have been assigned to the compartments a and b. The occurrence of $n_a!$ and $n_b!$ in the denominator of Eq. (8.1) ensures that each allotment of molecules to the two compartments is counted only once. The reader is encouraged to confirm that Eq. (8.1) gives the values of W given in Table 8.2.

We recall that once the molecules are chosen for compartment a, all of the remaining molecules must be assigned to compartment b. Eq. (8.1) thus gives the total number of distinct ways of choosing n_a molecules from six molecules. Denoting n_a simply by n and n_b by $6 - n$, Eq. (8.1) may be written

$$W = \frac{6!}{n!(6 - n)!}.$$

This last equation, which gives the number of distinct ways that n molecules can be chosen from six molecules, can easily be generalized to a collection of N molecules. The number of ways of choosing n molecules from a collection of N is

$$W = \frac{N!}{n!(N - n)!}. \tag{8.2}$$

The particular combination of integers appearing on the right-hand side of this equation has an important place in this chapter and will be denoted

$$\begin{pmatrix} N \\ n \end{pmatrix} = \frac{N!}{n!(N - n)!}. \tag{8.3}$$

The term appearing on the left-hand side of this equation is known as a *binomial coefficient*. According to Eqs. (8.2) and (8.3), the binomial coefficient with N appearing above n, which is referred to commonly as N choose n, gives the number of distinct ways of choosing n molecules from a collection of N molecules.

Example 8.1

Calculate the number of ways of choosing three molecules from a collection of six molecules.

Solution

Using Eq. (8.3), the number of ways of choosing three molecules from a collection of six molecules is equal to

$$\begin{pmatrix} 6 \\ 3 \end{pmatrix} = \frac{6!}{3!3!} = \frac{1 \cdot 2 \cdot 3 \cdot 4 \cdot 5 \cdot 6}{(1 \cdot 2 \cdot 3)(1 \cdot 2 \cdot 3)} = 4 \cdot 5 = 20.$$

There are thus 20 ways of choosing the three molecules, which is consistent with the appropriate entry in the fourth row of Table 8.2.

8.2 An ideal gas

The particles of a gas have a kinetic energy and also have long-range interactions with each other. If the long-range interaction of gas particles between collision is negligible in comparison with the translational kinetic energy of the particles, the gas will be said to be *perfect*. Between collisions, which occur on a very small time scale, the particles of a perfect gas move as though they were free particles. One can further differentiate between gases for which the particles are distinguishable or indistinguishable. If the density of the gas is low and the probability is very small that any single-particle state is occupied by more than one gas particle, one can in principle keep track of each of the gas particles. The particles can then be described as distinguishable and the arguments used to describe molecules in the previous section can be used to determine the statistical weight of the possible distributions of the particles. A perfect gas containing weakly interacting, distinguishable particles is known as an *ideal gas*. If the particles of a gas are weakly interacting and yet have a high enough density so that they are indistinguishable, the gas will be referred to as a *perfect quantum gas*. We shall describe the properties of an ideal gas in this section and discuss the properties of a perfect quantum gas in later sections of this chapter.

We shall now consider an ideal gas for which the long-range interactions of the particles is small and the gas particles are distinguishable. Since the interaction energy is negligible, the energy of each particle is approximately equal to its kinetic energy, and the total energy of the gas is equal to the sum of the energies of the individual gas particles. For simplicity, we shall assume that the energy of each particle of the gas belongs to the following discrete spectrum of energies

$$\epsilon_1 < \epsilon_2 < \cdots < \epsilon_r < \dots \quad ,$$

where the subscript r is used to label the possible energies of a gas particle. The Schrödinger equation for the gas particles will generally have a number of independent solutions for a given energy. We denote the number of single-particle states for a molecule with energy ϵ_r by g_r. Denoting the number of particles with energy ϵ_r by n_r, the total energy of the gas is

$$E(n_1, n_2, \dots, n_r, \dots) = \sum_r n_r \epsilon_r \quad , \tag{8.4}$$

and the total number of particles is

$$N = \sum_r n_r \quad . \tag{8.5}$$

The distribution of the particles among the single-particle states may be defined by specifying the number of particles with each energy: $n_1, n_2, \dots, n_r, \dots$. As before, the probability that a particular distribution will occur depends upon the number of distinct ways that the particles can be assigned to the single-particle states in order to achieve the distribution. The distributions that can be formed in the largest number of ways will have the highest statistical weights and will be the most likely to occur. Since the particles of a perfect gas move independently of each other, we can calculate the statistical weight for each single-particle energy separately. The number of ways of assigning n_1 particles to the states with the first energy ϵ_1 is equal to the number of ways n_1 particles can be chosen from N particles times the number of ways the n_1 particles can be assigned to the g_1 states with the first energy. Using Eq. (8.3), the number of distinct ways of choosing n_1 particles from N particles is given by the binomial coefficient N choose n_1. We thus have

$$\binom{N}{n_1} = \frac{N!}{n_1!(N - n_1)!} .$$

A particle with energy ϵ_1 can be assigned to any one of the g_1 states associated with that value of the energy. A second particle with energy ϵ_1 can also be assigned to any of the g_1 possible states making the total number of ways of assigning two particles to the states with energy ϵ_1 equal to $g_1 \times g_1$. Continuing in this way, the number of ways of assigning n_1 particles to the g_1 states with energy ϵ_1 is

$$g_1^{n_1} .$$

The number of ways of choosing n_1 particles and assigning them to the states with energy ϵ_1 can be obtained by multiplying the above two expressions to obtain

$$w_1 = \begin{pmatrix} N \\ n_1 \end{pmatrix} g_1{}^{n_1} = \frac{N!}{n_1!(N-n_1)!} g_1{}^{n_1} .$$

Similarly, the number of ways of choosing n_2 particles from the remaining $N - n_1$ particles and assigning them to the g_2 states with energy ϵ_2 is

$$w_2 = \begin{pmatrix} N - n_1 \\ n_2 \end{pmatrix} g_2{}^{n_2} = \frac{(N-n_1)!}{n_2!(N-n_1-n_2)!} g_2{}^{n_2} .$$

For the particles of a perfect gas, the assignment of the particles to the states associated with each value of the energy can be done independently and the total number of ways of assigning the particles to the single-particle states is equal to the product of the individual weights. We shall label the weight associated with a distribution by the numbers of particles with each value of the energy. We thus have

$$W(n_1, n_2, \ldots, n_r, \ldots) = w_1 \cdot w_2 \cdots w_r \cdots = \frac{N!}{n_1!(N-n_1)!} g_1{}^{n_1}$$
$$\cdot \frac{(N-n_1)!}{n_2!(N-n_1-n_2)!} g_2{}^{n_2} \cdot \ldots \cdot \frac{(N-n_1-n_2-\cdots-n_{r-1})!}{n_r!(N-n_1-n_2-\cdots-n_r)!} g_r{}^{n_r} \cdot \ldots \quad (8.6)$$

Canceling the common factors that occur in the above equation, we get

$$W(n_1, n_2, \ldots, n_r, \ldots) = N! \frac{g_1{}^{n_1}}{n_1!} \cdot \frac{g_2{}^{n_2}}{n_2!} \cdot \ldots \cdot \frac{g_r{}^{n_r}}{n_r!} \cdot \ldots .$$

This last equation may be written more simply by using the symbol $\prod$ to denote the product. We have

$$W(n_1, n_2, \ldots, n_r, \ldots) = N! \prod_r \frac{g_r{}^{n_r}}{n_r!} ,$$

where the product is over all of the possible energy states (denoted by the index r).

Each distribution has a statistical weight equal to the number of ways of assigning the particles to the single-particle states. The most probable distribution can be obtained by finding the values of n_r which maximize the function $W(n_1, n_2, \ldots, n_r, \ldots)$ while satisfying the subsidiary equations (8.4) and (8.5). As shown in Appendix HH, this leads to the following distribution law

$$\frac{n_r}{g_r} = \frac{N}{Z} e^{-\epsilon_r/k_B T} , \quad (8.7)$$

where Z is a normalization constant called the *partition function*. As before, n_r is the number of particles with energy ϵ_r, while g_r is the number of states with energy ϵ_r, and T is the temperature of the gas. The constant k_B is called the *Boltzmann constant* having the value 1.381×10^{-23} J/K or 8.617×10^{-5} eV/K. The ratio, n_r/g_r, is the average occupation number of the single-particle states with energy ϵ_r. Eq. (8.7) for the average occupation numbers is called the *Maxwell-Boltzmann distribution law*.

According to the Maxwell-Boltzmann distribution law, the average occupation number of a state with energy ϵ_r falls off exponentially with increasing energy. It is easy to understand in qualitative terms why the occupation number should decrease with increasing energy. For any given temperature, the total energy given by Eq. (8.4) is the same for all possible distributions. If any one state were given a large amount of energy, all of the other states would have a smaller amount of energy, and there would be fewer ways of partitioning out the remaining energy to the other single-particle states.

The normalization constant Z in Eq. (8.7) can be determined by multiplying the equation through with g_r and adding up the equations for the different values of r. Using Eq. (8.5), we obtain

$$N = \sum_r n_r = \frac{N}{Z} \sum_r g_r e^{-\epsilon_r/k_B T} .$$

This equation can be solved for Z to get

$$Z = \sum_r g_r e^{-\epsilon_r/k_B T} . \tag{8.8}$$

We conclude this section by describing our most important results. Eq. (8.7) gives the mean occupation numbers for the most probable distribution of the gas particles among the single-particle states. For macroscopic systems containing huge numbers of particles, the variation of the occupation numbers n_r from their most probable values are very small. The constant Z given by Eq. (8.8) is called the *partition function* of the system.

8.3 Applications of Maxwell-Boltzmann statistics

We now apply Maxwell-Boltzmann statistics to the motion of the particles of a gas and to black-body radiation. These problems show how statistical arguments may be applied, and they provide us some insight into the collective behavior of matter and radiation.

8.3.1 Maxwell distribution of the speeds of gas particles

The distribution of the speeds of gas particles was described by James Clerk Maxwell in 1860. While Maxwell obtained his distribution formula using classical arguments, we shall use a more modern approach which takes advantage of the ideas of de Broglie and Schrödinger discussed in Chapters 1 and 2.

We first note that Eqs. (8.7) and (8.8), were derived by supposing that the energy ϵ of the single-particle states could have discrete values which we labeled by an index r. The energy and number of states corresponding to a particular value of r were denoted ϵ_r and g_r respectively. In the following, we shall be dealing with the particles of a gas confined to a container. As we shall soon see, the possible values of the momentum and energy of particles confined to a container of macroscopic dimensions lie very close together. If the spacing of the energy levels is very small compared with $k_B T$, the momentum and energy of the particles can be regarded as continuous variables and both the number of particles n and the density of states g can be thought of as functions of the momentum or energy.

Since the kinetic energy ϵ of a gas particle is equal to $mv^2/2 = (mv)^2/2m$, the relation between the kinetic energy and momentum of a particle is

$$\epsilon = p^2/2m, \tag{8.9}$$

and we may use the momentum to label the single-particle states. The number of states between p and $p + dp$ will be denoted by $g(p)dp$, and the number of particles with a momentum between p and $p + dp$ will be denoted by $n(p)dp$. Using this notation the distribution law (8.7) can be written

$$n(p)dp = \frac{N}{Z} g(p)e^{-p^2/2mk_B T} dp , \tag{8.10}$$

and the partition function (8.8) becomes

$$Z = \int_0^\infty g(p) \, e^{-p^2/2mk_B T} dp . \tag{8.11}$$

An expression for the number of states, $g(p)dp$, in a particular momentum interval, p to $p + dp$, can be obtained by considering the de Broglie waves associated with the particles of gas. The de Broglie waves are solutions of the Schrödinger equation which satisfy boundary conditions at the walls of the vessel containing the gas. To be concrete, we imagine that the gas is contained in a large cubic box with walls having a length L. The condition upon the de Broglie wave associated with a particle is that an integral number of half wavelengths must fit within the box

$$n \times \frac{1}{2}\lambda = L, \tag{8.12}$$

where n is an integer. Solving this equation for the de Broglie wavelength, we get

$$\lambda = \frac{2L}{n} . \tag{8.13}$$

Each particle has a momentum related to the wavelength of the wave by the de Broglie relation

$$p = \frac{h}{\lambda}.$$

Substituting Eq. (8.13) into the de Broglie relation, we obtain the following equation for the possible values of the momentum of a particle

$$p = n\frac{h}{2L}. \tag{8.14}$$

An expression for the possible values of the kinetic energy can be obtained by substituting the value of p given by this last equation into Eq. (8.9). This leads to the following expression for the kinetic energy

$$\epsilon = \frac{n^2 h^2}{8mL^2},$$

which is identical to Eq. (2.17) for a particle in an infinite well. If the container is of macroscopic dimensions, the length L of the walls is large on an atomic scale and the allowed values of the momentum and the energy lie very close together.

The condition (8.12) for the de Broglie waves has led to the requirement that the momentum of the particles in the container can only have the values given by Eq. (8.14). To obtain an expression for the number of states in a particular range of momenta, we first invert Eq. (8.14) to obtain an expression for n in terms of p

$$n = \frac{2L}{h} p.$$

Differentiating this equation, we get

$$dn = \frac{2L}{h} dp.$$

The novel feature of modern thought is that a range of the momentum should contain a certain number of allowed states. Only for certain values of the momentum are the boundary conditions satisfied at the walls of the container. While this same effect occurred earlier for a particle moving in an infinite potential well, here the container of the gas particles is of macroscopic dimensions and so the allowed values of the momentum lie much closer together.

Denoting the number of states in the range between p and $p + dp$ by $g(p)dp$ this last equation can be written

$$g(p)dp = \frac{2L}{h} dp.$$

This result can easily be generalized to three dimensions. The number of states for which the x-component of the momentum is between p_x and $p_x + dp_x$, the y-component of the momentum is between p_y and $p_y + dp_y$, and the z-component of the momentum is between p_z and $p_z + dp_z$ is

$$g(p_x, p_y, p_z)dp_x dp_y dp_z = \left(\frac{2L}{h}\right)^3 dp_x dp_y dp_z.$$

The number of states within a volume element in momentum space is thus equal to $(2L/h)^3$ times the volume element.

Gas particles having a magnitude of the momentum in the range between p and $p + dp$ can be thought of as belonging to a spherical shell of momentum space with an inner radius p and an outer radius $p + dp$. Since the standing waves associated with a particle in a container have no direction associated with them, all of the states can be included by counting only those states for which the components of the momentum are positive. The number of standing waves corresponding to the momentum in the range between p and $p + dp$ can thus be calculated by finding the volume of one-eighth of a spherical shell with radius p and thickness dp

$$\frac{1}{8} \times \text{volume of spherical shell} = \frac{1}{8} \times 4\pi p^2 dp = \frac{\pi p^2 dp}{2}.$$

The number of states of a particle with momentum between p and $p+dp$ can be obtained by multiplying $(2L/h)^3 = 8V/h^3$ times the above volume to obtain

$$g(p)dp = \frac{4\pi V p^2 dp}{h^3} .$$

(8.15)

With this expression for $g(p)dp$, the distribution law (8.10) becomes

$$n(p)dp = \frac{N}{Z} \frac{V 4\pi p^2}{h^3} e^{-p^2/2mk_BT} dp ,$$

(8.16)

and the partition function (8.11) can be written

$$Z = \int_0^\infty \frac{V 4\pi p^2}{h^3} e^{-p^2/2mk_BT} dp .$$

(8.17)

Eq. (8.16) gives the number of gas particles with momentum between p and dp, and Eq. (8.17) gives the associated partition function. Integrals such as that appearing in Eq. (8.17) occur often in the kinetic theory of gases, and it is useful to have general formulas for evaluating them. If we make the substitutions $u = p$ and $a = 1/2mk_BT$, Eq. (8.17) can be written

$$Z = \frac{V 4\pi}{h^3} \int_0^\infty u^2 e^{-au^2} du .$$

We define the integral $I_n(a)$ by the equation

$$I_n(a) = \int_0^\infty u^n e^{-au^2} du .$$

(8.18)

The partition function can then be written

$$Z = \frac{V 4\pi}{h^3} I_2(a).$$

Formulas for the integral $I_n(a)$ and formulas for evaluating other integrals that commonly arise in statistical physics are given in Appendix G. The value of the integral $I_2(a)$ appearing in the expression for the partition function Z is

$$I_2(a) = \frac{1}{4a} \left(\frac{\pi}{a}\right)^{1/2} .$$

Substituting this expression for $I_2(a)$ and the value of a given above into the equation for the partition function gives

$$Z = V \frac{(2\pi mk_BT)^{3/2}}{h^3} .$$

The number of gas particles with a momentum in the range between p and $p + dp$ can be obtained by substituting this last equation into Eq. (8.16) giving

$$n(p)dp = N \frac{4\pi p^2}{(2\pi mk_BT)^{3/2}} e^{-p^2/2mk_BT} dp .$$

(8.19)

The probability that a gas particle has a momentum in the range between p and $p + dp$ is equal to the number of gas particles in this momentum range divided by the total number of gas particles

$$P(p)dp = \frac{n(p)dp}{N} .$$

Substituting Eq. (8.19) into this last equation, we obtain

$$P(p)dp = \frac{4\pi p^2}{(2\pi mk_BT)^{3/2}} e^{-p^2/2mk_BT} dp .$$

(8.20)

This equation, which gives the probability that a particle of the gas has a momentum in the range between p and $p + dp$, is one of our basic results. We shall use Eq. (8.20) to derive equations describing the probability that a gas particle has a speed in the range between v and $v + dv$ or an energy in the range between ϵ and $\epsilon + d\epsilon$. We leave as an exercise to show that the probability distribution (8.20) satisfies the normalization condition

$$\int_0^\infty P(p)dp = 1. \tag{8.21}$$

All of the probabilities should add up to one. (See Problem 4.)

An equation giving the probability that a particle has a speed between v and $v + dv$ can be obtained by substituting $p = mv$ into Eq. (8.20) to obtain

$$P(v)dv = 4\pi v^2 \left(\frac{m}{2\pi k_B T}\right)^{3/2} e^{-mv^2/2k_B T} dv . \tag{8.22}$$

This equation, which gives the probability that a particle of the gas has a speed in the range between v and $v + dv$ is known as the *Maxwell speed distribution* formula.

The Maxwell speed distribution formula (8.22) can be used to calculate the average values of quantities that depend upon the speed of the particles. As we have done for the hydrogen atom, the evaluation of integrals involving the distribution of velocities can be simplified by introducing a dimensionless variable to describe the velocity. To see how this should be done, we first note that the exponent of e in Eq. (8.22) must be dimensionless and therefore that the quantity,

$$\left(\frac{m}{2k_B T}\right)^{1/2},$$

has the dimensions of v^{-1}. For this reason, we introduce the change of variables

$$u = v \left(\frac{m}{2k_B T}\right)^{1/2}. \tag{8.23}$$

The dimensionless variable u is the speed of a particle measured in units of $(2k_B T/m)^{1/2}$. In terms of this new variable, the Maxwell distribution formula (8.22) may be written,

$$P(v)dv = F_1(u)du,$$

where

$$F_1(u) = \frac{4}{\sqrt{\pi}} u^2 e^{-u^2} .$$

The function $F_1(u)$, which gives the probability per speed interval that a particle has a speed between u and $u + du$, is illustrated in Fig. 8.2.

The speed for which the probability function has its maximum value is the *most probable* speed and will be denoted v_p. As indicated in Problem 5, the maximum of the Maxwell distribution occurs for $u = 1$. Setting u equal to one in Eq. (8.23), we find that the most probable speed of the particles is

$$v_p = \left(\frac{2k_B T}{m}\right)^{1/2}. \tag{8.24}$$

Example 8.2

Calculate the average speed of particles of an ideal gas at temperature T.

Solution

Following our discussion of average values in Section 2.2 of Chapter 2, the average speed is obtained by multiplying each speed v by the probability $P(v)dv$ and integrating over all possible speeds giving

$$v_{av} = \int_0^\infty v P(v)dv.$$

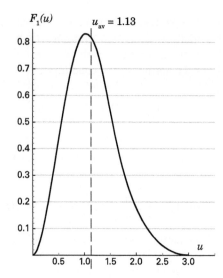

FIGURE 8.2 The function $F_1(u)$, which gives the probability per speed interval that a molecule in a gas has a speed between u and $u + du$.

Using Eqs. (8.22) and (8.23), the integrand in this last equation can be expressed in terms of the dimensionless variable u. We have

$$v_{av} = \frac{4}{\sqrt{\pi}} \left(\frac{2k_B T}{m} \right)^{1/2} \int_0^\infty u^3 e^{-u^2} du.$$

We may now use Eq. (8.18) to identify the above integral as $I_3(1)$ and express the average velocity as

$$v_{av} = \frac{4}{\sqrt{\pi}} \left(\frac{2k_B T}{m} \right)^{1/2} I_3(1).$$

According to the formula for $I_3(a)$ given in Appendix G, $I_3(1)$ has the value $1/2$. We thus have

$$v_{av} = \frac{2}{\sqrt{\pi}} \left(\frac{2k_B T}{m} \right)^{1/2}.$$

The average speed of the particles may be expressed in terms of the most probable speed v_p by using Eq. (8.24). We obtain

$$v_{av} = \frac{2}{\sqrt{\pi}} v_p .$$

The value u for the average speed is shown in Fig. 8.2. Since the Maxwell distribution curve is asymmetric declining more slowly to the right (as speed increases) than to the left (as speed decreases), the average value of the speed lies to the right of the maximum corresponding to the most probable speed.

The two properties of a planet that are most important in determining which constituents of its atmosphere can be retained are the escape velocity of the planet and the temperature of the outer layers of its atmosphere. The escape velocity is the velocity an object must attain to escape the gravitational field of the planet. For the Earth, the escape velocity is 11.2 km/s and the temperature of the upper atmosphere – called the thermosphere – can reach temperatures as high as 1770 K.

Example 8.3

Find the value of the dimensionless velocity variable u for a hydrogen atom moving with the escape velocity of the Earth in a region of the thermosphere for which the temperature is 1700 K and find the probability a hydrogen atom in this region of the thermosphere will escape into outer space.

Solution
Using Eq. (8.23) and the physical constants in Appendix A, we find

$$u = \left(\frac{1.027865 \times 1.6605402 \times 10^{-27} \text{ kg}}{2 \times 1.38066 \times 10^{-23} \text{ J/K} \times 1700 \text{ K}} \right)^{1/2} \times 11.2 \times 10^3 \text{ m/s} = 2.13564$$

The probability that a hydrogen atom will escape into outer space due to its thermal motion depends upon the probability a hydrogen atom will have a velocity greater than the escape velocity. This probability can be calculated using the following one-line MATLAB® program that integrates the probability from the value of u we have just calculated to a very large number.

```
quadl(@(u)u.^2.*exp(-u.^2),2.13564,10)
```

This program returned the answer $u = 0.0123$, which means that a hydrogen atom in this region of the thermosphere moving away from our planet has a slightly more than one percent chance of escaping into outer space. The process by which molecules of a planetary atmosphere escape into outer space due to their thermal motion is called *thermal escape*. As described in the classic book, "Theory of Planetary Atmospheres", cited at the end of this chapter, atmospheric escape is a complex process that includes many different processes. For example, recombination processes such as $O_2^+ + e \Longrightarrow O + O$ turns the binding energy of the O_2^+ ion into kinetic energy of the product O atoms that can lead to their escape into outer space. Recent attempts to find planets orbiting distant stars has renewed interest in planetary atmospheres.

Thus far, we have studied the distribution of the momenta and speeds of particles of a gas. We can obtain a distribution law for the kinetic energies associated with the translational motion of the particles by expressing the momentum distribution law (8.20) in terms of the kinetic energies. As before, we can express the kinetic energy of a particle as

$$\epsilon = \frac{p^2}{2m} .$$

Solving this equation for p^2 and taking the differential of the resulting equation, we obtain the following two equations

$$p^2 = 2m\epsilon,$$
$$2pdp = 2md\epsilon.$$

Using these two equations, we can write

$$p^2 dp = p \cdot pdp = \sqrt{2m\epsilon}\, md\epsilon = \frac{1}{2}(2m)^{3/2}\epsilon^{1/2}d\epsilon \tag{8.25}$$

and substitute this expression for $p^2 dp$ into Eq. (8.20) to obtain

$$P(\epsilon)d\epsilon = \frac{2}{\sqrt{\pi}} \frac{\epsilon^{1/2}d\epsilon}{(k_B T)^{3/2}} e^{-\epsilon/k_B T} . \tag{8.26}$$

The distribution law for the energy can be written very simply by introducing the following dimensionless variable to describe the energy

$$u = \epsilon/k_B T. \tag{8.27}$$

The new variable u gives the energy of a particle in units of $k_B T$. In terms of this new variable, Eq. (8.26) may be written,

$$P(\epsilon)d\epsilon = F_2(u)du,$$

where

$$F_2(u) = \frac{2}{\sqrt{\pi}} u^{1/2}e^{-u} .$$

The function $F_2(u)$, which gives the probability per energy interval that a particle has an energy between u and $u + du$, is illustrated in Fig. 8.3.

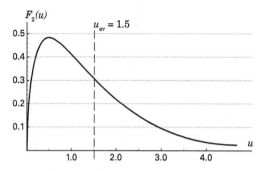

FIGURE 8.3 The function $F_2(u)$, which gives the probability per energy interval that a molecule has an energy between u and $u + du$.

Example 8.4

Calculate the average energy of particles of an ideal gas at temperature T.

Solution

The average kinetic energy of the particles is

$$\epsilon_{av} = \int_0^{\infty} \epsilon P(\epsilon) d\epsilon.$$

Using Eqs. (8.26) and (8.27) to express the above integrand in terms of the variable u, we obtain

$$\epsilon_{av} = \frac{2}{\sqrt{\pi}} k_B T \int_0^{\infty} u^{3/2} e^{-u} du.$$

The above integral, which may be evaluated using a formula in Appendix G, is equal to $3\sqrt{\pi}/4$. We thus obtain

$$\epsilon_{av} = \frac{3}{2} k_B T.$$

The value of u corresponding to the average kinetic energy is represented by a dashed line in Fig. 8.3.

The position of a gas particle in a container can be specified by giving its three Cartesian coordinates. We thus say that gas particles have three *degrees of freedom* associated with their motion in the three coordinate directions. The particle thus has an average kinetic energy $k_B T/2$ for each degree of freedom. This important result, which is called the principle of the *equipartition of energy*, is described by the following equation

$$\epsilon_{av} = \frac{1}{2} k_B T, \quad \text{for each degree of freedom.}$$

8.3.2 Black body radiation

The efforts of physicists to explain black body radiation had a very important impact upon our understanding of the interaction of radiation and matter. A black body is an idealization of a piece of matter that absorbs all incident radiation and is in equilibrium with radiation. Since a piece of matter in equilibrium with radiation emits the same amount of radiation it absorbs, a black body is said to be a 'perfect emitter of radiation'. The piece of matter is called a *black body* because a black piece of matter absorbs more radiation than matter of any other color. As illustrated in Fig. 8.4, a black body can be simulated by a cavity with a small hole through which radiation can pass. Since radiation incident upon the hole bounces around within the cavity until it is absorbed, the hole absorbs all radiation incident upon it and may therefore be used as a model of a black body.

In 1900, Max Planck presented the following empirical formula for the amount of energy per volume in a cavity for the frequency range between f and $f + df$

$$u(f)df = \frac{8\pi f^2}{c^3} \left(\frac{hf}{e^{hf/k_B T} - 1} \right) df. \tag{8.28}$$

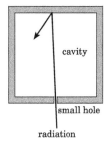

FIGURE 8.4 A cavity having a small hole through which radiation can pass.

This formula was found to fit observations better than one would expect of an empirical formula, and Planck was moved to find a fundamental justification for his formula. Although Planck considered electrons oscillating in the walls of the cavity, we shall here consider the standing waves within a cavity.

Each electromagnetic wave in a cavity has a wavelength and frequency and can be thought of as a harmonic oscillator. The average energy of all oscillators having a particular frequency can be obtained using Maxwell-Boltzmann distribution formula (8.7). Since the frequency is a continuous variable, we shall modify the Maxwell-Boltzmann distribution law as we have in describing the particles of a gas. The total number of waves with a frequency between f and $f + df$ will be denoted by $g(f)df$, and the number of waves oscillating in the cavity with a frequency between f and $f + df$ will be denoted by $n(f)df$. Using this notation, the Maxwell-Boltzmann distribution law (8.7) can be written

$$n(f)df = \frac{N}{Z} g(f)df e^{-\epsilon/k_B T} \ . \tag{8.29}$$

The arguments used in the previous section to find the number of de Broglie waves in a particular momentum interval may be applied to the electromagnetic waves within a cavity as well. The one new feature that must be included is that an electromagnetic wave in a cavity has two distinct polarizations. Using the term *mode* to describe an electromagnetic wave with a particular frequency and polarization, the number of modes per volume in the frequency range between f and $f + df$ may be obtained from Eq. (8.15) by using the de Broglie relation and including an extra factor of two for the two polarizations. We obtain

$$g(f)df = \frac{V 8\pi f^2}{c^3} df \ . \tag{8.30}$$

We now apply Maxwell-Boltzmann statistics to find the average energy of all oscillating waves having a particular frequency. We first note that the energy of an electromagnetic wave in classical electromagnetic theory depends upon its amplitude rather than its frequency. The coefficient $g(f)$ appearing in Eq. (8.30) depends only on the frequency and will have a single value for all of those waves having the same frequency. Eq. (8.29) thus implies that the number of waves having a particular frequency f and an energy between ϵ and $\epsilon + d\epsilon$ is proportional to the Boltzmann factor $e^{-\epsilon/k_B T}$, and the energy of the oscillators in this energy range is proportional to $\epsilon e^{-\epsilon/k_B T}$. If we suppose that the energy is a continuous variable, the total number of oscillators is proportional to $\int_0^\infty e^{-\epsilon/k_B T} d\epsilon$ and the total energy of the oscillators is proportional to $\int_0^\infty \epsilon e^{-\epsilon/k_B T} d\epsilon$. The average energy of the oscillators for a particular wavelength or frequency can thus be written

$$\bar{\epsilon} = \frac{\int_0^\infty \epsilon e^{-\epsilon/k_B T} d\epsilon}{\int_0^\infty e^{-\epsilon/k_B T} d\epsilon} \ .$$

The value of the integral in the numerator here is equal to $(k_B T)^2$, while the value of the integral in the denominator is $k_B T$. We thus have

$$\bar{\epsilon} = k_B T.$$

Applying Maxwell-Boltzmann statistics to the modes of oscillation corresponding to a single frequency thus leads to the conclusion that each oscillator has an average energy equal to $k_B T$. Since the electric field of a standing wave in the cavity has two distinct directions of polarization, this result is consistent with the principle of the equipartition of energy introduced in the previous section. Each mode of vibration of the waves in the cavity has an average energy $k_B T/2$.

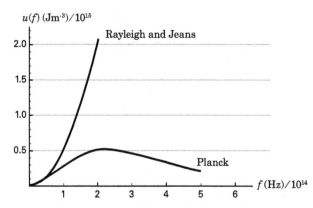

FIGURE 8.5 The function $u(f)$ predicted by Planck and by Rayleigh and Jeans.

Prior to Planck's work, Lord Rayleigh and James Jeans had studied the radiation field within a cavity. Using the fact that each mode of vibration should have an average energy $k_B T/2$ and Eq. (8.30) for the number of modes, they obtained the following expression for the energy per volume

$$u(f)df = \frac{8\pi f^2}{c^3} k_B T df. \tag{8.31}$$

The functions $u(f)$ given by Eqs. (8.28) and (8.31) are illustrated in Fig. 8.5. The formula (8.31) of Rayleigh and Jeans agrees with the empirical formula (8.28) of Planck for small frequencies but becomes infinite for large frequencies. The fact that the classical line of argument leads to the result that the energy per volume becomes infinite as the frequency approaches infinity is referred to as the *ultraviolet catastrophe*.

To avoid this unreasonable result, Planck made the remarkable assumption that the energy of an oscillator with a frequency f could only be an integral multiple of hf, where h is a constant. The energy ϵ of the oscillators is then given by the following formula

$$\epsilon = n hf,$$

where n is a positive integer. In order to understand the consequences of Planck's assumption, we again calculate the average energy of the oscillators having a single frequency f. For these oscillators, the allowed values of the energy are equal to $0, hf, 2hf, \ldots$. As before, the Boltzmann factor determines the number of oscillators with each possible energy. If we denote the number of oscillators with zero energy by A, then the number of oscillators with energy hf is $Ae^{-hf/k_B T}$, the number of oscillators with energy $2hf$ is $Ae^{-2hf/k_B T}$, and so forth. The total number of oscillators with frequency f is then

$$N = A + Ae^{-hf/k_B T} + Ae^{-2hf/k_B T} + \cdots$$
$$= A\left[1 + e^{-hf/k_B T} + (e^{-hf/k_B T})^2 + \cdots\right]. \tag{8.32}$$

Denoting the factor $e^{-hf/k_B T}$ by x, the infinite series appearing within square brackets can be written

$$1 + x + x^2 + \cdots .$$

Since $x = e^{-hf/k_B T}$ is always less than one, the above series converges having the value $1/(1-x)$. Using this result, Eq. (8.32) becomes

$$N = \frac{A}{1 - e^{-hf/k_B T}} . \tag{8.33}$$

The total energy of this group of oscillators is found by multiplying each term in the infinite series in Eq. (8.32) by the appropriate energy giving

$$E = 0 \cdot A + hf \cdot Ae^{-hf/k_B T} + 2hf \cdot Ae^{-2hf/k_B T} + \cdots$$
$$= A hf e^{-hf/k_B T}\left[1 + 2e^{-hf/k_B T} + 3(e^{-hf/k_B T})^2 + \cdots\right]. \tag{8.34}$$

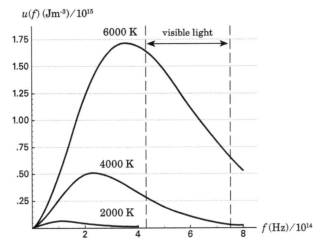

FIGURE 8.6 The function $u(f)$ given by Planck for three different temperatures.

Again denoting the factor e^{-hf/k_BT} by x, the series within square brackets can be written

$$1 + 2x + 3x^2 + \cdots .$$

This series converges having the value $1/(1-x)^2$. Using this result, Eq. (8.34) becomes

$$E = \frac{A h f e^{-hf/k_BT}}{(1 - e^{-hf/k_BT})^2} . \tag{8.35}$$

The average energy of the oscillator with frequency f may be obtained by dividing the total energy of the oscillators by the number of oscillators. Using Eqs. (8.35) and (8.33), we get

$$\bar{\epsilon} = \frac{h f e^{-hf/k_BT}}{1 - e^{-hf/k_BT}} = \frac{hf}{e^{hf/k_BT} - 1} . \tag{8.36}$$

Planck's formula (8.28) may now be obtained by multiplying the expression (8.30) for the number of oscillators per volume with frequencies in the range between f and $f + df$ by the expression (8.36) for the average energy of these oscillators. Although Planck himself felt uncomfortable with the hypothesis he made to obtain the correct formula, his bold hypothesis was soon confirmed by later developments. In 1905, Einstein successfully explained all of the observed features of the photoelectric effect by supposing that the radiation incident upon a metal surface consisted of quanta having an energy hf.

The function $u(f)$ given by Planck's formula (8.28) is the energy of the radiation per volume and per frequency. This function is plotted for three different temperatures in Fig. 8.6. The horizontal direction in this figure corresponds to frequency with the visible region marked. As the temperature increases, the curve representing $u(f)$ shifts toward higher frequencies and the area under the curve increases. The maximum of the function for a particular temperature can be found by setting the derivative of $u(f)$ with respect to f equal to zero. This leads to the equation

$$(3 - x)e^x = 3,$$

where $x = hf_{\max}/k_BT$. (See Problem 13.) Using this last equation, the frequency for which the radiation is a maximum and the temperature may be shown to be related by the equation

$$\frac{hf}{k_BT} = 2.8214. \tag{8.37}$$

We can thus see that the maximum of the curve representing $u(f)$ shifts to higher frequencies with increasing temperatures. Eq. (8.37) is known as the *Wien displacement law*.

The energy density $u(f)$ is related to the intensity of radiation emitted by a black body by the following equation

$$I(f) = \frac{c}{4} u(f),$$

where the intensity $I(f)$ is defined to be the energy emitted by a black body per frequency, per surface area, and per time. A derivation of this last equation can be found in the book by McGervey, which is cited at the end of this chapter. Substituting Eq. (8.28) into the above equation, we get

$$I(f)df = \frac{2\pi f^2}{c^2} \left(\frac{hf}{e^{hf/k_BT} - 1} \right) df. \tag{8.38}$$

The total intensity of light emitted by the black body per surface area and time is obtained by integrating Eq. (8.38) over all frequencies. This gives

$$I = \frac{2\pi h}{c^2} \int_0^\infty \frac{f^3}{e^{hf/k_BT} - 1} df.$$

By making the change of variable $u = hf/k_BT$, this last equation may be written

$$I = \frac{\hbar}{4\pi^2 c^2} \left(\frac{k_BT}{\hbar} \right)^4 \int_0^\infty \frac{u^3}{e^u - 1} du.$$

The integral occurring on the right-hand side of the above equation has the value, $\pi^4/15$, and the total intensity of radiation is described by the following equation

$$I = \sigma T^4, \tag{8.39}$$

where

$$\sigma = \frac{\pi^2 k_B^4}{60\hbar^3 c^2} = 5.67 \times 10^{-8} \text{ Jm}^{-2}\text{s}^{-1}\text{K}^{-4}.$$

Eq. (8.39) for the intensity of radiation per area is called the *Stefan-Boltzmann Law*. According to this law, the total intensity of the radiation emitted by a black body increases with the fourth power of the absolute temperature T.

There are many important applications of the theory of black-body radiation to astrophysics. The spectrum emitted by a star usually has the appearance of a black-body emission curve from which certain frequencies are missing. As light from layers just below the visible surface of the star pass by cooler atoms near the surface, certain well-defined frequencies are absorbed. These missing frequencies appear as dark lines in an otherwise continuous spectrum. From the color of a star, we obtain a direct indication of its surface temperature with the hottest stars appearing blue and the coolest stars appearing red. The relation between the color and luminosity of a star and its temperature can be understood from the theory of black-body radiation. We have already seen that the frequency for which a black body emits the most radiation is related to the temperature of the body. As an example of the application of the theory of black body radiation to the problem of understanding stellar spectra, we consider the star Betelgeuse in the constellation Orion. Betelgeuse is a red star that emits about one hundred thousand times more energy per second than the Sun. From the fact that Betelgeuse is red, we know it has a low surface temperature – about 3000 K. The Stefan-Boltzmann law then implies that each square meter of its surface area emits much less radiation than a square meter on the surface of our Sun which has a surface temperature of 5800 K. The fact that Betelgeuse is so much more luminous than our Sun then tells us that the star is huge. Betelgeuse belongs to a general class of stars called *red giants*.

Example 8.5

Calculate the radius of the star Betelgeuse.

Solution

Since the star Betelgeuse has a temperature about one half the temperature of the Sun, each square meter of Betelgeuse emits about one sixteenth the amount of energy per second as a square meter of the Sun. The fact that Betelgeuse is one hundred thousand times more luminous than the Sun then implies that it has $1,600,000$ times more surface area than the Sun. Since the surface area of a sphere is related to its radius by the formula, $A = 4\pi R^2$, the radius of a spherical object is proportional to the square root of its surface area. We may thus conclude that the radius of Betelgeuse is about equal to the square root of $1,600,000$

times the radius of the Sun. Using the fact that the radius of the Sun is 696,000 km, we find that the radius of Betelgeuse is $\sqrt{1,600,000} \times 696,000$ km $= 880$ million kilometers. In comparison, the distance of the Earth from the Sun is about 150 million kilometers, while the distance of Mars from the Sun is 228 million kilometers and the distance of Jupiter from the Sun is 778 million kilometers. If Betelgeuse were in the center of our solar system, its radius would reach out beyond the orbit of Jupiter.

8.4 Entropy and the laws of thermodynamics

We have thus far considered physical phenomena from a microscopic point of view. We have studied the motion of individual particles of a gas and the modes of vibration of the radiation field within a cavity. In this section, we take up the problem of describing physical systems with an appropriate set of macroscopic variables and try to understand how the microscopic and macroscopic descriptions of systems are related.

We consider again a perfect gas consisting of N gas particles in an insulated container of volume V. We suppose that the gas is in a state of equilibrium with the entire gas having a pressure P and temperature T. For a perfect gas, the long-range interaction between gas particles can be neglected and the energy of the gas is the sum of the kinetic energies of the individual gas particles.

The number of ways of assigning the particles of a gas to the single-particle states to achieve a particular macroscopic state is called the statistical weight W of the system. If we suppose as before that each assignment of the particles to single-particle states is equally likely to occur, then the probability that the system is in a particular macrostate is proportional to its weight. A system which is not in equilibrium will evolve until its weight attains its maximum value, and the system will then be in equilibrium. The *entropy* S can be defined by the equation

$$S(E, V, N) = k_B \ln W, \tag{8.40}$$

where k_B is the Boltzmann constant. Like the statistical weight, the entropy of a macrostate is a measure of the number of distinct ways the macrostate can be formed. It describes the disorder of the system. As the energy of the system increases, the number of ways the system can be formed and the entropy of the system also increase. The entropy of a system increases monotonically with energy, and the functional relation between the entropy and the energy may be inverted with the energy being expressed as a function of the entropy S, volume V and number of particles N

$$E = E(S, V, N).$$

We shall restrict ourselves here to processes for which the number of particles is constant and regard the energy to be a function of the entropy and the volume.

As the entropy of the gas increases with the volume held constant, the energy increases in the following way

$$dE|_V = \left(\frac{\partial E}{\partial S} \right)_V dS. \tag{8.41}$$

The partial derivative of E with respect to S occurring on the right-hand side of this equation gives the rate of change of the energy with respect to the entropy with the volume being held constant. We can use this partial derivative to define the temperature T of the gas

$$T = \left(\frac{\partial E}{\partial S} \right)_V. \tag{8.42}$$

To show that this equation provides an appropriate definition of the temperature, we consider the gas in a container that has been divided into two parts by a diathermal wall, which allows heat to flow freely between the two parts of the system without the exchange of particles. The system, which is shown in Fig. 8.7, has two parts with energy and volume labeled E_1, V_1 and E_2, V_2, respectively. We adopt the definition (8.42) of the temperature and show that the two parts will be in equilibrium with each other provided that the temperature T_1 is equal to the temperature T_2.

The number of ways of assigning gas particles to the single-particle states for the composite system is equal to the product of the number of ways of assigning the particles for the two systems separately. We have

$$W = W_1 W_2.$$

Taking the natural log of this equation and using Eq. (8.40), we obtain

$$S = S_1 + S_2.$$

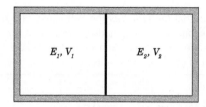

FIGURE 8.7 Gas in a container divided into two parts by a diathermal wall.

The entropy of the composite system is thus equal to the sum of the entropies of the two parts.

For a perfect gas, the energy is equal to the sum of the kinetic energies of the particles. Hence, the energy of the composite system is equal to the energies of the two parts

$$E = E_1 + E_2.$$

When the two parts are in thermal contact with each other, the energy of each part can change. However, the total energy will remain unchanged provided that the system as a whole is thermally and mechanically insulated from its environment. We must have

$$dE = dE_1 + dE_2 = 0.$$

This last equation may be written

$$dE_1 = -dE_2.$$

Applying Eqs. (8.41) and (8.42) to the two sides of this equation separately, we thus obtain

$$T_1 dS_1 = -T_2 dS_2. \tag{8.43}$$

This last equation can be simplified by considering more carefully the condition of a physical system in equilibrium. We have said previously that a physical system will tend to evolve into a state with the highest statistical weight, which can occur in more ways than other states of the system. According to Eq. (8.40), the state with the highest statistical weight will be the state of maximum entropy. We can thus describe the state of equilibrium of the system as a state of maximum entropy. The idea that the state of equilibrium of an isolated system is a state of maximum entropy has assumed the status of a physical law (called the second law of thermodynamics) because it has never been shown to be false.

Since the entropy of an isolated system has its maximum value at equilibrium, the entropy of the system at equilibrium is stationary with respect to small changes. We have

$$dS = dS_1 + dS_2 = 0,$$

which leads to the condition

$$dS_1 = -dS_2.$$

Substituting this last equation into Eq. (8.43), we obtain finally

$$T_1 = T_2.$$

The above definition of the temperature together with the condition that the two parts are in equilibrium has thus lead to the requirement that the temperatures of the two parts be equal. Using Eq. (8.42) to replace the partial derivative on the right-hand side of Eq. (8.41) with T, we get

$$dE|_V = T dS. \tag{8.44}$$

We may consider in a similar way the effect of an increase of the volume of the gas with the entropy held constant. We have

$$dE|_S = \left(\frac{\partial E}{\partial V}\right)_S dV. \tag{8.45}$$

The partial derivative of E with respect to V is equal to the rate of change of the energy with respect to the volume with the entropy being held constant. We can use this partial derivative to define the pressure P of the gas

$$-P = \left(\frac{\partial E}{\partial V}\right)_S. \tag{8.46}$$

To show that this equation provides an appropriate definition of the pressure, we consider an isolated system divided into two parts by a movable diathermal wall. The volume of the gas is equal to the sum of the volume of the two parts

$$V = V_1 + V_2.$$

We suppose that heat can flow through the diathermal wall and the wall can move freely until equilibrium is restored. If one adopts the definition of the pressure given by Eq. (8.46), one can show that the two parts will be in equilibrium provided that the pressure of the two parts are equal. The proof is left as an exercise. (See Problem 18.) Using Eq. (8.46) to replace the partial derivative on the right-hand side of Eq. (8.45) with $-P$, we get

$$dE|_S = -PdV. \tag{8.47}$$

The result of an increase of both the entropy and the volume can be obtained using Eqs. (8.41) and (8.45). We have

$$dE = \left(\frac{\partial E}{\partial S}\right)_V dS + \left(\frac{\partial E}{\partial V}\right)_S dV.$$

Using Eqs. (8.42) and (8.46), this equation can be written

$$dE = TdS - PdV.$$

8.4.1 The four laws of thermodynamics

Thermodynamics is a general theory of matter and radiation that depends upon four fundamental laws. These laws represent a synthesis of an enormous amount of empirical information. Einstein once said that if all of physics were somehow discredited, the last area of physics to fall would probably be thermodynamics because its empirical basis is so broad. We shall now discuss briefly the four laws of thermodynamics and discuss how these laws can be understood in microscopic terms. The results we shall obtain depend upon the pioneering work of Ludwig Boltzmann, who was the first to establish the connection between the microscopic and macroscopic worlds.

The *zeroth law* of thermodynamics says that a single number, the temperature, determines whether bodies will be in equilibrium with each other. Two bodies having the same temperature will be in equilibrium if they are brought in thermal contact with each other.

The *first law* of thermodynamics is a statement of the conservation of the energy. This law, which distinguishes between two forms of energy – heat and work – says that the energy of a system increases by an amount which is equal to the sum of the heat added to the system and the work done on the system

$$dE = dQ_{to} + dW_{on}. \tag{8.48}$$

Here we have labeled the element of heat with the word "to" to indicate that heat added to the system is taken to be positive, while heat lost by the system is negative. Similarly, the element of work is labeled by "on" to indicate that work done on the system is positive, while work done by the system is negative.

We can easily relate the first law of thermodynamics to the results obtained previously describing the changes of energy of the system. Suppose our system consists of a gas contained within a cylinder with a movable piston of cross sectional area A as shown in Fig. 8.8. Imagine that an applied force F moves the piston a distance dx. In keeping with our sign convention that work done on the system is positive, we will regard a displacement dx that compresses the gas to be positive. The amount of work done on the gas inside the cylinder is

$$dW = Fdx. \tag{8.49}$$

This equation can be expressed in terms of the pressure and the volume by using the following equations

$$P = \frac{F}{A},$$

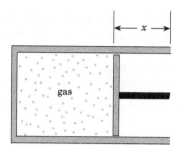

FIGURE 8.8 A cylinder containing a gas has a movable piston of cross sectional area A.

$$dV = -A dx.$$

The first of these equations gives the ordinary definition of pressure as force per unit area. The minus sign on the right-hand side of the second equation arises since we have taken dx to be positive if the gas is compressed. Using these two equations, Eq. (8.49) can be written

$$dW = (PA) \cdot \left(-\frac{dV}{A} \right)$$
$$= -P dV. \tag{8.50}$$

Eqs. (8.48) and (8.50) imply that the change of the energy of a system when work is done without heat being transferred to or from the system is given by the equation

$$dE|_Q = -P dV \tag{8.51}$$

Comparing this equation with Eq. (8.47) derived earlier, we see that the two results are identical provided that not adding heat to a system means that S is held constant.

Trying to understand how insulating a system from an influx of heat could effect the entropy of the system gives us an opportunity to consider how the microscopic and macroscopic descriptions are related. On a microscopic level, the states of gas particles are described by de Broglie waves that satisfy boundary conditions at the walls of the container. If the walls of the container slowly moved inward compressing the gas, all of the single-particle states and energies would change. However, if the compression were done slowly and no heat were added the system, each gas particle would stay in the same single-particle state. The number of ways of assigning the particles to the single-particle states would remain the same, and the entropy of the system would be unchanged. Eqs. (8.47) and (8.51) are thus equivalent.

Imagine now that energy is added to gas while the piston is held fixed. The cylinder may be heated or the gas may absorb light from an external source. Since the cylinder does not move, no external work is done and the first law (8.48) becomes

$$dE|_V = dQ|_V .$$

Comparing this last equation with Eq. (8.44), we see that the amount of heat absorbed by the gas under conditions of constant volume is related to the change of entropy of the system. When the volume of the gas remains constant, the single-particle states are unaffected by the influx of energy, and the energy absorbed by the gas excites the particles to higher-lying states. Just as the energy of the particles increases, the number of ways of distributing the energy among the single-particle states also increases, and the entropy of the gas increases.

According to the celebrated *second law* of thermodynamics, the entropy of a system not in equilibrium tends to increase. The macroscopic states which can occur in the most possible ways and hence have the highest statistical weights are more likely to occur. A system not in equilibrium may not be in a macrostate having a high statistical weight because of its past history. With the passage of time, all memory of how the system started out gets washed away by the multiplicity of possible events. Systems overcome the peculiarities of their origin evolving toward states with higher statistical weights and larger values of entropy.

The *third* and final *law* of thermodynamics states that the entropy of a body has a minimum at absolute zero. The minimum of entropy cannot be attained. An alternate formulation of the third law is that a body cannot be brought to absolute zero temperature by a finite sequence of operations.

In reviewing the laws of thermodynamics, the zeroth law is the touchstone which we used in introducing our definition of temperature. The first law interprets the concept of the conservation of energy at the macroscopic level, while the second law is formulated in terms of the entropy which has meaning only on the macroscopic level. The second law says that macroscopic processes evolve inevitably toward macrostates for which the energy is more or less evenly shared by the particles and the entropy of the system increases. The third law warns us to restrict thermodynamic arguments to nonzero temperatures.

8.5   A perfect quantum gas

We consider again a gas for which the long-range interaction of gas particles is negligible in comparison with the kinetic energy of the particles. This is the criterion for a perfect gas. In contrast to the case of an ideal gas considered previously, we now suppose that the temperature of the gas is sufficiently low and the density of the gas is sufficiently high that the mean occupation number of the single-particle states available to the gas particles is close to one. As discussed previously, the gas particles must then be regarded as indistinguishable, and the gas will be referred to as a perfect quantum gas. We shall suppose, as before, that the energy of each gas particle belongs to a discrete spectrum of energies $\epsilon_1 < \epsilon_2 < \cdots < \epsilon_r < \ldots$. The number of states corresponding to the single-particle energy ϵ_r will again be denoted by g_r, and we shall suppose n_r gas particles are assigned to the states with energy ϵ_r.

Two forms of statistics are necessary to describe a perfect quantum gas depending upon whether or not the particles satisfy the Pauli exclusion principle. The Pauli exclusion principle, which applies to the electrons in an atom, says that two particles cannot be in the same state. If the particles do not obey the Pauli exclusion principle, any number of particles can be in each single-particle state, and the form of statistics known as *Bose-Einstein statistics* (BE statistics for short). It was first introduced by Satyendra Bose in 1924 to derive the Planck radiation law, and Einstein, recognizing its full importance, applied it in the same year to particles with nonzero mass. Particles obeying Bose-Einstein statistics are called *Bosons*. Photons and a class of particles called *mesons* satisfy this form of statistics.

The other form of statistics applies to particles satisfying the Pauli-Exclusion Principle. The occupation number of the single-particle states can then only have the values zero or one. This kind of statistics was considered independently by Enrico Fermi and Paul Dirac in 1926 and is known as *Fermi-Dirac statistics* (FD statistics for short). Particles obeying Fermi-Dirac statistics are called *fermions*. Electrons, protons and neutrons satisfy this form of statistics.

An important connection has been established between the intrinsic angular momentum or spin of particles and the form of statistics they follow. Particles with zero or integral spin follow Bose-Einstein statistics, while particles with half-integer spin follow Fermi-Dirac statistics. The different statistics which bosons and fermions obey lead to departures from the ideal gas laws particularly at low temperatures. At high temperatures and low densities, the predictions of Bose-Einstein and Fermi-Dirac statistics reduce to the predictions of Maxwell-Boltzmann statistics considered earlier.

The starting point for the derivation of distribution laws is the enumeration of the number of ways that each distribution can occur. Since each assignment of the particles to the single-particle states is equally likely to occur, the most probable distribution is the one that can occur in the most possible ways. The fact that the particles we are now considering are indistinguishable has a fundamental effect upon our counting procedure. We begin by considering the problem of assigning three bosons (Pauli-Exclusion Principle does not apply) to two single-particle states. The possible assignments of the three particles are given in Table 8.3.

TABLE 8.3 Possible assignments of three bosons to two states.

Assignments	Illustration
3,0	○○○△
2,1	○○△○
1,2	○△○○
0,3	△○○○

Each of the rows of Table 8.3 corresponds to a possible assignment of the three particles to the two single-particle states. In the first assignment, three particles are assigned to the first state and no particles are assigned to the second state. In the illustrations on the right, the particles are represented by circles ○ and the partition between the two available states is represented by a triangle △. Since three particles have been assigned to the first state in the assignment shown in the first row, the three circles in the first entry on the right are followed by a triangle. In the second assignment shown in

Table 8.3, two particles are assigned to the first state and one particle is assigned to the second state. This is represented in the illustration on the right by two circles followed by a triangle and a circle. The two circles stand for the two particles in the first state, while the triangle serves as a partition between the two states and the final circle represents the one particle in the second state.

Each of the entries in the second column in Table 8.3 have four characters: three circles corresponding to the particles and one triangle serving as a partition between the two states. The four possible assignments of the three particles to the two states correspond to the four possible ways of assigning the one triangle to the four possible locations. These ideas can easily be generalized. An assignment of n_r particles to g_r single-particle states can be represented by a string containing n_r circles and $g_r - 1$ triangles. The number of possible assignments of the n_r particles to the g_r states is equal to the number of ways of assigning $g_r - 1$ triangles to $n_r + g_r - 1$ possible positions. This number can easily be calculated using the binomial coefficient (8.3) described near the end of the first section. We have

$$w_r = \left(\begin{array}{c} n_r + g_r - 1 \\ g_r - 1 \end{array} \right) = \frac{(n_r + g_r - 1)!}{(g_r - 1)!n_r!}. \tag{8.52}$$

The number of assignments shown in Table 8.3 can be obtained by setting $n_r = 3$ and $g_r = 2$ in Eq. (8.52) giving

$$w_r = \left(\begin{array}{c} 4 \\ 1 \end{array} \right) = \frac{4!}{1!3!} = 4.$$

In Fermi-Dirac statistics, the Pauli Exclusion Principle applies and each state may either be singly occupied or empty. Hence the number of ways of assigning n_r particles to g_r states is equal to the number of ways of selecting the n_r occupied states from the g_r states. We thus have

$$w_r = \left(\begin{array}{c} g_r \\ n_r \end{array} \right) = \frac{g_r!}{n_r!(g_r - n_r)!}. \tag{8.53}$$

Eqs. (8.52) and (8.53) are the basic equations we shall need for deriving the Bose-Einstein and Fermi-Dirac distribution laws.

As for the classical perfect gas considered earlier, the assignment of the particles to each of the energy states can be done and the total number of ways of assigning the particles to the single-particle states is equal to the product of the individual weights. Using Eq. (8.52) and the product notation as before, the total number of ways of assigning the particles to the single-particle states in Bose-Einstein statistics is

$$W(n_1, n_2, \ldots, n_r, \ldots) = \prod_r \frac{(n_r + g_r - 1)!}{n_r!(g_r - 1)!}. \tag{8.54}$$

Using Eq. (8.53), the total number of ways of assigning the particles to the single-particle states in Fermi-Dirac statistics is

$$W(n_1, n_2, \ldots, n_r, \ldots) = \prod_r^{\infty} \frac{g_r!}{n_r!(g_r - n_r)!}. \tag{8.55}$$

Example 8.6

Find the possible number of ways of assigning three particle to four states according to the Maxwell-Boltzmann, Bose-Einstein, and Fermi-Dirac statistics.

Solution

Each row of the following table gives possible distributions of three particles among the four states and gives the statistical weight of the distribution for the different kinds of statistics. The statistical weight associated with each distribution depends upon whether the particles are distinguishable and whether the Pauli exclusion principle applies.

Maxwell-Boltzmann statistics applies to distinguishable particles, while Bose-Einstein and Fermi-Dirac statistics apply to indistinguishable particles. For Maxwell-Boltzmann statistics, the weight of a particular distribution depends upon the

TABLE 8.4 Possible distributions of three particles among four states.

	Distributions			Weights		
n_a	n_b	n_c	n_d	*MB*	*BE*	*FD*
3	0	0	0	1	1	0
0	3	0	0	1	1	0
0	0	3	0	1	1	0
0	0	0	3	1	1	0
2	1	0	0	3	1	0
2	0	1	0	3	1	0
2	0	0	1	3	1	0
...	...	...	...	...	...	...
1	1	1	0	6	1	1
1	1	0	1	6	1	1
1	0	1	1	6	1	1
0	1	1	1	6	1	1
				64	20	4

number of ways of selecting the particles to be assigned to each state. Denoting the number of particles in the four states by n_a, n_b, n_c, and n_d, respectively, the number of possible ways of assigning the particles to the four states is

$$W(n_a, n_b, n_c, n_c) = \begin{pmatrix} 3 \\ n_a \end{pmatrix} \begin{pmatrix} 3 - n_a \\ n_b \end{pmatrix} \begin{pmatrix} 3 - n_a - n_b \\ n_c \end{pmatrix} \begin{pmatrix} 3 - n_a - n_b - n_c \\ n_d \end{pmatrix},$$

where the binomial coefficients appearing on the right-hand side of this last equation are defined by Eq. (8.3). Evaluating the binomial coefficients and simplifying the resulting equation, the weight of a particular distribution can be written

$$W(n_a, n_b, n_c, n_c) = \frac{3!}{n_a! n_b! n_c! n_d!}. \tag{8.56}$$

For the first distribution shown in Table 8.4, $n_a = 3$ and $n_b = n_c = n_c = 0$. Substituting these values into Eq. (8.56) gives $W(3, 0, 0, 0) = 1$. The assignment of all three particles to the first state uniquely defines this distribution. The second, third, and fourth distributions shown in the table have three particles in each of the other three states. Similarly, applying Eq. (8.56) to the fifth distribution in Table 8.4 gives $W(2, 1, 0, 0) = 3$. For the fifth distribution, the particle in the second state may be chosen in three possible ways. Once this particle is chosen, the remaining particles must be in the first state. In addition to the fifth, sixth, and seventh distributions, there are three additional distributions for which $n_b = 2$ and six other distributions for which $n_c = 2$ or $n_d = 2$. According to Maxwell-Boltzmann statistics, the total number of distributions of the three particles to the four states is

$$W_{\text{total}} = 4 \times 1 + 12 \times 3 + 4 \times 6 = 64.$$

The number of possible distributions of the 3 particles to the 4 states according to Bose-Einstein statistics is equal to the total number of ways of assigning 4-1=3 partitions to the $3 + (4 - 1) = 6$ positions available for particles and partitions. This gives

$$W(n_a, n_b, n_c, n_d) = \begin{pmatrix} 6 \\ 3 \end{pmatrix} = 20.$$

For Fermi-Dirac statistics the total number of distributions is equal to the number of ways of choosing three occupied states from the four states. We have

$$W(n_a, n_b, n_c, n_d) = \begin{pmatrix} 4 \\ 3 \end{pmatrix} = 4.$$

Notice that the weight of each distribution of the particles is equal to one for Bose-Einstein and Fermi-Dirac statistics. The two forms of quantum statistics apply to indistinguishable particles for which all possible ways of selecting the particles for a particular distribution are entirely equivalent. For Fermi-Dirac statistics, no more than one particle can be in each state.

The most probable distribution for Bose-Einstein and Fermi-Dirac statistics can be obtained by finding the values of n_r which maximize the expressions (8.54) and (8.55) for the weights W while satisfying the subsidiary equations (8.4) and (8.5). This leads to the following distribution laws for Bose-Einstein and Fermi-Dirac statistics

$$\frac{n_r}{g_r} = \frac{1}{e^\alpha e^{\epsilon_r/k_B T} - 1}, \quad \text{Bose-Einstein} \tag{8.57}$$

$$\frac{n_r}{g_r} = \frac{1}{e^\alpha e^{\epsilon_r/k_B T} + 1}, \quad \text{Fermi-Dirac} \tag{8.58}$$

As before, the ratio n_r/g_r is the average number of particles in the single-particle states with energy ϵ_r. A derivation of these distribution laws is given in Appendix HH. In this derivation, the parameter α is associated with the requirement that the total number of particles is constant.

As a first application of the quantum distribution laws, we return to the problem of black-body radiation, which we considered previously by applying the Maxwell-Boltzmann distribution laws to the electromagnetic waves within a cavity. We shall now regard the radiation field within a cavity as a photon gas. Photons have been found to have an intrinsic angular momentum or spin equal to one and are hence described by Bose-Einstein statistics. The energy of each photon is given by the relation $\epsilon = hf$ introduced originally by Planck. The number of photons in a cavity is not constant, and so the parameter α appearing in Eq. (8.57) may be taken to be equal to zero. Setting $\alpha = 0$ and using the continuous variable ϵ as a measure of the photon energy, Eq. (8.57) may be written

$$n(\epsilon)d\epsilon = g(\epsilon)f(\epsilon)d\epsilon, \tag{8.59}$$

where

$$f(\epsilon) = \frac{1}{e^{\epsilon/k_B T} - 1}. \tag{8.60}$$

The function $f(\epsilon)$ is called the *Bose-Einstein distribution function*. According to Eq. (8.59), the function $f(\epsilon)$ is equal to $n(\epsilon)d\epsilon/g(\epsilon)d\epsilon$ and may thus be interpreted as the average number of photons per state.

A photon with definite values of the momentum and energy can be in two independent states corresponding to the possible ways its spin can be oriented. The function $g(\epsilon)d\epsilon$ giving the number of states with energy between ϵ and $\epsilon + d\epsilon$ is thus twice the function (8.15) for the particles of a perfect classical gas. We have

$$g(\epsilon)d\epsilon = \frac{V 8\pi p^2 dp}{h^3}. \tag{8.61}$$

Using the de Broglie relation,

$$\lambda = \frac{h}{p},$$

and the equation,

$$\lambda f = c,$$

the momentum of a photon can be shown to be related to the frequency of the light as follows

$$p = \frac{hf}{c}.$$

Substituting this expression for the momentum into Eq. (8.61), we get

$$g(\epsilon)d\epsilon = \frac{8\pi V f^2 df}{c^3}. \tag{8.62}$$

The distribution function (8.60) may be expressed in terms of the frequency of the light by using the equation, $\epsilon = hf$, giving

$$f(\epsilon) = \frac{1}{e^{hf/k_B T} - 1}.$$ (8.63)

Substituting Eqs. (8.62) and (8.63) into Eq. (8.59), we get

$$n(\epsilon)d\epsilon = \frac{8\pi V f^2 df}{c^3} \frac{1}{e^{hf/k_B T} - 1}.$$

The energy density $u(f)df$ is equal to the product of the number of photons per unit volume $n(\epsilon)d\epsilon/V$ and the energy of each photon hf. We thus obtain

$$u(f)df = \frac{8\pi f^2 df}{c^3}\left(\frac{hf}{e^{hf/k_B T} - 1}\right).$$

This equation is identical to Eq. (8.28) obtained by Planck.

8.6   Bose-Einstein condensation

The application of quantum statistics to photons in a cavity was first carried out by Satyendra Bose in 1924. Shortly afterwards, Einstein applied Bose's form of statistics to material particles and realized that at low temperatures a substantial number of particles in a gas could precipitously fall into their lowest state. This transition in which the number of particles in the ground state suddenly and anomalously rises from a number near zero to a substantial fraction of the total number of particles is now referred to as *Bose-Einstein condensation*. Like the photons emitted from a laser, particles in their lowest quantum state would move coherently in step with each other. Quantum effects could then be visible on a large scale.

Einstein's idea was intriguing, but apparently unrealistic, because at the low temperatures at which such a condensation should occur, the particles would be in the form of a liquid or a solid. As we have seen, the Bose-Einstein distribution law applies to indistinguishable particles in a perfect gas for which the interaction between the particles is very small and can be neglected, while the interaction between particles held together in a liquid are substantial. The idea of Bose-Einstein condensation remained an unrealized possibility until 1938 when Fritz London pointed out that many of the peculiar properties of liquid ^4He could be understood if we thought of ^4He as a Bose-Einstein condensate.

Neutral ^4He atoms, which have an even number of nucleons and electrons, obey Bose-Einstein statistics. At atmospheric pressure, helium condenses into a liquid at 4.2 K. The temperature of helium can be reduced below this temperature by pumping the pressure down below atmospheric pressure. As the pressure and temperature are reduced, liquid helium boils until the temperature 2.186 K is reached. At this temperature, called the critical temperature T_C, boiling suddenly stops even though evaporation continues. The boiling of liquid helium below the phase transition temperature 4.2 K is due to a lack of uniformity of the temperature with bubbles forming at hot spots in the liquid. The cessation of boiling at T_C indicates a sharp rise in the thermal conductivity causing the entire liquid to be at a perfectly uniform temperature. Below the critical temperature 2.186 K, ^4He has other remarkable properties. While it passes freely through tiny capillaries indicating that it has zero viscosity, traditional measurements of the viscosity of ^4He performed by observing its drag on a flat surface immersed in the liquid show that its viscosity is not zero.

Fritz London suggested that liquid ^4He below the critical temperature consists of two interpenetrating fluids – a *normal fluid* and a *superfluid*. The atoms of the superfluid are in the ground state, while the atoms of the normal fluid are distributed among the other available states. The normal fluid causes bubbles to form as the temperature drops and causes the drag on immersed bodies. The superfluid, which consists of atoms in well-defined momentum states that are delocalized throughout space, has an infinite conductivity and zero viscosity. It maintains the liquid in thermal equilibrium and seeps through tiny capillaries. When heat is applied to the liquid, the superfluid is destroyed uniformly at all points in space.

As discussed in Chapter 6, Eric Cornell, Wolfgang Ketterle, and Carl Wieman received the Nobel Prize in 2001 for achieving Bose-Einstein condensation in cold clouds of alkali atoms. Three high-resolution images of Bose-Einstein condensation provided by Eric Cornell are shown in Fig. 8.9. The images were each produced by turning off the magnetic trap and viewing the density of the atomic cloud a short time afterwards. The atoms in the ground state remained stationary while the other atoms rapidly dispersed.

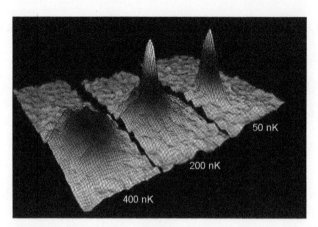

FIGURE 8.9 Three high-resolution images showing Bose-Einstein condensation at 400 nK, 200 nK, and 50 nK. These images were provided by Eric Cornell of the University of Colorado.

8.7   Free-electron theory of metals

The outer electrons of atoms in metals are not bound to individual atoms, but are free to move throughout the metal. These free electrons are called conduction electrons. Since electrons have a spin of one-half, the electrons in a metal are described by Fermi-Dirac statistics. Using the continuous variable ϵ for the energy of the electrons and writing the parameter α as $-\mu/k_BT$, the Fermi-Dirac distribution law (8.58) can be written

$$n(\epsilon)d\epsilon = g(\epsilon)d\epsilon f(\epsilon) \tag{8.64}$$

where

$$f(\epsilon) = \frac{1}{e^{(\epsilon-\mu)/k_BT} + 1} . \tag{8.65}$$

The parameter μ is known as the *chemical potential*, and the function $f(\epsilon)$ is called the *Fermi-Dirac distribution function*. We recall that in Fermi-Dirac statistics each single-particle state is either singly occupied or empty. Hence, the function $f(\epsilon)$, which is equal to $n(\epsilon)d\epsilon/g(\epsilon)d\epsilon$, may be interpreted to be the fraction of single-particle states that are occupied.

As in our earlier treatments of Bose-Einstein systems, we can use Eqs. (8.15) for the density of states in terms of the momentum of the electrons and then use Eq. (8.25) to express the density of states in terms of the energy. An extra multiplicative factor of two must be included to take into account the fact that electrons have two spin orientations. We obtain

$$g(\epsilon)d\epsilon = \frac{V4\pi}{h^3}(2m)^{3/2}\epsilon^{1/2}d\epsilon \tag{8.66}$$

The number of electrons having an energy between ϵ and $\epsilon + d\epsilon$ may be obtained by substituting Eqs. (8.65) and (8.66) into Eq. (8.64) to obtain

$$n(\epsilon)d\epsilon = \left[\frac{V4\pi}{h^3}(2m)^{3/2}\right]\epsilon^{1/2}d\epsilon\frac{1}{e^{(\epsilon-\mu)/k_BT} + 1} . \tag{8.67}$$

The chemical potential μ may be obtained by integrating Eq. (8.67) from 0 to ∞ giving

$$N = \left[\frac{V4\pi}{h^3}(2m)^{3/2}\right]\int_0^\infty \frac{\epsilon^{1/2}d\epsilon}{e^{(\epsilon-\mu)/k_BT} + 1} . \tag{8.68}$$

This last equation determines μ for a given number of electrons N, a given volume V and a given temperature T. For a system with a fixed number of fermions in a fixed volume, the ratio of N/V, which gives the number of fermions per volume, has a fixed value. The chemical potential μ appearing in the above equation varies with temperature to maintain the particle density. We may thus regard μ to be a function of the temperature

$$\mu = \mu(T).$$

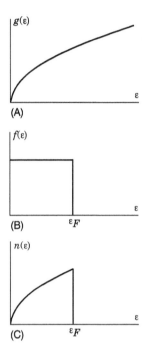

FIGURE 8.10 (A) The function $g(\epsilon)$, (B) the function $f(\epsilon)$, and (C) the function $n(\epsilon)$, which is the product of $g(\epsilon)$ and $f(\epsilon)$.

The *Fermi energy* ϵ_F is defined as the value of the chemical potential at absolute zero

$$\epsilon_F = \mu(0).$$

Since the chemical potential $\mu(T)$ is generally a slowly varying function near absolute zero, the chemical potential may be approximated by the Fermi energy ϵ_F for low temperatures.

Near absolute zero, the Fermi-Dirac distribution function (8.65) can be conveniently expressed in terms of the Fermi energy

$$f(\epsilon) = \frac{1}{e^{(\epsilon-\epsilon_F)/k_B T} + 1}. \tag{8.69}$$

According to Eq. (8.64), the number of electrons in the energy range between ϵ and $\epsilon + d\epsilon$ is equal to the product of $g(\epsilon)d\epsilon$ and $f(\epsilon)$. The function $g(\epsilon)$ given by Eq. (8.66) is illustrated in Fig. 8.10(A). For temperatures near absolute zero, the distribution function $f(\epsilon)$ given by Eq. (8.69) is equal to one for $\epsilon < \epsilon_F$ and equal to zero for $\epsilon > \epsilon_F$. This function is illustrated in Fig. 8.10(B). The function $n(\epsilon)$, which is the product of the two functions represented in Figs. 8.10(A) and 8.10(B), is illustrated in Fig. 8.10(C). At absolute zero, all of the states below the Fermi energy ϵ_F are occupied, and all of the states above the Fermi energy are unoccupied.

Since for temperatures near absolute zero the distribution function $f(\epsilon)$ is equal to one for $\epsilon < \epsilon_F$ and equal to zero for $\epsilon > \epsilon_F$, Eq. (8.68) assumes the following form for low temperatures

$$N = \left[\frac{V4\pi}{h^3}(2m)^{3/2}\right] \int_0^{\epsilon_F} \epsilon^{1/2} d\epsilon = \left[\frac{V4\pi}{h^3}(2m)^{3/2}\right] \frac{2}{3}\epsilon_F^{3/2}.$$

Solving this equation for ϵ_F, we obtain

$$\epsilon_F = \frac{h^2}{2m}\left[\frac{3}{8\pi}\frac{N}{V}\right]^{2/3}. \tag{8.70}$$

The Fermi energy depends upon the mass of the electrons m and upon the electron density N/V. Metals having one conduction electron per atom are said to be *monovalent*, while metals having two and three conduction electrons per atom are said to be *divalent* and *trivalent*, respectively. We leave as an exercise to calculate the ratio N/V and the Fermi energy for monovalent and divalent metals. (See Problems 28 & 29.)

The function $g(\epsilon)d\epsilon$ giving the number of states in the range between ϵ and $\epsilon + d\epsilon$ can be conveniently expressed using the Fermi energy ϵ_F. We first raise both sides of Eq. (8.70) to the 3/2 power giving

$$\epsilon_F^{3/2} = \frac{3}{2} \cdot \frac{h^3}{4\pi(2m)^{3/2}} \frac{N}{V},$$

and then rearrange terms to obtain

$$\frac{V4\pi(2m)^{3/2}}{h^3} = \frac{3}{2}N\epsilon_F^{-3/2}.$$

Using this last equation, Eq. (8.66) can be written

$$g(\epsilon)d\epsilon = \frac{3}{2}N\epsilon_F^{-3/2}\epsilon^{1/2}d\epsilon. \tag{8.71}$$

We may now obtain a more convenient equation for the number of electrons between ϵ and $\epsilon + d\epsilon$. Substituting Eqs. (8.71) and (8.69) into Eq. (8.64), we get

$$n(\epsilon)d\epsilon = N\frac{3}{2}\epsilon_F^{-3/2}\epsilon^{1/2}d\epsilon\frac{1}{e^{(\epsilon-\epsilon_F)/k_BT} + 1}. \tag{8.72}$$

Eq. (8.72), which applies for temperatures near absolute zero, may be used to calculate the average value of physical variables. The probability that an electron has an energy in the range between ϵ and $\epsilon + d\epsilon$ is equal to the number of electrons in this energy range divided by the total number of electrons N

$$P(\epsilon)d\epsilon = \frac{n(\epsilon)d\epsilon}{N}.$$

Substituting Eq. (8.72) into this last equation, we obtain

$$P(\epsilon)d\epsilon = \frac{3}{2}\epsilon_F^{-3/2}\epsilon^{1/2}d\epsilon\frac{1}{e^{(\epsilon-\epsilon_F)/k_BT} + 1}. \tag{8.73}$$

Example 8.7

Calculate the average energy of conduction electrons near absolute zero.

Solution
The average energy of the electrons may be obtained by multiplying each value of the energy ϵ by the probability $P(\epsilon)d\epsilon$ and integrating over all possible energies. We have

$$\epsilon_{av} = \int_0^\infty \epsilon P(\epsilon)d\epsilon.$$

Substituting Eq. (8.73) into the above equation gives

$$\epsilon_{av} = \frac{3}{2}\epsilon_F^{-3/2}\int_0^\infty \epsilon^{3/2}d\epsilon\frac{1}{e^{(\epsilon-\epsilon_F)/k_BT} + 1}.$$

This last equation may be simplified by again using the fact that near absolute zero the function $1/(e^{(\epsilon-\epsilon_F)/k_BT} + 1)$ is equal to one for $\epsilon < \epsilon_F$ and equal to zero for $\epsilon > \epsilon_F$. We thus obtain

$$\epsilon_{av} = \frac{3}{2}\epsilon_F^{-3/2}\int_0^{\epsilon_F} \epsilon^{3/2}d\epsilon.$$

Evaluating the integral, we obtain

$$\epsilon_{av} = \frac{3}{5}\epsilon_F.$$

Thus far, we have only considered the properties of the electron gas near absolute zero. To consider the value of the Fermi-Dirac distribution function for higher temperatures, it is convenient to factor μ from the parentheses in the denominator of Eq. (8.65) to obtain

$$f(\epsilon) = \frac{1}{e^{(\epsilon/\mu - 1)(\mu/k_B T)} + 1} \, .$$

We now define a dimensionless parameter u that gives the energy in units of the chemical potential

$$u = \epsilon/\mu. \tag{8.74}$$

The Fermi-Dirac distribution function $f(\epsilon)$ can then be written

$$f(\epsilon) = \frac{1}{e^{(u-1)(\mu/k_B T)} + 1} \, . \tag{8.75}$$

The chemical potential μ has the value ϵ_F for $T = 0$ K and for low temperatures is a slowing varying function of the temperature. For a temperature such that $K_B T = 0.1\mu$, Eq. (8.75) for the Fermi-Dirac distribution function becomes

$$f(\epsilon) = \frac{1}{e^{(u-1)10} + 1} \, . \tag{8.76}$$

The density of states (8.71) can also be expressed in terms of the variable u. Dividing and multiplying the expression of $g(\epsilon)$ in Eq. (8.71) by $\mu^{3/2}$ and using the definition of u provided by Eq. (8.74), we get

$$g(\epsilon)d\epsilon = \frac{3}{2}N\left(\frac{\mu}{\epsilon_F}\right)^{3/2} u^{1/2}du. \tag{8.77}$$

The probability that an electron has a particular value of the energy denoted by u can be obtained by multiplying the density of states given by Eq. (8.77) times the Fermi-Dirac distribution function $f(u)$ and dividing by the total number of particles. The resulting expression for the probability a particle has the energy associated with the variable u can be written

$$Prob(u)du = f_n p(u)du \tag{8.78}$$

where

$$f_n = \frac{3}{2}\left(\frac{\mu}{\epsilon_F}\right)^{3/2} \tag{8.79}$$

and

$$p(u)du = u^{1/2}f(u)du. \tag{8.80}$$

The most efficient way to evaluate the constant f_n is to note that the integral of the probability function over all non-negative values of u should be equal to one. After calculating the function $p(u)$, we can define a constant f_n which is equal to the integral of $p(u)$ from zero to infinity and define the function $prob(u)$ to be $(1/f_n)$ times $p(u)$. A MATLAB program which defines the Fermi-Dirac distribution function $f(u)$ for a temperature for which $k_B T = 0.1\mu$, plots this function $f(u)$ and calculates and plots the probability that an electron at that temperature will have the energy associate with u is given in MATLAB Program 8.1.

MATLAB Program 8.1

A MATLAB program which calculates and plots the Fermi-Dirac distribution function $f(u)$ for electrons with a temperature for which $k_B T = 0.1\mu$ and which calculates and plots the function $prob(u)$ which gives the probability an electron has an energy u.

```
u=linspace(0,1.5,300);
f = @(u) 1./(exp((u-1)*10)+1);
plot(u,f(u))
p = @(u) sqrt(u).*f(u);
fn = 1/quadl(p,0,1.5);
```

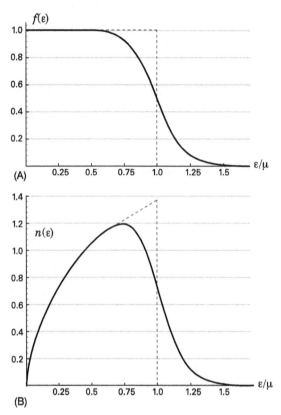

FIGURE 8.11 (A) The Fermi-Dirac distribution function $f(\epsilon)$ with electrons with a temperature for which $K_B T = 0.1\mu$. (B) The function $n(\epsilon)$ gives the number of electrons with an energy ϵ.

```
prob=@(u)fn*p(u);
plot(u,prob(u))
int = quadl(prob,1.0,1.5)
```

The first line of the MATLAB program defines an equally spaced grid with 300 points. The program then defines and plots the Fermi-Dirac distribution function $f(u)$ and uses this function to calculate the probability function $prob(u)$ that gives the probability an electron has an energy u. The plot of distribution function $f(u)$ is given in Fig. 8.11(A) and the probability function $prob(u)$ is given in Fig. 8.11(B). The last line of the program calculates the probability that an electron has an energy greater than the chemical potential μ.

The function $f(u)$ for the temperature for which $K_B T = 0.1\mu$ is shown in Fig. 8.11(A). Notice how different this function is from the function shown in Fig. 8.10(B). While the distribution function $f(u)$ shown in Fig. 8.10(B) falls immediately to zero at ϵ_F, which is the chemical potential μ when $T = 0$ K, the function $f(u)$ shown in Fig. 8.11(A) declines gradually to zero for values of energy above the chemical potential. The probability that an electron should have the energy u shown in Fig. 8.11(B) is also quite different than its counterpart shown in Fig. 8.11(B). Looking at Fig. 8.11(B) one can see clearly that there is a significant probability that the electrons have a value of the energy above the Fermi energy. The last line of the MATLAB Program 8.1 returned the value 0.107, which means that 10.7% of electrons are above the Fermi energy. Thermal excitations have populated levels above the Fermi energy leaving states below the Fermi energy vacant. These considerations will be very important in future chapters when we study semiconducting materials. While semiconductors generally have low densities of change carriers, they have small band-gaps making it possible for thermal excitations to populate the conducting states above the Fermi energy in which the electrons are free to move about the lattice.

For very low temperatures, electrons are packed into the lowest energy levels available. The Fermi energy gives the extent of the occupied levels for $T = 0$ K with all states below the Fermi energy being populated and all states above the Fermi energy being vacant. As the temperature rises, thermal excitations begin to populate energy levels above the Fermi energy. How responsive material are to changes in their temperature depends upon the Fermi energy. Materials with low Fermi energies respond more quickly to an increase in temperature. It is also convenient to define a Fermi temperature by the equation

$$\epsilon_F = k_B T_F,$$

where k_B is the Maxwell-Boltzmann constant and T_F is the Fermi temperature. The Fermi temperature for metals is in the range between 10^4 K and 10^5 K. The temperature in the example for which we wrote MATLAB Program 8.1 is about one tenth of the Fermi temperature. For temperatures T very small compared to T_F, the Fermi distribution function $f(\epsilon)$ and the function $n(\epsilon)$ differ very little from the behavior at $T = 0$ illustrated in Figs. 8.11(B) and 8.11(C).

In coming chapters, we will find that by doping crystals with impurities one can artificially raise or lower the position of the Fermi level and greatly increase the number of negative and positive charge carriers. The fundamental concepts of Fermi-Dirac statistics are as important for understanding recent developments as they are for understanding the intrinsic material considered in this chapter.

Suggestions for further reading

J. McGervey, *Introduction to Modern Physics* (New York: Academic Press, 1971).

D. Goodstein, *States of Matter* (New York: Dover Publications, 1985).

F. Mandl, *Statistical Physics*, Second Edition (New York: Wiley, 1988).

C. Kittel and H. Kroemer, *Thermal Physics*, Second Edition (San Francisco: W.H. Freeman, 1980).

C.J. Pethick and H. Smith, *Bose-Einstein Condensation in Dilute Gases* (Cambridge: Cambridge University Press, 2002).

Basic equations

Maxwell-Boltzmann statistics

The number of ways of choosing n particles from a collection of N

$$W = \frac{N!}{n!(N-n)!}$$

Binomial coefficient

$$\binom{N}{n} = \frac{N!}{n!(N-n)!}$$

The number of ways of assigning n_r particles to g_r states

$$w_r = \binom{N}{n_r} g_r^{n_r}$$

The Maxwell-Boltzmann distribution law

$$\frac{n_r}{g_r} = \frac{N}{Z} e^{-\epsilon_r/k_B T},$$

where the partition function is

$$Z = \sum_r g_r e^{-\epsilon_r/k_B T}$$

Applications of Maxwell-Boltzmann statistics

The number of particles with a momentum between p and p + dp

$$n(p)dp = \frac{N}{Z} g(p)dp \, e^{-\epsilon/k_B T}$$

where the partition function is

$$Z = \int_0^\infty g(p)dp \, e^{-p^2/2mk_B T}$$

and the number of states with momentum between p and $p + dp$ is

$$g(p)dp = \frac{V 4\pi p^2 dp}{h^3}$$

Planck's black-body radiation formula

$$u(f)df = \frac{8\pi f^2}{c^3}\left(\frac{hf}{e^{hf/k_BT}-1}\right)df$$

Wien displacement law

$$\frac{hf}{k_BT} = 2.8214$$

Stefan-Boltzmann law

$$I = \sigma T^4$$

Entropy and the laws of thermodynamics

Definition of the entropy S

$$S(E, V, N) = k_B \ln W$$

Definition of the temperature T and the pressure P

$$T = \left(\frac{\partial E}{\partial S}\right)_V \quad \text{and} \quad -P = \left(\frac{\partial E}{\partial V}\right)_S$$

The first law of thermodynamics

$$dE = dQ_{\text{to}} + dW_{\text{on}}$$

Quantum statistics

The weight according to Bose-Einstein statistics

$$w_r = \begin{pmatrix} n_r + g_r - 1 \\ g_r - 1 \end{pmatrix}$$

The weight according to Fermi-Dirac statistics

$$w_r = \begin{pmatrix} g_r \\ n_r \end{pmatrix}$$

The Bose-Einstein distribution law

$$\frac{n_r}{g_r} = \frac{1}{e^\alpha e^{\epsilon_r/k_BT} - 1}$$

The Fermi-Dirac distribution law

$$\frac{n_r}{g_r} = \frac{1}{e^\alpha e^{\epsilon_r/k_BT} + 1}$$

Free-electron theory of metals

The Fermi-Dirac distribution law

$$n(\epsilon)d\epsilon = g(\epsilon)d\epsilon f(\epsilon)$$

where

$$f(\epsilon) = \frac{1}{e^{(\epsilon - \epsilon_F)/k_B T} + 1},$$

and

$$g(\epsilon)d\epsilon = \frac{V 4\pi}{h^3}(2m)^{3/2}\epsilon^{1/2}d\epsilon$$

Summary

Statistical physics is based on the idea that the probability or statistical weight of a particular macroscopic state depends upon the number of possible ways the state can be constructed out of its microscopic constituents. The number of distinct ways a macroscopic state can be formed depends in turn upon whether or not the constituents are distinguishable. In Maxwell-Boltzmann's statistics, the constituents are distinguishable, and the number of ways a particular distribution of particles can be constructed depends upon the number of ways the particles can be selected for each energy range and how many different ways the particles can then be assigned to the microscopic states in that range. Maxwell-Boltzmann statistics leads to the distribution law

$$\frac{n_r}{g_r} = \frac{N}{Z}e^{-\epsilon_r/k_B T},$$

where n_r is the number of particles with energy ϵ_r and g_r is the number of microscopic states with energy ϵ_r. The partition function Z is

$$Z = \sum_r g_r e^{-\epsilon_r/k_B T}.$$

This form of statistics can be used to describe the velocity and energy distributions of the particles of a gas and to derive Planck's formula for black-body radiation.

Macroscopic system composed of indistinguishable components are described by two other kinds of statistics. Bose-Einstein statistics applies when the component parts of a macroscopic system do not satisfy the Pauli-exclusion principle and Fermi-Dirac statistics apply when the components do satisfy the Pauli-exclusion principle. These two forms of statistics lead to the following distribution laws

$$\frac{n_r}{g_r} = \frac{1}{e^{\alpha}e^{\epsilon_r/k_B T} - 1}, \qquad \text{Bose-Einstein}$$

$$\frac{n_r}{g_r} = \frac{1}{e^{\alpha}e^{\epsilon_r/k_B T} + 1}, \qquad \text{Fermi-Dirac.}$$

An important connection has been established between the intrinsic angular momentum or spin of particles and the form of statistics they follow. Particles with zero or integral spin follow Bose-Einstein statistics, while particles with half-integer spin follow Fermi-Dirac statistics. Bose-Einstein statistics can be used to describe the radiation field within a black-body and the collective phenomena associated with the condensation of a macroscopic system into its lowest quantum state. Fermi-Dirac statistics enables us to understand how the electronic characteristics of metals and semiconductors are related to the dynamical properties of their charge carriers.

Questions

1. Suppose that a drop of ink were to fall upon the surface of water contained in a jar. Would the ink eventually spread out uniformly throughout the water? Explain your answer.
2. Here is the scenario of a disaster: A group of students enter a classroom which has the dimensions, 3 m × 4 m × 4 m. At the time the students enter the room, all of the particles of air are located in a cube having the dimensions one cubic meter near one corner of the room. The students all suffocated and die. Does this strike you as a likely possible danger? Explain your answer.
3. What kind of statistics should be used to describe a collection of distinguishable particles?
4. How many different combinations of three particles could be chosen from a group of five distinguishable particles?
5. In how many different ways can five distinguishable particles be assigned to three states?

6. Suppose that a large collection of particles is described by Maxwell-Boltzmann statistics. Write down an expression for the fraction of the particles having an energy ϵ_r.

7. Suppose that a large collection of noninteracting particles described by Maxwell-Boltzmann statistics has a total energy E. Use general (nonmathematical) arguments to explain why it is exceedingly unlikely that any one particle would have an energy E.

8. Make a plot of the fractional number of occupied states as a function of energy according to Maxwell-Boltzmann statistics.

9. Write down an expression for the number of particles in the range p to $p + dp$ in terms of the total number of particles N, the partition function Z, the number of states, $g(p)dp$ and the momentum p.

10. What unlikely assumption was Planck forced to make in order to derive his equation describing the energy within a cavity?

11. Using a black body as a model, predict how the peak wavelength of the light emitted by a star will change as the star cools.

12. How is the entropy S of a gas related to the statistical weight W?

13. Suppose that a cloud of hydrogen atoms in their ground state absorbs a burst of radiation and the excited atoms transfer the energy they have received to other atoms in collision processes. How will the entropy of the cloud be affected by the absorption and collision processes?

14. Use the first law of thermodynamics to explain how the temperature of a gas changes as the gas expands adiabatically (no heat added or lost).

15. Would the pressure of a gas expanding isothermally (constant temperature) decrease more rapidly or more slowly than a gas expanding adiabatically (no heat added or lost).

16. If the universe were to continue to expand indefinitely, what would the second law of thermodynamics have to say about the fate of the universe.

17. What kind of statistics should be used to describe a collection of indistinguishable particles that do not obey the Pauli exclusion principle?

18. In how many different ways can three bosons be assigned to two states?

19. In how many different ways can three Fermions be assigned to five states?

20. Compare how the radiation field within a black body is described when the Planck's formula is derived using Maxwell-Boltzmann and Bose-Einstein statistics.

21. What kind of statistics should be used to describe the motion of electrons in a metal?

22. Make a plot of the Fermi-Dirac distribution function $f(\epsilon)$ for the conduction electrons for $T = 0$ K and for $T > 0$ K.

23. Make a plot of the number of conduction electrons in the energy range ϵ to $\epsilon + d\epsilon$ for $T = 0$ K and $T > 0$ K.

Problems

1. Use Eqs. (8.2) and (8.3) to calculate the statistical weights associated with the outcomes of flipping a coin four times.

2. How many distinct relationships involving two people are there in a family with four members?

3. The first excited energy level of a transition metal ion in a crystal field lies 0.025 eV above the ground state. The first excited level is doubly degenerate ($g_2 = 2$), while the ground state is nondegenerate ($g_1 = 1$). Using Maxwell-Boltzmann statistics calculate the relative population (n_2/n_1) of the two levels at room temperature $T = 298$ K.

4. Show that the function defined in Eq. (8.20) satisfies the normalization condition (8.21).

5. Show that the function $F_1(u)$ used to describe the speed distribution of particles has a maximum for $u = 1$.

6. Use the Maxwell speed distribution formula (8.22) to derive an expression for the average value of v^2 for the particles of a gas at a temperature T.

7. The root-mean-square speed v_{rms} is defined to be the square root of the average value of v^2. Using the result obtained in the previous problem, derive an equation describing the root-mean-square speed of the particles of a gas and compare this equation to the expressions for the most probable speed and the average speed. Show where the root-mean-square speed would appear in Fig. 8.2.

8. The temperature of the outer layers of the atmosphere of Mars is 240 K and the escape velocity of the planet is 5.0 km/s. Using MATLAB, calculate the percentage of C_2O molecules that will escape into outer space.

9. Write a MATLAB program to find the percentage particles with a velocity less than the maximum velocity v_p shown in Fig. 8.2 and find the percentage of the molecules with a velocity greater than the maximum velocity.

10. Calculate the average kinetic energy associated with each degree of freedom of a gas at room temperature $T = 298$ K and compare this energy to the energy necessary to excite a hydrogen atom to its first excited state.

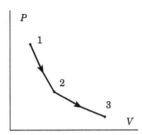

FIGURE 8.12 The change in the volume and pressure of an ideal gas as the gas expands. No heat is absorbed by the gas during the portion of the curve between 1 and 2, while the temperature is held constant during the portion of the curve between 2 and 3.

11. The concept of *molar mass* is useful for describing macroscopic samples of a substance. A *mole* of a substance is an amount of that substance having a mass in grams equal to its chemical weight measured in atomic mass units. One mole contains $N_A = 6.022 \times 10^{23}$ elementary units (atoms, particles, or whatever). The number N_A is referred to as *Avogadro's number*. Calculate the total kinetic energy of a mole of a gas at the temperature 298 K assuming that each particle of the gas has three degrees of freedom.

12. An individual atom has three degrees of freedom associated with its translational motion. By contrast, a diatomic particle has five degrees of freedom: three degrees of freedom associated with its translational motion and two degrees of freedom associated with the rotation of the axis of the particle. A polyatomic particle has six degrees of freedom. Using this information, estimate the specific heat of a mole of a gas consisting of individual atoms. Estimate the specific heat of a mole of a gas consisting of diatomic particles and a gas consisting of polyatomic particles.

13. Show that the condition

$$\frac{du(f)}{df} = 0,$$

where $u(f)$ is given by Eq. (8.28) leads to the equation, $(3 - x)e^x = 3$.

14. Use the Planck distribution law (8.28) to calculate the total energy density of a black-body at a temperature of 3000 K.

15. Derive an expression for the energy density of a black body as a function of wavelength $u(\lambda)$.

16. The surface temperature of the Sun is about 5800 K. Using the black-body model to represent the radiation emitted by the Sun, calculate the total intensity of light emitted by the Sun.

17. Considering again the surface of the Sun to be a black-body with a temperature of 5800 K, find the energy of photons corresponding to the most probable frequency of light emitted by the Sun. What is the probability that photons emitted from the surface of the Sun have an energy in the range 1.75 eV and 1.8 eV? What is the probability of their having an energy in the range 3.05 eV and 3.1 eV?

18. Use the definition of the pressure given by Eq. (8.46) to show that the two regions separated by a movable wall will be in equilibrium provided that the pressures of the two regions are equal.

19. The change in the volume and pressure of an ideal gas as the gas expands is illustrated in Fig. 8.12. No heat is absorbed by the gas during the portion of the curve between 1 and 2, while the temperature is held constant during the portion of the curve between 2 and 3. The volume, pressure and temperature of an ideal gas are related by the *ideal gas law*

$$PV = nRT,$$

where $R = N_A k_B$ and n is the number of moles. Using the ideal gas law, show that the work done by the gas during its expansion from the point in the curve indicated by 2 to the point indicated by 3 is

$$-nRT \ln \left(\frac{P_3}{P_2} \right).$$

During the portion of the curve between 2 and 3, the temperature of the gas remains constant. We may suppose that the energy E of the ideal gas remains constant during this portion of the expansion. Using the first law of thermodynamics (8.48), find the amount of heat added to the gas as the gas expands from 2 to 3.

20. Using the first law of thermodynamics, explain in qualitative terms why the volume of the gas decreases more sharply during the portion of the curve from 1 to 2 than it does during the portion from 2 to 3.

21. Calculate the critical temperature T_c for a gas with a density of 5.0×10^{14} atoms/cm^{-3}.

22. Derive an expression for the molar heat capacity of a gas by supposing that there are $N_{\epsilon>0}$ atoms in excited states and the energy of each is $3k_BT/2$.
23. Write a MATLAB program which calculates and plots the Fermi-Dirac distribution function $f(u)$ for electrons with a temperature for which $k_BT = 0.2\mu$ and which calculates and plots the function $prob(u)$ which gives the probability an electron has an energy u. Compare your results to the results obtained with MATLAB Program 8.1.
24. Using the MATLAB results from the preceding problems, calculate the probability that an electron has an energy greater than the chemical potential μ.
25. Show that the number of conduction electrons per volume is given by the following formula

$$\frac{N}{V} = \frac{n_c \rho N_A}{M},$$

where n_c is the number of conduction electrons per elementary unit, ρ is the density of the substance, N_A is Avogadro's number and M is the mass of a mole of the substance.
26. Using the formula given in the preceding problem, calculate the mean concentration of conduction electrons in gold. Gold is a monovalent metal having a density 19.32 g/cm^3 and an atomic weight of 197 g/mole.
27. Calculate the number of conduction electrons in a cube of gold 1 cm on an edge.
28. Sodium is a monovalent metal having a density of 0.971 g/cm^3 and an atomic weight of 23.0. Use this information to calculate (a) the number of conduction electrons per volume, (b) the Fermi energy, and (c) the Fermi temperature of sodium.
29. Magnesium is a divalent metal having a density of 1.7 g/cm^3 and an atomic weight of 24.3. Calculate the Fermi energy and the Fermi temperature of magnesium.

Chapter 9

Electronic structure of solids

Contents

9.1 The Bravais lattice	206	9.8 Classification of solids	227
9.2 Additional crystal structures	210	Suggestions for further reading	234
9.3 The reciprocal lattice	212	Basic equations	235
9.4 Lattice planes	215	Summary	236
9.5 Bloch's theorem	220	Questions	236
9.6 Diffraction of electrons by an ideal crystal	224	Problems	237
9.7 The band gap	226		

A crystal is a network of copies of a single, basic unit.

The crystalline nature of quartz is clear to the eye and hand. Light passing through it reflects from a myriad of interior planes, and it is sharp to the touch. These features depend upon the underlying structure of the material. On a microscopic level, the atoms of quartz are arranged in periodic arrays extending over vast distances.

The appearance of most metals is quite different. They do not have plane surfaces at sharp angles as does quartz, diamond, or rock salt. Their surfaces are dull and often quite malleable. Yet, most metals occur naturally in a crystalline state. The periodic arrangements of their atoms is as central to their nature as it is for the materials that we readily identify as crystalline.

We begin our study of condensed matter physics by studying the periodic structures that underlie most solids and have a decisive influence upon their electrical and optical qualities. Most metals and semiconductors have crystal lattices corresponding to a few commonly occurring structures that have very simple geometric properties. After studying the most important crystal structures, we shall see how the symmetry of the lattice is reflected in the form of single-electron wave functions. In Section 3.1, we found that a free particle (in one dimension) can be described by the wave function

$$\psi(x) = A e^{ikx},$$

where the wave vector k is related to the momentum of the particle by the equation

$$p = \hbar k.$$

We shall find that the free electron functions are periodic with respect to the crystal lattice if the wave vector k belongs to a lattice called the *reciprocal lattice*.

The reciprocal lattices of the important crystal structures are constructed in Section 9.4 and found to be related in an intuitive way to the original crystal lattices. According to Bloch's theorem considered in Section 9.6, the symmetry of the lattice implies that the wave function of any electron must have a wave vector associated with it. The wave vectors associated with electrons are vectors in the reciprocal lattice, which, like the crystal itself, consists of a multitude of basic cells. The unit cell of the reciprocal lattice is called the *Brillouin zone*. Using the concept of the wave vector and the Brillouin zone, one can characterize in simple terms how electrons interact within a crystal. Each electron having a wave vector $\mathbf{k}$ interacts with the electron with a wave vector $\mathbf{k}'$, which is the mirror image of $\mathbf{k}$ across the boundary of the Brillouin zone. This simple model of the interaction of electrons in a crystal will enable us to understand the band gap of solids and to classify materials into insulators, metals, and semiconductors.

The ideas formulated in this chapter form a basis for our treatment of semiconductors in Chapters 10 and 11.

Modern Physics with Modern Computational Methods. https://doi.org/10.1016/B978-0-12-817790-7.00016-0
Copyright © 2021 Elsevier Inc. All rights reserved.

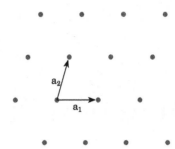

FIGURE 9.1 A two-dimensional Bravais lattice.

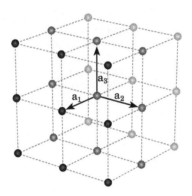

FIGURE 9.2 Simple cubic lattice.

9.1 The Bravais lattice

Imagine choosing some point in a crystal. It might be the center of an atom or a cluster of atoms. Then imagine identifying all points in the crystal that are equivalent to this point. The arrangement of atoms about each of these points is the same. The crystal appears the same from each vantage point. These equivalent points form a lattice known as a *Bravais lattice*. We may specify any point in the lattice by the formula

$$\mathbf{l} = l_1\mathbf{a}_1 + l_2\mathbf{a}_2 + l_3\mathbf{a}_3, \tag{9.1}$$

where l_1, l_2, and l_3 are integers. A two-dimensional lattice of this kind is shown in Fig. 9.1. By choosing the integers l_1 and l_2 associated with the vectors $\mathbf{a}_1$ and $\mathbf{a}_2$, we can identify any point in the array. The vectors, $\mathbf{a}_1$, $\mathbf{a}_2$, and $\mathbf{a}_3$, appearing in Eq. (9.1), and the vectors, $\mathbf{a}_1$ and $\mathbf{a}_2$, illustrated in Fig. 9.1 are called *primitive vectors*. All of the lattice points can be generated by forming integral multiples of these vectors.

A simple cubic crystal (scc) is shown in Fig. 9.2. The primitive vectors for this lattice are three perpendicular vectors of equal length. The points in a Bravais lattice that are closest to a given point are called its *nearest neighbors*. Due to the periodicity of the Bravais lattice, each point has the same number of nearest neighbors. The number of nearest neighbors is called the *coordination number* of the lattice. The coordination number of the simple cubic lattice is 6.

Another cubic lattice, which has many interesting properties, and which occurs quite frequently, is the *body-centered cubic* (bcc) structure formed from the simple cubic lattice by adding a single additional atom to the center of each cube. As can be seen in Fig. 9.3, this lattice can be thought of as consisting of two interpenetrating lattices with points **A** and **B**. Since each point is a center point in one structure and a corner point in the other structure, all of the points are clearly equivalent. We can choose the primitive vectors of the body centered cubic lattice in two different ways. One way is to choose a set of vectors that will generate all of the lattice points from a single point of the lattice. With the usual convention that $\hat{\mathbf{i}}$, $\hat{\mathbf{j}}$, and $\hat{\mathbf{k}}$ are unit vectors pointing along the x-, y-, and z-axes and that $a\hat{\mathbf{i}}$, $a\hat{\mathbf{j}}$, and $a\hat{\mathbf{k}}$ are the primitive vectors of the original simple cubic lattice, a possible choice of the primitive vectors of the body-centered cubic lattice is

$$\mathbf{a}_1 = a\hat{\mathbf{i}}, \quad \mathbf{a}_2 = a\hat{\mathbf{j}}, \quad \mathbf{a}_3 = \frac{a}{2}(\hat{\mathbf{i}} + \hat{\mathbf{j}} + \hat{\mathbf{k}}).$$

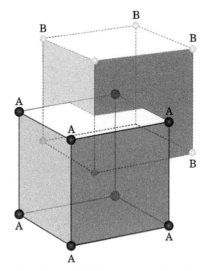

FIGURE 9.3 A body-centered cubic lattice. As shown in the figure, the body-centered cubic lattice can be thought of as consisting of two interpenetrating lattices with points **A** and **B**.

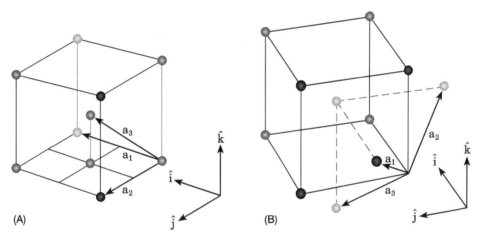

FIGURE 9.4 (A) The primitive vectors, $\mathbf{a}_1$, $\mathbf{a}_2$, $\mathbf{a}_3$ of the body-centered cubic lattice; (B) A more symmetrical choice of the primitive vectors of the body-centered cubic lattice.

These vectors are shown in Fig. 9.4(A). A more symmetrical choice of primitive vectors for the bcc lattice are the vectors,

$$\mathbf{a}_1 = \frac{a}{2}(\hat{\mathbf{j}} + \hat{\mathbf{k}} - \hat{\mathbf{i}}), \quad \mathbf{a}_2 = \frac{a}{2}(\hat{\mathbf{k}} + \hat{\mathbf{i}} - \hat{\mathbf{j}}), \quad \mathbf{a}_3 = \frac{a}{2}(\hat{\mathbf{i}} + \hat{\mathbf{j}} - \hat{\mathbf{k}}), \tag{9.2}$$

which are shown in Fig. 9.4(B). One may show both geometrically and analytically that the body-centered cubic Bravais lattice can be generated from either of these sets of primitive vectors by using Eq. (9.1) systematically. In either case, the entire lattice is generated as identical copies of a physical unit consisting of a single atom. These two sets of primitive vectors are compact, but each choice defines an irregular volume with parallel faces that does not display the cubic symmetry of the lattice.

The body-centered cubic lattice can also be generated by choosing a unit cell consisting of a corner point and a center point of a single cube

$$\mathbf{0}, \quad \frac{a}{2}(\hat{\mathbf{i}} + \hat{\mathbf{j}} + \hat{\mathbf{k}}). \tag{9.3}$$

The lattice, which is then generated by Eq. (9.1) with the primitive vectors of the original simple cubic lattice, consists of identical copies of a unit cell with two atoms. The two points given by Eq. (9.3) are called a *basis* for the lattice, and

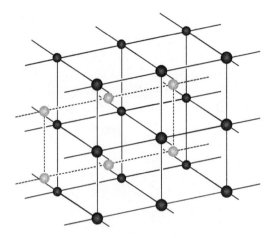

FIGURE 9.5 Cesium chloride lattice.

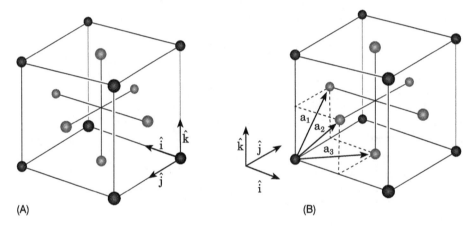

(A)   (B)

FIGURE 9.6 (A) A single cube in a face centered cubic lattice; (B) a set of primitive vectors, which generate the face-centered cubic lattice.

the lattice itself is called a *lattice with a basis*. This second way of representing the bcc lattice makes the symmetry more apparent.

If a body-centered cubic lattice consists of identical atoms, then either way of generating the lattice is entirely acceptable. However, if more than one kind of atom is present, it is necessary to choose a basic unit consisting of more than one atom. The crystal then consists of identical copies of this basis. The structure of the cesium chloride lattice is shown in Fig. 9.5. Each cesium atom occurs at the center of a cube with chlorine atoms on the corners. The atoms of each kind thus have eight of the other species as its nearest neighbors. In drawing Fig. 9.5, we have made the cesium and chloride ions smaller than they would be in an actual cesium chloride lattice. The extra space surrounding the ions in the figure allows one to see more clearly the geometrical form of the crystal. The translational symmetry of the cesium chloride lattice is that of a simple cubic lattice with the distance between two equivalent points being equal to the separation of two atoms of the same kind. We may choose a basis consisting of a chlorine atom at the origin $\mathbf{0}$ and a cesium atom at the cubic center $(a/2)(\hat{\mathbf{i}}+\hat{\mathbf{j}}+\hat{\mathbf{k}})$. The entire lattice is then generated by displacements with the primitive vectors of the cubic Bravais lattice $a\hat{\mathbf{i}}$, $a\hat{\mathbf{j}}$, and $a\hat{\mathbf{k}}$ where a is the distance between two adjoining chlorine atoms or two adjoining cesium atoms.

Another commonly occurring crystal structure is the face-centered cubic (fcc) lattice shown in Fig. 9.6(A). The lattice can be formed from the simple cubic lattice by adding an additional point to each square face. As for the body-centered cubic structure, this lattice can be thought of as consisting of interpenetrating lattices. For ease of description, one can think of each cube of the simple cubic lattice as consisting of top and bottom faces and four vertical sides facing north and south, and east and west. The points on the north and south faces of the original cubic structure form a new simple cubic lattice. The points of the original simple cubic lattice are now at the centers of the north and south faces of the new cubic structure, whereas the points that were added to the centers of the east and west faces of the original cube are in the centers of top and bottom faces of the new cubic structure. A set of primitive vectors that generates the face-centered cubic structure from a single point is shown in Fig. 9.6(B). A mathematical expression for these vectors is

$$\mathbf{a}_1 = \frac{a}{2}(\hat{\mathbf{j}} + \hat{\mathbf{k}}), \quad \mathbf{a}_2 = \frac{a}{2}(\hat{\mathbf{k}} + \hat{\mathbf{i}}), \quad \mathbf{a}_3 = \frac{a}{2}(\hat{\mathbf{i}} + \hat{\mathbf{j}}). \tag{9.4}$$

In keeping with Eq. (9.1), the points of the face-centered cubic lattice may be generated by taking integral multiples of these vectors. For instance, one may show both geometrically and analytically that the point on the upper left-hand corner of the cube in Fig. 9.8 is

$$a\hat{\mathbf{k}} = \mathbf{a}_1 + \mathbf{a}_2 - \mathbf{a}_3.$$

As for the body-centered cubic lattice, the points of the face-centered cubic lattice can be generated using a basis. A four-point basis of the fcc lattice is formed by adding the three primitive vectors (9.4) to the origin **0**

$$\mathbf{0}, \quad \mathbf{a}_1 = \frac{a}{2}(\hat{\mathbf{j}} + \hat{\mathbf{k}}), \quad \mathbf{a}_2 = \frac{a}{2}(\hat{\mathbf{k}} + \hat{\mathbf{i}}), \quad \mathbf{a}_3 = \frac{a}{2}(\hat{\mathbf{i}} + \hat{\mathbf{j}}).$$

The fcc lattice is obtained from this basis by making the displacement associated with the primitive vectors of the simple cubic lattice

$$\mathbf{l} = l_1 a\hat{\mathbf{i}} + l_2 a\hat{\mathbf{j}} + l_3 a\hat{\mathbf{k}}.$$

The number of nearest neighbors for the face-centered cubic lattice is 12. One can see this by considering the three mutually perpendicular planes passing through each corner point of the cube. Each plane contains four nearest neighbors, which are located at the centers of the faces of the eight cubes having that corner point.

From our discussions of the body-centered and face-centered lattices, one may note that the atoms in the two lattices have different numbers of nearest neighbors. The atoms of a body-centered cubic lattice have eight nearest neighbors, while the atoms of a face-centered cubic lattice have twelve nearest neighbors. It should also be apparent that the unit cells of face-centered cubic lattices have more lattice points than the unit cells of the body-centered cubic lattice. Before going further, we would like now to find the density of points in the three cubic lattices we have considered.

The simple cubic cell shown in Fig. 9.2 has eight lattice points located at the corners of each cube. Since each of these corner points is shared with eight other cubes, a simple cubic lattice has one lattice point per unit cell. The body-centered cubic lattice has one additional point at the center of the cube and thus has two lattice points for every cell of the cubic structure. In addition to the points at the corners of the cube, the face-centered cubic lattice has six lattice points on the faces of the simple cube. Since each point on the face of a cube is shared with one other cube, the points on the faces contribute three additional points to the cubic structure. Consequently, the face-centered cubic crystal has four lattice points for every cell of the cubic structure. We thus conclude that the face-centered cubic structure is the most dense and the simple cubic structure is the least dense of the three cubic Bravais lattices.

For questions concerning the relative density of different crystal structures, we note that for many structures the ions may be considered to be rigid spheres that are stacked together to form the crystals. This often provides a fairly good representation of ionic crystals and metals. In order to provide a criterion for judging how tightly a particular structure is packed, we define the *packing fraction* to be the volume of these spheres divided by the total volume of the crystal. The packing fraction is the fraction of the volume of space inside the spheres. If it were one, the spheres would be packed so tightly together that there was no empty space left, which is, of course, impossible. The packing fractions of the simple cubic, body-centered cubic, and face-centered cubic structures are 0.52, 0.68, and 0.74 respectively. (See Problem 11.) We thus find again that the face-centered cubic structure is the most dense and the simple cubic structure is the least dense of the three cubic structures. In the next section, we shall find that the face-centered cubic structure is closely related to the hexagonal close-packed structure. The two structures provide two very compact ways of stacking spheres.

Earlier in this section, we saw that an entire lattice can be generated by adding integral multiples of the primitive vectors of the Bravais lattice to the points in the unit cell. A region of space that entirely fills all of space when it is translated in this way is called a *primitive unit cell*. The choice is to some extent arbitrary. Two different choices of primitive cells in a two-dimensional lattice are shown in Fig. 9.7(A). The primitive cell that is used most widely is called the *Wigner-Seitz cell*. This cell is constructed by drawing perpendicular planes bisecting the lines joining the chosen center to the nearest equivalent lattice cites as illustrated in Fig. 9.7(B). The Wigner-Seitz cell entirely fills space when it is translated by the vectors of the Bravais lattice, and it has the symmetry of the lattice. An illustration of the Wigner-Seitz cell for a body-centered cubic lattice is given in Fig. 9.8(A).

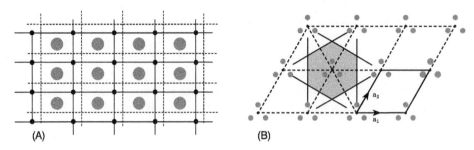

FIGURE 9.7 (A) Two different choices of primitive cells; (B) Wigner-Seitz cell in a two-dimensional lattice.

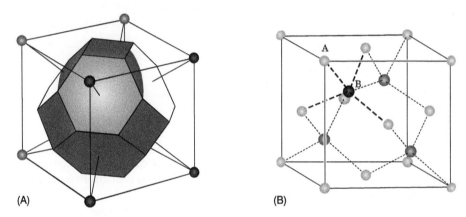

FIGURE 9.8 (A) Wigner-Seitz cell for a body-centered cubic lattice; (B) the structure of the diamond crystal.

9.2 Additional crystal structures

9.2.1 The diamond structure

Carbon atoms in naturally occurring diamond crystals occupy the sites of two interpenetrating face-centered cubic lattices. The diamond structure is shown in Fig. 9.8(B). In this figure, the sites A and B are corner points in the two different face-centered cubic lattices. B is situated one quarter of the way along the main diagonal of the cube with the corner point A. The lattice can be regarded as a face-centered cubic lattice with a two-point basis located at $\mathbf{0}$ and $(a/4)(\hat{\mathbf{i}} + \hat{\mathbf{j}} + \hat{\mathbf{k}})$. The elements carbon, silicon, and germanium crystallize in the diamond structure. An important variant of the diamond structure occurs for compounds involving two atomic species. The *zincblende structure* has equal numbers of zinc and sulfur atoms distributed on a diamond lattice in such a way that each atom has four of the atoms of the other species as its nearest neighbors. One species occupies the A site in Fig. 9.8(B) and the other species occupies the B site. Because two types of atoms occur, the zincblende structure must be described as a lattice with a basis as the cesium chloride structure.

9.2.2 The hexagonal close-packed structure

The *hexagonal close-packed* (hcp) structure ranks in importance with the body-centered cubic and face-centered cubic structures. About 30 elements crystallize in this form. As the name suggests, the hexagonal close-packed lattice can be constructed by stacking hard spheres so that each sphere rests on the spheres below it and touches the spheres next to it on the same level.

Fig. 9.9 shows two layers of hard spheres stacked in this way. From the figure, one can note that lines drawn from the centers of each sphere to the centers of neighboring spheres on the same level form a network of triangles. Looking more closely at the two layers of spheres in Fig. 9.9, one can also see that the spheres on the second level are placed in depressions in the first layer of spheres, and that not all of the depressions in the first layer are occupied by sphere of the second level. There are two alternate ways of placing spheres into the depressions of the second layer of spheres. If the spheres in the third layer are placed in the depressions directly above spheres in the first layer, as is the case for the site denoted by A in Fig. 9.9, and the spheres in the fourth level are placed above spheres in the second layer, and so forth, the resulting arrangement of spheres is the hexagonal close-packed or Wurtzite structure. This structure is shown in Fig. 9.10(A). Again

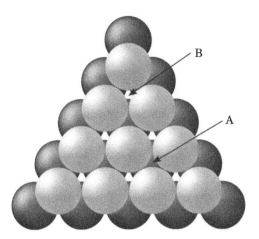

FIGURE 9.9 Two layers of stacked spheres.

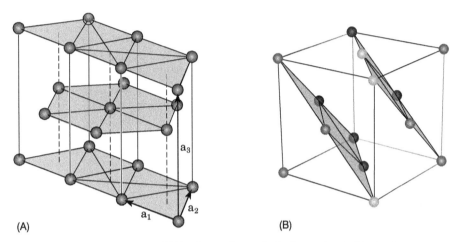

FIGURE 9.10 (A) The hexagonal close-packed structure; (B) Two layers of atoms in the face-centered cubic structure.

for clarity, small solid balls are drawn at each lattice point in the figure rather than full solid spheres. Once can readily see that the balls on the third layer in Fig. 9.10(A) lie above balls on the first layer.

The hexagonal close-packed structure is not the only way to close pack spheres. If the spheres in the third layer are placed in the depressions in the second layer lying above unused depressions in *both* the first and the second layers, and the fourth layer is placed in depressions in the third layer directly above spheres in the first, the fifth above the second, and so on, one generates another Bravais lattice. As suggested in Fig. 9.10(B), this second closed-packed structure is just the face-centered cubic lattice with the diagonal of the cube perpendicular to the planes of the solid spheres. The density of lattice points and the packing fractions are the same for the hexagonal and face-centered cubic structures.

Since each successive layer of the crystal can be placed in two alternate ways, there are very many other possible closed packed arrangements. For the hexagonal closed-packed structure, the third layer is the same as the first and the fourth is the same as the second, and so forth, and this structure may thus be denoted (...ABABAB...). Similarly, the face-center cubic structure may be denoted (...ABCABCABC...). There are other closed-packed structures. Certain rare earth metals crystallize in the form (...ABACABACABAC...).

9.2.3 The sodium chloride structure

The *sodium chloride* structure shown in Fig. 9.11 consists of equal numbers of sodium and chloride atoms placed at alternate points of a simple cubic lattice. Each ion has six ions of the other species as its nearest neighbors. The sodium chloride structure may also be described as two interpenetrating face-centered cubic structures, one consisting of sodium ions and the other of chloride atoms.

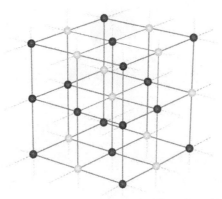

FIGURE 9.11 The sodium chloride structure.

9.3 The reciprocal lattice

The potential energy and the charge density of electrons in a crystal have the same periodicity as the lattice itself. A good deal of qualitative understanding of many solids can be obtained by supposing that the outer electrons of the atoms of the solid are moving in a weak periodic potential due to the ion cores. The electrons are then nearly free of their crystalline environment and can be approximated by free-electron wave functions. We begin our study of the properties of the valence electrons in crystals by considering the wave function of a free electron in a periodic environment.

In Section 3.1, we considered free particles that can be described by the wave function

$$\psi(x) = Ae^{ikx}, \tag{9.5}$$

where k is related to the momentum of the particle by the equation

$$p = \hbar k. \tag{9.6}$$

The requirement that the wave function $\psi(x)$ has the periodicity of the lattice can be satisfied by imposing the periodic boundary condition

$$\psi(a) = \psi(0). \tag{9.7}$$

Substituting the wave function (9.5) into Eq. (9.7) gives

$$Ae^{ika} = A.$$

This condition leads to the requirement

$$ka = n\,2\pi,$$

where n is an integer. Solving for k, we get

$$k = n\,\frac{2\pi}{a}. \tag{9.8}$$

The wave function (9.5) thus has the periodicity of the lattice provided that the wave vector k of the particle belongs to the discrete set of values given by Eq. (9.8).

Substituting the values of k given by Eq. (9.8) into the free-electron wave function (9.5), we obtain

$$\psi(x) = Ae^{2\pi inx/a}.$$

One may readily confirm that the wave function $\psi(x)$ above satisfies the periodic boundary condition (9.7). This wave function can be written in the form

$$\psi_g(x) = Ae^{igx}, \tag{9.9}$$

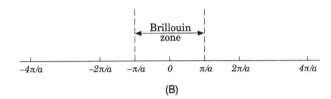

FIGURE 9.12 (A) A one-dimensional lattice; (B) the corresponding reciprocal lattice.

where the quantities g belong to the set of *reciprocal lattice lengths*

$$g = n\frac{2\pi}{a}. \tag{9.10}$$

The distance between two adjacent values of g is 2π times the inverse of the spacing in the original lattice. For this reason, the array of values of g is referred to as the reciprocal lattice. A one-dimensional lattice and the corresponding reciprocal lattice are illustrated in Fig. 9.12. The original Bravais lattice is often referred to as the *direct lattice*. Using this terminology, the one-dimensional lattice in Fig. 9.12(A) is the direct lattice, while the lattice in Fig. 9.12(B) is the reciprocal lattice. The concept of the reciprocal lattice may be used to state the condition for periodicity: *the free electron wave function (9.5) is a periodic function in the Bravais lattice if the value of k to which it corresponds is a member of the reciprocal lattice*. In Fig. 9.18(B), the Wigner-Seitz cell is also shown. The Wigner-Seitz cell in reciprocal lattice space is called the *Brillouin zone*.

The position of points in a one-dimensional lattice can be written

$$l = m\,a$$

where m is an integer and a is the lattice spacing. With this notation, the product of the position of a point in the reciprocal lattice (9.10) and a point in the direct lattice is

$$g\,l = 2\pi n\,m.$$

Here $n\,m$ is the product of two integers and is itself an integer. The product $g\,l$ is thus 2π times an integer. This leads immediately to the equation

$$e^{igl} = 1. \tag{9.11}$$

Eqs. (9.9) and (9.11) may be used to show that the functions $\psi_g(x)$ have the periodicity of the lattice. We have

$$\psi_g(x + l) = Ae^{ig(x+l)} = Ae^{igx}e^{igl} = Ae^{igx} = \psi_g(x).$$

The functions (9.9) serve as a basis that can be used to represent functions having the periodicity of the lattice. To show how a function may be expanded in terms of the functions (9.9), we use the ideas of Fourier analysis introduced in the Introduction and taken up again in Chapter 3. Any periodic function $f(x)$ can be represented as a Fourier series

$$f(x) = \sum_n F_n e^{2\pi inx/a}, \tag{9.12}$$

where n is an integer. This equation can be written more simply using Eq. (9.10) to obtain

$$f(x) = \sum_g F_g e^{igx}. \tag{9.13}$$

The coefficients F_g that appear in Eq. (9.13) are given by the formula

$$F_g = \frac{1}{a} \int_{\text{cell}} e^{-igx} f(x)dx,$$ (9.14)

where the range of integration extends over a cell of the lattice. A derivation of this result is sketched in Problem 22.

These ideas can readily be generalized to three dimensions. The wave function of a free electron in three dimensions can be written

$$\psi_{\mathbf{k}}(\mathbf{r}) = Ae^{i\mathbf{k}\cdot\mathbf{r}},$$ (9.15)

where $\mathbf{k}$ and $\mathbf{r}$ denote the wave and position vectors, respectively. The momentum of a free electron is related to the wave vector of the electron by the equation

$$\mathbf{p} = \hbar\mathbf{k}.$$ (9.16)

This last equation is a three-dimensional generalization of the relation $p = \hbar k$ discussed in Section 3.1.

The reciprocal lattice vectors $\mathbf{g}$ are the special wave vectors $\mathbf{k}$ for which the free electron wave function (9.15) is periodic. The wave vectors having this property will be said to belong to the reciprocal lattice. We shall now demonstrate that there are wave vectors for which the free-electron wave function is periodic by explicitly constructing these vectors. We shall then show how these wave vectors, which are called reciprocal lattice vectors, can be constructed for the body-centered and face-centered cubic crystals, and we shall show that these vectors provide a clear way of describing the periodic properties of these lattices.

The primitive vectors of the reciprocal lattice can be generated from the three primitive vectors $\mathbf{a}_1$, $\mathbf{a}_2$, and $\mathbf{a}_3$ of the direct lattice by the equations

$$\mathbf{b}_1 = 2\pi \frac{\mathbf{a}_2 \times \mathbf{a}_3}{\mathbf{a}_1 \cdot (\mathbf{a}_2 \times \mathbf{a}_3)}$$ (9.17)

$$\mathbf{b}_2 = 2\pi \frac{\mathbf{a}_3 \times \mathbf{a}_1}{\mathbf{a}_1 \cdot (\mathbf{a}_2 \times \mathbf{a}_3)}$$ (9.18)

$$\mathbf{b}_3 = 2\pi \frac{\mathbf{a}_1 \times \mathbf{a}_2}{\mathbf{a}_1 \cdot (\mathbf{a}_2 \times \mathbf{a}_3)},$$ (9.19)

and the reciprocal lattice vectors $\mathbf{g}$ can be written

$$\mathbf{g} = n_1\mathbf{b}_1 + n_2\mathbf{b}_2 + n_3\mathbf{b}_3,$$ (9.20)

where n_1, n_2, and n_3 are integers.

We leave it as an exercise (Problem 12) to show that the primitive vectors defined above have the property

$$\mathbf{b}_i \cdot \mathbf{a}_j = 2\pi \delta_{ij},$$ (9.21)

where δ_{ij} is the Kronecker delta symbol defined as follows

$$\delta_{ij} = \begin{cases} 1, & i = j \\ 0, & i \neq j. \end{cases}$$

Eq. (9.21) may then be used to show that the dot product of the reciprocal lattice vector (9.20) with the vector from the Bravais (direct) lattice defined by Eq. (9.1) gives

$$\begin{aligned} \mathbf{g} \cdot \mathbf{l} &= (n_1\mathbf{b}_1 + n_2\mathbf{b}_2 + n_3\mathbf{b}_3) \cdot (l_1\mathbf{a}_1 + l_2\mathbf{a}_2 + l_3\mathbf{a}_3) \\ &= 2\pi(n_1 l_1 + n_2 l_2 + n_3 l_3) \\ &= 2\pi \times \text{integer}. \end{aligned}$$ (9.22)

This leads immediately to the equation

$$e^{i\mathbf{g}\cdot\mathbf{l}} = 1.$$ (9.23)

As in the one-dimensional case, this result may be used to show that the free electron wave (9.15) is periodic in the original lattice provided that $\mathbf{k}$ is a vector of the reciprocal lattice. Allowing the wave vector $\mathbf{k}$ in Eqs. (9.15) to be a vector $\mathbf{g}$ of the reciprocal lattice and replacing $\mathbf{r}$ with $\mathbf{r} + \mathbf{l}$, we get

$$\psi_{\mathbf{g}}(\mathbf{r} + \mathbf{l}) = Ae^{i\mathbf{g}\cdot(\mathbf{r}+\mathbf{l})} = Ae^{i\mathbf{g}\cdot\mathbf{r}}e^{i\mathbf{g}\cdot\mathbf{l}} = Ae^{i\mathbf{g}\cdot\mathbf{r}} = \psi_{\mathbf{g}}(\mathbf{r}).$$

The free-electron function (9.15) is thus periodic if the wave vector $\mathbf{k}$ is a member of the reciprocal lattice.

A function $f(\mathbf{r})$ having the periodicity of the three-dimensional lattice can be represented by the Fourier series

$$f(\mathbf{r}) = \sum_{\mathbf{g}} F_{\mathbf{g}} e^{i\mathbf{g}\cdot\mathbf{r}}, \tag{9.24}$$

which is a natural generalization of (9.13) to three-dimensions. The coefficients $F_{\mathbf{g}}$ are given by the equation

$$F_{\mathbf{g}} = \frac{1}{v_{\text{cell}}} \int_{\text{cell}} f(\mathbf{r}) e^{-i\mathbf{g}\cdot\mathbf{r}} dV, \tag{9.25}$$

where v_{cell} is the volume of a primitive cell of the direct lattice and where, as before, the range of integration extends over a single cell.

Example 9.1

Construct the primitive cell of the reciprocal lattice for a simple cubic Bravais lattice.

Solution
As we have seen, the primitive vectors of the simple cubic lattice can be chosen to be

$$\mathbf{a}_1 = a\mathbf{i}, \quad \mathbf{a}_2 = a\mathbf{j}, \quad \mathbf{a}_3 = a\mathbf{k}.$$

The primitive vectors of the reciprocal lattice can be obtained immediately using Eqs. (9.17), (9.18), and (9.19). We have

$$\mathbf{b}_1 = \frac{2\pi}{a}\mathbf{i}, \quad \mathbf{b}_2 = \frac{2\pi}{a}\mathbf{j}, \quad \mathbf{b}_3 = \frac{2\pi}{a}\mathbf{k}.$$

The reciprocal lattice of the simple cubic lattice is itself a simple cubic lattice with the length of each side being $2\pi/a$.

Example 9.2

Show that the reciprocal lattice of the face-centered cubic lattice is the body-centered cubic lattice.

Solution
Again, substituting the expressions for the primitive vectors of the face-centered cubic lattice (9.4) into Eqs. (9.17), (9.18), and (9.19), we get

$$\mathbf{b}_1 = \frac{4\pi}{a}\frac{1}{2}(\hat{\mathbf{j}}+\hat{\mathbf{k}}-\hat{\mathbf{i}}), \quad \mathbf{b}_2 = \frac{4\pi}{a}\frac{1}{2}(\hat{\mathbf{k}}+\hat{\mathbf{i}}-\hat{\mathbf{j}}), \quad \mathbf{b}_3 = \frac{4\pi}{a}\frac{1}{2}(\hat{\mathbf{i}}+\hat{\mathbf{j}}-\hat{\mathbf{k}}). \tag{9.26}$$

By comparing these vectors with the vectors in Eq. (9.2), we see that the vectors (9.26) are the primitive vectors of a body-centered cubic lattice with the side of the cube equal to $2\pi/a$. As we shall find in Chapter 10, most semiconductors have a zincblende structure with two interpenetrating face-centered cubic lattices. The importance of the body-centered cubic lattice is largely due to its being the reciprocal lattice of the face-centered cubic lattice.

It is left as an exercise (Problem 16) to show that the reciprocal of the body-centered cubic lattice is the face-centered cubic lattice.

9.4 Lattice planes

In any Bravais lattice, one may identify *lattice planes* in a variety of different orientations. Each plane contains three or more lattice points that do not lie on a straight line. One can also identify groups of lattice planes, which together contain all of the points in a Bravais lattice. Two groups of lattice planes in a simple cubic lattice are shown in Fig. 9.13. Any lattice plane is a member of such a group or family of planes. We shall find that the vectors of the reciprocal lattice help us to

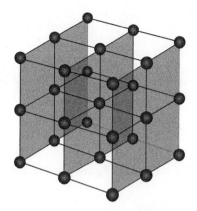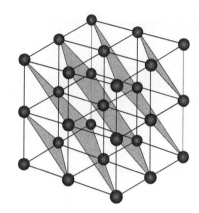

FIGURE 9.13 Two families of lattice plains in a simple cubic lattice.

identify and describe groups of lattice planes. We begin our discussion of lattice planes by demonstrating certain important features of lattice vectors and families of lattice planes.

(i) A vector extending from one lattice point to another is itself a lattice vector.

In keeping with Eq. (9.1), we denote the vectors pointing to the two lattice points

$$\mathbf{l} = l_1 \mathbf{a}_1 + l_2 \mathbf{a}_2 + l_3 \mathbf{a}_3$$

and

$$\mathbf{m} = m_1 \mathbf{a}_1 + m_2 \mathbf{a}_2 + m_3 \mathbf{a}_3,$$

where (l_1, l_2, l_3) and (m_1, m_2, m_3) are sets of integers. The vector extending from the first lattice point to the second is

$$\mathbf{m} - \mathbf{l} = (m_1 - l_1)\mathbf{a}_1 + (m_2 - l_2)\mathbf{a}_2 + (m_3 - l_3)\mathbf{a}_3.$$

This vector is of the form (9.1) and is also a member of the lattice.

(ii) Any plane determined by three points in the direct lattice is normal to some vector of the reciprocal lattice.

According to the property *(i)* above, a vector pointing from one lattice point to another will itself be a lattice vector. Using a similar notation as before, we denote the vector pointing from the first lattice point to the second lattice point

$$\mathbf{l} = l_1 \mathbf{a}_1 + l_2 \mathbf{a}_2 + l_3 \mathbf{a}_3,$$

and we denote the vector pointing from the first lattice point to the third lattice point

$$\mathbf{m} = m_1 \mathbf{a}_1 + m_2 \mathbf{a}_2 + m_3 \mathbf{a}_3.$$

A vector which is normal to the plane can be constructed by taking the vector product of these two vectors

$$\mathbf{n} = \mathbf{l} \times \mathbf{m}.$$

Substituting the above expressions for $\mathbf{l}$ and $\mathbf{m}$ into this last equation and using Eqs. (9.17)-(9.19), one may readily show that the vector $\mathbf{n}$ is parallel to a vector of the reciprocal lattice.

(iii) Each vector of the reciprocal lattice is normal to a set of lattice planes of the direct lattice.

In order to show this, we consider the reciprocal lattice vector

$$\mathbf{g} = n_1 \mathbf{b}_1 + n_2 \mathbf{b}_2 + n_3 \mathbf{b}_3$$

and the lattice vector

$$\mathbf{l} = l_1 \mathbf{a}_1 + l_2 \mathbf{a}_2 + l_3 \mathbf{a}_3,$$

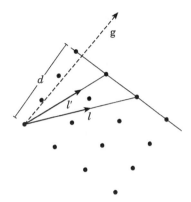

FIGURE 9.14 Two lattice vectors, $\mathbf{l}$ and $\mathbf{l}'$, which are directed to lattice points lying on a plane perpendicular to the reciprocal lattice vector, $\mathbf{g}$.

where (n_1, n_2, n_3) and (l_1, l_2, l_3) are sets of integers. Taking the inner product of these two vectors and using Eq. (9.21), we obtain

$$\mathbf{g} \cdot \mathbf{l} = 2\pi (n_1 l_1 + n_2 l_2 + n_3 l_3)$$
$$= 2\pi N. \tag{9.27}$$

In this last equation, N is the integer, $n_1 l_1 + n_2 l_2 + n_3 l_3$. The projection of the lattice vector $\mathbf{l}$ on the vector $\mathbf{g}$ is

$$d = \mathbf{l} \cdot \frac{\mathbf{g}}{|\mathbf{g}|},$$

where $|\mathbf{g}|$ is the length of the vector $\mathbf{g}$. Using Eq. (9.27) to evaluate the inner product, this last equation may be written

$$d = \frac{2\pi N}{|\mathbf{g}|}. \tag{9.28}$$

Any lattice vector $\mathbf{l}'$ satisfying the equation,

$$\mathbf{g} \cdot \mathbf{l}' = 2\pi N,$$

will have the same projection on $\mathbf{g}$ and must also be directed to a lattice point on the plane normal to $\mathbf{g}$, at a distance d from the origin. All the points having this property form a lattice plane. The vectors, $\mathbf{l}, \mathbf{l}'$, and $\mathbf{g}$, are illustrated in Fig. 9.14.

(iv) Consider now the reciprocal lattice vector,

$$\mathbf{g} = n_1 \mathbf{b}_1 + n_2 \mathbf{b}_2 + n_3 \mathbf{b}_3.$$

If the integers (n_1, n_2, n_3) have no common factors, the vector $\mathbf{g}$ is the shortest vector pointing in this direction, and the family of lattice planes normal to $\mathbf{g}$ is separated by the distance

$$d = \frac{2\pi}{|\mathbf{g}|}. \tag{9.29}$$

A derivation of this equation is sketched in Problem 23.

Three different lattice planes in a simple cubic structure are illustrated in Fig. 9.15. In describing these planes, we may superimpose in our own mind the reciprocal lattice space upon the ordinary lattice. The reciprocal lattice vectors then provide us an easy way of describing crystal planes. In Example 9.1, we found that the primitive vectors of the cubic lattice are

$$\mathbf{b}_1 = \frac{2\pi}{a}\mathbf{i}, \quad \mathbf{b}_2 = \frac{2\pi}{a}\mathbf{j}, \quad \mathbf{b}_3 = \frac{2\pi}{a}\mathbf{k}.$$

The lattice plane of the simple cubic lattice shown in Fig. 9.15(A) is perpendicular to the reciprocal lattice vector, $\mathbf{b}_2$, while the plane shown in Fig. 9.15(B) is perpendicular to the vector, $\mathbf{b}_1 + \mathbf{b}_2$, and the lattice plane in Fig. 9.15(C) is perpendicular to the vector, $\mathbf{b}_1 + \mathbf{b}_2 + \mathbf{b}_3$.

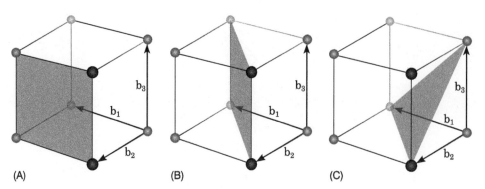

FIGURE 9.15 Three lattice planes in a simple cubic structure.

Example 9.3

Find the distance between the group of lattice planes that include the shaded plane shown in Fig. 9.21(C).

Solution

The lattice plane shown in Fig. 9.21(C) is perpendicular to the reciprocal lattice vector

$$\mathbf{g} = \mathbf{b}_1 + \mathbf{b}_2 + \mathbf{b}_3.$$

Since the three vectors, $\mathbf{b}_1$, $\mathbf{b}_2$, and $\mathbf{b}_3$, appear in this last equation with unit coefficients having no common multiplicative factors, the distance between the lattice planes can be obtained using Eq. (9.29). We note that the length of the reciprocal lattice vector is

$$\begin{aligned} |\mathbf{g}| &= \sqrt{(2\pi/a)^2 + (2\pi/a)^2 + (2\pi/a)^2} \\ &= \sqrt{3}(2\pi/a). \end{aligned} \tag{9.30}$$

Substituting this value for $|\mathbf{g}|$ into Eq. (9.29), we obtain

$$d = \frac{a\sqrt{3}}{3}. \tag{9.31}$$

The validity of this result can be confirmed by studying Fig. 9.21(C) more closely. Imagine drawing a diagonal of the cube passing from the lower right-hand corner to the upper left-hand corner of the cube. This diagonal passes through two lattice planes, the shaded plane and a plane passing through the three points above the shaded plane to the left. The points on the lower right-hand corner and the upper left-hand corner also lie on planes of the same family. The diagonal, whose length is $\sqrt{3}a$, is thus divided into three equal parts by the lattice planes, and the planes are separated by the distance $\sqrt{3}a/3$.

The correspondence between families of lattice planes and reciprocal lattice vectors provides the basis for the conventional way of describing lattice planes. One describes the orientation of families of lattice planes by giving the reciprocal lattice vector normal to the planes. To make the choice unique, one uses the shortest reciprocal lattice vector which has no common factors. The most prominent planes in the direct lattice are those which are most densely populated. Since each family of planes includes all of the points of the lattice, the most prominent planes are the planes which are most widely separated. According to Eq. (9.29), these planes correspond to the smallest reciprocal lattice vectors. For a particular set of primitive reciprocal lattice vectors, $\mathbf{b}_1$, $\mathbf{b}_2$, and $\mathbf{b}_3$, the *Miller indices* of a family of lattice planes are the integer components (n_1, n_2, n_3) of the shortest reciprocal lattice vector normal to the plane. The commas between the integral components are traditionally not written. In Example 9.3, we found that the reciprocal lattice vector, $\mathbf{g} = \mathbf{b}_1 + \mathbf{b}_2 + \mathbf{b}_3$, is normal to the shaded plane shown in Fig. 9.21(C). Since the components $(1, 1, 1)$ have no common factors, the Miller indices for the family of planes parallel to this plane are (111). The Miller indices for the family of planes parallel to the shaded plane in Fig. 9.21(A) are (010) and the Miller indices corresponding to the family of planes parallel to the shaded plane in Fig. 9.21(B) are (110). Some refinements of this simple notation are commonly used. For example, the Miller indices of the group of planes normal to the vector,

$$\mathbf{n} = 4\mathbf{b}_1 - 2\mathbf{b}_2 + \mathbf{b}_3,$$

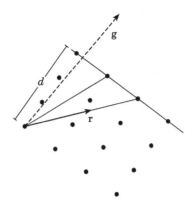

FIGURE 9.16 Three lattice planes in a simple cubic structure.

are $(4, -2, 1)$. However, the minus sign is removed by writing -2 as $\bar{2}$, and the designation of this group of planes thus becomes $(4\bar{2}1)$. To interpret the indices, one must know which set of primitive vectors is being used. A simple cubic basis is invariably used when the crystal has cubic symmetry.

Crystallographers usually introduce the Miller indices in a slightly different way. To make contact with their nomenclature, we consider a family of planes normal to the reciprocal lattice vector

$$\mathbf{g} = n_1\mathbf{b}_1 + n_2\mathbf{b}_2 + n_3\mathbf{b}_3 \qquad (9.32)$$

The condition that the position vector,

$$\mathbf{r} = r_1\mathbf{a}_1 + r_2\mathbf{a}_2 + r_3\mathbf{a}_3 , \qquad (9.33)$$

be directed to a point on a plane which is perpendicular to this vector can be written

$$\mathbf{g} \cdot \mathbf{r} = A, \qquad (9.34)$$

where A is a constant.

The vectors, $\mathbf{g}$ and $\mathbf{r}$, are illustrated in Fig. 9.16. All vectors $\mathbf{r}$ directed to points lying on the plane have the same scalar product with $\mathbf{g}$. Substituting Eqs. (9.32) and (9.33) into Eq. (9.34) and using Eq. (9.21), we obtain

$$2\pi(n_1 r_1 + n_2 r_2 + n_3 r_3) = A. \qquad (9.35)$$

The values of the components, r_1, r_2, and r_3, for which the plane intersects the line determined by the unit vector $\mathbf{a}_1$ may be denoted by $r_1 = x_1$, $r_2 = 0$, and $r_3 = 0$. Substituting these values into Eq. (9.35), we get

$$2\pi n_1 x_1 = A.$$

Solving this equation for n_1 gives

$$n_1 = \frac{A}{2\pi x_1}.$$

A similar line of argument may be used to find how the integers n_2 and n_3 are related to the points x_2 and x_3 where the plane crosses the lines determined by the unit vectors $\mathbf{a}_2$ and $\mathbf{a}_3$. The three integers are related to the intercepts by the equations

$$n_1 = \frac{A}{2\pi x_1}, \quad n_2 = \frac{A}{2\pi x_2}, \quad n_3 = \frac{A}{2\pi x_3}.$$

The Miller indices of a plane are thus inversely proportional to the intercepts of the lattice plane with the crystal axes. Crystallographers generally define the Miller indices to be a set of integers with no common factors that are inversely proportional to the intercepts of the lattice plane with the crystal axes

$$n_1 : n_2 : n_3 = \frac{1}{x_1} : \frac{1}{x_2} : \frac{1}{x_3}.$$

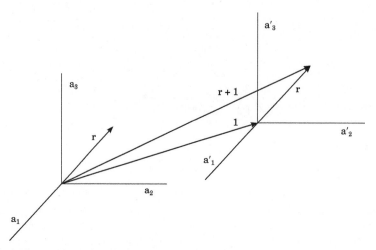

FIGURE 9.17 The translation of a coordinate frame by a lattice vector **l**.

9.5 Bloch's theorem

We have said that functions, such as the potential energy and the electron charge density, must have the periodicity of the lattice. We now turn our attention to the wave function of the electron. While the wave function need not itself be periodic, the form of the wave function is decisively influenced by the periodicity of the lattice.

To understand the conditions imposed upon the form of wave functions by the symmetry of the lattice, we consider the effect of a displacement of the coordinate system upon a wave function. Suppose an electron is described by a wave function $\psi(\mathbf{r})$, and we translate the origin of the coordinate system by the lattice vector **l**. A translation of this kind is illustrated in Fig. 9.17. The numerical values of the wave function are not affected by this displacement of the crystal axes, but the dependence of the function upon the new coordinates is different. We denote the new wave function by ψ'. The value of the new function ψ' at **r** is equal to the value of the old function ψ at $\mathbf{r}+\mathbf{l}$. We may express this mathematically by introducing a displacement operator $\mathbf{T_l}$ which transforms ψ into ψ'

$$\mathbf{T_l}\psi(\mathbf{r}) = \psi'(\mathbf{r})$$
$$= \psi(\mathbf{r}+\mathbf{l}). \tag{9.36}$$

The wave functions of the electrons in a crystal satisfy the Schrödinger equation

$$\left(-\frac{\hbar^2}{2m}\nabla^2 + V(\mathbf{r})\right)\psi(\mathbf{r}) = E\psi(\mathbf{r}). \tag{9.37}$$

The potential energy $V(\mathbf{r})$ appearing in this equation resembles the potential for an ion in the vicinity of an individual ion and flattens out in the region between ions. The condition that potential energy has the symmetry of the lattice can be written

$$V(\mathbf{r}+\mathbf{l}) = V(r), \tag{9.38}$$

where **l** is any displacement vector of the Bravais lattice.

A wave function which satisfies the Schrödinger equation with a periodic potential is known as a *Bloch function*. As a consequence of the periodicity of the potential $V(\mathbf{r})$, the wave functions have the following important property:

Bloch's theorem

For any wave function that satisfies the Schrödinger equation, there is a wave vector **k** such that translating the coordinate system by a lattice vector **l** is equivalent to multiplying the function by the phase factor $\exp(i\mathbf{k}\cdot\mathbf{l})$

$$\psi(\mathbf{r}+\mathbf{l}) = e^{i\mathbf{k}\cdot\mathbf{l}}\psi(\mathbf{r}).$$

This is a very strong condition imposed on the form of the wave function by the symmetry of the lattice. In the following, we shall explicitly denote solutions of the Schrödinger equation by the value of **k** to which they correspond. The condition

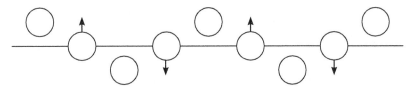

FIGURE 9.18 A chain of atoms vibrating about their equilibrium positions.

imposed by Bloch's theorem then becomes

$$\psi_{\mathbf{k}}(\mathbf{r} + \mathbf{l}) = e^{i\mathbf{k}\cdot\mathbf{l}}\psi_{\mathbf{k}}(\mathbf{r}). \tag{9.39}$$

A proof of Bloch's Theorem is given in Appendix II. In order to give an intuitive idea of the meaning of the theorem, we consider the motions of the atoms in a vibrating molecular chain. As illustrated in Fig. 9.18, the atoms in a chain of atoms vibrate independently about their equilibrium positions. Each atom vibrates up and down at its own rate. Looking at the relative position of the atoms, we see that the atoms are in different stages of the same motion. An atom at a certain position along the chain has a displacement which differs in phase from the displacements of the other atoms. Bloch's theorem makes a similar statement about the wave function of an electron in a periodic potential. The wave function in a cell specified by the lattice vector $\mathbf{l}$ differs from the wave function in the Wigner-Seitz cell near the origin by a phase factor $\exp(i\mathbf{k}\cdot\mathbf{l})$. The electron wave function is entirely determined by the behavior of the function in the unit cell about the origin.

Bloch's theorem can be stated a little differently. We define a function $u_{\mathbf{k}}(\mathbf{r})$ by the equation

$$u_{\mathbf{k}}(\mathbf{r}) = e^{-i\mathbf{k}\cdot\mathbf{r}}\psi_{\mathbf{k}}(\mathbf{r}). \tag{9.40}$$

Using Eqs. (9.40) and (9.39), we have

$$u_{\mathbf{k}}(\mathbf{r} + \mathbf{l}) = e^{-i\mathbf{k}\cdot(\mathbf{r}+\mathbf{l})}\psi_{\mathbf{k}}(\mathbf{r} + \mathbf{l}) = e^{-i\mathbf{k}\cdot\mathbf{r}}\psi_{\mathbf{k}}(\mathbf{r}) = u_{\mathbf{k}}(\mathbf{r}).$$

The function $u_{\mathbf{k}}(\mathbf{r})$ is thus periodic. We may solve Eq. (9.40) for $\psi_{\mathbf{k}}(\mathbf{r})$ obtaining

$$\psi_{\mathbf{k}}(\mathbf{r}) = e^{i\mathbf{k}\cdot\mathbf{r}}u_{\mathbf{k}}(\mathbf{r}), \tag{9.41}$$

and state Bloch's theorem as follows:

Alternate form of Bloch's theorem

The solutions of the Schrödinger equation of an electron in a Bravais lattice can be written as the product of a plane wave $e^{i\mathbf{k}\cdot\mathbf{r}}$ and a function $u_{\mathbf{k}}(\mathbf{r})$ having the periodicity of the lattice.

Example 9.4

Show that the wave function of a free electron,

$$\psi(\mathbf{r}) = A_{\mathbf{k}}e^{i\mathbf{k}\cdot\mathbf{r}}, \tag{9.42}$$

satisfies Bloch's theorem.

Solution
Using Eq. (9.42) and the properties of the exponential function, we may write

$$\psi(\mathbf{r} + \mathbf{l}) = A_{\mathbf{k}}e^{i\mathbf{k}\cdot(\mathbf{r}+\mathbf{l})} = e^{i\mathbf{k}\cdot\mathbf{l}}A_{\mathbf{k}}e^{i\mathbf{k}\cdot\mathbf{r}} = e^{i\mathbf{k}\cdot\mathbf{l}}\psi(\mathbf{r}).$$

The wave function (9.42) thus satisfies Eq. (9.39) corresponding to the first form of Bloch's theorem.

Since a constant is an example of a periodic function, the wave function (9.42) is equal to a periodic function times the exponential $\exp(i\mathbf{k}\cdot\mathbf{r})$. Eq. (9.42) is thus consistent with the alternate form of Bloch's theorem.

The form of the free-electron wave function (9.42) ensures that it transforms properly with respect to translations. We would expect that the perturbing effects of the ion cores should distort the free electron wave function. According to the

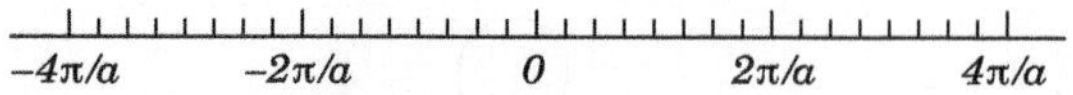

FIGURE 9.19 The points of the reciprocal lattice and the other observed values.

alternate form of Bloch's theorem, however, the distortion of the wave function due to the ion cores can only have the effect of replacing the constant $A_\mathbf{k}$ by a periodic function $u_\mathbf{k}(\mathbf{r})$. This is natural since we have supposed that the ion cores have the symmetry of the lattice.

Thus far we have imagined that the crystal lattice consists of an infinite array of atoms with complete translational symmetry. An infinite lattice of this kind can be thought of as the limiting case of a finite lattice as the number of lattice sites becomes very large. We might introduce boundaries as if they were real surfaces, at which the wave function vanishes. As we have seen in the third chapter, however, the wave function of an electron confined to a finite region by boundaries at which the function vanishes corresponds to a standing wave which in the one-dimensional case is represented by sine or cosine functions. Such a function differs considerably from a free electron wave function

$$\psi_\mathbf{k}(\mathbf{r}) = A_\mathbf{k} e^{i\mathbf{k}\cdot\mathbf{r}},$$

or a Bloch function defined by Eq. (9.41).

There is a mathematical device which serves to define the solutions of the Schrödinger equation (9.37) in a finite region, and which still does not introduce any direct physical effects due to the boundaries. This is to use *periodic* or *Born-von Karman* boundary conditions.

To clarify the issues involved, we consider a one-dimensional crystal having N cells, and require that the wave function be periodic over the length of the entire crystal. One can imagine that the ends of the crystal are joined together forming a circle. The periodic boundary condition can be expressed

$$\psi(x + Na) = \psi(x). \tag{9.43}$$

Here the integer N, which is the total number of unit cells in the lattice, is a very large number. According to Bloch's theorem, there is a wave vector k such that

$$\psi_k(x + Na) = e^{ikNa}\psi_k(x).$$

Substituting this last equation into Eq. (9.43) leads to the condition,

$$e^{ikNa} = 1,$$

which is satisfied by the following values of k

$$k = \frac{2\pi n}{Na}, \tag{9.44}$$

where n is an integer.

The points of the reciprocal lattice $n\,2\pi/a$ and the values of k given by Eq. (9.44) are represented in Fig. 9.19. We recall that the reciprocal lattice lengths are those values of k for which the free electron wave functions have the periodicity of the lattice. By contrast, Eq. (9.44) defines a much finer grid of k-values for which the wave function is periodic over the length of the entire crystal.

The free-electron wave function,

$$\psi_k(x) = A_k e^{ikx}, \tag{9.45}$$

may be normalized over a crystal of length $L = Na$ by imposing the condition

$$\int_0^L |\psi_k(x)|^2 dx = 1. \tag{9.46}$$

Substituting Eq. (9.45) into Eq. (9.46), we find that the constant A_k is equal to $1/L^{1/2}$. The wave function can thus be written

$$\psi(x) = \frac{1}{L^{1/2}} e^{ikx}. \tag{9.47}$$

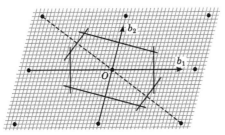

FIGURE 9.20 A two-dimensions unit cell (or Brillouin zone) that is filled by a fine grid of points. The allowed **k** vectors are directed to the points of intersection of families of lines parallel to the $\mathbf{b}_1$ and $\mathbf{b}_2$ axes.

Free-electron wave functions given by Eq. (9.47) satisfy the normalization condition (9.46). Wave functions corresponding to different values of k may also be shown to satisfy the following set of conditions

$$\int_0^L \psi_{k'}^*(x)\psi_k(x)dx = 0, \quad \text{for } k' \neq k. \tag{9.48}$$

These conditions are known as *orthogonality conditions*. (See Problem 22.)

We may extend our line of argument to three dimensions by requiring that the wave function be periodic along the three crystal axes corresponding to the unit vectors $\mathbf{a}_1$, $\mathbf{a}_2$, and $\mathbf{a}_3$. Denoting the number of sites along the crystal axes by N_1, N_2, and N_3 respectively, the boundary conditions in three dimensions are

$$\psi(\mathbf{r}+N_1\mathbf{a}_1) = \psi(\mathbf{r}) \quad \psi(\mathbf{r}+N_2\mathbf{a}_2) = \psi(\mathbf{r}) \quad \psi(\mathbf{r}+N_3\mathbf{a}_3) = \psi(\mathbf{r}). \tag{9.49}$$

For a wave function with wave vector $\mathbf{k}$, these conditions lead to the equalities

$$e^{i\mathbf{k}\cdot(N_1\mathbf{a}_1)} = e^{i\mathbf{k}\cdot(N_2\mathbf{a}_2)} = e^{i\mathbf{k}\cdot(N_3\mathbf{a}_3)} = 1,$$

which are satisfied by wave vectors $\mathbf{k}$ of the form

$$\mathbf{k} = \frac{n_1}{N_1}\mathbf{b}_1 + \frac{n_2}{N_2}\mathbf{b}_2 + \frac{n_3}{N_3}\mathbf{b}_3, \tag{9.50}$$

where n_1, n_2, and n_3 are integers, and $\mathbf{b}_1$, $\mathbf{b}_2$, and $\mathbf{b}_3$ are the unit vectors of the reciprocal lattice. (See Problem 25.)

The allowed wave vectors $\mathbf{k}$ are described by Eq. (9.50). For $\mathbf{k}$ vectors given by Eq. (9.50), the wave function $\psi_{\mathbf{k}}(\mathbf{r})$ satisfies the periodic boundary conditions (9.49). The $\mathbf{k}$ vectors may be thought of as being in the space of the reciprocal lattice vectors (9.20). The points, toward which the $\mathbf{k}$ vectors are directed, are obtained by dividing the unit cell of the reciprocal lattice in N_1 parts in the $\mathbf{b}_1$ direction, N_2 parts in the $\mathbf{b}_2$ direction and N_3 parts in the $\mathbf{b}_3$ direction. The unit cell (or Brillouin zone) is thus filled by a fine grid of points generated by families of equidistant crystal planes. This is illustrated for two dimensions in Fig. 9.20. In this figure, the allowed $\mathbf{k}$ vectors are directed to the points of intersection of families of lines perpendicular to the $\mathbf{b}_1$ and $\mathbf{b}_2$ axes.

For each allowed $\mathbf{k}$-value, the free-electron wave functions can be written

$$\psi_{\mathbf{k}}(\mathbf{r}) = \frac{1}{V^{1/2}}e^{i\mathbf{k}\cdot\mathbf{r}}. \tag{9.51}$$

The quantity V that occurs in Eq. (9.51) is the volume of a large irregular region with parallel sides $L_1 = N_1a_1$, $L_2 = N_2a_2$, and $L_3 = N_3a_3$. One may show that the wave function (9.51) is normalized in this large region and that wave functions corresponding to different allowed values of $\mathbf{k}$ are orthogonal

$$\int \psi_{\mathbf{k}'}^*(\mathbf{r})\psi_{\mathbf{k}}(\mathbf{r})dV = \begin{cases} 1, & \text{for } \mathbf{k}' = \mathbf{k} \\ 0, & \text{for } \mathbf{k}' \neq \mathbf{k}. \end{cases} \tag{9.52}$$

A derivation of these results are given in Appendix JJ. The free-electron functions (9.51) are thus normalized over the entire crystal, and they are orthogonal as are the wave functions of the one-dimensional crystal considered earlier.

9.6 Diffraction of electrons by an ideal crystal

In Chapter 1, we found that the scattering of X-rays by an ideal crystal is described by the equation,

$$2d\sin\theta = n\lambda.$$

This equation, which is known as Bragg's law, was derived using elementary properties of waves. Since considering Bragg's law, we have studied the ideas and methods of quantum mechanics, and we have studied in some detail the structure of crystals. In this section, we shall consider the elastic scattering of electrons by an ideal crystal using the methods that we have recently developed. There is an important lesson to be learned about the dynamics of the scattering process that we shall need to account for the energy levels of electrons in solids.

If a beam of electrons is directed into a crystal, the periodic potential of the ion cores will scatter the electrons into different states. We can describe the scattering process as we described atomic transitions in Chapter 4. Suppose that an electron is initially in a free-electron state described by a wave function $\psi_\mathbf{k}$, and at time $t = 0$, we allowed the potential $V(\mathbf{r})$ to interact with the electron. The interaction between the wave describing the electron and the ions of the crystal can be quite complicated. Imagine that at some later time, we turned off the potential function and considered the state of the electron. The wave function of the electron would then correspond to a superposition of plane waves moving in different directions. The probability that the electron made a transition from its initial state $\psi_\mathbf{k}$ to the state described by the wave function $\psi_{\mathbf{k}'}$ is proportional to the absolute value squared of the transition integral. We may write

$$P(\mathbf{k} \to \mathbf{k}') = \left[\mathbf{I}_{\mathbf{k}',\mathbf{k}}\right]^* \mathbf{I}_{\mathbf{k}',\mathbf{k}},$$

where

$$\mathbf{I}_{\mathbf{k}',\mathbf{k}} = \int \psi_{\mathbf{k}'}^*(\mathbf{r}) V(\mathbf{r}) \psi_\mathbf{k}(\mathbf{r}) dV. \tag{9.53}$$

All of the quantities that appear in the transition integral (9.53) can be written in terms of plane waves. The wave functions are given by Eq. (9.51), and the potential energy, being periodic, may be expanded in a Fourier series

$$V(\mathbf{r}) = \sum_\mathbf{g} V_\mathbf{g}\, e^{i\mathbf{g}\cdot\mathbf{r}}, \tag{9.54}$$

where the vectors $\mathbf{g}$ are members of the reciprocal lattice. Substituting Eqs. (9.51) and (9.54) into (9.53), we get

$$\mathbf{I}_{\mathbf{k}',\mathbf{k}} = \sum_\mathbf{g} V_\mathbf{g} \int \frac{1}{V} e^{i(\mathbf{k}+\mathbf{g}-\mathbf{k}')\cdot\mathbf{r}} dV$$

The integral that appears on the right-hand side of this last equation can be simplified by using Eq. (9.51) to rewrite the functions inside the integral as a product of the complex conjugate of the wave function $\psi_{\mathbf{k}'}$ times the wave function $\psi_{\mathbf{k}+\mathbf{g}}$

$$\mathbf{I}_{\mathbf{k}',\mathbf{k}} = \sum_\mathbf{g} V_\mathbf{g} \int \psi_{\mathbf{k}'}^*(\mathbf{r}) \psi_{\mathbf{k}+\mathbf{g}}(\mathbf{r}) dV.$$

We may then take advantage of the orthogonality and normalization conditions of the plane wave states summarized by Eq. (9.52) to obtain the following result

$$\mathbf{I}_{\mathbf{k}',\mathbf{k}} = \begin{cases} V_\mathbf{g}, & \text{if } \mathbf{k}' = \mathbf{k} + \mathbf{g} \\ 0, & \text{otherwise.} \end{cases}$$

For a particular value of $\mathbf{k}$, diffracted electrons will be observed in directions corresponding to wave-vectors $\mathbf{k}'$ satisfying

$$\mathbf{k}' = \mathbf{k} + \mathbf{g}. \tag{9.55}$$

The vectors $\mathbf{k}$, $\mathbf{k}'$, and $\mathbf{g}$ are shown in Fig. 9.21. For elastic scattering, the energy of the diffracted electrons is the same as the energy of the incident electrons. The vectors $\mathbf{k}$ and $\mathbf{k}'$ thus have the same length. The angle between $\mathbf{k}$ and $\mathbf{k}'$ in Fig. 9.21 is 2θ, and the length of $\mathbf{g}$ is equal to twice the projection of $\mathbf{k}$ upon $\mathbf{g}$

$$|\mathbf{g}| = 2|\mathbf{k}|\sin\theta. \tag{9.56}$$

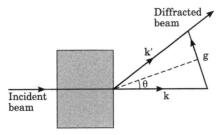

FIGURE 9.21 Elastic scattering of electron by a crystal. The wave vector of the incident electron is denoted by **k** and the wave vector of the scattered electron is denoted by **k**′.

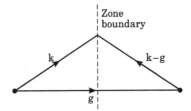

FIGURE 9.22 The reciprocal lattice vector **g** and the wave vectors **k** and **k**′ = **k** − **g**.

Eq. (9.56) can be written more simply by taking advantage of the properties of the reciprocal lattice vectors. The length of the reciprocal lattice vector |**g**| is related to the distance between adjacent lattice planes by Eq. (9.28), where N is an integer. We also recall from Chapter 1 that the magnitude of the wave vector **k** is related to the wave length of the incident wave λ by the equation

$$|\mathbf{k}| = \frac{2\pi}{\lambda} \tag{9.57}$$

Substituting values of |**g**| and |**k**| obtained from Eqs. (9.28) and (9.57) into Eq. (9.56) leads to the Bragg diffraction Law

$$N\lambda = 2d\sin\theta,$$

which was derived in Chapter 1 by requiring that waves scattered from successive lattice planes constructively interfere.

Eq. (9.55) is a condition upon the initial wave vector **k** and the final wave vector **k**′ which must be satisfied for scattering to occur. This equation was derived by requiring that the transition integral (9.53) be nonzero. A condition upon the initial wave vector **k** and the final wave vector **k**′, which we shall find to be more useful, can be derived by requiring that the complex conjugate of the transition integral be nonzero. As indicated in Problem 26, this leads to the equation

$$\mathbf{k} = \mathbf{k}' + \mathbf{g}, \tag{9.58}$$

which can be obtained from Eq. (9.55) by interchanging **k** and **k**′. Eq. (9.58) can be written

$$\mathbf{k}' = \mathbf{k} - \mathbf{g}.$$

The vectors **k**, **g**, and **k**′ = **k** − **g** are illustrated in Fig. 9.22. The requirement that the scattering process be elastic is equivalent to the condition

$$|\mathbf{k}| = |\mathbf{k} - \mathbf{g}|.$$

This implies that **k** lies on the perpendicular bisector of the reciprocal lattice vector **g**. At this point, we recall that the Wigner-Seitz cell or Brillouin zone is constructed by drawing planes that bisect the reciprocal lattice vectors in the different crystalline directions. We thus conclude that elastic scattering of electrons will occur when **k** lies on the boundary of the Brillouin zone. The wave vector of the diffracted electrons **k** − **g** is then the mirror image of **k** across the zone boundary, and the incident and diffracted states $\psi_{\mathbf{k}}(\mathbf{r})$ and $\psi_{\mathbf{k}-\mathbf{g}}(\mathbf{r})$ have the same kinetic energy.

A qualitatively correct description of many solids can be obtained by adding the states corresponding to the mirror images of the wave vectors across the zone boundaries to the wave functions of free electrons. This approach is called the *nearly free electron model.* A single-electron wave function, which includes the scattering at the zone boundaries, can be written

$$\psi(\mathbf{r}) = \sum_{\mathbf{g}} \alpha_{\mathbf{k-g}} e^{i(\mathbf{k-g})\cdot\mathbf{r}}. \tag{9.59}$$

We note that the sum on the right-hand side of this equation includes the free-electron wave function $\psi_{\mathbf{k}}$ and all wave functions $\psi_{\mathbf{k-g}}$ into which the state $\psi_{\mathbf{k}}$ can be scattered by the crystal field.

9.7 The band gap

We begin this section by considering the changes that occur in sodium atoms when they condense to form a solid. Each sodium atom contains 11 electrons. Ten electrons fill the inner $1s$, $2s$, and $2p$ states, and one electron is in the outer $3s$ state. The outer electron is loosely bound to the atoms. When sodium atoms are brought together to form a solid, the electrons in the inner lying states remain localized around the nuclear centers, while the $3s$ electrons become conduction electrons described by Block functions that are extended over the entire solid. The wave functions of the outer electrons then satisfy the Schrödinger equation of an electron moving in the potential $V(\mathbf{r})$ of the ion cores.

To the extent that the perturbing effects of the ion cores can be neglected, the conduction electrons can be described by the free-electron wave functions

$$\psi_{\mathbf{k}}(\mathbf{r}) = \frac{1}{V^{1/2}} e^{i\mathbf{k}\cdot\mathbf{r}},$$

and the energy of the electrons is given by the formula

$$\mathcal{E}^0(\mathbf{k}) = \frac{\hbar^2 k^2}{2m}. \tag{9.60}$$

A sketch of free-electron energy function $\mathcal{E}^0(\mathbf{k})$ in one dimension is given in Fig. 9.23(A).

In the previous section, we found that electrons are diffracted at the zone boundaries in reciprocal lattice space. The diffraction of the electrons can be understood as a mixing effect involving the two free-electron states, $e^{i\mathbf{k}\cdot\mathbf{r}}$ and $e^{i(\mathbf{k-g})\cdot\mathbf{r}}$. These two free-electron states have the same energy, $\mathcal{E}^0(\mathbf{k}) = \mathcal{E}^0(\mathbf{k-g})$. At the zone boundary, the electrostatic field of the crystal combines the two free-electron states with wave vectors $\mathbf{k}$ and $\mathbf{k-g}$ into two other states with energy $\mathcal{E}^+$ and $\mathcal{E}^-$. The separation between these two energy values, which is known as a *band gap*, is shown in Fig. 9.23(B). A derivation of formulas describing the splitting of the $\mathcal{E}^+$ and $\mathcal{E}^-$ levels is given in Appendix KK. Here we shall describe the underlying physical causes of the band gap.

As we have suggested, the wave functions of an electron at the zone boundary is a linear combination of the free-electron states, $e^{i\mathbf{k}\cdot\mathbf{r}}$ and $e^{i(\mathbf{k-g})\cdot\mathbf{r}}$, and is thus of the general form

$$\psi_{\mathbf{k}}(\mathbf{r}) = \alpha_{\mathbf{k}} e^{i\mathbf{k}\cdot\mathbf{r}} + \alpha_{\mathbf{k-g}} e^{i(\mathbf{k-g})\cdot\mathbf{r}}.$$

For the lower value of the energy $\mathcal{E}^-$, the coefficients $\alpha_{\mathbf{k-g}}$ and $\alpha_{\mathbf{k}}$ are equal, and the wave function is

$$\psi_{\mathbf{k}}^-(\mathbf{r}) = \alpha_{\mathbf{k}} \left[e^{i\mathbf{k}\cdot\mathbf{r}} + e^{i(\mathbf{k-g})\cdot\mathbf{r}} \right].$$

Factoring $e^{i(\mathbf{k}-(1/2)\mathbf{g})\cdot\mathbf{r}}$ from the exponential terms on the right-hand side gives

$$\psi_{\mathbf{k}}^-(\mathbf{r}) = \alpha_{\mathbf{k}} e^{i(\mathbf{k}-(1/2)\mathbf{g})\cdot\mathbf{r}} \left[e^{(1/2)\mathbf{g}\cdot\mathbf{r}} + e^{-(1/2)\mathbf{g}\cdot\mathbf{r}} \right].$$

Using the expression for the cosine function in terms of exponentials given in the first section of Chapter 3, we may then identify the expression within square brackets to be $2\cos(\frac{1}{2}\mathbf{g}\cdot\mathbf{r})$. The equation for the wave function of the lower state thus becomes

$$\psi_{\mathbf{k}}^-(\mathbf{r}) = \alpha_{\mathbf{k}} e^{i(\mathbf{k}-\frac{1}{2}\mathbf{g})\cdot\mathbf{r}} \, 2\cos(\frac{1}{2}\mathbf{g}\cdot\mathbf{r}).$$

The probability of finding an electron at a position $\mathbf{r}$ in space is given by the absolute value squared of the wave function

$$|\psi_{\mathbf{k}}^-(\mathbf{r})|^2 = |\alpha_{\mathbf{k}}|^2 \, 4\cos^2(\frac{1}{2}\mathbf{g}\cdot\mathbf{r}).$$

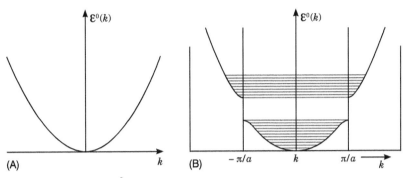

FIGURE 9.23 (A) The free-electron energy function $\mathcal{E}^0$ as a function of the angular wave vector k; (B) the single-electron levels and the band gaps that occur at the zone boundaries.

The function $|\psi_{\mathbf{k}}^-|^2$ is large near $\mathbf{r} = 0$ and at other sites of the Bravais lattice. Hence, the energy is lowered by the attraction of the electron to the ions of the crystal. Similarly, the wave function $\psi_{\mathbf{k}}^+(\mathbf{r})$ of the upper state is

$$\psi_{\mathbf{k}}^+(\mathbf{r}) = \alpha_{\mathbf{k}} e^{i(\mathbf{k}-\frac{1}{2}\mathbf{g})\cdot\mathbf{r}} \, 2i \, \sin(\frac{1}{2}\mathbf{g}\cdot\mathbf{r}).$$

The absolute value square of $\psi_{\mathbf{k}}^+$ is

$$|\psi_{\mathbf{k}}^+(\mathbf{r})|^2 = |\alpha_{\mathbf{k}}|^2 \, 4\sin^2(\frac{1}{2}\mathbf{g}\cdot\mathbf{r}).$$

$|\psi_{\mathbf{k}}^+|^2$ is small near the origin and large in regions of positive potential. This has the effect of raising $\mathcal{E}^+$ above $\mathcal{E}^-$. The appearance of a band gap may thus be traced to the form of the single-electron wave functions.

9.8   Classification of solids

We have discussed the electronic structure of solids in such detail because it enables us to classify solids into broad groups having similar macroscopic properties. In this section, we shall first distinguish the various kinds of solids using the idea of electronic bands introduced recently, and we shall then study more fully how the atoms of the solid are bound together in a crystal.

9.8.1   The band picture

As we have suggested before, the energy levels of a crystal consist of dense bands. The energy bands that occur above 0 eV are called *conduction bands*, while the bands occurring below zero are known as the *valence bands*. Electrons in the conduction band are generally free to move about the solid while electrons in the valence band are localized near particular atoms. The conduction and valence bands are separated by a band gap, which, as we have seen, is associated with scattering at the Bragg planes. The distribution of electron states is usually symmetric with respect to the origin in $\mathbf{k}$-space. The two states with wave vectors $\mathbf{k}$ and $-\mathbf{k}$ have the same energy and both states are either occupied or unoccupied. If the ground state of a solid is symmetric and a sizable gap separates the occupied and unoccupied levels, no net electric current can flow in the crystal. The current flowing in one direction will entirely balance the current flowing in the other direction. For there to be a current, electrons must be excited into unoccupied states. An electron excited into a conduction band is called a *conduction electron*, while an empty state in the valence band is called a *hole*.

Insulators

If there is a large band gap $\Delta\mathcal{E}$ between the occupied and unoccupied states, a prohibitively large amount of energy must be supplied to the electrons to carry them up over the gap into the next band. The solid would then be an *insulator*. Several solid elements to the right of the periodic table serve as good insulators. For instance, sulfur and iodine are good insulators. Other material that require a considerable amount of energy to dislodge electrons are composed of more than one element. Fused quartz, which is a crystal made up of silicon and oxygen, is a good insulator. So are rubber, teflon, and many plastics.

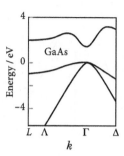

FIGURE 9.24 The lowest conduction band and the highest two valence bands of $GaAs$.

Semiconductors

Suppose, however, the band gap $\Delta\mathcal{E}$ is small. As we found in the previous chapter, the distributions of electrons between the energy levels of a solid is described by Fermi-Dirac statistics. At a nonzero temperature T, a small density of electrons will be excited by thermal fluctuations into the upper band. Since these electrons can carry a current, the material then has an observable conductivity, which will increase as the temperature increases. Such a material is called a *semiconductor*. The most common semiconductors occur in the center-right portion of the periodic table: silicon, germanium, gallium, arsenic, indium, antimony, and tellurium. These elements form crystals with small band gaps and a very small number of conduction electrons at room temperature. Crystals formed from two of these elements generally have the same properties.

As an example of a semiconductor, we consider gallium arsenide ($GaAs$), which has a band gap of 1.424 eV at room temperature (300 K). The lowest conduction band and the two highest valence bands of $GaAs$ are shown in Fig. 9.24. Notice that the lowest conduction band of $GaAs$ has a local minimum with the two sides slopping upward as for the free-electron energy illustrated in Fig. 9.23(A). From Eq. (9.60) for the free-electron energy $\mathcal{E}^0(\mathbf{k})$, we see that the second derivative of the free-electron energy is inversely proportional to the mass of the electron. In Fig. 9.24, we see that two bands converge at the valence band maximum at the center of the Brillouin-zone center. These bands are known as the *heavy-* and *light-*hole bands; the flatter one, with the smaller value of the second derivative, is the heavy hole band, and the steeper one is the light hole band. The underlying cause of their being two valence bands at the same point of the reciprocal lattice is that valence electrons moving in one direction in the crystal have an effective mass that is different from the effective mass of a valence electron moving in another direction.

Metals

If a band is not filled, there will be current carrying states available just above the top of the occupied levels. The material will then readily conduct a current, and the conductivity will not depend upon temperature except in so far as temperature affects the scattering of electrons by the ion cores. The solid would then be described as a *metal*.

Graphene

In recent times, a number of new materials have been developed which offer the promise of producing very fine electrical devices. Since these materials cannot be classified as pure metals or pure semiconductors, we call them here *complex materials*. The two new kinds of materials we will consider are *graphene* and *carbon nanotubes*.

Graphene is the name given to a flat monolayer of carbon atoms joined together in a two-dimensional honeycomb lattice. As can be seen in Fig. 9.25, each unit cell of the two-dimensional structure consist of six carbon atoms bound together in a compact unit. Graphene was discovered in 2004 on top of noncrystalline substrates, in liquid suspensions, and in suspended membranes bringing to an end more than 70 years of theoretical arguments that such materials would be thermodynamically unstable and could not exist. Graphene is in fact very stable, and is made typically today by chemical vapor deposition of a carbon monolayer on a nickel surface. Another film can then be deposited on top and nickel etched away leaving a graphene monolayer on an insulating substrate.

To see how the properties of graphene compare with the properties of other materials, we consider first how metals respond to electric fields. When a metal is placed in an external electric field, the charge carriers in the metal acquire a *drift velocity*. The *mobility* of the charge carriers in the metal is defined to be the radio of the drift velocity to the strength of the electric field. Graphene waivers have been produced with carrier mobilities higher than the mobility of any other known material at room temperature. The high mobility shows that the motion of charge carriers in graphene is not disturbed by scattering processes. The charge carriers of graphene can travel thousands of interatomic distances without scattering.

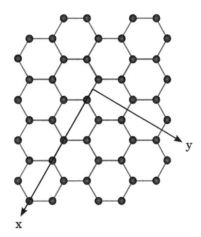

FIGURE 9.25 The two-dimensional graphene lattice.

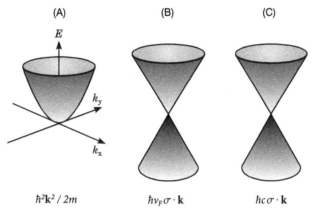

FIGURE 9.26 (A) The energy surface for nonrelativistic charge carriers whose wave functions satisfy the Schrödinger equation; (B) the energy surface for charge carriers in graphene; (C) the energy surface for ultra relativistic electrons with a velocity near the speed of light.

Another remarkable property of graphene is that the electrons and holes that serve as charge carriers have entirely lost their mass. The charge carriers are Fermions with zero mass, which cannot be described by the Schrödinger equation we have used thus far but must be described by the relativistic Dirac equation to be studied in Chapter 12. While we do not yet have the technical details of the Dirac theory available to us, we can immediately see some of the consequences of the electron having zero mass. When we described the Compton effect in Chapter 1, we said that the energy of a photon, which has zero mass, is related to its momentum by the equation

$$E = cp,$$

where c is the velocity of light. Since the momentum of a particle is equal to $\hbar k$, we see immediately that the energy of the massless charger carriers of graphene should be linearly related to the wave vector k. Fig. 9.26(B) shows the two-dimensional energy surface of graphene. For charge carriers with a **k** vector having components, k_x and k_y, the energy surface is a cone that rises linearly in each direction. For comparison, the energy surface of a Schrödinger electron is shown in Fig. 9.26(A) and the energy surface of a Dirac electron with a velocity near the speed of light is shown in Fig. 9.26(C). The energy surface for fast-moving Dirac electrons has the same cone-like shape but rises more steeply than the energy surface for the charge carriers for graphene.

To describe a charge carrier moving in graphene, one must also take into account the fact that the honeycomb lattice of graphene is made up of two equivalent carbon sublattices, A and B. Like the electron spin, which can either be up or down, an electron in graphene can be at A or B. When we study the Dirac theory in Chapter 12, we will find that the spin of the electron can be described by the Pauli matrices

$$\sigma_x = \begin{bmatrix} 0 & 1 \\ 1 & 0 \end{bmatrix}, \quad \sigma_y = \begin{bmatrix} 0 & -i \\ i & 0 \end{bmatrix}, \quad \sigma_z = \begin{bmatrix} 1 & 0 \\ 0 & -1 \end{bmatrix}. \tag{9.61}$$

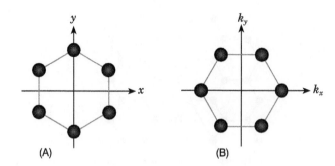

FIGURE 9.27 (A) A unit cell of the graphene lattice and (B) the corresponding unit cell of the reciprocal lattice.

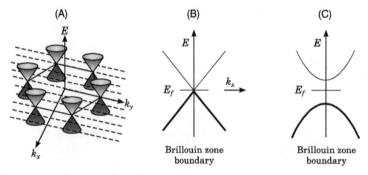

FIGURE 9.28 (A) an upward-directed and a downward-directed cone in k-space attached to each site of the reciprocal lattice, (B) a two-dimensional figure produced when a vertical plane passing through one site of the reciprocal lattice slices through the center of the cone at an adjacent site, (C) the two-dimensional figure produced by a vertical plane that slices through a cone some distance from the central axis of the cone.

For motion in the xy-plane, the electrons in graphene can be described by the model Hamiltonian

$$H = \hbar v_F \begin{bmatrix} 0 & k_x - ik_y \\ k_x + ik_y & 0 \end{bmatrix} = \hbar v_F \sigma \cdot \mathbf{k}, \tag{9.62}$$

where σ refers to the Pauli matrices, and v_F is the velocity of the charge carrier. By comparison, the Hamiltonian of an ultra-relativistic electron with a velocity near the speed of light c is

$$H = \hbar c \sigma \cdot \mathbf{k}. \tag{9.63}$$

The velocity v_F of charge carriers in graphene is about equal to the velocity of light c divided by 360. The close analogy between graphene electronics and relativistic quantum mechanics has made it possible to test the premises of relativistic theories in simple table top experiments.

The band structure of graphene has the unusual quality that the material is metallic for motion of the charge carriers in specific directions and is a semiconductor for other directions. To be in a position to describe the motion of charge carriers of graphene, we first construct the reciprocal lattice. The direction of the k-vectors of the reciprocal lattice is defined by the condition that they be perpendicular to the vectors of the original graphene lattice. A unit cell of the graphene lattice is illustrated in Fig. 9.27(A). The corresponding unit cell of the reciprocal lattice can be obtained by rotating the lattice shown in Fig. 9.27(A) by 90° in a counterclockwise direction to obtain the lattice shown in Fig. 9.27(B). Notice that lines drawn between neighboring sites of the lattice in Fig. 9.27(B) are perpendicular to the sides of the figure shown in Fig. 9.27(A).

We can imagine an upward-directed and a downward-directed cone in k-space attached to each site of the reciprocal lattice as shown in Fig. 9.28(A). A vertical plane passing through a site of the reciprocal lattice and through the center of the cone on an adjacent site produces the two-dimensional figure shown in Fig. 9.28(B), while a vertical plane that passes some distance from the center of a cone on an adjacent site of reciprocal lattice produces the two-dimensional figure shown in Fig. 9.28(C). Fig. 9.28(B) is the band structure associated with a zero-gap metal, while Fig. 9.28(C) is the band structure of a semiconductor. Graphene is a metal for electrons or holes directed from one site to another of the reciprocal lattice and a semiconductor for electrons or holes passing by the sites of the reciprocal lattice.

In addition to the properties of the material we have just discussed, graphene is very responsive to electrical biasing potentials. Charge carriers can be tuned continuously between electrons and holes in concentrations as high as 10^{13} cm^{-2}.

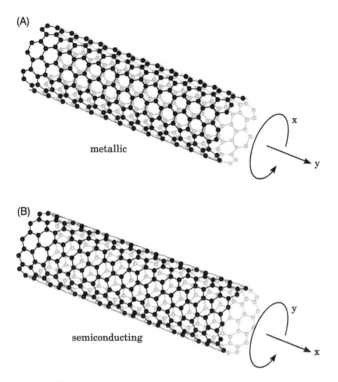

FIGURE 9.29 (A) A metallic carbon nanotube; (B) A semiconducting carbon nanotube.

This property of graphene can be understood by understanding the position of the Fermi level for the cones of states at the sites of the reciprocal lattice. A positive biasing potential raises the Fermi level with the states available near the Fermi level being electron states. Similarly, a negative biasing potential lowers the Fermi level with the states near the Fermi level being hole states. The sensitivity of graphene to biasing potential together with the very high mobility of its charge carriers at room temperature make it very likely that graphene will continue to make important contributions to electronics.

Carbon nanotubes

Single-walled carbon nanotubes are nanometer-diameter cylinders consisting of a single graphene sheet wrapped up to form a tube. Experiment and theory show that nanotubes can be either metals or semiconductors, and their electrical properties rival or exceed the properties of the best metals or semiconductors known. The high quality of nanotubes is due to the remarkable electronic properties of graphene from which the nanotubes are made. The band structure of nanotubes depends upon the orientation of the nanotube. If the axis of the nanotube points in one of the metallic directions of the graphene layer, the nanotube will be a metal, while if the axis of the nanotube points in another direction, the nanotube will be a semiconductor. A metallic nanotube is illustrated in Fig. 9.29(A), and a semiconducting nanotube is illustrated in Fig. 9.29(B). Notice that the orientation of the unit cell of the metallic nanotube shown in Fig. 9.29(A) is the same as the orientation of unit cell shown in Fig. 9.27(A). The direction of the current in the nanotube is the same as the direction of the k_x axis in Fig. 9.27(B). The current flows in the direction of the axis of the carbon nanotube.

9.8.2   The bond picture

Solids can also be classified according to the way in which the atoms are bound together to their neighbors. We shall distinguish five different types of bonding: covalent bonding, ionic bonding, molecular bonding, hydrogen bonding, and metallic bonding.

Covalent bonding

We begin by considering the germanium semiconductor. The crystal of this element has the diamond structure shown in Fig. 9.8(B). As we have seen, this structure consists of two interpenetrating face-centered cubic lattices. A germanium atom at point B in the figure is bonded to four neighboring atoms. Bonding of this kind is easily described using the concept of

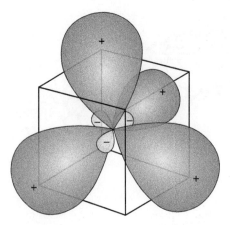

FIGURE 9.30 Four $s - p$ hybrid wave functions directed to the corners of a tetrahedron.

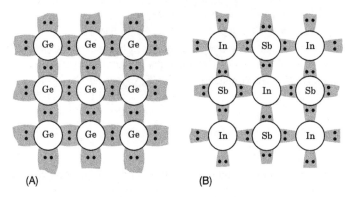

(A)   (B)

FIGURE 9.31 (A) Schematic representation of the bounding in the germanium crystal; (B) Schematic representation of the bonding in a $InSb$ crystal.

hybridization employed in modern chemistry. Germanium has one $4s$ wave function and three $4p$ wave functions which combine to form four hybrid wave functions directed to the corners of a tetrahedron centered at B. Hybrid orbitals of this kind are illustrated in Fig. 9.30. The covalent bond between neighboring germanium atoms is due to the sharing of electron pairs in each of these bonding orbitals. Looking again at Fig. 9.8(B), we see that the atom at B has four nearest neighbors, arranged at the vertices of a tetrahedron. Each of these atoms is also joined in a similar fashion to its neighbors.

Fig. 9.31(A) shows a schematic representation of the electronic structure of the germanium crystal in which each Ge atom shares eight electrons with its neighbors. This picture is overly simplified. The charge density of the valence electrons is spread around the ion cores; however, there is a large charge density corresponding to a pair of electrons with opposite spins in the regions corresponding to the electronic bonds.

It is instructive to compare the crystal structure of Ge to that of $InSb$. While germanium is in column IV of the periodic table, indium (In) is in column III and antimony (Sb) is in column V. The $InSb$ semiconductor forms the same diamond structure as germanium with indium atoms occupying all of the A sites of the crystal and antimony atoms occupying all of the B sites. A schematic representation of the $InSb$ structure is shown in Fig. 9.31(B). The electronic structure of $InSb$ is very similar to that of germanium. In each case, eight valence electrons are distributed among the four bonds. However, whereas the four electron pairs occur midway between the germanium atoms in the germanium structure, the electron pairs are drawn more closely to the Sb atom in the $InSb$ lattice. This can be explained with the concept of electronegativity used in modern chemistry. The electronegativity of an atomic species is a measure of the attractive power of the atom for an electron pair. If two atoms, which are bonded together, have the same electronegativity, the bonding electrons are shared equally between them. If, however, one atom is more electronegative than the other, the electron pair is drawn more closely to the atom that is more electronegative and the chemical bond has a polar character. More negative charge is concentrated near the atom which is more electronegative and the other atom is more electrically positive. Linus Pauling originally devised the scale for electronegativity based on a value of 2.1 for hydrogen. Fluorine (4.0) has the highest electronegativity on this scale, and cesium (0.8) has the lowest. The nonmetallic elements toward the right in the periodic table have the highest values of electronegativity. These elements attract electrons helping them fill their electron shells. The metals, appearing on the left side of the periodic table, give up their electrons easily and have lower values of electronegativity. The element Sb has an electronegativity of 1.9, while In has an electronegativity of 1.6. One thus expects the electron pairs in the bonds between In and Sb to be drawn toward the Sb atom.

There are many other compounds for which the electron pairs associated with the bonds are displaced more toward one of the atomic centers. Compounds such as BeO and MgO which have atoms in the II and VI columns of the periodic table are generally more polar than $InSb$ because the electronegativity of II column atoms differ by more from VI column atoms than do the electronegativities of In and Sb. The most polar compounds are those involving atoms of the I and VII columns of the periodic table. $NaCl$ is a common example of a compound of this kind. Such compounds can most simply be described using the concept of ionic bonding described in the next section.

Ionic bonding

In an ionic compound, one or more electrons from one of the species is transferred to the other species. In a $NaCl$ crystal, for instance, the outer $3s$ electron of each sodium atom is drawn over to a chlorine atom with the additional electron completing the outer $3p$ shell of chlorine. The resulting ionic crystal is composed of Na^+ and Cl^- ions. One can imagine that an ionic crystal is composed of charged ions that are stacked together like impenetrable charged spheres. The crystal is held together by the electrostatic force of attraction between these oppositely charged spheres, and is prevented from collapsing by their impenetrability.

The ideal ionic crystal of spherical, charged spheres is most nearly realized by the alkali halides. These crystals are all cubic at normal pressures. The positive spheres consist of singly ionized alkali atoms (Li^+, Na^+, K^+, Rb^+, or Cs^+) and the negative spheres are negatively charge halogen atoms (F^-, Cl^-, Br^-, or I^-). If their atomic numbers are comparable, the positive ions are usually quite a bit smaller than the negative ions. The alkali halides all crystallize under normal conditions in the sodium chloride structure except $CsCl$, $CsBr$, and CsI which are most stable in the cesium chloride structure shown in Fig. 9.5.

Doubly charged ions from columns II and column VI of the periodic table also form ionic crystals. However, there are a number of tetrahedrally coordinated II-VI compounds which are primarily covalent in character. Crystals formed from ions of the III and V columns of the periodic table are generally covalent and almost all assume the zincblende structure characteristic of covalent crystals.

The distinguishing feature of ionic crystals is that all of the electronic charge is localized near the individual ions. Since the electron charge is well localized, most ionic crystals are insulators. In covalent crystals, an appreciable part of the electron charge occurs in the region between neighboring ions and serves as a kind of glue holding the ions together. Most covalent crystals have tetrahedrally coordinated lattices such as the zincblende structure. Depending on the size of their band gap, covalent crystals can be either insulators or semiconductors.

Molecular crystals

In contrast to covalent and ionic crystals, molecular crystals consist of individual atoms or molecules drawn together by the weak van der Waals force. The solid phase of the noble gases from column VIII of the periodic table are the best examples of molecular crystals. Except for helium, the noble gases all crystallize in monatomic face-centered cubic lattices. The ground state of each atom in the lattice corresponds to filled shells that are little deformed in the solid. Other examples of molecular crystals are provided by solid oxygen and nitrogen, which consist of nonpolar diatomic molecules. The interaction between the units of the molecular crystal is due to weak perturbations of the entities that occur when the crystal is formed. Because the force holding the units together is so weak, molecular crystals generally have very low melting and boiling points.

The elements from columns V, VI, and VII of the periodic table form crystals that to varying degrees have both molecular and covalent character. Solids composed of nonpolar molecules such as oxygen or nitrogen generally have covalent bonding within the individual molecules. These crystals are thus held together by both covalent and molecular bonds. There are also substances that cannot be simply categorized in this way. Sulfur and selenium, for instance, have elaborate and intricate structures that cannot be fit reasonably into any of these classification schemes.

Hydrogen-bonded crystals

The hydrogen atom has unique properties that give the bonds which it forms with other elements a special character. The ionization energy of hydrogen (13.59 eV) is much higher than for the other elements from column I of the periodic table. Lithium with an ionization energy of 5.39 eV has the next highest value. Since it is so difficult to remove an electron completely from hydrogen, the atom does not form ionic crystals as do the other first column elements. Also, since hydrogen only lacks one electron to form a closed shell, it can form only one covalent bond by sharing an electron pair. Hydrogen

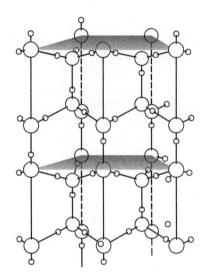

FIGURE 9.32 Hydrogen bonding in one of the phases of ice.

can thus not form the tetrahedrally coordinated structures characteristic of the other covalent elements. Finally, because the proton is so much smaller than any other ion core, the nucleus of hydrogen can come very close to large negative ions forming structures unattainable with other positive ions. As an illustration of hydrogen bonding, one of the phases of ice is shown in Fig. 9.32. Each oxygen atom in the figure is joined to two hydrogen atoms by covalent bonds and to two other hydrogen atoms by electrostatic attraction in much the same way as ions are joined to each other in an ionic crystal. The resulting lattice has a tetrahedral bond structure similar to that of covalent crystals.

Metals

As one moves toward the left of the periodic table, covalent bonds expand until there is an appreciable density of electrons throughout the interstitial regions between the positive ions, and the overlap of the energy bands in k-space increases. The material then exhibits the properties of a metal, in which valence electrons are free of their ions, forming a nearly uniform gas. The purest examples of metallic crystals are the alkali metals of column I. Other examples are provided by the alkali earths of column II and by the transition metal elements.

The way in which the atoms of a crystal are bonded together is important for studying the macroscopic properties of the crystal. One important macroscopic property of a crystal which depends upon how the units of the crystal are bonded together is the *cohesive energy*, which is the energy required to break the crystal up into its component parts. The cohesive energy has the same significance for solids as ionization potentials have for atoms and molecules.

The most weakly bonded solids are the molecular crystals for which the cohesive energy is of the order of 0.1 eV per unit. Hydrogen bonded crystals have somewhat higher cohesive energies–about 0.5 eV per molecule. The weakness of the bonding of these two types of crystals is responsible for their having low melting points. Metals are much more strongly bonded, having cohesive energies between 1 and 5 eV per atom, while the ionic and covalent crystals are the most strongly bonded crystals with cohesive energies of around 10 eV per atom.

Calculations of the cohesive energy of a solid depend upon an understanding of how the solid is bonded together. In order to calculate the cohesive energy of a molecular crystal, one must estimate the weak van der Waals interactions between the atoms or molecules from which the crystal is formed. Similarly, to calculate the cohesive energy of an ionic crystal one must calculate the electrostatic energy of all pairs of ions in the crystal and estimate the Coulomb repulsion between the ion cores. A fairly good approximation of the cohesive energy of covalent crystals can usually be attained by adding up the energies of the individual covalent bonds. For metals, the cohesive energy is a balance between the energy due to the attraction of the ion cores for the conduction electrons and the energy due to the repulsive interaction between the conduction electrons.

Suggestions for further reading

N. Ashcroft and N. Mermin, *Solid State Physics* (New York: Harcourt Brace, 1976).
J. Ziman, *Principles of the Theory of Solids*, Second Edition (Cambridge: Cambridge University Press, 1972).

A.K. Geim, *Graphene: Status and Prospects*, Science, Vol. 324, June, 2009.

A.K. Geim and S. Novoselov, *The rise of graphene*, Nature, Vol. 6, March, 2007.

P.l. McEuen, M. Fuhrer, and H. Park, *Single-Walled Carbon Nanotube Electronics*, The inaugural issue of IEEE Transactions on Nanotechnology, 2002.

Basic equations

Bravais lattice

Lattice points

$$\mathbf{l} = l_1\mathbf{a}_1 + l_2\mathbf{a}_2 + l_3\mathbf{a}_3$$

Basis vectors for simple cubic lattice

$$\mathbf{a}_1 = a\hat{\mathbf{i}}, \qquad \mathbf{a}_2 = a\hat{\mathbf{j}}, \qquad \mathbf{a}_3 = a\hat{\mathbf{k}},$$

Basis vectors for body-centered cubic lattice

$$\mathbf{a}_1 = \frac{a}{2}(\hat{\mathbf{j}} + \hat{\mathbf{k}} - \hat{\mathbf{i}}), \quad \mathbf{a}_2 = \frac{a}{2}(\hat{\mathbf{k}} + \hat{\mathbf{i}} - \hat{\mathbf{j}}), \quad \mathbf{a}_3 = \frac{a}{2}(\hat{\mathbf{i}} + \hat{\mathbf{j}} - \hat{\mathbf{k}})$$

Basis vectors for face-centered cubic lattice

$$\mathbf{a}_1 = \frac{a}{2}(\hat{\mathbf{j}} + \hat{\mathbf{k}}), \quad \mathbf{a}_2 = \frac{a}{2}(\hat{\mathbf{k}} + \hat{\mathbf{i}}), \quad \mathbf{a}_3 = \frac{a}{2}(\hat{\mathbf{i}} + \hat{\mathbf{j}})$$

Reciprocal lattice

One dimensional lattice

$$k = n\,\frac{2\pi}{a}$$

Primitive vectors of the reciprocal lattice

$$\mathbf{b}_1 = 2\pi\,\frac{\mathbf{a}_2 \times \mathbf{a}_3}{\mathbf{a}_1 \cdot (\mathbf{a}_2 \times \mathbf{a}_3)}$$

$$\mathbf{b}_2 = 2\pi\,\frac{\mathbf{a}_3 \times \mathbf{a}_1}{\mathbf{a}_1 \cdot (\mathbf{a}_2 \times \mathbf{a}_3)}$$

$$\mathbf{b}_3 = 2\pi\,\frac{\mathbf{a}_1 \times \mathbf{a}_2}{\mathbf{a}_1 \cdot (\mathbf{a}_2 \times \mathbf{a}_3)}$$

Reciprocal lattice vectors

$$\mathbf{g} = n_1\mathbf{b}_1 + n_2\mathbf{b}_2 + n_3\mathbf{b}_3$$

Distance separating lattice planes

$$d = \frac{2\pi}{|\mathbf{g}|}$$

Bloch's theorem

Standard form of Bloch's theorem

$$\psi_{\mathbf{k}}(\mathbf{r} + \mathbf{l}) = e^{i\mathbf{k}\cdot\mathbf{l}}\psi_{\mathbf{k}}(\mathbf{r})$$

Alternate form of Bloch's theorem

$$\psi_{\mathbf{k}}(\mathbf{r}) = e^{i\mathbf{k}\cdot\mathbf{r}}u_{\mathbf{k}}(\mathbf{r})$$

Scattering of electrons by a crystal

Relation between final vector $\mathbf{k}'$, initial vector $\mathbf{k}$, and reciprocal lattice vector $\mathbf{g}$

$$\mathbf{k}' = \mathbf{k} - \mathbf{g}$$

Single-electron wave function

$$\psi(\mathbf{r}) = \sum_{\mathbf{g}} \alpha_{\mathbf{k}-\mathbf{g}} e^{i(\mathbf{k}-\mathbf{g})\cdot\mathbf{r}}$$

Free-electron energy

$$\mathcal{E}^0(\mathbf{k}) = \frac{\hbar^2 k^2}{2m}$$

Summary

Metals and semiconductors have crystal lattices with a well-defined translational symmetry. Most important among the crystal lattices are the simple cubic, body-centered cubic, and face-centered cubic structures, the diamond structure, and the hexagonal close-packed structure.

The structure of the crystal of which a substance is composed determines the properties of the wave functions of electrons moving through the crystal. Free-electron wave functions are periodic with respect to the crystal lattice if the wave vector k belongs to the reciprocal lattice, which may be constructed using the primitive vectors of the original crystal lattice. The reciprocal lattice of the simple cubic lattice is itself a cubic lattice, while the reciprocal lattice of the body-centered cubic lattice is a face-centered cubic lattice and the reciprocal lattice of the face-centered cubic lattice is a body-centered cubic lattice.

According to Bloch's theorem, the translational symmetry of the lattice implies that each wave function, which is a solution of Schrödinger equation, has an associated wave vector $\mathbf{k}$. The vector $\mathbf{k}$ is a vector in the reciprocal lattice, which, like the crystal itself, consists of a multitude of basic cells. The unit cell of the reciprocal lattice is called the Brillouin zone. Each electron having a wave vector $\mathbf{k}$ interacts with the electron with a wave vector $\mathbf{k}'$, which is the mirror image of $\mathbf{k}$ across the boundary of the Brillouin zone. This simple model of the interaction of electrons in a crystal allows us to understand the band gap of solids and to classify materials into insulators, metals, and semiconductors.

Graphene and nanotube, which have appeared recently, are complex materials with properties of metals and semiconductors.

Questions

1. Write down a formula specifying the points in a Bravais lattice.
2. Give the number of nearest neighbors for the sites in the simple cubic, body-centered cubic, and face-centered cubic lattices.
3. Give an example of a crystal lattice consisting of two interpenetrating face-centered cubic lattices.
4. Describe the hexagonal close-packed structure.
5. Describe another type of close-packed structure and say how it differs from the hexagonal close-packed structure.
6. Describe the sodium chloride and cesium chloride crystals.
7. What is the zincblende structure?
8. What is the Wurtzite structure?
9. Which of the cubic lattices has the highest density of points?
10. Write down a formula specifying the points in a one-dimensional reciprocal lattice.
11. Write down formulas expressing the primitive vectors of the reciprocal lattice in terms of the primitive vectors of the direct lattice.
12. Write down a formula for the dot product of a member of the reciprocal lattice $\mathbf{g}$ with a member of the direct lattice $\mathbf{l}$.
13. If $\mathbf{g}$ is a member of the reciprocal lattice, what distinctive property does the function, $\psi_{\mathbf{g}}(\mathbf{r}) = e^{i\mathbf{g}\cdot\mathbf{r}}$, have?
14. Make a drawing of the Wigner-Seitz cell of the reciprocal lattice (the Brillouin zone) for a simple cubic lattice with lattice constant a.
15. Make a drawing of the Brillouin zone of the face-centered cubic lattice.
16. Sketch the lattice planes with Miller indices, (100) and (011) for a simple cubic lattice.

17. Using Bloch's theorem, write down a formula relating the functions, $\psi_k(\mathbf{r} + \mathbf{l})$ and $\psi_k(\mathbf{r})$.
18. Write down a formula expressing the alternate form of Bloch's theorem.
19. Make a drawing of the vectors, $\mathbf{k}$, $\mathbf{g}$, and $\mathbf{k} - \mathbf{g}$.
20. Write down an expression for a single-electron wave function which includes the scattering at the zone boundaries.
21. Explain why the energy $\mathcal{E}^-$ of the lower state at the band gap is less that the energy $\mathcal{E}^+$ of the upper state.
22. Where would the band gap for a material that crystallizes in a face-center cubic lattice occur?
23. What is the basic unit of the graphene lattice?
24. What distinctive property do the charge carriers of graphene have?
25. Why is it useful to introduce spin to describe the interaction of the charge carriers of graphene?
26. For which direction of the charge carriers does graphene behave as a metal?
27. For which direction of the charge carriers does graphene behave as a semiconductor?
28. What determines whether a carbon nanotube behaves like a metal or a semiconductor?
29. How are the atoms bonded together in the NaF crystal?
30. The methane molecule (CH_4) is very nonpolar since it has four hydrogen atoms arranged symmetrically around a central carbon atom. How would you expect molecules of methane to be bonded together in a crystal?
31. How are water molecules bonded together in ice crystals?
32. Do you think the boiling point of methane will be lower or higher than the boiling point of water? Explain your answer.
33. Why do materials that are held together by molecular bonds have lower melting points than covalently bonded materials?
34. Describe the forces responsible for the cohesive energy of ionic crystals and metals.

Problems

1. The nearest neighbors of a cesium ion in the cesium chloride lattice are chloride ions and the next nearest neighbors of the ion are other cesium ions. How many nearest neighbors and how many next nearest neighbors does a cesium ion have?
2. A Bravais lattice has a set of primitive vectors

$$\mathbf{a} = (\frac{a}{2})\hat{\mathbf{i}} + (\frac{a}{2})\hat{\mathbf{j}}, \quad \mathbf{b} = a\hat{\mathbf{j}}, \quad \mathbf{c} = \frac{a}{\sqrt{2}}\hat{\mathbf{k}},$$

where $\hat{\mathbf{i}}, \hat{\mathbf{j}}$, and $\hat{\mathbf{k}}$ are unit vectors pointing along the x-, y-, and z-axes. Make a drawing of those lattice points lying in the x-y plane and say what kind of lattice it is.
3. Show that the angle between any two of the bonds joining a carbon atom to its neighbors in the diamond lattice is $\cos^{-1}(-1/3) = 109°28'$.
4. Suppose that identical ions are stacked upon each other to form a body-centered cubic lattice. The nearest neighbors of a point in this lattice are located one half the way along the main diagonals of the adjoining cubes. Show the nearest neighbor distance for the body-centered cubic lattice is $(\sqrt{3}/2)a$, where a is the length of the sides of the cubes.
5. The nearest neighbors of a point in a face-centered cubic lattice are located at the midpoints of the adjoining faces of the cubes. Show that the nearest neighbor distances for the simple cubic and face-centered cubic lattices are a and $(\sqrt{2}/2)a$ respectively, where a is the length of the sides of the cubes.
6. Sodium metal, which has a body-centered cubic structure, has a density of 0.971 g/cm^3 and a molar mass of 23.0 g. Calculate the center–to–center distance for neighboring sodium ions in the crystal.
7. Copper metal, which has a face-centered cubic structure, has a density of 8.96 g/cm^3 and a molar mass of 63.5 g. Calculate the center–to–center distance for neighboring copper ions in the crystal.
8. A set of primitive vectors that generates the face-centered cubic lattice is defined by Eq. (9.4) and shown in Fig. 9.6(B). As can be seen from Fig. 9.6(B), these three vectors each point to a lattice point at the center of a face of a single cube. Express the lattice vectors pointing to the centers of the other three faces of this cube as a linear combination of these primitive vectors.
9. Express the lattice vectors pointing to the four upper corners of the cube shown in Fig. 9.4(B) as linear combinations of the primitive vectors given by Eq. (9.4).
10. Express the lattice vectors pointing to the four upper corners of the cube shown in Fig. 9.5 as linear combinations of the primitive vectors given by Eq. (9.2).
11. Ionic crystals and metals may be thought of as being comprised of ionic spheres. A criterion for judging how tightly the ionic spheres are packed is provided by the packing fraction (F) which is the volume of the spheres divided by the

total volume of the crystal. The packing fraction is given by the following equation

$$F = n\frac{4/3\pi R^3}{a^3},$$

where n is the number of ionic spheres per unit cell and R is the radius of the spheres. As we have seen, the values of n for the simple cubic, body-centered cubic and face-centered cubic lattices are 1, 2, and 4 respectively. The radius (R) of the spheres is related to the nearest neighbor distances (d) by the equation $2R = d$. Using the nearest neighbor distances found in Problems 4 and 5, show that the packing fractions for the three cubic systems we have considered are

$$\frac{\pi}{6} = 0.52 \ \text{simple cubic}$$

$$\frac{\sqrt{3}\pi}{8} = 0.68 \ \text{body-centered cubic}$$

$$\frac{\sqrt{2}\pi}{6} = 0.74 \ \text{face-centered cubic.}$$

12. Use Eqs. (9.17)-(9.19) defining the primitive vectors of the reciprocal lattice to derive Eq. (9.21).

13. A crystal lattice has a set of primitive vectors

$$\mathbf{a}_1 = (a/2)\hat{\mathbf{i}} + (a/2)\hat{\mathbf{j}}$$

$$\mathbf{a}_2 = a\hat{\mathbf{j}}$$

$$\mathbf{a}_3 = (a/\sqrt{2})\hat{\mathbf{k}},$$

where $\hat{\mathbf{i}}, \hat{\mathbf{j}}$, and $\hat{\mathbf{k}}$ are unit vectors pointing along the x, y, and z axes. Calculate the primitive vectors of the reciprocal lattice and identify the type of crystal to which the reciprocal lattice belongs.

14. A crystal lattice has a set of primitive vectors

$$\mathbf{a}_1 = \left(\frac{\sqrt{2}a}{2}\right)\hat{\mathbf{i}} + \left(\frac{\sqrt{2}a}{2}\right)\hat{\mathbf{j}}$$

$$\mathbf{a}_2 = -\left(\frac{\sqrt{2}a}{2}\right)\hat{\mathbf{i}} + \left(\frac{\sqrt{2}a}{2}\right)\hat{\mathbf{j}}$$

$$\mathbf{a}_3 = -a\hat{\mathbf{k}}.$$

Calculate the primitive vectors of the reciprocal lattice and identify the type of crystal to which the reciprocal lattice belongs.

15. Using Eq. (9.2), which defines the primitive vectors for the body-centered cubic lattice, calculate the primitive vectors of the reciprocal lattice and identify the kind of lattice defined by the primitive vectors of the reciprocal lattice vectors.

16. Using Eq. (9.2), which defines the primitive vectors for the body-centered cubic lattice, calculate the primitive vectors of the reciprocal lattice and show that these vectors could serve as primitive vectors of a face-centered cubic lattice.

17. The primitive vectors of the hexagonal closed-packed structure are

$$\mathbf{a} = (\sqrt{3}a/2)\hat{\mathbf{i}} + (a/2)\hat{\mathbf{j}}, \quad \mathbf{b} = -(\sqrt{3}a/2)\hat{\mathbf{i}} + (a/2)\hat{\mathbf{j}}, \quad \mathbf{c} = c\hat{\mathbf{k}},$$

where a and c are lattice constants. Calculate the primitive vectors of the reciprocal lattice and the volume of the unit cell in reciprocal lattice space. Identify the reciprocal lattice and say in qualitative terms how the reciprocal lattice is related to the direct lattice.

18. The hexagonal close-packed and face-centered cubic structures are closely related. The first two primitive vectors of a face-centered cubic lattice can be chosen to be the vectors **a** and **b** defined in the previous problem with the constant a being equal to the length of a side of a cubic unit cell. How must the third vector **c** be chosen then for the lattice to be face-centered cubic?

19. Show that the reciprocal of the reciprocal lattice is the corresponding direct lattice.

20. Find the distance between the family of lattice planes with Miller indices [110] for a simple cubic lattice.
21. Find the distance between the family of lattice planes with Miller indices [210] for a simple cubic lattice.
22. (a) If g and g' are reciprocal lattice lengths given by Eq. (9.10), show that the functions e^{igx} and $e^{ig'x}$ satisfy the equation

$$\int_0^a e^{-ig'x} e^{igx} dx = \begin{cases} a, & g = g' \\ 0, & g \neq g'. \end{cases}$$

(b) Multiply Eq. (9.13) from the left with $e^{-ig'x}$, integrate the left- and right-hand sides from 0 to a, and use the result of part (a) to obtain the following equation

$$\int_0^a e^{-ig'x} f(x) dx = a F_{g'}.$$

This last equation is equivalent to Eq. (9.14).
23. Suppose

$$\mathbf{l} = l_1 \mathbf{a}_1 + l_2 \mathbf{a}_2 + l_3 \mathbf{a}_3,$$

is a lattice vector, and

$$\mathbf{g} = n_1 \mathbf{b}_1 + n_2 \mathbf{b}_2 + n_3 \mathbf{b}_3,$$

is a reciprocal lattice vector which is normal to a family of lattice planes. If the set of integers (n_1, n_2, n_3) have no common factors and $\mathbf{l}$ satisfies the equation

$$\mathbf{g} \cdot \mathbf{l} = 2\pi N,$$

then one may show that there is a vector $\mathbf{l}'$ satisfying the condition

$$\mathbf{g} \cdot \mathbf{l}' = 2\pi (N + 1).$$

Using this result, show that the lattice planes containing the points $\mathbf{l}$ and $\mathbf{l}'$ are separated by a distance

$$d = \frac{2\pi}{|\mathbf{g}|}.$$

24. Show that the free-electron wave functions given by Eq. (9.45) satisfy the orthogonality condition (9.48).
25. (a) Show that the Born-von Karman boundary condition

$$\psi(\mathbf{r} + N_1 \mathbf{a}_1) = \psi(\mathbf{r})$$

and Bloch's theorem (9.39) lead to the condition

$$e^{i\mathbf{k}\cdot(N_1\mathbf{a}_1)} = 1.$$

The corresponding conditions for the other lattice directions are

$$e^{i\mathbf{k}\cdot(N_2\mathbf{a}_2)} = 1, \quad e^{i\mathbf{k}\cdot(N_3\mathbf{a}_3)} = 1.$$

(b) Show that the conditions given in part (a) are satisfied by the $\mathbf{k}$ vector (9.50).
26. The complex conjugate of the transition integral (9.53) may be written

$$\mathbf{I_{k',k}}^* = \int \psi_{\mathbf{k}'}(\mathbf{r}) V(\mathbf{r}) \psi_{\mathbf{k}}^*(\mathbf{r}) dV.$$

By substituting Eqs. (9.51) and (9.54) into this integral and using the orthogonality properties of the free-electron wave functions, show that the integral can be nonzero only if the vectors, $\mathbf{k}$, $\mathbf{k}'$, and $\mathbf{g}$, satisfy Eq. (9.58).
27. Explain why many semiconductors are opaque to visible light but transparent to infrared light.

Chapter 10

Charge carriers in semiconductors

Contents

10.1 Density of charge carriers in semiconductors	241	Suggestions for further reading	250
10.2 Doped crystals	244	Summary	251
10.3 A few simple devices	245	Questions	251

The conductivity of a semiconductor is dependent upon the small number of electrons in the conduction band and holes in the valence band.

When atoms condense to form a solid, the wave functions of electrons in the different atomic shells respond differently to their environment. The wave functions of the inner electrons, which are only weakly perturbed by the solid state environment, are very similar to the wave functions of isolated atoms. The electronic states of the inner electrons are called *core states*. By contrast, the wave functions of more loosely bound electrons become more distorted by the presence of other neighboring atoms. Electron states, which are significantly distorted by the solid state environment and yet remain localized about the atomic center are called *valence states*, while other states, which become delocalized over the entire solid, are called *conduction states*. Unlike the lowest energy levels of atoms, which are generally well separated, the energy levels of the valence and conduction electrons of a solid occur in dense bands with many different levels close to each other. As we have seen in the previous chapter, there is generally an energy gap called a *band gap* between the highest valence level and the least energetic conduction electrons.

10.1 Density of charge carriers in semiconductors

The distinctive property of semiconductors is that they have small bandgaps separating valence and conduction states. While electrons can be excited thermally from the valence band into the conduction band of these materials, the number of electron-hole pairs created in this way is quite small. The small number of charge carriers limits the conductivity of these materials. For example, *Si* at room temperature has only 10^{10} free electrons and holes per cubic centimeter even though the *Si* crystal has more than 10^{23} atoms per cubic centimeter. Increasing the temperature of a semiconductor increases the number of electrons excited thermally into the conduction band.

In a filled band, all available states are occupied. For every electron moving in one direction, there is another electron moving in the opposite direction and the net current is zero. However, if an electron is excited from a filled shell, a hole is created and the resulting imbalance between the velocities of the electrons makes it possible for the material to conduct a current. Even though the current is due to an electron moving in a direction opposite to that of the missing electron, it is convenient to imagine that the hole carries the electric change. An electron with negative charge moving in one direction is equivalent to a hole with positive charge moving in the opposite direction.

A perfect semiconductor with no impurities or lattice defects is called an *intrinsic semiconductor*. The generation of a hole in an intrinsic semiconductor is accomplished by breaking a covalent bond of the crystal producing a free electron and a vacancy in the lattice. Since free electrons and holes are created in pairs, the number of free electrons in the conduction band is always equal to the number of holes in the valence band. Fig. 10.1 gives a simple illustration of electrons and holes in an intrinsic semiconductor. As shown in Fig. 10.1, the Fermi energy is between the highest occupied level and the lowest unoccupied level. For intrinsic semiconductors, the Fermi energy occurs near the middle of the gap between the valence and conduction bands. In the next section, we shall find that it is possible to introduce additional electrons or holes into a semiconductor by adding impurities. Semiconductors with additional charge carriers due to impurities are called *extrinsic semiconductors* and said to be *doped*. For extrinsic semiconductors, the Fermi energy may be very close to the bottom of the conduction band or very close to the top of the valence band.

Modern Physics with Modern Computational Methods. https://doi.org/10.1016/B978-0-12-817790-7.00017-2

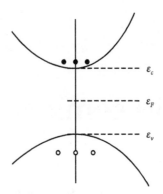

FIGURE 10.1 Illustration of electrons and holes in an intrinsic semiconductor. The top of the valence band is denoted by ϵ_v and the bottom of the conduction band is denoted by ϵ_c. ϵ_F is the Fermi energy.

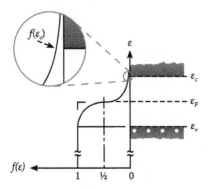

FIGURE 10.2 The function $f(E)$ for intrinsic semiconductors drawn to show which of the levels in the conduction band are occupied.

Many of the optical and electrical properties of semiconductors depend upon the density of charge carriers. The number of conduction electrons in a particular energy range can be calculated using the free-electron model which we applied to the electrons of a metal in Chapter 8. Eq. (8.66) of Chapter 8 gives the number of states available to a free electron in the range between ϵ and $\epsilon + d\epsilon$. For the conduction electrons in a semiconductor, this equation can be written

$$g(\epsilon)d\epsilon = \frac{V4\pi}{h^3}(2m_c)^{3/2}(\epsilon - \epsilon_c)^{1/2}d\epsilon, \tag{10.1}$$

where m_c is the effective mass of a conduction electron and ϵ_c is the energy of the bottom of the conduction band.

The fraction of single-electron states that are occupied is given by the Fermi-Dirac distribution function,

$$f(\epsilon) = \frac{1}{e^{(\epsilon - \epsilon_F)/k_B T} + 1}, \tag{10.2}$$

where k_B is Boltzmann's constant and ϵ_F is the Fermi energy. The function $g(\epsilon)$ is illustrated in Fig. 8.10(A), while the function $f(\epsilon)$ for a particular temperature $T > 0$ is illustrated in Fig. 8.11(A). Using Eqs. (10.1) and (10.2), the density of electrons in the range between ϵ and $\epsilon + d\epsilon$ can be written

$$n(\epsilon)d\epsilon/V = f(\epsilon)g(\epsilon)d\epsilon/V = \frac{4\pi}{h^3}(2m_c)^{3/2}\frac{(\epsilon - \epsilon_c)^{1/2}d\epsilon}{e^{(\epsilon - \epsilon_F)/k_B T} + 1}. \tag{10.3}$$

The density of occupied states can best be visualized by turning the diagram of $f(\epsilon)$ versus ϵ shown in Fig. 8.11(A) on its side so that the axis corresponding to the energy variable in Fig. 8.11(A) is the same as in a band diagram. An illustration of this kind is shown in Fig. 10.2. The function $f(\epsilon)$ appearing on the left in this figure shows the fraction of the states in the conduction band that are occupied. One can see that the fraction of states that are occupied decreases with increasing energy and that the function $f(\epsilon)$ declines exponentially for large energies. Due to the exponential decline of $f(\epsilon)$ only energies within $k_B T$ of the band edge contribute significantly to the carrier density.

The total density of carriers in the conduction band is

$$n = \frac{4\pi}{h^3}(2m_c)^{3/2}\int_{\epsilon_c}^{\infty}\frac{(\epsilon-\epsilon_c)^{1/2}d\epsilon}{e^{(\epsilon-\epsilon_F)/k_BT}+1}.$$

The above integral can be evaluated by making the substitutions, $y = (\epsilon-\epsilon_c)/k_BT$ and $v = (\epsilon_F-\epsilon_c)/k_BT$, and the expression for the density of charge carriers can be written

$$n = N_c\frac{2}{\sqrt{\pi}}F_{1/2}(v),\tag{10.4}$$

where

$$N_c = 2\left(\frac{2\pi m_c k_B T}{h^2}\right)^{3/2}\tag{10.5}$$

and

$$F_{1/2}(v) = \int_0^{\infty}\frac{\sqrt{y}\,dy}{e^{y-v}+1}.$$

For an intrinsic semiconductor, the Fermi energy is near the center of the energy gap and $k_BT \ll \epsilon_c - \epsilon_F$. The exponential term in the denominator of the integral $F_{1/2}$ is then very much larger than one, and the integral becomes

$$F_{1/2}(v) = e^v\int_0^{\infty}\sqrt{y}\,e^{-y}dy = e^v\frac{\sqrt{\pi}}{2}.$$

Substituting this expression for $F_{1/2}(v)$ back into Eq. (10.4), we obtain

$$n = N_c e^{-(\epsilon_c-\epsilon_F)/k_BT}.\tag{10.6}$$

The number of electrons in the conduction band thus decreases exponentially as the distance between ϵ_c and ϵ_F increases.

The theory of charge carriers in crystals is important for studying carrier loss in semiconductor devices such as semiconductor lasers. Electrons in crystals typically have an effective mass which is different from the mass ordinarily associated with the electron. While electrons in a silicon crystal have an effective mass equal to 0.96 times the electron mass for motion in a direction perpendicular to the boundary of the Brillouin zone, the electrons have a mass equal to 0.19 times the electron mass for motion parallel to the zone boundary. Denoting the electron mass by m_0, the *longitudinal mass* for motion perpendicular to the face of the Brillouin zone is $0.96\,m_0$ and the *transverse mass* for motion parallel to the zone boundary is $0.19\,m_0$. Electrons in silicon are less mobile in the longitudinal direction than in the transverse directions.

Example 10.1

Find the density of transverse charge carriers in a silicon crystal at $T = 300$ K.

Solution

The density of charge carriers for silicon can be calculated using Eq. (10.6). We shall first find the value of N_c and then evaluate the other constants appearing in this equation. Using the physical constants given in Appendix A with the mass m equal to 0.19 times the electron mass, one finds that the constant N_c given by Eq. (10.5) is equal to 4.00×10^{20} cm^{-3}. Silicon has a band gap of approximately 1.17 eV between the valence and the conduction bands. Since for an intrinsic semiconductor such as a pure silicon crystal, the Fermi energy ϵ_F is midway between the valence and conduction bands, the energy difference, $\epsilon_c - \epsilon_F$, is equal to 0.585 eV, and, for $T = 300$ K, the product $k_B T$ is equal to 0.02585 eV. The value of $(\epsilon_c - \epsilon_F)/k_B T$ is thus equal to 22.6 and $e^{-(\epsilon_c-\epsilon_F)/k_BT}$ is equal to 1.49×10^{-10}. Substituting this value into Eq. (10.6) and taking $N_c = 4.00 \times 10^{20}$ cm^{-3} we find that the density of transverse charge carriers in silicon at $T = 300$ K is 6.0×10^{10} cm^{-3}. This value, which is consistent with the value given at the beginning of his section, is many of orders of magnitude less than the number of atoms per cubic centimeter.

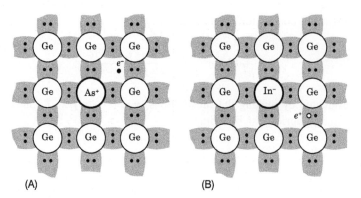

FIGURE 10.3 (A) Arsenic donor atoms and (B) indium acceptor atoms in a germanium lattice.

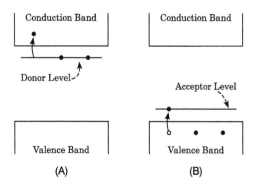

FIGURE 10.4 (A) A donor level just below the conduction band in a n-doped semiconductor, (B) an acceptor level just above the valence band in a p-doped semiconductor.

10.2 Doped crystals

As noted earlier, the concentration of free electrons and holes in a crystal can be modified by adding impurities to the crystal. To give some idea of how this is done, we consider adding silicon impurities to GaP. Crystals of GaP can be made by melting gallium and phosphorus in a crucible and adding a seed GaP crystal to the melt. GaP crystallizes on the seed, and if a small amount of silicon is added to the melt, a small fraction of the gallium atoms in the GaP crystal will be replaced by silicon atoms. Silicon is in the fourth (IV) column of the periodic table, while Gallium is a third (III) column element. Since silicon has one more valence electron than gallium, each silicon site contribute an additional electron to the conduction band. Impurities that contribute electrons to the conduction bands are called *donors*, while impurities that contribute holes to the valence bands are called *acceptors*. A semiconductor for which extra free electrons or holes have been introduced by adding impurities is called an *extrinsic semiconductor*, and the process of adding impurities to modify the concentration of free electrons or holes is called *doping*. Through doping, a crystal can be modified so that it has a predominance of either electrons or holes.

As a second example of doping a crystal, we consider the addition of arsenic impurities to a germanium crystal. Germanium belongs to column IV of the periodic table having two electrons in the $4s$ shell and two electrons in the $4p$ shell, while arsenic belongs to the column V of the periodic table having two electrons in the $4s$ shell and three electrons in the $4p$ shell. As shown in Fig. 10.3(A), the arsenic atoms contributes an additional electron which can easily be excited to the conduction band. The germanium crystal with arsenic impurities is said to be *n-doped* because it has additional negative charged carriers due to arsenic. Fig. 10.3(B) shows a germanium crystal with indium impurities. Indium is in the column III in the periodic table with two electrons in the $5s$ shell and one electron in the $5p$ shell. Since indium has one less electron in the valence shell, the germanium crystal with indium impurities has additional holes in the valence band and is said to be *p-doped*.

Adding donor or acceptor impurities to a crystal has the effect of adding additional energy levels to the band structure and changes the position of the Fermi level. The additional energy levels are usually within the bandgap. For example, an As atom added as an impurity to a Ge crystal has an additional electron which can easily be excited to the conduction band. Since only a small amount of energy is required to excite the extra electron of the As atom, the additional donor state is slightly under the conduction band as shown in Fig. 10.4(A). The additional occupied levels added to the Ge crystal by

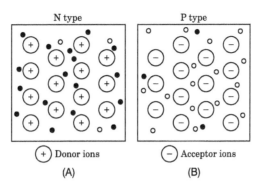

FIGURE 10.5 (A) An n-doped semiconductor. The ion cores of donor states are represented by open circles with plus signs. Free electrons are represented by small dark circles, and holes are represented by small open circles. (B) A p-doped semiconductor. Acceptor ions are represented by open circles with minus signs.

As impurities raises the Fermi energy so that it is just below the conduction band. Similarly, an indium atom added to a *Ge* crystal needs an additional electron to complete the four bonds joining the atom to its four *Ge* neighbors and is thus inclined to take an electron from the valence band of germanium leaving a hole. As shown in Fig. 10.4(B), the empty states due to indium atoms lie slightly above the valence band. The new unoccupied states lower the Fermi energy so that it is just above the valence band.

10.3 A few simple devices

We shall now briefly discuss the properties of *n*- and *p*-type materials and then consider the kinds of semiconductor devices that can be formed by joining *n*- and *p*-type materials together. Fig. 10.5(A) shows an *n*-doped material with a relatively small number of donor impurity atoms. The open circles with plus signs in the figure represent the ion cores of the donor atoms, while the small dark circles represent free electrons. The small open circles in Fig. 10.5(A) correspond to holes which are due to thermal excitation of electrons from the valence band. In a *n*-doped material, electrons are referred to as the *majority carriers* and holes are referred to as *minority carriers*.

Similarly, Fig. 10.5(B) shows a *p*-doped material with a relatively small number of acceptor atoms. The negative ions formed when electrons combine with acceptor atoms are represented by open circles with minus signs, and the holes formed when atoms lose valence electrons are represented by small open circles. In a *p*-doped material, the holes are the majority carriers and the electrons are the minority carriers.

10.3.1 The *p-n* junction

A p-n junction is formed by depositing a material doped with either donor or acceptor atoms upon another material with the opposite kind of doping. The junction has two-terminals and is thus referred to as a *p-n* diode.

Once the *p-n* junction is formed, electrons in the *n*-material diffuse across the boundary leaving behind positively charged donor ions, and holes in the *p*-material diffuse across the boundary leaving behind negatively charged acceptor ions.

The majority charge carriers that cross over the boundary recombine with majority carriers in the other material and disappear. Electrons diffusing into the *p*-material encounter a great number of holes in the *p* region and immediately recombine, while holes diffusing into the *n*-material immediately recombine with electrons. As shown in Fig. 10.6, only fixed ions remain near the junction. Positive ions remain on the *n*-side, while negative ions remain on the *p*-side. Because of the lack of charge carriers, this region near the interface between the two materials is called the *depletion zone*.

The charge ions in the depletion zone create an electrostatic potential which eventually becomes strong enough to resist the flow of majority charge carriers across the boundary. Electrons in the *n*-material are hindered by the electric field due to the negative ions in the *p*-region, while holes in the *p*-region are hindered by positive ions in the *n*-region. Because of the charged ions near the junction, this region is also called the *space-charge region*.

The extent of the depletion zone or space-charge region can be modified by applying a biasing potential. Before considering the effect of biasing potentials we first note that electrons in the *n*-material will naturally flow to the left into the *p*-material and holes will naturally flow from the *p* material to the right into the *n*-material. Adopting the ordinary convention of associating the direction of the current with the motion of the positive charge carriers, the natural flow across

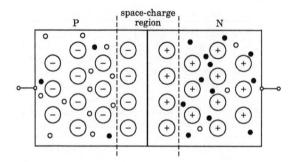

FIGURE 10.6 A p-n junction after majority carriers have diffused across the boundary producing a space-charge region.

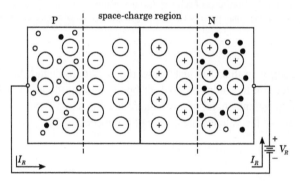

FIGURE 10.7 A reversed-biased p-n junction.

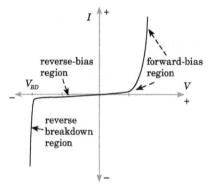

FIGURE 10.8 A plot of the current (I) versus potential difference (V) for a p-n junction. The breakdown voltage is denoted by V_{BD} on the negative horizontal axis.

the boundary of the $p - n$ junction is from left to right. Current flow to the left is entirely due to the minority charge carriers. Fig. 10.7 shows the effect upon the junction of a *reverse bias* causing the flow of current from right to left and hence attracting electrons and holes away from the junction between the n- and p-materials. As can be seen by comparing Fig. 10.7 to Fig. 10.6 such a voltage the applied to a $p - n$ junction increases the size of the depletion zone. In contrast, a *forward bias* causes electrons and hole to flow in the opposite direction and reduces the size of the depletion zone.

Having discussed the nature of the $p - n$ junction, we now examine the properties of current flow through the junction. Fig. 10.8 shows a plot of the current (I) versus potential difference (V) for the p-n junction. The potential being positive corresponds to forward biasing, while the potential being negative corresponds to reverse biasing. As the potential difference increases from zero, the current through the device remains very small until the forward bias exceeds the electrostatic potential due to the presence of positive and negative ions at the boundary between the n- and p-materials. The potential difference needed to overcome the barrier potential is typically around 0.3 Volts for germanium diodes and 0.7 Volts for silicon diodes. For larger values of the potential difference, the forward current then increases rapidly.

The current through the p-n junction for small positive values of the potential difference and for negative values of the potential difference is due entirely to the minority carriers and remains very small for a considerable range of the applied potential. While a positive barrier potential restricts the flow of majority carriers across the junction, minority carriers are

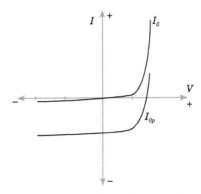

FIGURE 10.9 A plot showing the current (I_0) of a p-n junction without optical excitations and a current I_{Op} that includes the effect of optical excitations of electrons from the valence band.

accelerated by the barrier potential as they cross the junction. If the reverse voltage is high enough, the minority carriers move so rapidly that they knock valence electrons from atoms creating more charge carriers. The value of the potential for ionization processes to occur is called the *reverse breakdown voltage*. The current in the reverse direction increases rapidly for more negative values of the applied potential.

A basic characteristic of *p-n* junctions is that they conduct current much more readily in the forward direction than the reverse direction unless the breakdown voltage is exceeded. The current in the forward direction is limited only by the resistance in series with the junction, while the current in the reverse direction is generally very small. $p - n$ junctions can thus be used to produce *rectifiers*, which convert alternating currents into pulsating direct currents.

10.3.2   Solar cells

We now consider solar cells which are an alternate source of electrical energy that is of growing importance with the present concern about global warming and the need of finding alternates to the burning of fossil fuels. The amount of solar power striking the earth far exceeds the world's energy needs. The current energy consumption of the United States could be supplied by solar power if farmers in Kansas converted just 4% of existing farmland into solar installations.

The density of black body radiation for different temperatures is illustrated in Fig. 8.6 of Chapter 8. The intensity of the radiation emitted by the black body is proportional to the energy density. Using the fact that the surface temperature of our Sun is approximately 6000 K, we see that the maximum intensity of solar radiation occurs for a frequency of about 3.5×10^{14} Hz. Using the equation

$$\lambda f = c,$$

we find that the maximum intensity of the solar spectrum occurs for $\lambda = 1200$ nm which is in the infrared portion of the spectra. The visible spectra extend from 400 nm to 700 nm.

A semiconductor will absorb a photon if the energy of the photon is sufficiently large to excite an electron from the valence band to the conduction band. If the photon energy is less than the band gap, the semiconductor can not absorb the photon and the semiconductor will be transparent. Silicon is commonly used for solar cells because it is relatively inexpensive and has a band gap of 1.1 eV that is well matched to the Sun's spectrum. A photon emitted by the Sun at the solar maximum (3.5×10^{14} Hz) would have an energy equal to $hf = 1.44$ eV. So, the band gap of silicon occurs just before the maximum of the solar radiation curve. The transfer of an electron from the valence band to the conduction band creates a nonequilibrium situation. In a fraction of a second, the electron would ordinarily drop back into the valence band, giving off heat and not doing any useful work. For a solar cell to be effective, the electron and hole produced must be quickly separated from each other.

The design of a solar cell takes these basic properties of electron-hole pairs into account. Solar cell generally consist of n-doped and p-doped silicon interfaces that form a p-n junction. The junction has a space charge region with few charge carriers that could combine with a new electron or hole and has an electric field that accelerates the electron-hole pairs that are formed rapidly away from each other. The electric field within the space charge region points from the n-side to the p-side of the junction. When an electron and hole are produced in the space charge region, the conduction electron is accelerated towards the n-side, and the hole is accelerated towards the p-side. The current generated by electrons and the holes are both in the direction from the n-side to the p-side. Fig. 10.9 shows the current (I_0) of a p-n junction without optical excitations and a current I_{Op} that includes the effect of optical excitations of electrons from the valence band.

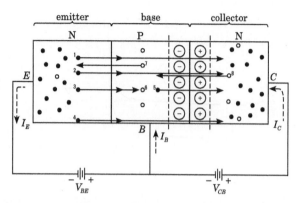

FIGURE 10.10 A NPN bipolar transistor with a forward bias between the emitter and base and a reverse bias between the collector and base.

The current generated by the solar cell can power an external circuit. In this context, we note that potential V and the current I of the p-n junction shown in Fig. 10.8 are both negative or both positive. So the electric power P, which is the product of V and I, is always positive indicating that power is dissipated. In contrast, the power produced by the current I_{Op} of a solar cell shown in Fig. 10.9 is positive in the region where V and I_{Op} are both negative and negative in the region where V is positive and I_{Op} is negative. The regions for which the power is negative corresponds to the solar cell generating power.

Solar cell is used in circuits for which a resistance is attached to the cell. In an implementation where the solar cell delivers power to the electrical grid, many solar cells are connected together and an inverter converts the solar cell's DC voltage to an AC voltage matching the grid voltage.

10.3.3 Bipolar transistors

We next consider devices called *bipolar transistors*, which have two $p-n$ junctions. The most widely used applications of the transistor occur when one p-n junction has a forward bias, and the other p-n junction has a reverse bias. Fig. 10.10 shows a transistor that is biased in this way. Some of the electrons and holes shown in the figure have been numbered to facilitate the following description of charge flow.

With the forward biasing on the p-n junction on the left, electrons flow freely across this junction into the central region. For this reason, the n-region on the left is called the *emitter*. Notice that the space-charge region of the p-n junction on the left has been overcome by forward biasing, while the p-n junction on the right, which is reverse biased, has a space-charge region. In Fig. 10.10, electrons number 1, 2, and 4 make it through the central region to the region on the right called the *collector*. Electron number 3 passes through the junction on the left only to recombine with hole number 6 in the central p-region, while hole number 7 passes from the central region into the region on the left. The central region called the *base* is generally lightly doped and very narrow so that electrons diffusing into it usually reach the collector before encountering a hole.

The currents shown in Fig. 10.10 are again drawn using the standard convention that the direction of the current corresponds to the direction of the positively charged holes. The emitter current is denoted by I_E and the collector current by I_C. The reverse biasing on the p-n junction on the right causes minority carriers to flow across this junction. Notice that electron number 5 passes from the central region to the collector, while hole number 8 passes from the collector into the central region. The current of the minority carriers across the B-C junction will be denoted by I_M.

The currents flowing through the terminals of the transistor are related by the equation

$$I_E = I_B + I_C . \tag{10.7}$$

Of the two currents appearing on the right side of this last equation, the collector current is the sum of the emitter current that reaches the collector and the current due to minority carriers passing through the C-B junction. We may write

$$I_C = \alpha_{EC} I_E + I_M ,$$

where α_{EC} is the portion of the emitter current reaching the collector and I_M is the current due to minority carriers. In most modern silicon transistors, I_M is much smaller than $\alpha_{EC} I_E$ and the collector current is approximately given by the equation

$$I_C \approx \alpha_{EC} I_E . \tag{10.8}$$

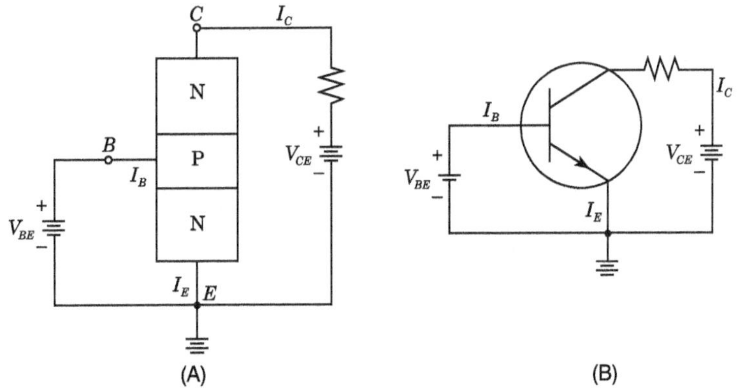

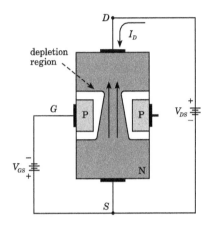

FIGURE 10.11 (A) A NPN transistor with its emitter grounded, and (B) the symbolic representation of this circuit.

FIGURE 10.12 An N-channel JFET.

An approximate expression for I_B can be obtained by substituting Eq. (10.8) into Eq. (10.7) and solving for I_B

$$I_B \approx (1 - \alpha_{EC})I_E .$$

(10.9)

Eqs. (10.8) and (10.9) can now be used to obtain an expression for I_C in terms of I_B. We have

$$I_C \approx \beta_{EC} I_B,$$

where the *current gain* factor β_{DC} is given by the equation

$$\beta_{EC} = \frac{\alpha_{EC}}{1 - \alpha_{EC}}.$$

Since α_{EC} is close to one, the current gain β_{EC} can be quite large. For example, if $\alpha_{EC} = 0.95$, $\beta_{EC} = 19$. This explains the usefulness of dipole transistors in producing current amplifiers.

Fig. 10.11(A) shows a *NPN* transistor with its emitter grounded, and Fig. 10.11(B) shows the symbolic representation of this circuit. For a three-terminal device, one terminal is generally grounded with the other two terminals serving as input and output.

10.3.4 Junction field-effect transistors (JFET)

Fig. 10.12 shows the structure of a simple *N*-channel *JFET*. The metal contact at one end of the transistor is called the *source* (S), while the metal contact at the other end is called the *drain* (D). Midway between the source and the drain, two *p*-doped regions extend into the *n*-doped interior from either side. In the normal operation of the transistor, electrons are injected into the *n*-doped region from the source, pass between the two *p*-doped regions called the *gate* and are collected by the drain.

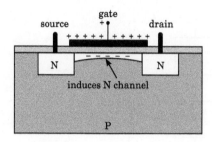

FIGURE 10.13 An N-channel MOSFET.

As shown in Fig. 10.12, the gate is biased negatively with respect to the source, and the drain is biased positively with respect to the source. The electric potential in the N-channel declines continuously from the value of the potential at the drain to its value at the source. Denoting the potential difference between the drain and the source by V_{DS} and the potential difference between the gate and the source by V_{GS}, the potential difference between the drain and the gate is $V_{DS} + |V_{GS}|$. For example, if $V_{DS} = 20$ V and $V_{GS} = -5$ V, the reverse bias between the gate and the drain is 25 V.

The breadth of the channel through which the electrons pass depends upon the size of the depletion zones surrounding the two gates. The reverse bias applied to the positive charge carriers in the gate causes the depletion zones to be larger and more obtrusive than they would be with no bias. The depletion regions are wedge-shaped, being larger at the drain end than at the source end. This is because the gate-to-channel reverse bias at the drain end of the gate is larger than the reverse bias at the source end of the gate.

The behavior of *JFET*s resembles the behavior of bipolar transistors in many respects. Since the drain current of the *JFET* is very sensitive to the voltage between the gate and the source, *JFET*s can also be used to make amplifiers. The JFET is a majority carrier device with a high input impedance (Z) which functions like a variable resistor. The effective resistance of the JFET is determined by the gate as it limits the number of carriers in the channel.

10.3.5 MOSFETs

The N-channel *MOSFET*, which is shown in Fig. 10.13, is also a majority carrier device. The MOSFET is built on a high-resistivity p-type semiconductor with two n-type regions diffused into the substrate. The upper surface of the device is covered with a layer of insulating silicon dioxide, and a metal contact area is placed over the silicon dioxide, covering the entire area between source and drain. There is no electrical contact between the metal plate, which serves as the gate terminal, and the p-substrate due to the insulation afforded by the silicon dioxide.

The silicon dioxide can be regarded as the dielectric medium between the two plates of a capacitor formed by the metal gate and the p-substrate. In these terms, the device can be thought of as a metal-oxide semiconductor (MOS) capacitor. A positive voltage applied to the metal gate leads to the build up of positive charge on the gate and an induced negative charge on the surface of the p-type semiconductor creating under the silicon dioxide an n-type region, which forms a channel between the source and the drain. Current can now be made to flow from source to drain through this induced channel by biasing the drain positive relative to the source.

Making the gate voltage more positive increases the number of conduction electrons that pass through the n-channel and thus *enhances* the current flow. The device will have no current with a zero or negative gate voltage, and is hence referred to as a *normally off MOSFET*. A p-channel *MOSFET* can be built in the same way by diffusing p-type regions in an n-substrate.

A MOSFET dissipates far lower power than either a JFET or bipolar transistor because the input to the MOSFET, the gate, is connected to an insulator and thus no current flows. In a JFET, the gate is a reverse biased diode with a small amount of current. The bipolar transistor has a forward biased diode (base to emitter), and thus the input current is very large compared to either a JFET or MOSFET. Modern computer chips with billions of transistors are made almost exclusively with MOSFETs because of the small amount of current they draw.

Suggestions for further reading

Ronald J. Tocci and Mark E. Oliver, *Fundamentals of Electronic Devices*, Fourth Edition (Macmillan, 1991).
Ben G. Streetman and Sanjay Banerjee, *Solid State Electronic Devices*, Fifth Edition (Prentice Hall, 2000).
Larry A. Coldren and Scott W. Corzine, *Diode Lasers and Photonic Integrated Circuits* (Wiley, 1995).

Shamus McNamara, *Operating Principles of Semiconductor Devices* (Department of Electrical and Computer Engineering, University of Louisville, 2016).

Summary

The distinctive property of semiconductors is that they have small band gaps separating the valence and conduction bands. In these materials, a relatively small number of electrons are excited thermally from the valence band into the conduction band creating electron-hole pairs that can conduct a current. The density of free electrons and holes can be modified by adding impurities to the crystal. A semiconductor having impurities that contribute electrons to the conduction band is said to be n-doped or to have donor atoms, while a semiconductor having impurities that contribute holes to the valence band is said to be p-doped or to have acceptor atoms.

A p-n junction is formed by depositing a material with either donors or acceptors upon another material with the opposite kind of doping. Interfaces produced in this way have a preferred direction of current and can be used to produce rectifiers. Solar cells which consist typically of n-doped and p-doped silicon interfaces forming a p-n junction provide a reliable alternative to the burning of fossil fuels. Transistors with two $p-n$ junctions are commonly used to produce amplifiers. The amplification of a signal by a transistor depends upon the fact that the voltage cross two terminals (emitter and collector), which serve as the output terminals, is very sensitive to the voltage across two other terminals (base and emitter), which serve as the input terminals. In contrast to bipolar transistors in which both the majority and minority charge carriers are responsible for current flow, the JFET and MOSFET are majority carrier devices. The JFET and MOSFET function like variable resistors with the effective resistance of the device being determined by the gate as it adjusts the number of carriers in the channel. Modern computer chips with billions of transistors are made almost exclusively with MOSFETs because of the small amount of current they draw.

Questions

1. Describe the general features that are characteristic of the band structure of semiconductors.
2. Why does the conductivity of a semiconductor increase as the temperature increases?
3. Make a sketch showing which of the energy levels of an intrinsic semiconductor are occupied.
4. What is the significance of the Fermi energy?
5. Where does the Fermi energy occur for an intrinsic semiconductor?
6. How does the addition of donor and acceptor dopants affect the position of the Fermi level?
7. The elements *Al* and *Ga* are in column III of the periodic table, while *Si* and *Ge* belong to column IV of the periodic table, and *P* and *As* belong to column V. How would the density of charge carriers in silicon be affected if a small number of *Si* atoms were replaced by *Al* or *P*?
8. How would the density of charge carriers in a *Si* or *Ge* crystal be affected by the introduction of an *As* impurity?
9. Make a sketch showing how electrons and holes move across a p-n junction.
10. What is the preferred direction of current across a p-n junction?
11. Make a sketch showing how the current through in a p-n junction varies when an external voltage is applied.
12. What effect does reverse and forward biasing have upon the size of the space-charge region in a p-n junction?
13. Why does breakdown occur if the reverse biasing voltage across a p-n junction becomes too large?
14. Why do electrons and holes formed in the space-charge region of a solar cell not immediately recombine to produce heat?
15. Make a sketch of a NPN bipolar transistor showing how electrons and hole move when a forward bias is applied between the emitter and the base and a reverse bias is applied between the collector and the base.
16. What property of NPN bipolar transistors makes them suitable for amplifiers?
17. Make a sketch of a N-channel JFET labeling the source, drain, and gate.
18. No current flows through a MOSFET when they are closed. Why is it not possible to eliminate the leakage current from bipolar transistors?

Chapter 11

Semiconductor lasers

Contents

11.1 Motion of electrons in a crystal	253	Suggestions for further reading 275
11.2 Band structure of semiconductors	255	Basic equations 275
11.3 Heterostructures	258	Summary 277
11.4 Quantum wells	262	Questions 277
11.5 Quantum barriers	265	Problems 278
11.6 Phenomenological description of diode lasers	270	

The electrons in the active region of a semiconductor laser are contained within a quantum well in one direction while being able to move freely in the other two directions.

Most semiconductors used to make electronic and optical devices have either the face-centered cubic or hexagonal structures. The face-centered cubic lattice, which is the structure of the gallium arsenide crystal and most other semiconductors used in optical devices, is shown in Fig. 9.6. Aluminum nitride can actually crystallize in either the face-centered cubic or hexagonal structures; however, the semiconductor used in blue lasers has the hexagonal structure.

11.1 Motion of electrons in a crystal

The motion of an electron in a crystal is similar to the motion of a free electron. In one dimension, the momentum and energy of a free electron is given by the equations

$$p = \hbar k$$

and

$$\mathcal{E}^0(k) = \frac{p^2}{2m} = \frac{\hbar^2 k^2}{2m} . \tag{11.1}$$

Since the energy $\mathcal{E}^0(k)$ depends upon the square of k, the energy of a free electron as a function of k has the form of a parabola. This function is illustrated in Fig. 9.23.

According to Bloch's theorem described in Section 9.6, an electron moving in a periodic potential has a wave vector $\mathbf{k}$ associated with it. In terms of $\mathbf{k}$, the wave function of the electron can be written

$$\psi_{\mathbf{k}}(\mathbf{r}) = e^{i\mathbf{k}\cdot\mathbf{r}} u_{\mathbf{k}}(\mathbf{r}),$$

where $u_{\mathbf{k}}(\mathbf{r})$ is a function having the periodicity of the lattice. The wave function describing an electron moving through a lattice is thus very similar to the wave function of a free electron,

$$\psi_{\mathbf{k}}(\mathbf{r}) = A_{\mathbf{k}} e^{i\mathbf{k}\cdot\mathbf{r}} .$$

A free electron is described by a plane wave with amplitude $A_{\mathbf{k}}$, whereas the amplitude of the wave function of an electron in a lattice is modulated by the function $u_{\mathbf{k}}(\mathbf{r})$ having the periodicity of the lattice.

The energy of an electron in a lattice depends upon the vector $\mathbf{k}$ which can be thought of as a vector in the reciprocal lattice. In one-dimension, the proximity cell in reciprocal lattice space consists of the points that are closest to the origin. These k-values are in the range

$$-\pi/a \le k \le \pi/a,$$

Modern Physics with Modern Computational Methods. https://doi.org/10.1016/B978-0-12-817790-7.00018-4

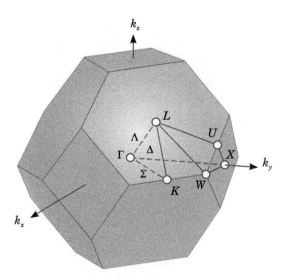

FIGURE 11.1 The Brillouin zone of the face-centered cubic lattice with conventional notation indicating special points and directions.

where the constant a called the lattice constant is the distance between neighboring points in the original lattice. The proximity cell in the reciprocal lattice is called the *Brillouin zone*. As illustrated in Fig. 9.23(B), the energy levels of an electron in a periodic structure have band gaps that occur at the zone boundaries.

Many of the semiconductors we shall consider have a zincblende crystal structure with equal numbers of two atomic species at the sites of two interpenetrating face-centered cubic lattices. The reciprocal lattice of the face-centered cubic lattice is the body-centered cubic lattice illustrated in Fig. 9.3. As discussed in Chapter 9, the proximity cell or Wigner-Seitz cell of the body-centered cubic lattice contains the points which are closer to the origin than to any other lattice point. The Wigner-Seitz cell of the body-centered cubic lattice, which is the Brillouin zone for a face-centered cubic lattice, is shown in Fig. 11.1.

Fig. 11.1 also gives the standard notation for the points of high symmetry of the Brillouin zone. The point Γ is at the origin of **k**-space, while the points indicated by Δ lie along the positive k_y-axis. The k_y-axis meets the zone boundary at the point X, which is in the middle of a square face of the polyhedron. If the k_x- and k_y-axes are denoted [100] and [010], respectively, the points indicated by Σ are in the [110] direction meeting the boundary at K, which is in the middle of an edge shared by two hexagonal faces. The points Λ are along a line in the [111] direction. The [111] line passes through the middle of a hexagonal face of the zone at L.

All of the crystal structures we have considered have a high degree of symmetry, and the symmetry of the crystal can be used to characterize its structure. The face-centered and body-centered crystals have cubic structures, which are invariant with respect to the rotations and reflections that would leave a cube unchanged. One can readily confirm that the face-centered cubic lattice shown in Fig. 9.6 and the Brillouin zone of the face-centered crystal shown in Fig. 11.1 are transformed into themselves, for example, by a rotation by 90° about a line pointing in the [001] direction, by a rotation by 120° about a line pointing in the [111] direction, or by a reflection through a plane passing through the origin perpendicular to the [100] direction. The set of symmetry operations, for which at least one point remains fixed and unchanged in space, is called the *point group* of the crystal. Using this terminology, the zincblende structure and the Brillouin zone of its reciprocal lattice both have cubic point groups. In addition to their point symmetry, a Bravais lattice is also invariant with respect to translations along the directions of its primitive vectors. The set of all symmetry operations – translations, as well as rotations and reflections – that leave a Bravais lattice unchanged is called the *space group* of the lattice.

The symmetry of crystals has important implications for their band structure. If one wave vector in the Brillouin zone can be transformed into another wave vector by a symmetry operation of the crystal, then the electronic energy for these two wave vectors must be the same. For example, the energy corresponding to the wave vector at the point X in Fig. 11.1 must be the same as the energy for the point which is at the center of a square face perpendicular to the x-axis. The point X and the other point just mentioned can be transformed into each other by rotations by 90° about the z-axis.

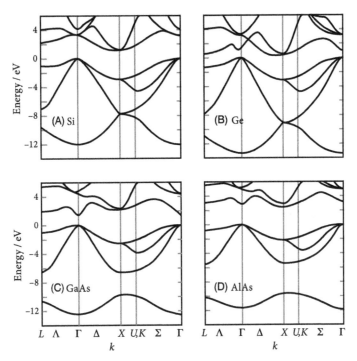

FIGURE 11.2 The band structure for (A) Si, (B) Ge, (C) GaAs, and (D) AlAs. The notation for special points and directions is that shown in Fig. 11.1.

11.2 Band structure of semiconductors

The band structures for Si, Ge, $GaAs$, and $AlAs$ are shown in Fig. 11.2. The horizontal axes of these figures are labeled by the symmetry points of the Brillouin zone shown in Fig. 11.1, and the vertical axes give the energy of an electron at the given point. The energy bands that occur above 0 eV are the *conduction bands*, while the bands occurring below zero are known as the *valence bands*. The valence bands for Si, Ge, $GaAS$, and $AlAs$ shown in Fig. 11.2 all have maxima at the Γ point, and the valence bands of these four semiconductors are all very similar. For the conduction bands, however, small shifts relative to one another change the nature of the lowest minima. Only for $GaAs$ does the minimum in the conduction band occur at the Γ point. For Si and $AlAs$ the minimum in the conduction band occurs in the Δ direction near X, while the minimum in the conduction band of Ge occurs at L.

Information about the energy bands of semiconductors can be obtained by using the open source suite of programs called ABINIT which is distributed under the GNU General Public License. The ABINIT package, which uses a simplified form of the Hartree-Fock equations obtained by approximating the exchange terms in the equations, is sufficiently accurate to provide a realistic description of the band gaps of commonly occurring semiconductors.

11.2.1 Conduction bands

The conduction band of $GaAs$ shown in Fig. 11.2(C) has a minimum at Γ. The form of the band near the minimum is similar to the form of the function for the free-electron energy illustrated in Fig. 9.23 and described by Eq. (11.1). Approximate agreement between the free-electron energy and the calculated band structure near the minimum can be obtained by using an effective mass for the electron. In terms of the effective mass m_e, which is generally measured in units of the true electron mass m_0, the energy of conduction electrons near the minimum at Γ can be written

$$\mathcal{E}(k) = \epsilon_c + \frac{\hbar^2 k^2}{2m_0 m_e}, \tag{11.2}$$

where ϵ_c is the energy of an electron at Γ. For $GaAs$, m_e is equal to 0.067 at $T = 0$ K. The low effective mass of the conduction electrons of $GaAs$ makes them very responsive to changes in the electric field and explains the suitability of $GaAs$ for high-frequency transistors.

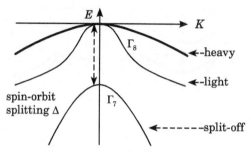

FIGURE 11.3 States near the top of the valence band of a semiconductor with the zincblende structure. The conventional notation is used to designate the states.

The minima in the conduction bands of *Si* and *AlAs* shown in Fig. 11.2(A) and (D) occur in the Δ direction at a point whose k_y value we shall denote by k_0. The minimum at k_0 can be described by the anisotropic function

$$\mathcal{E}(k) = \epsilon_c + \frac{\hbar^2}{2m_0}\left[\frac{(k_y - k_0)^2}{m_l} + \frac{k_x^2}{m_t} + \frac{k_z^2}{m_t}\right], \tag{11.3}$$

where m_l is the *longitudinal* effective mass for motion in the Δ direction perpendicular to the square face of the Brillouin zone at X, and m_t is the *transverse* effective mass for motion in the x- and z-directions parallel to the face of the zone. For *Si*, the effective masses are $m_l = 0.98$ and $m_t = 0.19$. So, electrons near the minimum are much less mobile in the longitudinal y-direction than in the transverse x- and z-directions.

11.2.2 Valence bands

The semiconductors, *Si*, *Ge*, *GaAs*, and *AlAs*, shown in Fig. 11.2 all have valence bands with maxima at the point Γ in the Brillouin zone. In semiconductor devices, electrons are typically excited from the upper part of the valence band into the conduction bands leaving holes in the valence bands. For this reason, the states in the valence bands are called hole states and the states in the conduction band are called electron states. Fig. 11.3 shows a schematic drawing of the states near the top of the valence band of a semiconductor with the zincblende structure. Near Γ, the energy of the two upper branches are roughly described by the equation

$$\mathcal{E}(k) = \epsilon_v - \frac{\hbar^2 k^2}{2m_0 m_e}, \tag{11.4}$$

where m_e is the effective mass of a hole and m_0 is the electron mass. The upper curve in Fig. 11.3, which is denoted by the word *heavy*, corresponds to holes with a large effective mass, while the lower curve, which is denoted by the word *light*, corresponds to holes with a small effective mass. For *GaAs*, the effective mass of *light holes* is $m_l = 0.082$, while the mass of the *heavy holes* is $m_h = 0.51$. Since the effective mass appears in the denominator of Eq. (11.4), the energy of light holes changes more rapidly with k than does the energy for heavy holes. Notice that in Fig. 11.3, the curve for light holes turns more rapidly downward as one moves away from the central point.

We have said previously that each atom in a zincblende structure is joined to four neighboring atoms by covalent bonds and that the orbitals for the electrons forming these bonds are linear combinations of *s*- and *p*-orbitals. The separation between the upper two bands and the *split-off* band shown in Fig. 11.3 is due to the spin-orbit interaction of the *p*-electrons. As seen in Chapter 5, the spin-orbit interaction increases with increasing atomic number Z. The spin-orbit splitting is 0.044 eV for *Si* with $Z = 14$, while for *Ge* with $Z = 32$ the spin-orbit splitting is 0.19 eV. The book *Semiconductors – Basic Data*, which is cited at the end of this chapter, gives a good compilation of electronic and structural data for semiconductors.

11.2.3 Optical transitions

An absorption process in a semiconductor can be thought of as a scattering event involving a photon and an electron in which momentum and energy are conserved. In such a scattering process, we shall denote the initial and final momentum of the electron by $\hbar\mathbf{k}_i$ and $\hbar\mathbf{k}_f$ and the initial and final energy by E_i and E_f. The momentum and energy of the photon is

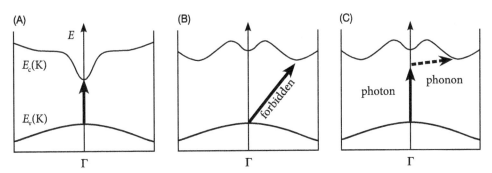

FIGURE 11.4 Optical absorption across the band gap in different semiconductors. (A) A vertical transition for a semiconductor for which the lowest minimum in the conductor occurs at Γ, (B) a forbidden transition across an indirect band gap, and (C) a transition across an indirect band gap with absorption of a photon and a phonon.

given by the equations

$$\mathbf{p} = \hbar\mathbf{q}, \qquad E = hf,$$

where $\mathbf{q}$ is the wave vector of the photon and f is the frequency of the light. The condition that the momentum and energy be conserved in an absorption process leads to the equations

$$k_f = k_i + q, \qquad E_f = E_i + E.$$

The range of the wave vectors corresponding to optical light may be found by using the definition of the wave vector

$$q = \frac{2\pi}{\lambda}.$$

Using this equation, the magnitude of the wave vector of violet light with wave length equal to 400 nm is found to be 1.57×10^{-2} nm^{-1}, while the magnitude of the wave number of red light with wave length equal to 700 nm is 8.98×10^{-3} nm^{-1}. In contrast, the wave vector of an electron near the zone boundary is

$$k = \frac{\pi}{a} \approx \frac{\pi}{0.5 \text{ nm}} \approx 6.3 \text{ nm}^{-1}.$$

Since the wave vector of a photon of optical light is tiny compared to the size of the Brillouin zone, the momentum of the electron involved in an optical transition changes very little and the transition corresponds to a vertical line in the band diagram. Fig. 11.4(A) shows a vertical transition for a semiconductor such as $GaAs$, for which the lowest minimum in the conduction band occurs at Γ. Most light-emitting devices are composed of materials such as $GaAs$ or InP which have vertical band gaps.

We found previously that the lowest minima in the conduction bands of Si, Ge, and $AlAs$ do not occur above the maximum in the valence band at Γ. The lowest point in the conduction bands of Si and $AlAs$ occurs in the Δ-direction near X, while the lowest point in the conduction band of Ge occurs at L. For the semiconductors, Si, Ge, and $AlAs$, transitions between the highest maximum of the valence band and the lowest minima of the conduction band cannot occur since transitions of this kind would involve large changes in the momentum of the electron. A forbidden optical transition involving a change of the momentum is shown in Fig. 11.4(B).

Nonvertical, optical transitions are generally less efficient than vertical transitions and they can only occur if some mechanism provides the change of momentum. One possibility is for the vibrational motion of the crystal to transfer the necessary momentum. Just as photons are quanta of the electromagnetic interaction, *phonons* are quanta of the vibrational energy of the lattice. Optical transitions, which are not vertical, can occur if a phonon changes the momentum of the electron. A transition of this kind for which both a photon and a phonon are absorbed is illustrated in Fig. 11.4(C).

The atoms in a crystal vibrate about their equilibrium positions as would small masses attached to springs. While lattice vibrations disrupt the flow of charge carriers in a semiconductor, thermal energy also excites electrons causing them to populate the conduction band and thus increases the conductivity of the material. Lattice vibrations can also play a significant role in optical transitions.

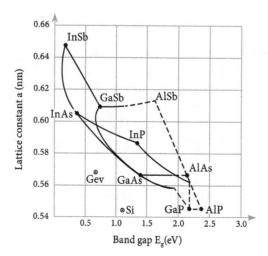

FIGURE 11.5 The lattice constant (a) of a number of semiconductors plotted against the band gap (E_g).

11.3 Heterostructures

Many important semiconductor devices have different layers of materials grown on a common substrate so that the crystal structure is continuous across the interfaces. A wide variety of complex multilayered crystals of this kind– called *heterostructures* – have been grown and are used to make electronic and optical devices.

11.3.1 Properties of heterostructures

For crystals composed of distinct layers to be free of defects, it is necessary that the different layers have the same crystal structure and the separation between neighboring atoms in each layer be similar. The lattice constant (a) of a number of semiconductors is plotted against band gap (E_g) in Fig. 11.5. All of the semiconductors shown in Fig. 11.5 have the zincblende structure. We note that $GaAs$ and $AlAs$ have lattice constants that are very similar and hence heterostructures can be formed from these two semiconductors. The two semiconductors, $GaSb$ and $AlSb$, are also very similar.

The range of possible heterostructures can be increased considerably by using alloys of the various compounds. Crystals composed of the alloy, $Al_xGa_{1-x}As$, for example, have Al or Ga at one set of sites in the zincblende crystal and As at the other sites. The indices, x and $1 - x$, give the fractional number of atoms of the corresponding species in the crystal. For an alloy, the lattice constant can usually be found by linear interpolation. Denoting the lattice constant of $AlAs$ by $a(AlAs)$ and the lattice constant of $GaAs$ by $a(GaAs)$, the lattice constant of $Al_xGa_{1-x}As$ is

$$a(Al_xGa_{1-x}As) = xa(AlAs) + (1 - x)a(GaAs).$$

This is known as Vegard's law. Since the lattice constants of $AlAs$ and $GaAs$ differ by less than 0.15%, it is possible to grow any $Al_xGa_{1-x}As$ alloy on a $GaAs$ substrate without a significant amount of strain.

Other alloys can be grown on an InP substrate. The two alloys, $Ga_{0.47}In_{0.53}As$ and $Al_{0.48}In_{0.52}As$, have the same lattice constant as InP. This can be understood by drawing a line between $InAs$ and $GaAs$ in Fig. 11.5 and also a line between $InAs$ and $AlAs$. Both lines intersect a horizontal line passing through InP. This indicates that one $GaInAs$ alloy and one $AlInAs$ alloy have the same lattice constants as InP.

For heterostructures to perform well in optical and electrical devices, the layers must match each other and be free of impurities. The methods which have been used most successfully in producing high-quality heterostructures are *molecular-beam epitaxy* (MBE) and *metal-organic chemical vapor deposition* (MOCVD). Simplified illustrations of these two methods of producing multilayered surfaces are given in Figs. 11.6(A) and (B). In the MBE apparatus shown in Fig. 11.6(A), the substrate is placed on a rotating holder in a vacuum chamber, and the elements to be incorporated into the layers of the heterostructure are vaporized in furnaces with outlets directed toward the substrate. The molecules emerging from the furnaces form low-density molecular beams in which the molecules travel in straight lines without collisions. In the MOCVD apparatus shown in Fig. 11.6(B), the substrate is placed in a heated chamber through which different gases are passed in a carrier of hydrogen. Group III elements form molecules with methane groups CH_3 in the containers shown near the bottom of the illustration and the resulting molecules then combine with metal hydrides of group V element as in the

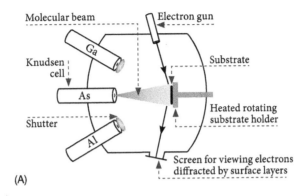

$$(CH_3)_3Ga + AsH_3 \xrightarrow{650°C} GaAs\downarrow + 3CH_4$$

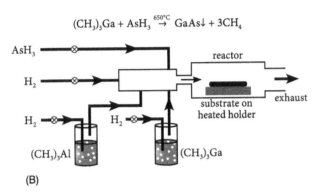

FIGURE 11.6 Simplified drawings of (A) a molecular-beam epitaxy apparatus and (B) a metal-organic chemical vapor deposition apparatus.

following reaction

$$(CH_3)_3Ga + AsH_3 \rightarrow GaAs + 3CH_4 \,.$$

The molecules of group III and group V elements formed in this way are then deposited on the substrate. While both the MBE and MOCVD techniques have produce heterostructures of high quality, the MOCVD technique has a reputation for producing better optoelectronic devices, and has been successfully scaled up for commercial production.

Each layer of a heterostructure has a band structure which depends upon the location of atoms within the layer and also upon the surrounding layers. The structure of the valence and conduction bands for a layer is usually very similar to the structure of the bands in a bulk sample; however, the position of the valence and conduction bands of a layer will generally be shifted at the interface to another layer. Determining the valence and conduction band offsets at interfaces and the ability to tune the offsets to particular optimal values is essential for developing new devices. The band offsets at a junction between two layers are said to be *intrinsic* if they depend only upon the properties of the bulk materials, and they are said to be *extrinsic* if they depend on specific properties of the interface. Certain classes of interfaces have band offsets that can be considered to be intrinsic to a high degree of accuracy, while other classes of interfaces have band offsets that can be modified by manipulating the interfacial structure.

11.3.2   Experimental methods

We consider now the interface between two lattice-matched semiconductors, A and B, with bandgaps, E_g^A and E_g^B. The valence and conduction bands of these two semiconductors are discontinuous across the interface as shown in Fig. 11.7. The maximum of the valence band and the minimum of the conduction band for semiconductor A are denoted by E_v^A and E_c^A, while the maximum of the valence band and the minimum of the conduction band for semiconductor B are denoted by E_v^B and E_c^B. The bending of the bands occurs over distances corresponding to many layers of atoms.

A large number of experimental techniques have been developed to study how the energy levels of semiconductors change at the junction between two materials. Fig. 11.8(A) gives a simplified illustration of an experiment performed by Coluzza and his collaborators to determine the shift of the conduction band at the junction between $AlGaAs$ and $GaAs$. Notice that the bottom of the conduction band, which is indicated by CB in the figure, rises and then dips down at the

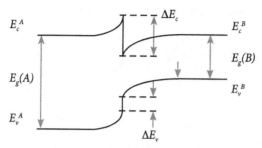

FIGURE 11.7 The interface between two lattice-matched semiconductors, A and B, with band gaps, $E_G{}^A$ and $E_G{}^B$. The maximum of the valence band and the minimum of the conduction band for semiconductors A and B are denoted by $E_v{}^A$ and $E_c{}^A$ and by $E_v{}^B$ and $E_c{}^B$, respectively.

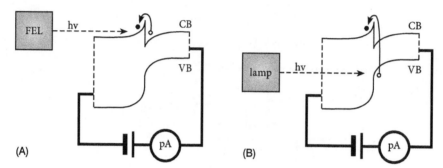

FIGURE 11.8 (A) A illustration of an experiment performed by Coluzza et al. to determine the shift of the conduction band at the junction between AlGaAs and GaAs. (B) A photoemission experiment performed by Haase et al. and Dahmen et al. using conventional light sources.

junction between the two materials just as the conduction band in Fig. 11.7. Electrons to the right of the junction find themselves in a potential well, and these electrons are excited from the energy levels in the well by the light from a free-electron laser operating in the 2–10 μm wavelength range. The change in the energy at the junction between the two materials was found by measuring the external current to determine the maximum wavelength of light for which absorption of the laser light occurred.

Fig. 11.8(B) illustrates a photoemission experiment performed by Haase el al. and Dahmen et al. using a conventional light source. In these experiments, photons with a sufficiently long wavelength travel through the wider-gap material with little absorption and are absorbed by electrons at the top of the valence band of the lower-gap material. Electrons that have absorbed photons makes transitions to the bottom of the conduction band of the wider-gap material. As in our treatment of the photoelectric effect in Chapter 1, the difference between the top of the valence band and the bottom of the conduction band at the junction between the two materials is found by the determining the bias voltage necessary to cut off optical transitions.

Photoemission spectroscopy, which provides very accurate information about the surface properties of solids, is an ideal tool for determining band offsets. While various light sources can be used, synchrotron radiation is particularly well suited to photoemission spectroscopy because it has a well-defined energy profile. If the experimental energy resolution is smaller than the valence-band offset, the valence-band maxima of the two semiconductors in a heterostructure can be identified in the photoelectron energy distribution curves.

Fig. 11.9 shows the photoemission spectra obtained by Katnani and Margaritondo in 1983 using a beam of synchrotron radiation directed upon CdS surfaces on which Si layers of various thicknesses have been deposited. The lowest curve in Fig. 11.9 for a "clean" surface without any silicon shows the distribution of electrons emitted from the valence band of CdS. The other curves for heterostructures with a layer of Si upon a CdS substrate have elongated tails extending beyond the energy range for the clean surface. In the curve corresponding to the 20 Å layer of Si, one can discern double maxima corresponding to photoemission from the valence bands of CdS and Si.

11.3.3 Theoretical methods

Suppose that two semiconductors, A and B, with bandgaps, E_g^A and E_g^B, are brought together and suppose the top of the valence band is displaced at the interface by an amount ΔE_v and the bottom of the conduction band is displaced by an amount ΔE_c. These discontinuities and also the average electrostatic potentials, $\overline{V}_A$ and $\overline{V}_B$, are shown in Fig. 11.10.

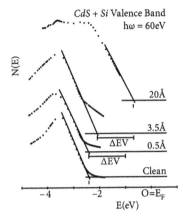

FIGURE 11.9 The photoemission spectra obtained by Katnani and Margaritondo using directed beams of synchrotron radiation upon CdS surfaces on which Si layers had been deposited.

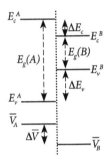

FIGURE 11.10 The top of the valence band and the bottom of the conduction band for two semiconductors, A and B.

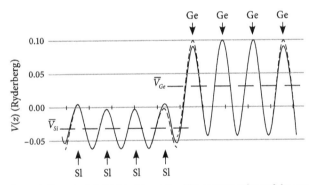

FIGURE 11.11 The oscillation of the electrostatic potential $V(z)$ in Si and Ge. The average values of the potential in the two materials is denoted by $\overline{V}_{Si}$ and $\overline{V}_{Ge}$. The vertical scale is given in units of the binding energy of hydrogen, which is equal to about 13.6 eV.

Since we are only interested now in a region on the order of a few atomic distances, the band bending described earlier is very small, and we have depicted the bands on either side of the interface as being flat.

The electrons in a semiconductor move in a potential field due to the ion-cores and the other electrons. The dependence of the electrostatic potential upon the coordinate corresponding to the direction perpendicular to the interface can be obtained by averaging the potential on the lattice planes parallel to the interface

$$V(z) = \frac{1}{S} \int_S V(x, y, z) dx dy,$$

where the integration extends over the surface S of a lattice plane.

As illustrated in Fig. 11.11 for a Si/Ge interface, the function $V(z)$ rapidly recovers its bulk behavior in each of the two semiconductors as one moves away from the interface. In each of the two regions, the average value of this one-dimensional potential function is denoted by $\overline{V}$. As for the top of the valence band E_v and the bottom of the conduction band E_c, the

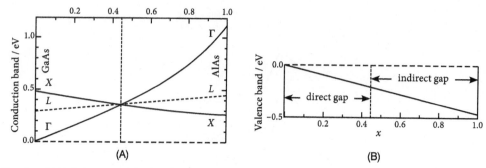

FIGURE 11.12 Energies of (A) the bottoms of the three lowest conduction bands and (B) the top of the valence band of $Al_x Ga_{1-x} As$ as a function of x giving the fraction of Al present. The band gap of GaAs at room temperature is 1.424 eV.

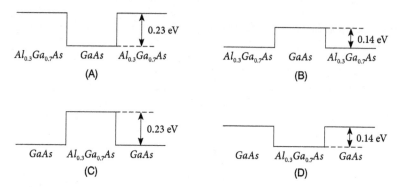

FIGURE 11.13 (A) Quantum well for electrons, (B) quantum well for holes, (C) quantum barrier for electrons, and (D) quantum barrier for holes.

average potential $\overline{V}$ changes discontinuously at the interface, and it is just this discontinuous change in the average potential that is needed to calculate the shift in the valence and conduction bands. Once the shift of the average potential is known, the position of the valence and conduction bands in each material can be obtained by separate bulk calculations.

11.4 Quantum wells

The motion of electrons and holes in heterostructures depends upon the relative position of the conduction and valence bands of adjoining layers. Our developing ability to manipulate the behavior of electrons and holes in heterostructures by growing multilayered surfaces with predictable band structures is called *band engineering*.

Consider the heterostructure formed by sandwiching $GaAs$ as a filling between two layers of $Al_{0.3}Ga_{0.7}As$. For these two materials, the relative values of E_c and E_v can be obtained using Fig. 11.12(A) and (B). Fig. 11.12(A) gives the energies of the three lowest conduction band minima in $Al_x Ga_{1-x} As$ as a function of the relative number of Al atoms specified by the value of x, while Fig. 11.12(B) gives the top of the valence band in $Al_x Ga_{1-x} As$ for any value of x. These two figures were drawn using data obtained by Yu, McCaldin, and McGill in 1992 and Adachi in 1985. For $Al_{0.3}Ga_{0.7}As$ with the fraction x of Al in the alloy $Al_x Ga_{1-x} As$ equal to 0.3, we find that the position of the lowest Γ band goes up by approximately 0.23 eV while the position of the valence band goes down by approximately 0.14 eV. The presence of the Al impurity thus increases the bandgap by $\Delta E_g \approx 0.37$ eV.

For $Al_{0.3}Ga_{0.7}As$–$GaAs$–$Al_{0.3}Ga_{0.7}As$, the alignment of the lowest conduction bands are shown in Fig. 11.13(A) and the alignment of the valence bands are shown in Fig. 11.13(B). Notice that the electrons and holes in $GaAs$ are confined by the surrounding layers of $AlGaAs$. Electrons near the bottom of the conduction band and holes near the top of the valence band in $GaAs$ are not free to move because there are no available states of the same energy in $AlGaAs$. We can describe this situation by saying that the electrons and holes in $GaAs$ are confined within a *quantum well*.

Consider now the bands shown in Fig. 11.13(C) and (D). These bands, which were obtained by interchanging the band minima and maxima in the previous example, correspond to $Al_{0.3}Ga_{0.7}As$ being a filling between two layers of $GaAs$. Since the electrons near the bottom of the conduction band and holes near the top of the valence band in $GaAs$ cannot move into available states in the central region, the $GaAs$–$Al_{0.3}Ga_{0.7}As$–$GaAs$ structure provides an example of a *quantum barrier*. For the two structures shown in Figs. 11.13(A) and (B) and in (C) and (D), the electrons and holes are treated the same. The central region in Figs. 11.13(A) and (B) are quantum wells for both electrons and holes, while

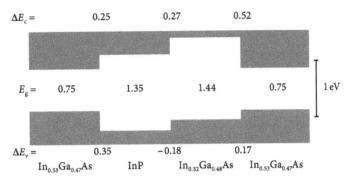

$\Delta E_c =$ 0.25   0.27   0.52

$E_g =$ 0.75   1.35   1.44   0.75   1 eV

$\Delta E_v =$ 0.35   −0.18   0.17

$In_{0.53}Ga_{0.47}As$ InP $In_{0.52}Ga_{0.48}As$ $In_{0.53}Ga_{0.47}As$

FIGURE 11.14 Junctions between the semiconductors, $In_{0.53}Ga_{0.47}As$, InP, and $In_{0.52}Al_{0.48}As$, and $In_{0.53}Ga_{0.47}As$.

the central region in Fig. 11.13(C) and (D) are quantum barriers for both electrons and holes. Alignments of this kind, for which the energy at the top of the valence band goes up and the energy at the top of the valence band goes down or visa versa, are called *Type I* or *straddling alignments*.

Other possible alignments are shown in Fig. 11.14. Notice that the semiconductors, $In_{0.53}Ga_{0.47}As$ and InP, are related to each other in the same way as the semiconductors, $GaAs$ and $Al_{0.3}Ga_{0.7}As$. The sandwich, InP-$In_{0.53}Ga_{0.47}As$-InP, is a quantum well for both electrons and holes, while the sandwich, $In_{0.53}Ga_{0.47}As$-InP-$In_{0.53}Ga_{0.47}As$, is a barrier for both electrons and holes. The junction between the semiconductors $In_{0.53}Ga_{0.47}As$ and InP in Fig. 11.14 is thus also an example of type I or straddling alignment. Another example of straddling alignment is the junction of the two semiconductors, $In_{0.53}Ga_{0.47}As$ and $In_{0.52}Al_{0.48}As$.

The junction between the semiconductors, InP and $In_{0.52}Al_{0.48}As$, in Fig. 11.14 has different characteristics. Electrons near the bottom of the conduction band of InP are not free to move into available states in $In_{0.52}Al_{0.48}As$ and holes near the top of the valence band of $In_{0.52}Al_{0.48}As$ are not free to move into available states in InP. Electrons and holes are thus not on the same footing at this junction. The sandwich, $In_{0.52}Al_{0.48}As$-InP-$In_{0.52}Al_{0.48}As$, is a quantum well for electrons and a quantum barrier for holes, while the sandwich, InP-$In_{0.52}Al_{0.48}As$-InP, is a quantum barrier for electrons and a quantum well for holes. Alignments of this kind, for which the energy at the bottom of the conduction band and the energy at the top of the valence band both go up or both go down, are called *type II* or *staggered* alignments.

The remaining possibility is that the two band gaps do not overlap at all giving *type III* or *broken-gap alignment*. The interface between $InAs$ and $GaSb$ provides an example of an alignment of this kind. For an $InAs$-$GaSb$ junction, the conduction band of $InAs$ overlaps the valence band of $GaSb$. Electrons and holes then move freely across the interface until they are restrained by the electrostatic field of the accumulated charge as for $p-n$ junctions.

The quantum wells used in the fabrication of semiconductor devices consist of the layers of heterostructures. We shall denote the distance measured along a line perpendicular to the planes of atoms by the coordinate z and regard z as a dimensionless coordinate giving the distance in nanometers. The finite well described in Chapters 2 and 3, consists of a layer of $GaAs$ sandwiched between two layers of $Al_{0.3}Ga_{0.7}As$. The well extends from $x = -5$ nm to $x = +5$ nm and has the depth $V_0 = 0.3$ eV. Because we want to solve the Schrödinger equation for values inside and outside the well, the z-coordinate extends beyond the edge to the well.

We shall denote the free electron mass by m_0. We then denote the effective mass of an electron within the well by $m_0 m_W$ and the effective mass of an electron in the barrier forming the wall of the well is $m_0 m_B$. The Schrödinger equation for an electron inside the well is

$$-\left(\frac{\hbar^2}{2m_0 m_W}\right)\frac{d^2\psi}{dz^2} = E\psi$$

and the Schrödinger equation for an electron outside the well is

$$-\left(\frac{\hbar^2}{2m_0 m_B}\right)\frac{d^2\psi}{dz^2} + V_0\psi = E\psi.$$

As we have done in Chapters 2 and 3, we can multiply the above equations from the left by $-2m_0/\hbar^2$ and express the equations in dimensionless units in which z gives the distance normal to the planes of the lattice in nanometers and the energy is in electron Volts. The Schrödinger equation for an electrons inside the well may then be written

$$-\frac{1}{m_W}\frac{d^2\psi}{dz^2} = \epsilon\psi \tag{11.5}$$

where $m_W m_0$ is the effective mass of the electron in the well and ϵ is the energy in electron volts. The Schrödinger equation in the $Al_{0.3}Ga_{0.7}As$ barrier can be written

$$-\frac{1}{m_B}\frac{d^2\psi}{dz^2} + E_0\mathcal{V}\psi = E_0\epsilon\psi, \tag{11.6}$$

where $m_B m_0$ is the effective mass in the barrier. ϵ and $\mathcal{V}_0$ are the energy and the depth of the well in electron volts. The value of the constant E_0 is given Eq. (2.28) of Chapter 2. With the factor $m_W = 0.067$ included E_0 has the value 1.759. However, in Eqs. (11.5) and (11.6) with the factors m_W and m_B written separately

$$E_0 = \frac{1.759}{0.067} = 26.25$$

As before, we shall require that the wave function be continuous across the edge of the well.

$$\psi(z=L_-) = \psi(z=L_+) \tag{11.7}$$

To find the appropriate continuity condition involving the derivatives, we anticipate a result obtained in Section 13.3 of Chapter 13. The Schrödinger equation may be used to obtain a continuity equation for the probability current and density, which requires that the flow of probability into a volume is equal to the rate the probability that the particle is inside the volume increases. In one-dimension, the probability current is

$$j(z) = \frac{\hbar}{2mi}(\psi^*\frac{d\psi}{dz} - \psi\frac{d\psi^*}{dz}).$$

The correct continuity condition involving the derivatives to impose is that the probability current $j(z)$ be continuous across the interface. Otherwise the interface will absorb or emit electrons. If the wave functions are continuous across the boundary, the current will be continuous if the derivatives satisfy the following condition

$$\frac{1}{m_W}\frac{d\psi}{dz}\Big|_{z=L-} = \frac{1}{m_B}\frac{d\psi}{dz}\Big|_{z=L+}. \tag{11.8}$$

The eigenvalue equation defined by Eq. (11.5) and (11.6) with the boundary condition (11.8) can be solved using the inverse-iteration method. We first discretize Eqs. (11.5) and (11.6) using the spline collocation method described in Chapter 3 to obtain a matrix equation of the form

$$\mathbf{C}_{mat}\mathbf{u} = \epsilon E_0\mathbf{B}_{mat}\mathbf{u}. \tag{11.9}$$

We then subtract a term $\sigma E_0\mathbf{B}_{mat}\mathbf{u}$ from both sided of Eq. (11.9) to obtain

$$[\mathbf{C}_{mat} - \sigma E_0\mathbf{B}_{mat}]\mathbf{u} = (\epsilon - \sigma)E_0\mathbf{B}_{mat}\mathbf{u}. \tag{11.10}$$

The constant σ is typically chosen to be slightly less that the eigenvalue of interest. The eigenvalue, $\epsilon - \sigma$, will then be the smallest eigenvalue and the inverse power method can be expected to converge rapidly.

Using an approximate eigenfunction $\mathbf{u}$ and the corresponding eigenvalue

$$\epsilon = \int \mathbf{u}^*\mathbf{C}_{mat}\mathbf{u}dz,$$

we form the right-hand side of Eq. (11.10) and solve the equation to obtain a new approximation of the eigenfunction. In doing this the boundary condition (11.8) is taken into account by bringing the coefficients for the spline functions at the edge of the well over to the right-hand side of Eq. (11.10). The equation is then solved and the process continued until the calculated eigenvalues produced by the iterations approach a limiting value.

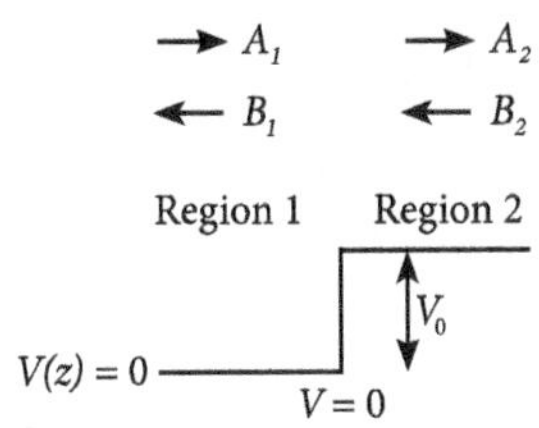

FIGURE 11.15 A potential step at $z = 0$ with potential energy $V = 0$ for negative z and $V = V_0$ for positive z. The arrows directed toward and away from the step correspond to waves moving in the positive and negative z directions.

11.5 Quantum barriers

We shall now consider how the plane waves associated with free electrons and with electromagnetic radiation are scattered by barriers. The transport and reflection of electrons is essential for understanding tunneling phenomena, while the transport and reflection of electromagnetic waves are of fundamental importance for the applications to lasers we shall soon consider.

11.5.1 Scattering of electrons by potential barriers

Many of the general features of electron scattering can be found in the scattering of an electron by the potential step illustrated in Fig. 11.15. As in Chapter 3, we shall call the region to the left of the step region 1, and the region to the right of the step region 2, and we first consider the case for which the energy E is greater than the step height V_0. Unlike the treatment given in Chapter 3, though, we shall use the variable z to denote the coordinate of the electron.

The solution of the Schrödinger equation in the regions to the left and right of the barrier were described in Chapter 3. To the left of the step, we shall here denote the wave function

$$\psi_1(z) = A_1 e^{ik_1 z} + B_1 e^{-ik_1 z}, \tag{11.11}$$

where

$$k_1 = \sqrt{\frac{2mE}{\hbar^2}}, \tag{11.12}$$

where A_1 and B_1 denote the amplitudes of the two waves in Region 1. The first term on the right-hand side of Eq. (11.11) describes the incident electron moving in the positive z-direction, while the second term describes the reflected electron moving in the negative z-direction. The parameter k_1 depends upon the energy of the electron. To the right of the step, the electron is described by the wave function

$$\psi_2(z) = A_2 e^{ik_2 z} + B_2 e^{-ik_2 z}, \tag{11.13}$$

where A_2 and B_2 gives the amplitudes of the waves in region 2 and k_2 is defined by the equation

$$k_2 = \sqrt{\frac{2m(E - V_0)}{\hbar^2}}. \tag{11.14}$$

The relations between the coefficients, A_1 and B_1 for the wave in Region 1 and the coefficients, A_2 and B_2 for the waves in Region 2 can be expressed in terms of matrices. As discusses in Chapter 3, the advantage of expressing the relation between the incoming and outgoing amplitudes in matrix form is that the transmission and reflection coefficients for complex systems can then be calculated by multiplying the matrices for the individual parts.

The T-matrix is defined by the equation

$$\begin{bmatrix} A_1 \\ B_1 \end{bmatrix} = \begin{bmatrix} T_{11} & T_{12} \\ T_{21} & T_{22} \end{bmatrix} \begin{bmatrix} A_2 \\ B_2 \end{bmatrix}. \tag{11.15}$$

Equations for the coefficients, A_1 and B_1, in terms of the elements of the T-matrix and the coefficients, A_2 and B_2, can be obtained by multiplying the matrix in Eq. (11.15) times the column vector to its right and equating each component of the resulting vector with the corresponding component of the vector on the left. We obtain

$$A_1 = T_{11} A_2 + T_{12} B_2 \tag{11.16}$$

$$B_1 = T_{21}A_2 + T_{22}B_2 \qquad (11.17)$$

With the definition (11.15) of the T-matrix, the order of the matrix-vector operation is the same as the order of the corresponding quantities in Fig. 11.15. The incident amplitudes are to the left in both Fig. 11.15 and Eq. (11.15), and the out-going amplitudes are to the right in both the figure and the equation. This is a convenient convention for describing complex scattering processes.

The transmission coefficient for the barrier can be expressed in terms of the elements of the T-matrix. For the potential step, we denote the amplitudes of the incoming and outgoing waves on the left by A_1 and B_1, and we denote the amplitudes of the waves leaving and approaching the barrier on the right by A_2 and B_2. To obtain the transmission and reflection coefficients, we set the incoming amplitude on the right B_2 equal to zero in Eqs. (11.16) and (11.17) to obtain

$$A_1 = T_{11}A_2. \qquad (11.18)$$
$$B_1 = T_{21}A_2. \qquad (11.19)$$

Eq. (11.18) may be used to obtain an equation for the transmission amplitude t, which is the ratio of the transmitted and incident amplitudes. We have

$$t = \frac{A_2}{A_1}\bigg|_{B_2=0} = \frac{1}{T_{11}}. \qquad (11.20)$$

An equation for the reflection amplitude r can be obtained by dividing Eq. (11.19) by Eq. (11.18) to get

$$r = \frac{B_1}{A_1}\bigg|_{B_2=0} = \frac{T_{21}}{T_{11}} \qquad (11.21)$$

Eqs. (11.20) and (11.21) give the transmission and reflection amplitudes for a barrier in terms of the elements of the T matrix.

11.5.2 Light waves

The radiation field within a laser cavity is due to traveling plane waves of the form

$$\mathbf{E} = \hat{\mathbf{e}}E_0 e^{i(\bar{k}z - \omega t)},$$

where E_0 is the amplitude of the oscillations of the electric field, $\hat{\mathbf{e}}$ is a unit vector pointing in the direction of the field, and $\bar{k}$ is the wave vector. We denote the wave vector by $\bar{k}$ to allow for the possibility of absorption or amplification of light. The wave vector can be written

$$\bar{k} = k - ik_i, \qquad (11.22)$$

where k and k_i are the real and imaginary parts of the wave vector $\bar{k}$. As described in the introduction, waves traveling in the positive and negative directions have characteristic dependences upon the spatial coordinate, which we here denote by z. A wave moving in the positive z-direction may be described by the function

$$u_+(z) = Ae^{i\bar{k}z}$$

while a wave moving in the negative z-direction is described by the function

$$u_-(z) = Be^{-i\bar{k}z}.$$

If the wave vector $\bar{k}$ is real, the functions, $u_+(z)$ and $u_-(z)$, corresponds to waves of constant amplitude. The effect of the imaginary part of $\bar{k}$ can be understood by substituting Eq. (11.22) into the above equation for $u_+(z)$, to obtain

$$u_+(z) = Ae^{k_i z}e^{ikz}.$$

A positive value of k_i amplifies a wave moving in the positive x-direction and attenuates waves moving in the negative direction, while a negative value of k_i attenuates waves moving in the positive x-direction and amplifies waves moving in the negative direction.

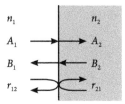

FIGURE 11.16 A dielectric interface between two materials showing the amplitudes of the waves in the two regions and the indices of refraction.

11.5.3 Reflection and transmission by an interface

Light-emitting diodes and semiconductor lasers typically have gratings or etched ridges in which light is reflected or transmitted many times at the interface between different materials. Fig. 11.16 illustrates the amplitudes of light at the interface of two semiconductors with indices of refraction, n_1 and n_2. In this section, we shall derive expressions for the transmission and reflections amplitudes, t and r, for a single interface between two dielectric materials, and we shall then consider more complex semiconductor devices with several dielectric interfaces.

The relation between the transmission and reflection amplitudes and components of the T-matrix were already obtained during the course of our discussion of the square barrier. In applying these equations to a problem in which light traveling in the semiconductor denoted by 1 is incident upon the interface of a second semiconductor denoted by 2, we denote the transmission and the reflection amplitudes by t_{12} and r_{12}, respectively. According to Eqs. (11.20) and (11.21), the transmission and reflection amplitudes are related to the elements of the T-matrix by the equations

$$t_{12} = \frac{A_2}{A_1}|_{B_2=0} = \frac{1}{T_{11}}$$

$$r_{12} = \frac{B_1}{A_1}|_{B_2=0} = \frac{T_{21}}{T_{11}}.$$

These equations were derived by setting the amplitude (B_2) of the wave incident from the right equal to zero and finding the ratio of the transmitted and reflected amplitudes to the amplitude of the wave incident from the left. Similarly, equations relating the amplitudes of transmitted and reflected waves to the amplitude of light incident upon the interface from the right can be derived by setting the amplitude (A_1) of the wave incident from the left equal to zero. This leads to the equations

$$t_{21} = \frac{B_1}{B_2}|_{A_1=0} = \frac{\det \mathbf{T}}{T_{11}} \tag{11.23}$$

$$r_{21} = \frac{A_2}{B_2}|_{A_1=0} = -\frac{T_{12}}{T_{11}}. \tag{11.24}$$

A discussion of the properties of T-matrices can be found in the book by Davies cited at the end of this chapter. Here we mention only that the determinant of the T-matrix can be shown to be equal to one.

Solving the above equations for the individual elements of the T-matrix, we obtain the following representation of the T-matrix in terms of the transmission and reflection coefficients

$$\mathbf{T} = \frac{1}{t_{12}} \begin{bmatrix} 1 & -r_{21} \\ r_{12} & t_{12}t_{21} - r_{12}r_{21} \end{bmatrix} \tag{11.25}$$

The boundary conditions upon the electromagnetic fields at the interface between the two materials may be used to derive equations for the ratios of the scattering amplitudes, and the resulting equations in turn help us simplify the T-matrix. The ratio of the reflected and the incident amplitudes of light incident upon the interface between the first and second materials is

$$\frac{B_1}{A_1}|_{B_2=0} = \frac{n_1 - n_2}{n_1 + n_2},$$

where n_1 and n_2 are the indices of refraction in the two materials. Denoting the ratio of the two amplitudes by r_{12}, the above equation becomes

$$r_{12} = \frac{n_1 - n_2}{n_1 + n_2}.$$

This equation is antisymmetric with respect to 1 and 2 and hence implies

$$r_{21} = -r_{12}.$$

For light that is normally incident upon the interface between the first and second materials, the transmission coefficient is

$$t_{12} = \frac{2(n_1 n_2)^{1/2}}{n_1 + n_2}.$$

Since the right-hand side of this last equation is symmetric with respect to an interchange of 1 and 2, the transmission coefficient for the two media are equal

$$t_{21} = t_{12}.$$

We shall generally suppose that power is conserved at the interface implying that

$$r_{12}^2 + t_{12}^2 = r_{21}^2 + t_{21}^2 = 1.$$

The above relations between the transmission and reflection coefficients enable us to simplify the form of the T-matrix given by Eq. (11.25). We thus obtain

$$\mathbf{T} = \frac{1}{t_{12}} \begin{bmatrix} 1 & r_{12} \\ r_{12} & 1 \end{bmatrix}. \tag{11.26}$$

To describe devices made up of layers of semiconducting materials, we must be able to describe the effect of a translation upon the waves. Our treatment here will be analogous to our discussion in Chapter 3 of the effect of translations upon the waves associated with electrons. Within a block of a dielectric material, an electromagnetic wave can be described by functions of the form

$$\psi(z) = A e^{i\bar{k}z} + B e^{-i\bar{k}z}, \tag{11.27}$$

where A is the amplitude of a wave propagating to the right through the media and B is the amplitude of a wave propagating to the left. To find the effect upon the above function (11.27) of a translation by a distance L, we substitute $z + L$ for z in Eq. (11.27) to obtain

$$\psi(z + L) = \left(A e^{i\bar{k}L} \right) e^{i\bar{k}z} + \left(B e^{-i\bar{k}L} \right) e^{-i\bar{k}z}.$$

The translated wave function has $A e^{i\bar{k}L}$ in place of A and $B e^{-i\bar{k}L}$ in place of B. The effect of the translation upon the wave function can thus be described by the diagonal matrix

$$T_L(\bar{k}) = \begin{bmatrix} e^{i\bar{k}L} & 0 \\ 0 & e^{-i\bar{k}L} \end{bmatrix}.$$

11.5.4 The Fabry-Perot laser

The simplest kind of laser, which can be built of semiconductor materials, has a cavity consisting of a layer of one semiconductor material sandwiched between layers of a second semiconductor material. A laser of this kind is called a *Fabry-Perot* laser because it resembles a Fabry-Perot interferometer with an optical cavity closed off on both ends by mirrors.

The T-matrix for the cavity can be obtained as for the barrier considered previously by multiplying the T-matrix for the interface on the left, the T-matrix for transmission across the cavity, and the T-matrix for the interface on the right. We have

$$\mathbf{T} = \frac{1}{t_{12}} \begin{bmatrix} 1 & r_{12} \\ r_{12} & 1 \end{bmatrix} \begin{bmatrix} e^{i\bar{k}L} & 0 \\ 0 & e^{-i\bar{k}L} \end{bmatrix} \frac{1}{t_{21}} \begin{bmatrix} 1 & r_{21} \\ r_{21} & 1 \end{bmatrix}. \tag{11.28}$$

According to Eq. (11.20), the transmission amplitude of the cavity is equal to the inverse of the T_{11} matrix element, which may be obtained by evaluating the matrix product in Eq. (11.28). MATLAB® Function 11.1 given below evaluates the transmission coefficient of a Fabry-Perot laser for a particular value of the reflection coefficient r_{21} and a particular wavelength of light λ. For this program, the wave vector $\bar{k}$ is assumed to be real.

MATLAB Function 11.1

This MATLAB function calculates the transmission coefficient of a Fabry-Perot laser for a particular value of the reflection coefficient r_{21} and a particular wavelength of light λ.

```
function T = laserfp(r21,lambda)

L=200;
k=2*pi/lambda;
r12=-r21;
t21=sqrt(1-r21^2);
t12=t21;
A=(1/t12)*[1 r12; r12 1];
B=(1/t21)*[1 r21; r21 1];
C=[exp(i*k*L) 0; 0 exp(-i*k*L)];
M=A*(C*B);
t=1/M(1,1);
T = abs(t)^2;T = abs(t)^2;
```

The first line of the program gives the length to the Fabry-Perot cavity, which we have taken to be 200 nm. The second line of the program uses the value of λ given in the input to calculate the wave number of the light, and the following three lines of the program uses the value of r_{21} given in the input to calculate the values of r_{12}, t_{21}, and t_{12}. As for the transmission of electrons through a potential barrier, the matrix A determines how the light wave is effected by the interface between the first and the second semiconducting materials, the matrix B determines how the wave is effected by the interface between the second and the first materials, and the matrix C translates the wave along the length of the cavity. The program evaluates the product of the three matrices and evaluate the transition amplitude, which is equal to the inverse of the T_{11} matrix element. The last line of the program calculates transmission coefficient by taking the absolute value squared to the transmission amplitude. MATLAB Program 11.1 given below plots the transmission coefficient of the Fabry-Perot laser as a function of $2kL/2\pi$.

MATLAB Program 11.1

This MATLAB program plots the transmission coefficient of a Fabry-Perot laser as a function of $2kL/2\pi$ for the reflection coefficients, $r_{21} = 0.3, r_{21} = 0.565$, and $r_{21} = 0.9$.

```
for n=1:500
    lambda=100+n-1;
    X(n)=400/lambda;;
    T1(n)=laserfp(0.3,lambda);
    T2(n)=laserfp(0.565,lambda);
    T3(n)=laserfp(0.9,lambda);
end
plot(X,T1,X,T2,X,T3);
```

Fig. 11.17 gives a plot of the power transmitted through the Fabry-Perot cavity as function of $x = 2kL/2\pi = 2L/\lambda$. Notice that the power has district maximas for integral values of x. When x is equal to an integer, the length of the cavity is an integral multiple of half wavelengths and a wave transmitted along the cavity is in phase with a wave that passes through the cavity a second time after reflections from the back and front surfaces. These two waves are in phase and the power of the laser then has a maximum.

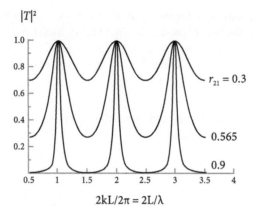

$|T|^2$

$r_{21} = 0.3$

0.565

0.9

$2kL/2\pi = 2L/\lambda$

FIGURE 11.17 The power transmitted through the Fabry-Perot cavity as a function of kL. The horizontal scale is $2kL/2\pi$ and the vertical scale is transmitted power $|T|^2$.

11.6 Phenomenological description of diode lasers

Light emitting processes were considered in some detail in Chapters 4. Fig. 4.12 show the three kinds of radiative processes that can occur for atoms: spontaneous emission in which an atom makes a transition from an upper level to a lower level emitting a quantum of radiation, absorption in which light is absorbed from the radiation field and an atom makes a transition from a lower to an upper level, and induced or stimulated emission in which an atom is induced by the radiation field to make a transition from an upper to a lower level. As discussed in the first section of Chapter 6, laser amplification can occur when the number of atoms in the upper state is greater than the number of atoms in the lower state and the rate of stimulated emission is greater than the rate of absorption. Such a population inversion is created for atoms by pumping the atoms up to higher levels with another radiative source and allowing them to fall down to the lower of the two laser levels.

The active region of both light-emitting diodes and lasers is sandwiched between n- and p-doped layers. As shown in Fig. 11.18, forward biasing causes electrons from the n-doped layer and holes from the p-doped layer to enter the active region. Light is then emitted as electrons near the bottom of the conduction band recombine with holes near the top of the valence band. Population inversion is thus created very efficiently in semiconductor lasers by the ordinary movement of charge carriers.

All light-emitting devices have some way of confining the current within the active region. The current flow in the simple stripe laser shown in Fig. 11.19(A) is limited by the size of the contact area, while the charge carriers and the photons in the more sophisticated device shown in Fig. 11.19(B) are confined to a ridge etched into the heterostructure. Current is confined to the ridge by reverse biasing and by doping the regrown material around the ridge to be semiinsulating.

As for atomic transitions studied in Chapters 4 and 6, electrons in the active region of a semiconductor device can combine spontaneously with holes to emit light, or the electrons can be stimulated to recombine with holes by radiation. Whether spontaneous emission or stimulated emission dominates depends upon the density of photons in the active region.

11.6.1 The rate equation

We now derive a coupled set of equations for the density of charge carriers and photons. The rate of change of the density of charge carriers in the active region is equal to the difference between the rate charge carriers are injected into the active region (R_{gen}) and the rate charge carriers recombine (R_{rec}). Denoting the density of charge carriers by N, we have

$$\frac{dN}{dt} = R_{gen} - R_{rec}. \tag{11.29}$$

The current flowing into the active region will be denoted by I. The fractional number of charge carriers arriving in the active region will be denoted by η_i, and the charge of the carriers will be denoted by q. With this notation the number of charge carriers injected into the active region per time and per volume V is

$$R_{gen} = \frac{\eta_i I}{q V}.$$

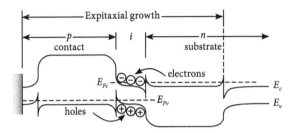

FIGURE 11.18 Schematic drawing of a light-emitting diode or a laser in which the active region is sandwiched between *n*- and *p*-doped layers.

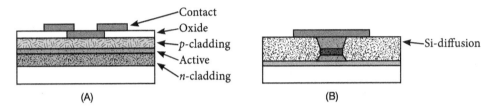

FIGURE 11.19 (A) A simple strip laser in which the current flow is limited by the size of the contact area. (B) A more sophisticated device in which the charge carriers and the photons are confined to a ridge etched into the heterostructure.

The recombination rate depends upon a number of different recombination processes. In Chapter 4, we found that the rate transitions occur from an initial state to a final state depends upon the number of electrons in the initial state. The various recombination processes are thus all functions of the carrier density (N). We may write

$$R_{rec} = R_{sp}(N) + R_{nr}(N) + R_l(N) + R_{st}(N), \qquad (11.30)$$

where $R_{sp}(N)$ is the rate of spontaneous recombination, $R_{nr}(N)$ is the rate of nonradiative recombination, $R_l(N)$ is the rate of carrier leakage, and $R_{st}(N)$ is the rate of stimulated recombination. The first three terms on the right-hand side of Eq. (11.30) correspond to processes in which charge carriers are lost without contributing very much to the laser output. As discussed in our treatment of lasers in Chapter 6, spontaneous emission processes are random in nature making only a small contribution to the light emitted by a laser beam. Nonradiative recombination only serves to heat the medium, while the current, which leaks out, is simply lost. For this reason, we shall refer to the three processes of spontaneous recombination, nonradiative recombination, and current leakage, as *dissipative processes* and characterize them by the mean *carrier lifetime* τ defined by the equation

$$\frac{N}{\tau} = R_{sp}(N) + R_{nr}(N) + R_l(N). \qquad (11.31)$$

The term $R_{st}(N)$ in Eq. (11.30) describes absorption and stimulated emission. In Chapter 4, we found that the rates of absorption and stimulated emission are proportional to the density of the radiation field. Denoting the density of photons by N_p, the total rate of absorption and stimulated emission can be written

$$R_{st}(N) = v_g g(N) N_p, \qquad (11.32)$$

where v_g is the group velocity of radiation within the laser cavity and $g(N)$ is a function of the carrier density that characterizes the laser gain. Since $R_{st}(N)$ has units of $s^{-1}m^{-3}$, and the units of v_g and N_p are m/s and m^{-3}, respectively, the function $g(N)$ has units of m^{-1}. $g(N)$ is the fractional gain of laser light along the laser cavity in units of inverse meters.

Using Eqs. (11.31) and (11.32), Eq. (11.30) can be written

$$R_{rec} = \frac{N}{\tau} + v_g g N_p,$$

and Eq. (11.29) becomes

$$\frac{dN}{dt} = \frac{\eta_i I}{q V} - \frac{N}{\tau} - v_g g N_p. \qquad (11.33)$$

This last equation, which describes how the density of charge carriers changes with time, is called the *rate equation*.

The gain of the laser is a measure of how much the number of photons in the laser cavity is increased by the process of electron/hole recombination. Since each charge carrier can combine with a hole to increase the number of photons in the cavity, the gain of the laser depends upon the density of charge carriers N. As the current increases more and more charge carriers are injected into the active region of the laser, the gain increases. However, one would expect there to be a limit to the number of charge carriers the laser cavity can accommodate. When that limit is reached, one would expect the density of charge carriers and the gain of the laser to remain constant.

The density of photons in a laser cavity depends upon the same physical processes that determine the rate equation for charge carriers. The rate equations for photons can be written

$$\frac{dN_p}{dt} = \left[\Gamma v_g g(N) - \frac{1}{\tau_p} \right] N_p + \Gamma \beta_{sp} R_{sp}(N). \tag{11.34}$$

The first term in this equation is the contribution to the density of photons in the active region due to stimulated recombination. Each time an electron-hole pair is stimulated to recombine, an electron is lost and a photon is generated. This explains the difference in signs of the stimulated recombination terms in Eqs. (11.33) and (11.34). Since the volume occupied by photons V_p is usually larger than the volume of the active region occupied by electrons V, the rate the density of photons in the active region increases due to stimulated recombination is V/V_p times R_{st}. The factor Γ, which is equal to V/V_p, is generally called the *confinement factor*.

The second term on the right-hand side of Eq. (11.34) describes the loss of photons in the active region due to miscellaneous scattering and absorption processes. In the absence of generating terms in the rate equation, photons would decay with a rate that is proportional to the number of photons present. Hence, the decay of photons is exponential with a lifetime τ_p. The last term on the right-hand side of Eq. (11.34) describes the small contribution to the density of photons in the active region due to spontaneous recombination. The proportionality constant β_{sp} is the reciprocal of the number of optical modes in the bandwidth of the spontaneous emission.

The steady-state solutions of Eqs. (11.33) and (11.34) can be obtained by setting the time derivatives in these equations equal to zero. The requirement that the number of photons in the active region of the laser is constant is achieved by setting the derivative on the left-hand side of Eq. (11.34) equal to zero and solving for N_p to obtain

$$N_p = \frac{\Gamma \beta_{sp} R_{sp}(N)}{1/\tau_p - \Gamma v_g g(N)}. \tag{11.35}$$

The density of photons present under steady-state conditions thus depends upon the number of photons produced by spontaneous recombination ($\Gamma \beta_{sp} R_{sp}$) and the difference between the photon decay rate ($1/\tau_p$) and the rate photons are produced by stimulated emission ($\Gamma v_g g$). As a semiconductor laser approaches threshold, the denominator in Eq. (11.35) becomes small and the density of photons and the laser power builds up.

Similarly, the requirement that the number of charge carriers in the active region of the laser is constant can be achieved by setting the left-hand side of Eq. (11.33) equal to zero and rearranging terms to obtain

$$\frac{\eta_i I}{qV} - \frac{N}{\tau} = v_g g(N) N_p. \tag{11.36}$$

The left-hand side of this last equation is the difference between the rate charge carriers are injected into the active region and the rate charge carriers are lost due to spontaneous and nonradiative recombination and to leakage, while the right-hand side of the equation equals the rate charge carriers are lost due to stimulated recombination. At equilibrium, the net rate charge carriers are injected into the active region is equal rate charge carriers are consumed in the stimulated recombination process.

To gain some understanding of the steady-state solutions of the rate equations, we now consider the limiting form of the equations below threshold and then see how the carrier density and laser output change as the current increases.

11.6.2 Well below threshold

When the rate current is injected into the active region is very small, the rate photons are produced by stimulated recombination will be much less that the photon decay rate, and the supply of photons is entirely due to spontaneous recombination.

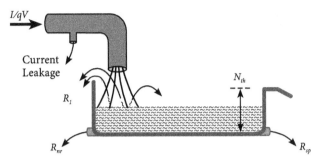

FIGURE 11.20 Reservoir with continuous supply and leakage as an analogy of the steady-state condition of charge carriers in the active region of a laser cavity. Charge carriers are continuously injected into the active region only to be lost due to different recombination processes and leakage.

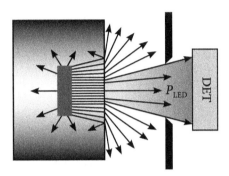

FIGURE 11.21 Schematic drawing of a light-emitting diode showing how only a small portion of the generated light reaches a desired detector.

An equation for the current below threshold can be obtained by setting the term corresponding to gain on the right-hand side of Eq. (11.36) equal to zero and using Eq. (11.31) to evaluate the term involving τ in this equation. We obtain

$$\frac{\eta_i I}{q V} = R_{sp}(N) + R_{nr}(N) + R_l(N). \tag{11.37}$$

For a laser well below threshold, the rate charge carriers are injected into the active region is equal to the rate charge carriers are lost due to spontaneous and nonradiative recombination and due to current leakage.

The steady-state condition of charge carriers in the active region of a laser cavity is analogous to the process of establishing a certain water level in the reservoir shown in Fig. 11.20. Just as water flows into a reservoir only to run off in various ways, the charge carriers entering the active region are lost due to the different recombination processes and leakage. The density of carriers eventually established depends upon the rate that charge carriers enter the active region and the rate they are lost.

A laser operating below threshold has properties that are very similar to those of a light-emitting diode (LED). For the laser below threshold as well as the LED, light is produced mainly by spontaneous recombination with generated photons not being amplified. As shown in Fig. 11.21, much of the light emitted by an LED is reflected back toward the active region rather than passing out of the semiconductor chip.

11.6.3 The laser threshold

If the rate charge carriers are injected into the active region exceeds the rate charge carriers are lost by recombination and charge leakage, the density of charge carriers in the active region N and the gain $g(N)$ will increase. A laser reaches threshold when the rate photons are produced by stimulated emission, $\Gamma v_g g(N) N_p$, approaches the rate photons are lost by scattering and absorption, N_p/τ_p. According to Eq. (11.35), the density of photons then rises substantially and the process of stimulated emission begins to dominate the other radiation processes. With an excess of photons in the active region of the lasers, electrons and holes recombine with photons being emitted. The increase in the rate of simulated emission reduces the rate at which the density of charge carriers increases and a new equilibrium is established with the density of charge carriers and the gain being held constant at their threshold values.

Fig. 11.22 again shows our reservoir analogy when the water level is higher. The processes of spontaneous recombination $R_{sp}(N)$, nonradiative recombination $R_{nr}(N)$, and carrier leakage $R_l(N)$ are still shown contributing to the loss

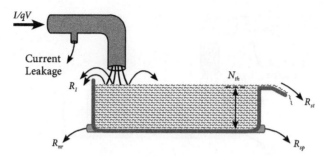

FIGURE 11.22 Reservoir analogy above threshold when the water level has risen to the spillway so that an increased input results in an increased output (R_{st}) but no further increase in carrier density (N).

of charge carriers. The process of stimulated recombination $R_{st}(N)$ is represented in Fig. 11.22 by the spillway of the reservoir, which limits the rise of the water above a certain level. It is important to realize that the recombination rates on the right-hand side of the steady stated conditions (11.35) and (11.36) depend upon the carrier density. As $\Gamma v_g g(N) N_p$ approaches N_p/τ_p, and the carrier density becomes stable at its threshold value, all of the terms in the two equations also become stationary except the denominator of Eq. (11.35).

11.6.4 Above threshold

To obtain an expression for the density of photons above threshold, we first estimate the current at threshold I_{th} by supposing that Eq. (11.37) can be applied almost at threshold

$$\frac{\eta_i I_{th}}{qV} = \left(R_{sp}(N) + R_{nr}(N) + R_l(N)\right)_{th}.$$

The right-hand side of this last equation may be identified as the rate of carrier loss at threshold. With this value for the rate of carrier loss, N/τ, and with the value of gain at threshold, g_{th}, Eq. (11.36) becomes

$$\frac{\eta_i I}{qV} - \frac{\eta_i I_{th}}{qV} = v_g g_{th} N_p.$$

This equation can be then solved for N_p to obtain

$$N_p = \frac{\eta_i(I - I_{th})}{qV v_g g_{th}}. \tag{11.38}$$

To obtain the power output of the laser, we first find the *stored optical energy in the cavity*, E_{os}, by multiplying the density of photons times the energy per photon and the volume occupied by the photons. We have

$$E_{os} = N_p hf V_p.$$

The optical power output can then be obtained by multiplying stored optical energy times the *energy loss rate through the mirrors*, $v_g \alpha_m$. We obtain

$$P_0 = v_g \alpha_m N_p hf V_p.$$

Substituting Eq. (11.38) into this equation and using $\Gamma = V/V_p$, the equation for the power can be written

$$P_0 = \eta_d \frac{hf}{q}(I - I_{th}),$$

where

$$\eta_d = \frac{\eta_i \alpha_m}{g_{th} \Gamma}.$$

The carrier density and gain of a semiconductor laser operating with a current in the range between zero and 2 mA is shown in Fig. 11.23. Notice that the carrier density and gain rises until the threshold is reached and then level off. The rates of spontaneous and nonradiative recombination and carrier leakage all become "clamped" at threshold.

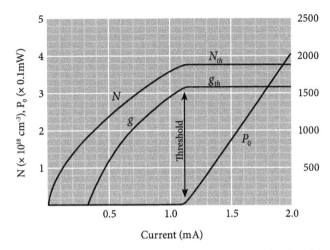

FIGURE 11.23 The carrier density N, gain g, and output power P_0 of a semiconductor laser as a function of the injected current. The horizontal scale gives current, while the vertical scale on the left gives carrier density N and power P_0, and the vertical scale on the right gives gain g.

Suggestions for further reading

John H. Davies, *The Physics of Low-Dimensional Semiconductors* (Cambridge University Press, 1998).

M.J. Kelly, *Low-Dimensional Semiconductors, Materials, Physics, Technology, Devices* (Clarendon Press, 1995).

Aflonso Franciosi and Chris G. Van de Walle, *Heterojunction band offset engineering*, Surface Science Reports, Vol. 25 (1996) pages 1–140.

Larry A. Coldren and Scott W. Corzine, *Diode Lasers and Photonic Integrated Circuits* (Wiley, 1995).

Otfried Madelung, Editor, *Semiconductors – Basic Data*, Second Edition (Springer, 1996).

Neil W. Ashcroft and N. David Mermin, *Solid State Physics* (Harcourt Brace, 1976).

Basic equations

Quantum wells in heterostructures

Schrödinger equation inside the well

$$-\frac{1}{m_W}\frac{d^2\psi}{dz^2} = \epsilon\psi$$

Schrödinger equation inside the barrier

$$-\frac{1}{m_B}\frac{d^2\psi}{dz^2} + E_0\mathcal{V}\psi = E_0\epsilon\psi$$

Boundary conditions

$$\psi(z = L_-) = \psi(z = L_+)$$

$$\frac{1}{m_W}\frac{d\psi}{dz}\Big|_{z=L-} = \frac{1}{m_B}\frac{d\psi}{dz}\Big|_{z=L+}$$

Quantum barriers

Potential step
To the left of the step, the wave function is

$$\psi_1(z) = A_1 e^{ik_1z} + B_1 e^{-ik_1z},$$

where

$$k_1 = \sqrt{\frac{2mE}{\hbar^2}}.$$

To the right of the step, the wave function is

$$\psi_2(z) = A_2 e^{ik_2 z} + B_2 e^{-ik_2 z},$$

where

$$k_2 = \sqrt{\frac{2m(E - V_0)}{\hbar^2}}.$$

The T-matrix, which is defined by the equation,

$$\begin{bmatrix} A_1 \\ B_1 \end{bmatrix} = \begin{bmatrix} T_{11} & T_{12} \\ T_{21} & T_{22} \end{bmatrix} \begin{bmatrix} A_2 \\ B_2 \end{bmatrix},$$

relates the incident and reflected amplitudes, A_1 and B_1, on the left side of a junction to the transmitted and incident amplitudes, A_2 and B_2, on the right side of the junction.

The transmission amplitude t

$$t = \frac{A_2}{A_1}\Big|_{B_2=0} = \frac{1}{T_{11}}$$

The reflection amplitude r

$$r = \frac{B_1}{A_1}\Big|_{B_2=0} = \frac{T_{21}}{T_{11}}$$

Reflection and transmission of light

The reflection coefficient

$$r_{12} = \frac{n_1 - n_2}{n_1 + n_2}$$

The transmission coefficient

$$t_{12} = \frac{2(n_1 n_2)^{1/2}}{n_1 + n_2}$$

The **T**-matrix may be written

$$\mathbf{T} = \frac{1}{t_{12}} \begin{bmatrix} 1 & r_{12} \\ r_{12} & 1 \end{bmatrix}.$$

Within a block of a dielectric material, an electromagnetic wave can be described by functions of the form

$$\psi(z) = A e^{i\bar{k}z} + B e^{-i\bar{k}z}, \tag{11.39}$$

where A is the amplitude of a wave propagating to the right through the media and B is the amplitude of a wave propagating to the left. The wave vector is

$$\bar{k} = k - ik_i.$$

The imaginary part of $\bar{k}$ is responsible for absorption and amplification of a signal.

The effect of the translation upon the wave function can be described by the diagonal matrix

$$T_L(\bar{k}) = \begin{bmatrix} e^{i\bar{k}L} & 0 \\ 0 & e^{-i\bar{k}L} \end{bmatrix}.$$

Phenomenological description of diode lasers

The rate equation

$$\frac{dN}{dt} = \frac{\eta_i I}{qV} - \frac{N}{\tau} - v_g g N_p$$

The rate the number of charge carriers increases is equal to the difference of the rate charge carriers are injected into the active region and the rate charge carriers are lost by dissipative processes and by stimulated recombination.

Well below threshold

$$\frac{\eta_i I}{qV} = R_{sp}(N) + R_{nr}(N) + R_l(N)$$

The rate charge carriers are injected into the active region is equal to the rate charge carriers are lost due to spontaneous and nonradiative recombination and due to current leakage.

Photon density above threshold

$$N_p = \frac{\eta_i (I - I_{th})}{qV v_g g_{th}}.$$

The photon density increases with increasing current above threshold. The carrier density and gain rises until the threshold is reached and then level off.

Summary

The energy ($\mathcal{E}$) available to electrons in semiconductors occur in broad bands that depend upon the magnitude and direction of the wave vector (k). The conduction bands correspond to energies above 0 eV, while the valence bands correspond to negative energies. For Si, Ge, $GaAS$, and $AlAs$, the maxima of the valence bands occurs at the symmetry point denoted by Γ. $GaAs$ is one of only a few semiconductors for which the minimum in the conduction band also occurs at Γ and vertical transitions between the valence and conduction bands can occur.

A discontinuous jump occurs in the potential energy of electrons at the juncture between two semiconductors. Discontinuities of this kind make it possible to create quantum wells and potential barriers for conduction electrons by forming multilayer structures called heterostructures. For multilayered crystals to perform well in optical and electrical devices, the layers must have the same lattice constants and be free of impurities.

Our discussion of the single-electron states of quantum wells and barriers is followed by a phenomenological description of semiconductor lasers.

Questions

1. Describe the outer surfaces of the Brillouin zone of the face-centered cubic crystal.
2. What special property of the $GaAs$ crystal makes it useful for optical devices?
3. How is the mobility of electrons related to their effective mass?
4. Which interaction is responsible for the separation between the upper two valence bands and the split-off band of semiconductors with the zincblende structure?
5. How does the magnitude of the wave vector of a photon compare with the magnitude of the wave vector of an electron near the boundary of the Brillouin zone?
6. What condition must be satisfied for crystals composed of distinct layers to be free of strain?
7. Sketch the maximum of the valence and the minimum of the conduction band as a function of the distance from the interface between two lattice matched semiconductors.
8. Sketch the alignment of the conduction bands for the quantum well formed by the layers $Al_{0.3}Ga_{0.7}As–GaAs–Al_{0.3}Ga_{0.7}As$.
9. The T-matrix used to describe electron scattering phenomena relates the amplitudes of the wave function to the left of a junction, A_1 and B_1, to the amplitudes of the wave function to the right of the junction, A_2 and B_2. Write down the equation defining the T-matrix.
10. What advantage is there in expressing the relation between the outgoing and incoming amplitudes in matrix form?
11. The T-matrix for a cavity can be written as the product of three matrices. Describe the physical significance of these three matrices.

12. Write down an equation expressing the rate of change of the density of charge carriers in the active region of a semiconductor laser in terms of the rate charge carriers are injected into the active region (R_{gen}) and the rate charge carriers recombine (R_{rec}).

13. Write down an equation expressing the fact that the rate charge carriers are injected into the active region of a semiconductor laser below threshold is equal to the rate charge carriers are lost.

14. How does the number of photons in the active region of a semiconductor laser change as the rate photons are produced by stimulated emission approaches the rate photons are lost by dispersion.

15. The steady-state condition of charge carriers in a laser cavity is analogous to a reservoir with water being supplied and lost continuously. To what property of the laser does the water level of the reservoir correspond?

16. Show how the reservoir corresponding to a laser appears as the laser approaches threshold.

17. How do the density of charge carriers and the gain of a semiconductor laser vary with increasing current as the laser approaches threshold?

Problems

1. The band gap of $GaAs$ is 1.519 eV. Calculate the wavelength of light emitted when an electron near the bottom of the conduction band of $GaAs$ makes a direct transition to a state near the top of the valence band.

2. The lattice constant a for $GaAs$ is equal to 5.65 eV and the effective mass is $0.067m_0$. Using Eq. (11.2) calculate the energy difference $\mathcal{E} - E_c$ of an electron with $k = 0.1(a/\pi)$.

3. The lattice constant a is 5.869 Å for InP, 6.058 Å for $InAs$, 5.653 Å for $GaAs$, and 5.660 Å for $AlAs$. Find the compositions of $In_xGa_{1-x}As$ and $In_xAl_{1-x}As$ that can be grown without strain on a substrate of InP.

4. The collapse of the Tacoma Narrows Bridge showed that engineers who do not know Fourier theory are a threat to themselves and society. The same could be said of physicists and engineers who do not know how to multiply matrices. Evaluate the following matrix times vector multiplications

$$\begin{bmatrix} 1 & 2 & 0 \\ 1 & 1 & 2 \\ 1 & 3 & 1 \end{bmatrix}\begin{bmatrix} 1 \\ 1 \\ 0 \end{bmatrix}, \quad \begin{bmatrix} 1 & 0 & 1 \\ 1 & 2 & 1 \\ 1 & 1 & 3 \end{bmatrix}\begin{bmatrix} 1 \\ 0 \\ 1 \end{bmatrix}, \quad \begin{bmatrix} 1 & 2 & 0 \\ 1 & 1 & 2 \\ 1 & 3 & 1 \end{bmatrix}\begin{bmatrix} 2 \\ 1 \\ 1 \end{bmatrix}, \quad \begin{bmatrix} 2 & 1 & 1 \\ 1 & 0 & 1 \\ 1 & 1 & 0 \end{bmatrix}\begin{bmatrix} 1 \\ 2 \\ 1 \end{bmatrix}.$$

5. Evaluate the following matrix times matrix multiplications

$$\begin{bmatrix} 0 & 1 \\ 1 & 0 \end{bmatrix}\begin{bmatrix} 0 & -i \\ i & 0 \end{bmatrix}, \quad \begin{bmatrix} 0 & 1 \\ 1 & 0 \end{bmatrix}\begin{bmatrix} 2 & 0 \\ 0 & -2 \end{bmatrix}, \quad \begin{bmatrix} 1 & 2 & 0 \\ 1 & 1 & 2 \\ 1 & 3 & 1 \end{bmatrix}\begin{bmatrix} 2 & 1 & 1 \\ 1 & 0 & 1 \\ 1 & 1 & 0 \end{bmatrix}.$$

6. By evaluating the product of the matrices in Eq. (11.28), obtain an explicit expression for the T_{11} matrix element.

7. A Fabry Perot laser consists of two strips of length 200 nm. Write a MATLAB function that calculates the transmission coefficient for a two-strip Fabry-Perot laser for particular values of the reflection coefficients, r_{21} and r_{32} and a particular wavelength of light λ.

8. Write a MATLAB program that plots the transmission coefficient of a two-strip Fabry-Perot laser described in the preceding problem as a function of $2kL/2\pi$ for the reflection coefficients, $r_{21} = 0.565$, and $r_{32} = 0.9$.

9. A reservoir of area A is filled by water at a rate of R_f (in ft^3/min) and simultaneously drained from two ports with the flow rate depending upon the height of water h. The drain rates are, $R_{d1} = C_1h$, and $R_{d2} = C_2h$, respectively.
   (a) Write down the rate equation for the reservoir.
   (b) Derive an expression for the steady-state height of the water.

Chapter 12

The special theory of relativity

Contents

12.1 Galilean transformations	279	Suggestions for further reading	305
12.2 The relative nature of simultaneity	282	Basic equations	305
12.3 Lorentz transformation	284	Summary	307
12.4 Space-time diagrams	295	Questions	307
12.5 Four-vectors	301	Problems	308

The laws of physics must be the same in all frames of reference moving at a constant velocity with respect to each other.

The special theory of relativity has its origins in the work of Einstein in the early part of the twentieth century. More recent work has been concerned with relativistic wave equations and with the quantum mechanical description of interacting fields. We begin this chapter by considering two frames of reference moving with respect to each other. Before the emergence of relativity theory, the transformations used to describe the relation between the coordinates of such moving frames of reference were called the *Galilean transformations*. After discussing the Galilean transformations, we shall describe briefly how the classical perspective failed to provide a framework in which optical phenomena could be understood and how this failure led Einstein to formulate the special theory of relativity.

12.1 Galilean transformations

Two coordinate frames S and S' moving with respect to each other are shown in Fig. 12.1. The position of a particle can be specified by giving its coordinates x, y, z in the frame S or the coordinates x', y', z' in the frame S'. We suppose that the origins of the two frames are in the same position for $t = 0$, and that S' is moving with a constant speed u along the x-axis of S. After a time t, the origins of the two frames will be separated by a distance ut. According to the conception of space derived from our usual experience, the two sets of coordinates are related by the equations

$$
\begin{aligned}
x' &= x - ut \\
y' &= y \\
z' &= z.
\end{aligned}
\qquad (12.1)
$$

The x' coordinate of a point is equal to the x coordinate of the point minus the separation of the two coordinate frames, which is equal to ut. Since S' is moving along the x-axis, the y and z coordinates of the two systems are equal. One also supposes that time is the same in both frames of reference. The transformation equations (12.1) are known as the *Galilean transformations*.

The Galilean transformations (12.1) gives the coordinates x', y', z' of a point in the frame S' that correspond to the coordinates x, y, z of the point in S. The inverse transformation, which gives the coordinates x, y, z in terms of the coordinates x', y', z' can be obtained by solving Eq. (12.1) for x, y, z giving

$$
\begin{aligned}
x &= x' + ut \\
y &= y' \\
z &= z'.
\end{aligned}
$$

We note that the inverse transformation can be obtained from the Galilean transformation (12.1) by replacing the velocity u with $-u$ and interchanging the primed and unprimed coordinates. This way of obtaining the inverse transformation can be

Modern Physics with Modern Computational Methods. https://doi.org/10.1016/B978-0-12-817790-7.00019-6

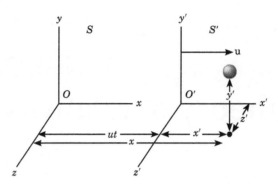

FIGURE 12.1 Two coordinate frames S and S'. The coordinate frame S' is moving with velocity u along the x-axis of the coordinate frame S.

understood in intuitive terms. Just as the reference frame S' moves with a velocity u with respect to S, the frame S moves with a velocity $-u$ with respect to S'.

We now consider a moving particle and think how observers in the coordinate frames S and S' will describe its motion. Equations relating the components of the velocity of the particle in the two reference frames can be obtained by differentiating Eq. (12.1). We obtain

$$\frac{dx'}{dt} = \frac{dx}{dt} - u$$
$$\frac{dy'}{dt} = \frac{dy}{dt}$$
$$\frac{dz'}{dt} = \frac{dz}{dt} \ .$$

Denoting the derivatives of the coordinates in the two reference frames by v_x, v_y, v_z and v'_x, v'_y, v'_z respectively, these equations become

$$v'_x = v_x - u$$
$$v'_y = v_y$$
$$v'_z = v_z \ .$$

If the particle is moving along the x-axis, the v_y and v_z components are zero, and the x-component of the velocity may be denoted by v. The equation relating the velocity of the particle in the two coordinate frames then becomes simply

$$v' = v - u \ . \tag{12.2}$$

Here u denotes the velocity of the coordinate frame S' with respect to the coordinate frame S, while v and v' are the velocities of the particle with respect to the S and S' frames, respectively. Eq. (12.2) says that the velocity of the particle in the coordinate frame S' is equal to its velocity in the frame S minus the velocity of S' with respect to S. This result is what one would expect. Suppose, for instance, that a train is moving through a station and a person standing on the ground throws a ball along the train in the same direction the train is moving. To an observer on the train, the ball should appear to be moving with a velocity equal to its velocity with respect to the station minus the velocity of the train.

The correctness of the Galilean coordinate and velocity transformations appears to be intuitively clear. They provide a very accurate description of objects moving with the velocities that we would associate with trains and balls. However, the Galilean velocity transformation led to conceptual difficulties when it was applied to light. Suppose that a light signal is traveling with velocity c along the x axis of the coordinate frame shown on the left of Fig. 12.1. According to Eq. (12.2), the velocity of light with respect to the primed frame would be $c - u$. If the primed frame was moving to the right, u would be positive and the velocity of light in the primed frame could be written $c - |u|$. For negative values of u, the velocity of light in the primed frame would be $c + |u|$. It was difficult for physicists in the latter part of the nineteenth century to reconcile this result with prevailing ideas in electromagnetic theory.

In 1863, James Clerk Maxwell had consolidated all that was then known about electricity and magnetism in four differential equations – known today as Maxwell's equations. Maxwell showed that in order for the equations to be internally

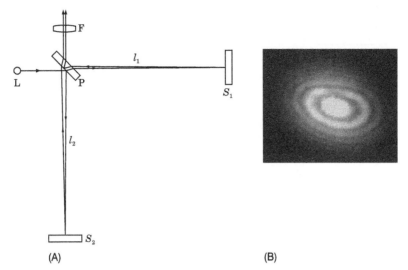

FIGURE 12.2 (A) A schematic drawing of the Michelson and Morley experiment. (B) Interference pattern produced with the interferometer of Michelson and Morley. Source: http://commons.wikimedia.org/wiki/User:Falcorian.

consistent, it was necessary for him to add a single time-dependent term to one of the equations. He realized that the resulting equations implied that electromagnetic waves would be propagated in a vacuum. The ideas of Maxwell were confirmed in 1890 when Heinrich Hertz produced radio waves in his laboratory.

Maxwell's theoretical work and the experiments of Hertz thirty years later led to a view of light we have today. Light can be thought of as an electromagnetic disturbance which propagates through space as a wave. The velocity of light, which we denote by c, can be determined by purely electrical measurements. If the velocity of light were to have different values in reference frames moving with respect to each other, Maxwell's equations could not be the same in all frames of reference.

Physicists in the 1890's and in the early years of the twentieth century attempted to resolve these difficulties using ideas that had been fruitful for them in the past. Many of the advances of physics had come from successfully constructing mechanical models. Kepler and Newton had constructed a mechanical model of the solar system. Optical Interference patterns, which were successfully explained by Huygens and Young using a wave model of light, could be appreciated intuitively by considering the patterns of waves moving across the surface of a body of water. Physicists of that time attributed the oscillations associated with light waves to a substance called the ether which permeated the universe. They speculated that the privileged frame in which the laws of Maxwell applied was a frame that was stationary with respect to the ether. In this frame of reference, a light wave spreads out uniformly with a velocity equal to c in all directions.

Around the turn of the century, several experiments were designed to measure the velocity of the Earth with respect to the frame of reference in which the ether is at rest. The Earth itself could not be this frame because the velocity of the Earth changes as it moves in its orbit about the Sun.

The experiment, which produced the clearest result, was performed by Albert Michelson and Edward Morley. A schematic drawing of the apparatus is shown in Fig. 12.2(A). Light from a light source L is split into two beams by a partially silvered glass plate P. One beam passes through the plate P and along the path of length l_1. This beam is reflected by the mirror S_1 and the partially silvered plate and passes through the lens F. The other beam is reflected by the plate P, passes down along the path of length l_2 and is reflected by the mirror S_2. As it passes upward, the second beam passes through the partially silvered plate and the lens F. The two beams of light come together to produce interference fringes such as those shown in Fig. 12.2(B). The position of the fringes depends upon the phase difference of the two beams after they have passed along the two paths.

We shall suppose that the velocity of light is c in the frame of reference in which the ether is at rest and that the velocity of light in a frame of reference attached to the Earth can be calculated using the Galilean transformations. The number of oscillations the waves make along the two paths will then depend upon the velocities of light and the Earth in the frame of reference in which the ether is at rest. We shall suppose that the apparatus on the Earth has a velocity V_E with respect to the frame of reference in which the ether is at rest and that this motion is directed along the path l_1.

Fig. 12.3(A) shows the velocity of the Earth V_E and the velocity of light moving toward and away from mirror S_1. If the light ray is moving in a direction opposite to the direction of the velocity of the Earth, the velocity of the light relative to the reflecting mirror S_1 will be $c + V_E$, while if the light is moving in the same direction as the Earth, the velocity of the

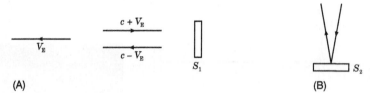

FIGURE 12.3 (A) The path of light rays moving toward and away from mirror S_1 in the Michelson and Morley experiment. The velocity of the Earth (V_E) is also shown. (B) The path of a light ray moving toward and away from mirror S_2.

light relative to the mirror S_1 will be $c - V_E$. Fig. 12.3(B) show a light ray moving toward and away from mirror S_2. Due to the motion of the Earth, this ray moves along a triangular path in the frame of reference in which the ether is at rest. Notice the motion of the Earth affects the relative velocity of the light moving toward and away from mirror S_1 and also the length of the path of light moving toward and away from mirror S_2.

In their experiment, Michelson and Morley observed the interferences fringes. They then rotated the apparatus by 90° about its axis and noted that the interference fringes did *not* move. Michelson and Morley carefully estimated ordinary sources of error, such as stresses and thermal variations. If the velocity of the Earth relative to the frame of reverence in which the ether was at rest were comparable to the velocity of the Earth relative to the Sun, the effect of rotating the apparatus should be clearly observable. Yet, the rotation of the apparatus had no discernible effect upon the interference fringes. Michelson and Morley performed their experiment repeatedly during different times in the year when the motion of the Earth in its orbit was quite different. They could see no effect.

12.2 The relative nature of simultaneity

A number of other experiments were devised by physicists to measure the velocity of the Earth with respect to the ether. Several decades of experimental research failed to provide any evidence that the velocity of light depended upon the motion of the Earth in any way. FitzGerald and Lorentz succeeded in explaining the negative result of these experiments by supposing that physical bodies moving through the ether contracted in the direction of motion. Einstein, on the contrary, accepted the experiments as decisive evidence that the laws of electrodynamics as well as the laws of mechanics are the same in every inertial frame of reference. In a paper published in 1905, Einstein proposed two postulates that form the basis of the special theory of relativity

Postulate 1. The laws of physics are the same in all inertial frames of reference.

Postulate 2. The speed of light in a vacuum is equal to the value c, independent of the motion of the source or the observer.

With these principles in mind, Einstein directed his efforts toward modifying the Galilean transformations so that the velocity of light would be the same in each frame. He emphasized the need of specifying how we would test experimentally the concepts that lie at the root of our physical description. For us to define a time scale, we must be able to say in principle how we would decide if two events occurred simultaneously. This could be done by setting up a coincidence counter at the midpoint of a straight line connecting the sites of the two events as illustrated in Fig. 12.4. If a light signal is emitted from the site of each event as it occurs and if the two light signals arrive at the midpoint at the same time, we shall say that the two events occurred simultaneously.

We now consider whether the definition of simultaneity depends upon our choice of coordinate frame. To answer this question, we shall use an imaginary experiment suggested by Einstein. Consider two frames of reference: one attached to the Earth and the other frame attached to a train moving along a straight track with a velocity u. Imagine that there is an observer stationed on the ground alongside the railroad tracks and another observer riding on the train. Each observer has a coincidence counter and a measuring stick.

Suppose now that two lightning bolts strike the train and the tracks leaving permanent marks as illustrated in Fig. 12.5(A). The marks on the track are located at A and B, and the marks on the train are located at A' and B'. At the instant the lightning strikes A coincides with A' and B coincides with B'. Coincidence counters are located at C and C'. Suppose that the observer on the ground can confirm that his coincidence counter at C is located midway between the marks at A and B and the observer on the train can confirm that the coincidence counter at C' is halfway between the marks at A' and B'. The two observer would say that the two lightning bolts struck simultaneously if the light signals emitted from the bolts when they struck arrived at their coincidence counters at the same time.

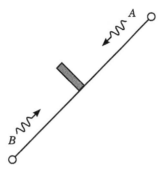

FIGURE 12.4 A coincidence counter at the midpoint of a straight line connecting the sites of two events, A and B.

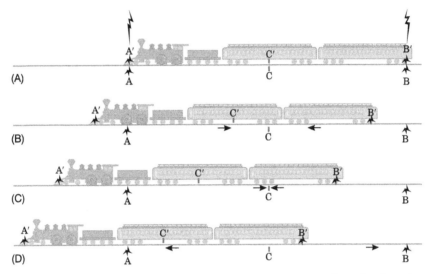

FIGURE 12.5 Two lightning bolts strike a moving train leaving permanent marks on the tracks at A and B and marks on the train at A' and B'.

Fig. 12.5(B) to (D) show the movement of the train after the lightning has struck. Secondary signals emitted from A and A' and from B and B' are also shown. In drawing these figures, we have assumed that the lightning bolts struck simultaneously with respect to the observer on the ground. The light signal emitted from A and A' passes through C' in Fig. 12.5(B) and arrives at the point C in Fig. 12.5(C). Similarly, the light signal from B and B' passes through C in Fig. 12.5(C) and arrives at C' in Fig. 12.5(D). Since the two light signals arrive at C at the same time, the observer on the ground would say that the two lightning bolts struck simultaneously. However, since the light signal from A and A' arrives at C' before the light signal from B and B', the observer on the train would say that the one lightning bolt struck A' before the other lightning bolt struck B'. We thus conclude that two events which are simultaneous with respect to one frame of reference are not always simultaneous with respect to another frame of reference.

This conclusion affects length measurements. We can suppose that the observer on the ground and the observer on the train have measuring sticks which they can use to measure the distance between two points in their coordinate frames. The length of an object moving with respect to an observer can be measured by recording the position of its two end points *at the same time*. Since the two lightning bolts struck simultaneously with respect to the ground observer, the observer on the ground can obtain the length of the train with respect to his frame of reference by measuring the distance between the two burnt marks on the track. However, the observer on the train would find that the lightning struck B and B' after it struck A and A', and he would think that the distance between the two marks on the tracks is shorter than the length of the train. If the lightning bolt had struck B' at the earlier time when the lightning struck the train at A', the train would not have moved so far along the track, and the distance between the two marks would be greater.

Before making time measurements, one must ensure that clocks placed in different positions are synchronized with each other. Suppose that the observer on the ground and the observer on the train each had a large assortment of clocks, and all of the clocks whether on the ground or on the train were identical to each other. Each observer could place a clock at any point in his coordinate frame where he intended to measure time and synchronize the clocks in his frame of reference to a single clock at the origin. The hands on the face of a clock at some distance from the origin could be adjusted so that

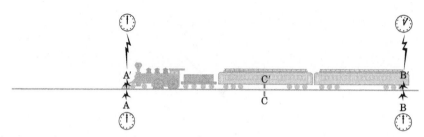

FIGURE 12.6 Clocks at A and A' and at B and B' which record the time the lightning struck at each point. The face of the clocks at A, A', and B show the time to be twelve o'clock while the clock at B' shows the time to be after twelve.

if a light signal were emitted from the clock and the clock at the origin when both clocks showed the same time, the two signals would arrive at a coincidence counter at the midpoint between them simultaneously. Suppose that the clock at A on the track and the clock at A' on the train agree with each other when the lightning bolt strikes at A and A'. Since the two lightning bolts strike simultaneously with respect to the observer on the ground, the clocks at A and B on the ground will show the same time when the two lightning bolts strike. As we have said, however, the observer on the train found that the one lightning bolt struck the train at A' before the other struck the train at B'. This means that the hand of the clock at B' on the train will show a later time when the lightning struck at B' than the clock showed at B. Fig. 12.6 shows clocks at A and A' and at B and B' which record the time when the lightning struck at each point. The face of the clocks at A, A', and B show the time to be twelve o'clock while the clock at B' shows the time to be five minutes past twelve.

The illustration we have given of two lightning bolts striking a train shows that the concept of simultaneity has no absolute meaning. An observer in any frame of reference must see whether or not events occur at the same time. Two events that occur simultaneously in one frame of reference need not be simultaneous in another. We have also learned that our measurements of moving objects may be shortened in the direction of their motion and our measurement of time intervals may be different.

12.3 Lorentz transformation

The above considerations help us resolve the apparent contradiction between the law of propagation of electromagnetic waves and the Michelson-Morley experiment. If the time and location of events depend upon the reference frame in which they are measured, it is possible that the speed of light is the same with respect to frames of reference moving relatively to each other.

We would now like to replace the Galilean transformations (12.1) by new equations that would enable us to relate the space and time coordinates of events. Our criterion will be that a light wave spreads out isotropically with the same speed in all frames of reference. The requirement that events that occur at equally spaced points in one reference frame should be equally spaced in other frames and that a clock running steadily in one reference frame run steadily in another will be satisfied if the coordinate transformations are linear.

12.3.1 The transformation equations

We consider again the two coordinate frames, S and S', shown in Fig. 12.1. As before, we suppose that the origins of the two frames are in the same position for $t = 0$, and that the frame S' is moving with a constant velocity u along the x-axis of the frame S. A condition upon the transformations can be obtained by considering a particle which is at rest relative to S'. A particle which is at rest in S' moves with a velocity u along the x-axis of S. The requirement that the x' coordinate of the particle should not vary is satisfied if the transformation of the x-coordinates is of the form

$$x' = \lambda(x - ut), \tag{12.3}$$

where λ is a constant that we will determine.

The length of two measuring sticks that are parallel to each other and perpendicular to the direction of motion are unaffected by the motion. So, we require that the y and z coordinates in the two systems be the same

$$y' = y$$
$$z' = z. \tag{12.4}$$

To complete this set of equations, we need an equation relating the time t' in S' to the time t in S. From our example of the train given in the previous section, we consider a particular clock beside the track having a position corresponding to a value of x. For a particular time t, the clock in the train above that point on the track would have a definite time t'. This suggests a transformation of the form

$$t' = \mu t + \nu x. \tag{12.5}$$

For the relativistic transformations of coordinates to be well-defined, we must evaluate the constant λ in Eq. (12.3) and the constants μ and ν in Eq. (12.5). These constants can be found by requiring that the speed of light is the same with respect to S and S' and that the relativistic transformations reduce to the Galilean transformations (12.1) when the relative velocity of the two systems u is small compared to the speed of light c.

We suppose that a spherical wave is emitted from the origin of the frames S and S' at $t = 0$ when the two frames coincide, and the wave spreads out uniformly in both frames with a velocity c. The coordinates of a point on the surface of the wave satisfy the equations

$$x^2 + y^2 + z^2 = c^2 t^2 \tag{12.6}$$
$$x'^2 + y'^2 + z'^2 = c^2 t'^2.$$

Substituting Eqs. (12.3), (12.4), and (12.5) into this last equation gives

$$\lambda^2 (x - ut)^2 + y^2 + z^2 = c^2 (\mu t + \nu x)^2.$$

The terms in this equation can be rearranged giving

$$(\lambda^2 - c^2 \nu^2) x^2 + y^2 + z^2 - 2(u\lambda^2 + c^2 \mu \nu) xt = (c^2 \mu^2 - u^2 \lambda^2) t^2. \tag{12.7}$$

Eq. (12.7) will be identical to Eq. (12.6) if the coefficients of x^2 and t^2 in Eq. (12.7) are equal to $+1$ and c^2, respectively, and if the coefficient of xt in Eq. (12.7) is equal to zero. This leads to the following equations for the constants λ, μ, and ν

$$\lambda^2 - c^2 \nu^2 = 1,$$
$$c^2 \mu^2 - u^2 \lambda^2 = c^2,$$
$$u\lambda^2 + c^2 \mu \nu = 0.$$

The last of these equations may be solved for λ^2 giving

$$\lambda^2 = -\frac{c^2 \mu \nu}{u}. \tag{12.8}$$

Substituting this value of λ^2 into the first two equations, we obtain

$$c^2 \nu (\mu + u\nu) = -u, \tag{12.9}$$
$$\mu (\mu + u\nu) = 1. \tag{12.10}$$

Dividing Eq. (12.9) by Eq. (12.10) gives

$$\frac{c^2 \nu}{\mu} = -u,$$

which may be used to solve for ν in terms of μ. We obtain

$$\nu = -\frac{u\mu}{c^2}. \tag{12.11}$$

Substituting Eq. (12.11) into Eq. (12.10) and solving for μ^2, we obtain

$$\mu^2 = \frac{1}{1 - u^2/c^2}. \tag{12.12}$$

The constant μ appearing in Eq. (12.5) is not equal to one as in the classical theory; however, we can make it nearly equal to one for small values of u^2/c^2 by choosing the positive square root of Eq. (12.12). The constant μ is then

$$\mu = \frac{1}{\sqrt{1 - u^2/c^2}}. \tag{12.13}$$

Substituting this equation into Eq. (12.11), we obtain

$$v = -\frac{u}{c^2} \frac{1}{\sqrt{1 - u^2/c^2}}. \tag{12.14}$$

The constant λ can then be obtained by substituting Eqs. (12.13) and (12.14) into Eq. (12.8) and again taking the positive square root to obtain

$$\lambda = \frac{1}{\sqrt{1 - u^2/c^2}}.$$

The relativistic transformation equations can now be obtained by substituting these values of the constants, λ, μ, and v, into Eqs. (12.3), (12.4) and (12.5). We obtain

$$\begin{aligned}
x' &= \gamma(x - u\,t) \\
y' &= y \\
z' &= z \\
t' &= \gamma(t - u\,x/c^2),
\end{aligned} \tag{12.15}$$

where

$$\gamma = \frac{1}{\sqrt{1 - u^2/c^2}}. \tag{12.16}$$

These equations are called the *Lorentz transformations*. For small values of u/c, the constant γ in the above equation is close to one and Eqs. (12.15) reduce to the Galilean transformations (12.1) considered previously.

The Lorentz transformations (12.15) relate the time and the space coordinates of an event in the frame S' to the time and space coordinates of the event in S. Suppose that at time t an event occurs in the reference frame S at a point with coordinates (x, y, z). The transformation equations (12.15) gives the time t' and space coordinates (x', y', t') of the event in the frame S'. As for the Galilean transformations, the inverse transformation can be obtained from Eq. (12.15) by replacing the velocity u with $-u$ and interchanging the primed and unprimed coordinates

$$\begin{aligned}
x &= \gamma(x' + u\,t') \\
y &= y' \\
z &= z' \\
t &= \gamma(t' + u\,x'/c^2).
\end{aligned} \tag{12.17}$$

Before discussing the kinematic consequences of the Lorentz transformations, we shall make a few qualitative observations about the nature of the transformations. Underlying the derivation of the transformation equations is the idea that all frames of reference having a constant velocity with respect to each other are equivalent. The laws of nature are the same in all such frames, and, in particular, the speed of light is the same. Our expectation is that in each frame of reference, an observer has a measuring stick and a number of clocks which he has distributed at different points where he might want to measure the time. All of the clocks in each frame are synchronized as we have described before so that they give consistent time measurements. Each observer makes his own time and space measurements and reports the result. In the Lorentz transformations (12.15), the time and space coordinates are inextricably tied together. The basic entity is an event which occurs at a particular time at a particular location. We shall think of the time as a coordinate on the same footing as the space coordinates. An event, which has time and space coordinates (t, x, y, z) in the frame S, has coordinates (t', x', y', z') in the frame S'. These two sets of coordinates are related by the Lorentz transformations.

12.3.2 Lorentz contraction

Suppose now that we wish to determine the length of an object moving past us. As we have with the train in the previous section, we can measure a moving object in our frame of reference by recording the position of its two end points at the same time. We associate the moving object with the reference frame S' and ourselves with the frame S. The two end points in the frame S are denoted by x_1 and x_2 and the time in this frame is denoted by t. The length of the object in the frame S is equal to the difference between x_1 and x_2. Since the object is moving with respect to the frame S, we denote the length of the object in this frame by L_M. We thus have

$$L_M = x_2 - x_1. \tag{12.18}$$

In order to find a relationship between the length of the object in our frame of reference and its length in a frame moving with the object, we apply the first of the Lorentz transformations (12.15) to the two end points

$$x'_2 = \gamma(x_2 - u\,t) \tag{12.19}$$
$$x'_1 = \gamma(x_1 - u\,t). \tag{12.20}$$

Since the position of the points x_1 and x_2 are determined at the same time in the reference frame S, the same time variable t appears in Eqs. (12.19) and (12.20). Subtracting the second of these equations from the first, we obtain

$$x'_2 - x'_1 = \gamma(x_2 - x_1). \tag{12.21}$$

The distance $x'_2 - x'_1$ that appears on the left of Eq. (12.21) is the length of the object in the frame S'. Since the object is at rest with respect to frame S', we denote this distance by L_R

$$L_R = x'_2 - x'_1. \tag{12.22}$$

Using the definitions of L_M and L_R provided by (12.18) and (12.22), Eq. (12.21) can be written

$$L_R = \gamma L_M.$$

We now use Eq. (12.16) to express the length of the moving object in terms of its velocity

$$L_M = \sqrt{1 - u^2/c^2}L_R. \tag{12.23}$$

Since the factor $\sqrt{1 - u^2/c^2}$ is always less than one, the object is shorter than it would be at rest. This effect is known as *Lorentz contraction*. A moving body is contracted in the direction of its motion by the factor $\sqrt{1 - u^2/c^2}$, while its dimensions perpendicular to the direction of motion are unaffected. Note that in deriving this result, we used the ordinary Lorentz transformation (12.15) rather than the inverse transformation (12.17). Our reason for doing this was that the measurement of the position of the two end points was taken at the same time in the frame S. The values of t appearing in Eqs. (12.19) and (12.20) are thus the same, and the time drops out when we take the difference of x'_2 and x'_1.

Example 12.1

Find how much a spaceship, which is traveling at half the speed of light (c/2), will be contracted for an observer on the Earth.

Solution
The contraction of the spaceship can be obtained using Eq. (12.23). Setting $u = c/2$, the quantity $\sqrt{1 - u^2/c^2}$ appearing in this equation is found to be equal to $\sqrt{3}/2$, and the length of the moving spacecraft is $L_M = \sqrt{3}L_R/2$. So, the spaceship is shortened by about 13.4 %.

Fig. 12.7(A) shows a contracted spaceship as viewed by an observer on the Earth, while Fig. 12.7(B) shows that the Earth appears contracted from the view point of an observer in the spaceship. A basic premise of relativity theory is that all frames of reference moving at a constant velocity with respect to each other are equivalent. Observations made in one frame of reference are as valid as observations made in any other.

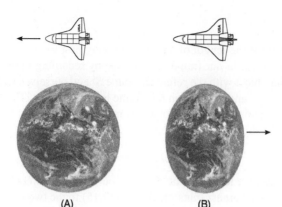

FIGURE 12.7 (A) An observer on the Earth views a passing, contracted spaceship. (B) On the spaceship, an observer sees that the Earth appears contracted.

12.3.3 Time dilation

We now consider the effect of the Lorentz transformations on time measurements. Suppose that a clock is attached to the moving frame S', and the spatial coordinates of the clock are (x', y', z'). We would like to compare time intervals of this clock in S' to time intervals in S. For this purpose, we use the last of the inverse Lorentz transformations

$$t = \gamma (t' + u\, x'/c^2).\tag{12.24}$$

Since the clock we have mentioned is stationary in S', the coordinate x' will be the same for all of our time measurements. It will therefore cancel out for measurement of time intervals just as t dropped out previously in our consideration of space intervals. Using Eq. (12.17), the interval $t_2 - t_1$ between two times in the S frame can be related to readings t'_2 and t'_1 of the clock in S' by the equation

$$t_2 - t_1 = \gamma (t'_2 - t'_1).$$

The dependence of the term on the right upon the velocity of the clock can be made explicit using Eq. (12.16) to obtain

$$t_2 - t_1 = \frac{(t'_2 - t'_1)}{\sqrt{1 - u^2/c^2}}.\tag{12.25}$$

Again, since the clock is moving with respect to S and at rest with respect to S', we use the definitions,

$$\Delta t_M = t_2 - t_1$$
$$\Delta t_R = t'_2 - t'_1,$$

to write Eq. (12.25)

$$\Delta t_M = \frac{\Delta t_R}{\sqrt{1 - u^2/c^2}}.\tag{12.26}$$

The time interval measured by an observer in S is thus longer than the time interval for an observer in S'. This effect is called *time dilation*.

To show that Eq. (12.26) for time dilation is correct, we consider a simple experiment. Suppose that the oscillator shown in Fig. 12.8(A) is on a train moving with velocity v past the stationary platform in a railroad station. As shown in Fig. 12.8(A), a pulse of light leaves a light source (B), travels vertically upward a distance (D), is reflected by the mirror and then travels vertically downward to the source where it is detected. The length of time for the light pulse to travel up to the mirror and then back down to the detector is

$$\Delta t_0 = \frac{2D}{c}.$$

Solving this last equation for D, we obtain

$$D = \frac{c\Delta t_0}{2}.\tag{12.27}$$

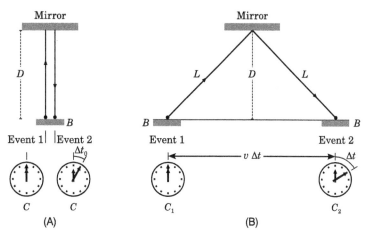

FIGURE 12.8 (A) An observer on the train measures the time interval Δt_0 between events 1 and 2 using a single clock C on the train. The clock is shown twice: first for event 1 and then for event 2. (B) An observer on a stationary platform in a railroad station uses two synchronized clocks, C_1 at event 1 and C_2 at event 2, to measure the time interval between the two events. This measured time interval is denoted Δt.

We consider now how the flight of the pulse would appear to an observer on the train platform. Because the experiment moves with the train, the observer on the platform sees the light follow the path shown in Fig. 12.8(B). According to relativity theory, the speed of light is the same for the observer on the stationary platform as it is for an observer on the train. Hence, the time for the pulse to travel along the path shown in Fig. 12.8(B) is

$$\Delta t = \frac{2L}{c} , \tag{12.28}$$

where

$$L = \sqrt{\left(v\frac{\Delta t}{2}\right)^2 + D^2} . \tag{12.29}$$

Solving Eq. (12.28) for L, we obtain

$$L = \frac{c\Delta t}{2} .$$

We may now substitute the above value of L and the value of D given by Eq. (12.27) into Eq. (12.29) and solve for Δt to obtain

$$\Delta t = \frac{\Delta t_0}{\sqrt{1 - v^2/c^2}},$$

which is equivalent to Eq. (12.26).

Example 12.2

A muon, which decays in 2.2×10^{-6} seconds when it is at rest, is moving with a velocity of $0.95\,c$ in the upper atmosphere. How far will the particle travel before it decays?

Solution

We shall call the frame of reference moving with a particle the *rest frame* of the particle. The lifetime of the muon in its own rest frame is 2.2×10^{-6} s. To find the length of the muon's flight, we first calculate the lifetime of the muon in a reference frame which is stationary with respect to the upper atmosphere. Since the muon is moving through the upper atmosphere, its lifetime as viewed by a stationary observer in the upper atmosphere is denoted Δt_M. Applying Eq. (12.26) to the muon, we obtain

$$\Delta t_M = \frac{2.2 \times 10^{-6} \text{ s}}{\sqrt{1 - (0.95)^2}} = 7.0 \times 10^{-6} \text{ s}.$$

The length of the muon's flight is equal to $\Delta t_M \times 0.95c = 7.0 \times 10^{-6}$ s $\times 0.95 \times 2.998 \times 10^8$ m/s $= 2.0$ km.

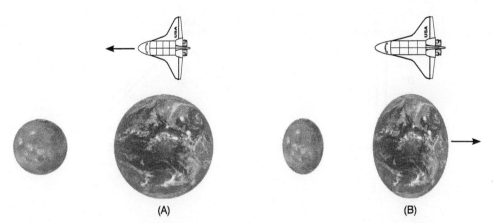

FIGURE 12.9 A spaceship travels from the Earth to a distant planet. (A) As viewed from the Earth, the Earth and the distant planet are round, while the spaceship is contracted. (B) To the astronaut on the spaceship, the spaceship appears to be of ordinary length, while the Earth and distant planet are contracted.

Example 12.3

A space traveler has been ordered to journey to a distant planet which is 100 light-years from Earth. The Earth, the planet, and the spaceship are illustrated in Figs. 12.9(A) and (B). Fig. 12.9(A) shows the Earth, the planet, and the spaceship as viewed by an observer stationary with respect to the Earth, while Fig. 12.9(B) shows the Earth, the planet, and the spaceship as viewed by the traveler in the spaceship. Adopting a frame of reference which is stationary with respect to the Earth, find how fast the astronaut must travel to reach the planet when he is ten years older. Do the same problem using instead a frame of reference stationary with respect to his spaceship.

Solution

The View From Planet Earth

During his flight, the astronaut becomes ten years older in his own frame of reference. Since he will be moving with respect to the Earth, we must use the time dilation formula to find out how much older he will be in Earth years. Denoting the velocity of the astronaut by v, the duration of his flight in Earth years will be

$$\Delta t_M = \frac{10 \text{ yr}}{\sqrt{1 - (v/c)^2}} .$$

The distance to the planet is 100 light years, which is the distance light would travel in a hundred years. Denoting this distance by $100\,c \times$ yr, the condition that the astronaut reach the planet when he is ten years older is

$$v \frac{10 \text{ yr}}{\sqrt{1 - (v/c)^2}} = 100\,c \times \text{yr}. \tag{12.30}$$

Dividing this equation through by $100\,c \times$ yr and multiplying by $\sqrt{1 - (v/c)^2}$, we get

$$(0.1)\frac{v}{c} = \sqrt{1 - (v/c)^2}$$

This last equation may be solved for v to obtain

$$v = 0.995\,c.$$

The View From the Spaceship

In his own frame of reference the astronaut will become 10 years older during the flight; however, the distance to the planet will become shorter due to space contraction. Using Eq. (12.23), the condition that the astronaut reach the planet in ten years is

$$v \times 10 \text{ yr} = 100\,c \times \text{yr} \times \sqrt{1 - (v/c)^2}. \tag{12.31}$$

We immediately see that this last equation can be obtained by multiplying Eq. (12.30) through by $\sqrt{1 - (v/c)^2}$. It is entirely equivalent to Eq. (12.30). We thus obtain the same result.

Time dilation caused the time of the flight to be longer in the reference frame of the Earth, while space contraction caused the distance to be shorter in the reference frame of the spaceship. The result is the same.

12.3.4 The invariant space-time interval

A Lorentz transformation generally changes the difference in time and space between two events. We can describe this by saying that time intervals and space intervals are not invariant with respect to Lorentz transformations. However, we shall now show that the difference between the square of c times the time separation between two events and the square of the space separation between the two events is invariant with respect to Lorentz transformations. The invariant space-time interval is defined by the equation

$$\text{(invariant interval)}^2 = (c \times \text{time separation})^2 - \text{(space separation)}^2 . \tag{12.32}$$

Since the product of the speed of light and the time separation, which occurs in the first term on the right-hand side of this equation, is a distance, all of the terms in this last equation will have the units of distance squared.

To show that the space-time interval defined by this last equation is invariant with respect to Lorentz transformations, we consider two events whose time and space coordinates in S are (t_1, x_1, y_1, z_1) and (t_2, x_2, y_2, z_2). Using these time and space coordinates to express the time and space separations in the frame S, we obtain

$$\text{(invariant interval)}^2 = [c(t_2 - t_1)]^2 - [(x_2 - x_1)^2 + (y_2 - y_1)^2 + (z_2 - z_1)^2]. \tag{12.33}$$

In the reference frame S' moving with respect to S, the corresponding linear combination of space and time coordinates is

$$\text{(invariant interval)}^2 = [c(t'_2 - t'_1)]^2 - [(x'_2 - x'_1)^2 + (y'_2 - y'_1)^2 + (z'_2 - z'_1)^2].$$

The above expression for the interval in the frame S' may be expressed in terms of coordinates of the frame S by using Eqs. (12.15) and (12.16) to obtain

$$\text{(invariant interval)}^2 = \frac{[c(t_2 - t_1) - (u/c)(x_2 - x_1)]^2}{1 - u^2/c^2} - \frac{[(x_2 - x_1) - u(t_2 - t_1)]^2}{(1 - u^2/c^2)} - (y_2 - y_1)^2 - (z_2 - z_1)^2. \tag{12.34}$$

The terms involving $(t_2 - t_1)^2$ and $(x_2 - x_1)^2$ on the right-hand side of this last equation may be grouped together and the cross terms due to the square of the two terms within square brackets may be shown to cancel. In this way, one may show that the interval squared in the frame S' is identical to the interval squared in S. The space-time interval is invariant with respect to Lorentz transformations.

Eq. (12.33) for the invariant interval reduces to Eq. (12.6) used to derive the Lorentz transformations if one replaces the coordinates (t_2, x_2, y_2, z_2) in Eq. (12.33) by (t, x, y, z) and lets the coordinates (t_1, x_1, y_1, z_1) correspond to the origin at $t = 0$. The invariance of the interval is thus equivalent to our original condition that a light wave should spread out isotropically in reference frames moving relative to each other. Notice that Eq. (12.32) is very similar to the Pythagorean theorem except that the two squares on the right-hand side of the equation are separated by a minus sign rather than a plus sign.

The invariant interval may be used in describing events which occur at the same spatial location. If two events take place at the same location in the frame S, then x_2 equals x_1, y_2 equals y_1, and z_2 equals z_1. According to Eq. (12.33), the invariant interval is then equal to c times the time separation of the two events. When two events occur in a reference frame at the same point, the time interval in that frame is called the *proper time*. As an example of the use of the concept of proper time, we consider again the decay of a particle. Suppose that a particle is formed at the time t_1 and decays at the time t_2 in the laboratory. The time interval that we would associate with the particle in the laboratory is $\Delta t = t_2 - t_1$. The proper time interval denoted by τ is the lifetime of the particle in its rest frame. As we have seen, the lifetime of a moving particle is longer than the lifetime of the particle in its rest frame.

12.3.5 Addition of velocities

We consider now how velocities transform with respect to Lorentz transformations. The additive nature of the Galilean velocity transformations played a central role in the chain of events that led Einstein to break with tradition and formulate relativity theory. Suppose that a particle is moving in the x-direction with a velocity v in the reference frame S and that we

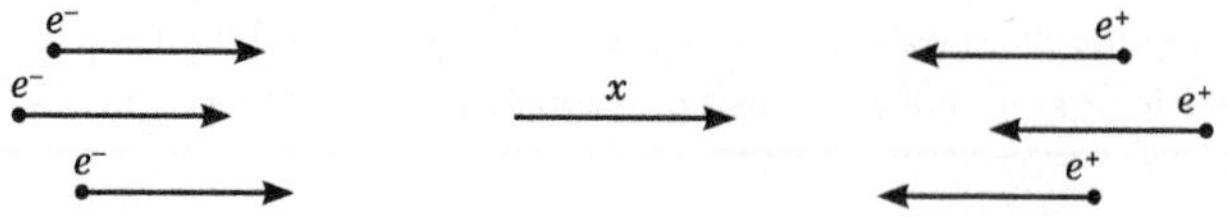

FIGURE 12.10 Electrons and positrons approaching each other in a colliding beam experiment. The electrons e^- move in the positive x-direction, while the positrons e^+ move in the negative direction.

wish to describe the motion of the particle with respect to a frame S', which is moving with a velocity u along the x-axis of S. The velocity of the particle in the frame S' can be obtained by using the chain rule

$$v' = \frac{dx'}{dt'} = \frac{dx'}{dt}\frac{dt}{dt'}. \tag{12.35}$$

To evaluate the derivative of x' with respect to t, we take the derivative with respect to t of the first equation of the Lorentz transformations (12.15) to obtain

$$\frac{dx'}{dt} = \gamma(v - u). \tag{12.36}$$

The derivative of t with respect to t' in Eq. (12.35), can be evaluated by taking the derivative with respect to t' of the last equation of the inverse Lorentz transformations (12.17) to obtain

$$\frac{dt}{dt'} = \gamma(1 + \frac{u}{c^2}v'). \tag{12.37}$$

Substituting Eqs. (12.36) and (12.37) into Eq. (12.35), we get

$$v' = \gamma(v - u) \cdot \gamma(1 + \frac{u}{c^2}v'). \tag{12.38}$$

The relativistic velocity transformation can be obtained by solving this last equation for v' and using Eq. (12.16) to obtain

$$v' = \frac{v - u}{1 - uv/c^2}. \tag{12.39}$$

The derivation of this result is left as an exercise (see Problem 15). We note that the numerator of Eq. (12.39) is the same as the Galilean velocity transformation (12.2). The denominator provides a relativistic correction that is small unless u and v are close to c. If the velocity v in Eq. (12.39) is equal to c, then v' is also equal to c. To see this, we substitute $v = c$ into Eq. (12.39) to obtain

$$v' = \frac{c - u}{1 - (u/c)} = \frac{c - u}{(c - u)/c} = c.$$

The speed of light is the same in all frames of reference moving relative to each other.

Example 12.4

Electrons and positrons interact in a colliding beam experiment with the absolute value of the velocity of both kinds of particles being $0.95c$. Find the velocity of the electrons in the frame of reference of the positrons.

Solution

As illustrated in Fig. 12.10, we choose the positive x-direction to correspond to the direction of the electrons. The velocity of the electrons is then $v = +0.95c$, and the velocity of the frame of reference moving with the positrons is $u = -0.95c$. Substituting these values into Eq. (12.39), we obtain

$$v' = \frac{0.95c + 0.95c}{1 - (-0.95c)(0.95c)/c^2} = 0.9987c.$$

While the velocity of the electrons is now much closer to the speed of light, it does not exceed the velocity of light.

FIGURE 12.11 A light source moves with velocity v toward an observer at A while receding from an observer B.

12.3.6 The Doppler effect

We shall now consider the shift which is observed in the frequency of an electromagnetic wave when the source of the wave is moving with respect to the observer. This shift is called the *Doppler Effect*.

Consider a light source moving with velocity v as shown in Fig. 12.11. The source, which is in the reference frame S', is approaching an observer at A and receding from an observer at B. Both A and B are in the reference frame S. First, we consider the train of waves approaching A. Suppose that during the time Δt the source emits N waves. During this time, the first wave emitted will have moved a distance $c\Delta t$ and the source will have moved a distance $v\Delta t$. Since the N waves seen by the observer at A occupy a distance $c\Delta t - v\Delta t$, their wavelength is

$$\lambda = \frac{c\Delta t - v\Delta t}{N},$$

and their frequency is

$$f = \frac{c}{\lambda} = \frac{cN}{(c-v)\Delta t}.$$

Using the notation $\beta = v/c$, this last equation for the frequency can be written

$$f = \frac{1}{1-\beta}\frac{N}{\Delta t}. \tag{12.40}$$

The quantity Δt appearing in the denominator of Eq. (12.40) is the period of time in the reference frame S during which N waves are emitted. We denote by $\Delta t'$ the period of time in the reference frame S' during which N waves are emitted. Since the source is at rest in the reference frame S', these two time periods are related by the equation

$$\Delta t = \gamma \Delta t'.$$

Substituting this equation for Δt into Eq. (12.40) and using the fact that the frequency f_0 in the frame of reference of the source is equal to $N/\Delta t'$, we obtain

$$f = \frac{f_0}{1-\beta}\frac{1}{\gamma}.$$

Finally, using Eq. (12.16), this last equation can be written

$$f = \frac{\sqrt{1-\beta^2}}{1-\beta}f_0 = \sqrt{\frac{1+\beta}{1-\beta}}\,f_0, \quad \textit{for approaching light source.}$$

Notice that the frequency f will be greater than f_0 if the source is approaching. Since for visible light this corresponds to a shift to the blue part of the spectra, the frequency shift is called a *blue shift*.

Consider now the case when the light source is moving away from the observer, as for observer B in Fig. 12.11. The above line of argument must then be adjusted to take into account the fact that the observer B sees N waves occupying a distance $c\Delta t + v\Delta t$, and the following equation is obtained

$$f = \frac{\sqrt{1-\beta^2}}{1+\beta}f_0 = \sqrt{\frac{1-\beta}{1+\beta}}\,f_0, \quad \textit{for receding light source.} \tag{12.41}$$

In this case, the frequency f is less than f_0. Since red light has the lowest frequency for visible light, the frequency shift of a receding light source is called a *red shift*.

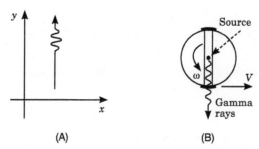

FIGURE 12.12 The transverse Doppler effect. (A) A light source moves upward along the y-axis, while the observer is located at a position along the positive x-axis. (B) Kündig confirmed the relativistic prediction for the transverse Doppler effect by placing a gamma-ray source at the center of an ultracentrifuge and a detector on the rim of the centrifuge.

Example 12.5

In 1963 Maarten Schmidt found that the fractional red shift $(\lambda - \lambda_0)/\lambda_0$ of the galaxy 3C 273 was 0.158. From this information, calculate the speed with which this galaxy is receding from the Earth.

Solution

Eq. (12.41) gives the frequency of light for a known value of $\beta = v/c$. Solving this equation for β, we get

$$\beta = \frac{1 - (f/f_0)^2}{1 + (f/f_0)^2} \,. \tag{12.42}$$

We can thus calculate the value of β and the velocity of the galaxy if we know the ratio f/f_0.

Using the information given, the fractional red shift can be written

$$\frac{\lambda - \lambda_0}{\lambda_0} = 0.158.$$

This equation can be written

$$\lambda - \lambda_0 = 0.158\lambda_0 \,,$$

which in turn may be used to show

$$\frac{\lambda}{\lambda_0} = 1.158 \,.$$

We can now find the ratio f/f_0. Using the relation $\lambda = c/f$, we write

$$\frac{f}{f_0} = \frac{cf}{cf_0} = \frac{c/f_0}{c/f} = \frac{\lambda_0}{\lambda} = 0.864.$$

Substituting this value of f/f_0 into Eq. (12.42) we find that $\beta = 0.15$. The velocity of the galaxy is thus $v = 0.15c = 45,000$ km/s. This remarkable result was used to identify the first quasar.

The Doppler effect can be thought of as consisting of two different kinds of physical effects. The first of these effects is the compression or attenuation of waves emitted by a light source due to its relative motion, and the other effect is the relativistic time-dilation effect. The time-dilation effect is independent of the direction of motion of the light source. To understand this, consider again the simple experiment illustrated in Fig. 12.8. If this experiment were moved from a train to a helicopter, and the helicopter were flown in any direction with the orientation of the oscillator kept perpendicular to the direction of flight, the result would be the same. Time dilation is isotropic in nature.

This fact leads to a striking prediction. The time-dilation part of the Doppler effect still occurs if the motion of the light source is transverse as shown in Fig. 12.12(A). In this figure, the light source moves upward along the y-axis, while the

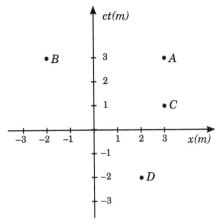

FIGURE 12.13 Space-time diagrams for A, B, C, and D. Two of the space dimensions (y and z) are suppressed. The horizontal axis gives the x coordinate measured in meters, while the vertical axis gives the product of the velocity of light c and the time of the event t. Like the x coordinate, ct is measured in meters.

observer is located at a position along the positive x-axis. The shift of the frequency is then given by the equation

$$f = \frac{f_0}{\gamma}.$$

As predicted by Einstein's theory, the relative motion of the light source causes its oscillations to become slower.

Following a suggestion made by Einstein in 1907, Kündig confirmed the relativistic prediction for the transverse Doppler effect in 1962. In his experiment, which is illustrated in Fig. 12.12(B), Kündig placed a gamma-ray source at the center of an ultracentrifuge and a detector on the rim of the centrifuge. Using a sensitive measuring technique, Kündig was able to confirm the relativistic prediction to $\pm 1\%$ over a wide range of relative speeds.

12.4 Space-time diagrams

The Lorentz transformations (12.15) relates the time and space coordinates of two reference frames. The difficulties one may have in understanding this transformation in intuitive terms stem partly from the fact that four space and time variables are involved, x, y, z, t. We cannot imagine events in four dimensions. This difficulty can be removed by characterizing events by their x and t coordinates. Since we have chosen the direction of relative motion of reference frames to be along the x-axis, y is equal to y' and z is equal to z'. Nothing essential is lost by using the x and t coordinates to describe an event. In this section, we shall give x-positions and time in meters. A distance in meters can be associated with time by multiplying all times by the speed of light c. Time in meters is the distance a flash of light will travel in that time.

Four events, A, B, C, and D, are shown in Fig. 12.13. We note that A and B occur at the same time with $ct = 3m$. The x coordinate of A is 3 m, while the x coordinate of B is -2 m. The events, A and C, have the same x coordinate and hence occur at the same point along the x-axis at different times. The ct coordinate is 3 m for A and 1 m for C. The coordinates of D are $x = 2$ m and $ct = -2$ m. Diagrams of this kind which give the x and ct coordinates of events are called *space-time diagrams*. As we shall see, these diagrams can help us to understand intuitively many of the special features of relativity theory.

12.4.1 Particle motion

A moving particle traces out a line in the space-time diagram called the *worldline* of the particle. The worldline is the trajectory of the particle in a graph of x versus ct. As an example, we consider four particles moving with a constant speed during 3 m of time. The displacement of the particles during this time are shown in Fig. 12.14(A). During the time interval, $ct = 3.0$ m, the particles, #1, #2, #3, and #4, move distances, -1.0 m, 0.0 m, 1.5 m, and 3.0 m, respectively. The worldlines of these four particles, which are shown in Fig. 12.14(B), correspond to straight lines. In this figure, the slope of each line is equal to the ratio of its rise $\Delta(ct)$ and its displacement Δx. The slope of a worldline and the velocity of the corresponding particle are thus related by the equation

$$\Delta(ct)/\Delta x = c\,(1/(\Delta x/\Delta t)) = c/v.$$

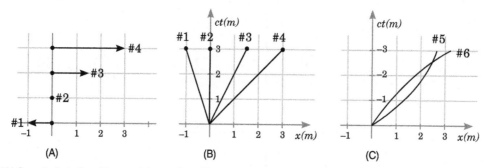

FIGURE 12.14 (A) Space trajectories of four particles moving with constant speeds. During the time intervals, $ct = 3.0$ m, the particles, #1, #2, #3, and #4, move distances, -1.0 m, 0.0 m, 1.5 m, and 3.0 m, respectively. (B) The world lines of these four particles. The constant slopes are a consequence of their constant speeds. (C) The curved world lines of two accelerating particles. The velocity of particle #5 is decreasing, while the velocity of particle #6 is increasing.

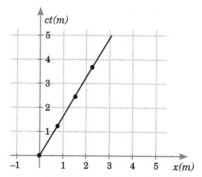

FIGURE 12.15 The worldline of a particle moving with a constant speed in the positive x-direction. The solid dots on the worldline indicate the times when the particle passes by clocks equally spaced along the x-axis.

Fig. 12.14(C) shows the curved worldlines of two particles with velocities that are changing. The velocity of particle #5 is decreasing, while the velocity of particle #6 is increasing.

Example 12.6

Find the speed of particles #3 and #4 in Fig. 12.14(B).

Solution

For particle #3, we find that $\Delta(ct) = 3.0$ m $- 0.0$ m $= 3.0$ m and $\Delta x = 1.5$ m $- 0.0$ m $= 1.5$ m. So, the slope of the worldline of this particle is equal to 3.0 m/1.5 m $= 2$. Since the slope of the worldline is equal to c/v, we set

$$\frac{c}{v} = 2.$$

The speed v of particle #3 is thus equal to $0.5c$. Similarly, the slope of the worldline of particle #4 is $\Delta(ct)/\Delta x = 3.0$ m/3.0 m $= 1$, and the velocity of particle #4 is equal to c. Only particles with zero mass can travel with the velocity of light.

The worldline of a particle is a record of its path through space-time. Fig. 12.15 shows a particle that travels a distance of $3m$ in $5m$ of time. This particle starts at the origin when the clock there records zero time and moves with constant speed in the positive x-direction passing by an array of equally spaced clocks in the laboratory. Each incident of a particle passing a clock is an event that could be recorded by an observer in the laboratory. The solid dots on the worldline representing these events are like beads on a string.

In the previous section, we found that it is possible to define an invariant space-time interval, which is the same for all observers. The space-time interval is given by Eq. (12.32). For the particle with the worldline shown in Fig. 12.15, this

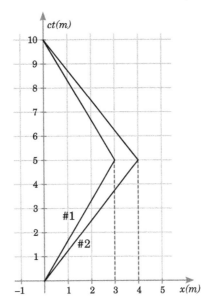

FIGURE 12.16 The worldline of two particles which travel with constant speed in the positive x-direction and then return to the origin.

formula leads to the following result

$$(\text{invariant interval})^2$$
$$= (5\text{m})^2 - (3\text{m})^2 = (4\text{m})^2 .$$

The invariant space-time interval is thus 4m. As discussed in the previous section, this is equal to the proper time of the particle or the length of time that would elapse on a clock moving in the reference frame of the particle. The fact that 4m of time elapses in the moving frame while 5m passes in the laboratory system is an example of time dilation.

This result can easily be generalized to a particle moving with speed v in the laboratory frame. The proper time for the particle in meters is given by the formula

$$(\text{proper time})^2 = (\text{lab time})^2 - (\text{distance traveled})^2 .$$

The distance traveled by the particle in the laboratory frame is

$$(\text{distance traveled}) = (\text{speed}) \times (\text{lab time}).$$

Substituting the above expression for the distance traveled into the equation for the proper time, we obtain

$$(\text{proper time})^2 = (\text{lab time})^2[1 - (\text{speed})^2].$$

This leads to the following expression for the ratio of the lab time and the proper time

$$\frac{(\text{lab time})}{(\text{proper time})} = \frac{1}{\sqrt{1 - (\text{speed})^2}},$$

which is consistent with the time dilation formula obtained earlier. Notice that the ratio of the laboratory and proper time is larger for a faster moving particle.

Fig. 12.16 shows the "kinked" worldlines of two particles (#1 and #2) which travel a distance in the positive x-direction and then return to the origin. When each particle and the clock traveling with it turns around to return to the origin, it experiences a period of rapid acceleration, which corresponds to the change in the slope of the worldline of the particle. Clocks behave differently when accelerated with small clocks such as wristwatches faring better than large pendulums. We shall assume the ideal limit of (acceleration-proof) clocks.

The total length of time elapsed on laboratory clocks during the outward and return motions of particles #1 and #2 is 10m. Since the lower portion of the worldline of particle #1 is identical to the worldline of the particle shown in Fig. 12.15,

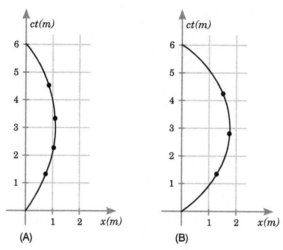

FIGURE 12.17 (A) The curved worldline of a particle that moves in the positive x-direction and returns to the origin. (B) The longer worldline of a particle that moves in the positive x-direction and returns to the origin in a shorter amount of proper time.

we know that the total proper time for particle #1 is equal to 2×4m $= 8$m. The proper time corresponding to the lower portion of the worldline of particle #2 can be found using the equation

$$(\text{proper time})^2 = (5\text{m})^2 - (4\text{m})^2 = (3\text{m})^2 .$$

The proper time for the outward motion of particle #2 is 3m, and the total proper time of particle #2 is equal to 2×3m $= 6$m. For particle #2, which travels faster and has a longer worldline, the total proper time is less. This is due to the minus sign in the expression for the invariant interval (12.32). For particles traveling through space-time, longer path length means less proper time.

The observations we have made about the worldlines in Fig. 12.16 apply as well to particles whose velocity and direction of motion change continuously. Figs. 12.17(A) and (B) show the curved worldlines of two particles that move in the positive x-direction and eventually find there way back to the origin. Each particle has a beacon which flashes at regular intervals according to the time in its own reference frame. These flashes, which are indicated by solid dots on the world lines of the particle, represent the passage of proper time for each particle. Notice that the world line of the particle shown in Fig. 12.17(A) is longer than the portion of the time axis corresponding to the motion of the particle and the elapsed proper time is less than the elapsed time measured along the vertical axis. The distance along the vertical axis corresponding to the motion of the particle is equal to the elapsed time in the laboratory frame of reference. Also notice that the total proper time for the particle moving along the longer worldline shown in Fig. 12.17(B) is less than the proper time of the other particle.

12.4.2   Lorentz transformations

The coordinates of two reference frames – one moving relative to the other – can be displayed in a single space-time diagram. Consider again the two coordinate frames, S and S', shown in Fig. 12.1. As before, we suppose that the origins of the two frames are in the same position for $t = 0$, and that the frame S' is moving with a constant velocity u along the x-axis of the frame S. Using the x-axis and clocks of frame S, we may determine the x and ct coordinates of events and record points for them describing the events in our space-time diagram. To assign the ct' and x' coordinates of events, we must first draw the ct' and x' axis in our diagram.

The ct' axis may be identified as the locus of points for which $x' = 0$. We set $x' = 0$ in the first of the Lorentz transformations (12.15), to obtained

$$x' = \gamma(x - u\,t) = 0.$$

This last equation may be divided by γ and the resulting equation rewritten giving

$$x - (u/c)ct = 0.$$

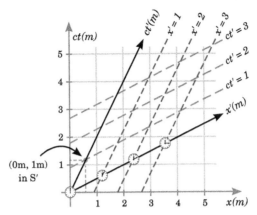

FIGURE 12.18 A space-time diagram for the reference frame S showing the coordinate axes x' and ct' of the S' reference frame. S' moves with speed $u = 0.5\,c$ along the x-axis of S. The coordinate lines for $x' = 1, 2, 3$ and $ct' = 1, 2, 3$ are represented by dashed lines. The point $(0\,m, 1\,m)$ in S' is indicated.

The factor u/c may then be denoted by β and the equation solved for ct to obtain

$$ct = (1/\beta)x,$$

This equation corresponds to a straight line passing through the origin with slope equal to $(1/\beta)$.

In the same way, the x' axis may be defined as the locus of points for which $ct' = 0$. Setting $t' = 0$ in the fourth Lorentz transformation (12.15), we get

$$t' = \gamma(t - u\,x/c^2) = 0,$$

which leads to the equation

$$ct = \frac{u}{c}x = \beta x.$$

The slope of the x' axis is β.

Fig. 12.18 gives the space-time diagram of S showing S' moving at the speed $u = 0.5\,c$. In this figure, the x' and ct' axes of S' are shown and also the coordinate lines for $ct' = 1, 2, 3$ and $x' = 1, 2, 3$. These coordinate lines can be determined in the same way we determined the coordinate axes. Fig. 12.18 may be used to see how the space-time coordinates of S and S' are related.

One feature of Fig. 12.18 to notice is that both the ct' axis and the x' axis are inclined with respect to the corresponding axes in S. As we have mentioned, the points along the ct' axis correspond to the origin ($x' = 0$) in the S' frame. The ct' axis is the worldline of the origin of the S' frame in the space-time diagram of an observer in the S frame. We have found previously that the slope of a worldline is equal to the velocity of light divided by the velocity of the particle or $1/\beta$. Fig. 12.18 is drawn to describe the case for which the velocity of the S' frame is $0.5\,c$. The parameter β would then be equal to 0.5 and $1/\beta = 2.0$, which is consistent with the orientation of the ct' axis shown in Fig. 12.18. For the case in which the reference frame S' is moving with a velocity $0.5\,c$, the slope of the x' axis is equal to $1/2$. Of course, if the vertical axis is pitched forward and the horizontal axis tilted up, the two axes will no longer be orthogonal. One must keep in mind that Fig. 12.18 illustrates a space-time diagram – not two coordinate frames in space.

Example 12.7

Use the space-time diagram shown in Fig. 12.18 to transform the space-time coordinates $(0m, 1m)$ in S' into coordinates in S.

Solution

The point $(0m, 1m)$ in S' corresponds to the coordinates, $x' = 0$ and $ct' = 1m$. In Fig. 12.18, this point is at the intersection of the ct' axis with the $ct' = 1$ coordinate line. A vertical line coming down from this point intersects the x axis at a point slightly beyond the midpoint of the first interval. We estimate that $0.5m \leq x \leq 0.7m$. Similarly, a horizontal line passing through the point intersects the ct axis slightly above the grid line $ct = 1.0m$. We estimate that $1.1m \leq ct \leq 1.3m$. The exact values of the x and ct coordinates can be found using the inverse Lorentz transformations (12.17). To obtain the value of x, we use the first equation, which can be

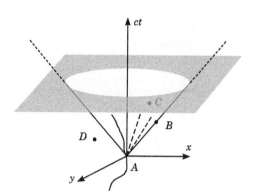

FIGURE 12.19 The worldline of a particle moving in two-dimensions with a cone (called the *light cone*) joined to the worldline at point A.

written

$$x = \gamma(x' + \beta\, ct').$$

As we have said, Fig. 12.18 is drawn with the velocity of S' being $0.5\,c$. Then, $\beta = 0.5$ and $\gamma \approx 1.155$. Since $x' = 0$ and $ct' = 1$ m, we have

$$x = \gamma(0.5m) \approx 0.5775 \text{ m}.$$

To obtain the value of ct, we multiply the fourth of the inverse transformations through with c to obtain

$$ct = \gamma(ct' + \beta x').$$

Using the value of γ obtained previously and the coordinates $x' = 0$ and $ct' = 1.0m$, we obtain

$$ct = \gamma(1.0m) \approx 1.155 \text{ m}.$$

The values of x and ct we obtained using the inverse Lorentz transformations are consistent with the approximate values obtained from the space-time diagram shown in Fig. 12.18. Other examples of the use of space-time diagrams can be found in the problems.

12.4.3 The light cone

Thus far we have only been concerned with motion in one-dimension. We can add another spatial dimension to our space-time diagrams by making our drawings three-dimensional. Fig. 12.19 shows the worldline of a particle moving in two dimensions. The vertical axis again corresponds to ct, while the two horizontal axes correspond to the x- and y-coordinates. At a particular point along the worldline of the particle denoted by A, a cone is drawn with a base angle of 45 degrees. Suppose light is emitted by the particle at the point A. In a time interval Δt, light will move a distance $c\Delta t$ in the ct direction and also a distance $c\Delta t$ in space. The light signal will thus move from A along the surface of the cone. For this reason, the cone shown in Fig. 12.19 is called the *light cone*. Two dashed lines moving upward from the point A in Fig. 12.19 correspond to other possible trajectories of the worldline of the particle. Because the speed of a particle with nonzero mass must be less than c, the trajectory of a massive particle must lie within the light cone.

The passage of the particle through point A is an event that would be recorded by an observer at that location. Three other events denoted by B, C, and D are also shown in Fig. 12.19. The event B lies on the light cone, while event C lies within the light cone and event D lies outside the light cone. To describe the relation of event A with other events in space-time, we define three different space-time separations

$$(\text{timelike separation})^2 = (c \times \text{time separation})^2 - (\text{distance})^2 > 0$$
$$(\text{spacelike separation})^2 = (\text{distance})^2 - (c \times \text{time separation})^2 > 0$$
$$(\text{lightlike separation})^2 = (c \times \text{time separation})^2 - (\text{distance})^2 = 0.$$

Two events are said to have a lightlike separation if the invariant separation between them is equal to zero. The existence of this possibility is a unique feature of the Lorentz geometry due to the minus sign in the definition of the space-time interval. The fact that B lies on the light cone implies that a light signal emitted from A in the direction of B would arrive at event B and, hence, that the separation of A and B is lightlike. In contrast, the event C in the upper portion of the light cone is timelike, and event D outside the light cone are spacelike.

The motion of a particle with nonzero mass provides an example of timelike events. Imagine a massive particle emitting flashes of light at regular time intervals as it moves through space. Since a massive particle has a velocity less than the velocity of light, the spatial separation between two flashes will be less that c times the time separation of the flashes. The separation of the flashes will thus be timelike. If the particle were traveling with a constant velocity, all of the flashes would occur at the same point in the rest frame of the particle. This property of events with a timelike separation is generally true. For any two events having a timelike separation, one can make a Lorentz transformation to a moving frame in which the two event occur at the same location.

Two events have a spacelike separation when the spatial separation of the events dominates over the time separation. The distance between two spacelike events is greater than c times the time separation. This implies that signal would have to travel faster than the speed of light to pass from one event to the other event. Since no signal can travel faster than the speed of light, an event cannot *cause* or *influence* another event with a spacelike separation. One can always make a Lorentz transformation to a moving frame in which the events with a spacelike separation occur at the same time. Of course, one event cannot cause another event occurring at the same time. An event, which causes another event, must occur before the event it causes.

12.5 Four-vectors

We can put the Lorentz transformations in a more symmetric form by multiplying the equation transforming the time coordinates in Eq. (12.15) by c and writing these equations before the others to obtain

$$ct' = \gamma \left[ct - \beta x \right]$$
$$x' = \gamma \left[x - \beta ct \right]$$
$$y' = y$$
$$z' = z,$$

where $\beta = u/c$. In the above equations, the quantities ct and ct' appear on the same footing as the x and x' coordinates.

We define the position-time four-vector x^μ with components

$$x^0 = ct$$
$$x^1 = x$$
$$x^2 = y$$
$$x^3 = z.$$

The position-time vector may be written so that the components appear explicitly

$$x^\mu = (ct, x, y, z), \tag{12.43}$$

or more simply

$$x^\mu = (ct, \mathbf{r}).$$

The Lorentz transformations can be written in terms of the components of x^μ giving

$$x'^0 = \gamma(x^0 - \beta x^1)$$
$$x'^1 = \gamma(x^1 - \beta x^0)$$
$$x'^2 = x^2$$
$$x'^3 = x^3, \tag{12.44}$$

where β and γ are defined by the equations

$$\beta = u/c \tag{12.45}$$

and

$$\gamma = \frac{1}{\sqrt{1 - u^2/c^2}}. \tag{12.46}$$

We shall use a single Greek letter Λ to denote the coefficients in the Lorentz transformations (12.44) writing the equation

$$
\begin{aligned}
x'^0 &= \Lambda^0{}_0 x^0 + \Lambda^0{}_1 x^1 \\
x'^1 &= \Lambda^1{}_0 x^0 + \Lambda^1{}_1 x^1 \\
x'^2 &= \Lambda^2{}_2 x^2 \\
x'^3 &= \Lambda^3{}_3 x^3.
\end{aligned}
\tag{12.47}
$$

In this notation, the upper index of Λ refers to the component of the vector on the left-hand side of the equation, while the lower index refers to component of the vector on the right. The coefficients $\Lambda^\mu{}_\nu$ may be identified by comparing Eqs. (12.44) and (12.47). The coefficients $\Lambda^0{}_0$ and $\Lambda^1{}_1$ are equal to γ, $\Lambda^0{}_1$ and $\Lambda^1{}_0$ are equal to $-\beta\gamma$, and $\Lambda^2{}_2$ and $\Lambda^3{}_3$ are equal to 1. All other coefficient are zero. The transformation coefficients may be displayed using matrix notation

$$
[\Lambda^\mu{}_\nu] =
\begin{bmatrix}
\gamma & -\beta\gamma & 0 & 0 \\
-\beta\gamma & \gamma & 0 & 0 \\
0 & 0 & 1 & 0 \\
0 & 0 & 0 & 1
\end{bmatrix},
$$

where μ labels the rows of the Λ-matrix and ν labels the columns.

The Lorentz transformations may be written in the following compact form using vector notation

$$x'^\mu = \sum_{\nu=0}^{3} \Lambda^\mu{}_\nu x^\nu. \tag{12.48}$$

Notice that any one of Eq. (12.47) can be obtained by setting μ equal to $0, 1, 2,$ or 3 in this last equation. A further simplification of the transformation equations can be obtained by adopting a convention originally suggested by Einstein. According to this convention called the *summation convention*, repeated Greek indices are summed from 0 to 3. The summation over ν in Eq. (12.48) is then understood and the Lorentz transformations become simply

$$x'^\mu = \Lambda^\mu{}_\nu x^\nu. \tag{12.49}$$

For the concept of a four-vector to be useful, we must be able to define the length of a vector. We recall that the length of a vector in three-dimensional space is invariant with respect to rotations. It is thus natural for us to choose a definition of the length of a four vector which is invariant with respect to the relativistic transformation of coordinates given by Eqs. (12.44) and (12.49). We define the square of the length of the vector x^μ by the equation

$$|x|^2 = (x^0)^2 - (x^1)^2 - (x^2)^2 - (x^3)^2. \tag{12.50}$$

Notice how this definition of length differs from the definition used in three dimensions

$$|\mathbf{r}|^2 = (x^1)^2 + (x^2)^2 + (x^3)^2 .$$

In three dimensions, the length of a vector is equal to the sum of the squares of its components. In contrast, the length of a four vector given by Eq. (12.50) is equal to the square of its zeroth component *minus* the squares of its spatial components. Of course, the form of our definition of the length of a four-vector is similar to the definition of the invariant space-time interval considered earlier.

In Eq. (12.50), the coordinates (x^0, x^1, x^2, x^3) are defined in the reference frame S. The square of the length of the four vector x'^μ in the reference frame S' is

$$|x'|^2 = \left(x'^0\right)^2 - \left(x'^1\right)^2 - \left(x'^2\right)^2 - \left(x'^3\right)^2 .$$

Substituting the values of the primed coordinates given by Eq. (12.44) into the above equation, we obtain

$$|x'|^2 = \gamma^2(x^0 - \beta x^1)^2 - \gamma^2(x^1 - \beta x^0)^2 - (x^2)^2 - (x^3)^2 \, .$$

In this last equation, the cross terms, which result from the first two terms in parentheses, cancel and the remaining terms can be rearranged to give

$$|x'|^2 = \gamma^2(1 - \beta^2)(x^0)^2 - \gamma^2(1 - \beta^2)(x^1)^2 - (x^2)^2 - (x^3)^2 \, . \tag{12.51}$$

Using Eqs. (12.46) and (12.45), the quantity $\gamma^2(1 - \beta^2)$ may be shown to be equal to one. It then follows immediately that the right-hand side of Eq. (12.51) is equal to the right-hand side of Eq. (12.50). We thus have

$$|x'|^2 = |x|^2. \tag{12.52}$$

The above equation says that the square of the length of a vector is independent of the reference frame. This is generally described by saying that the square of the length of a vector is *Lorentz invariant*. The idea of Lorentz invariance is of central importance to relativity theory. Einstein emphasized the point that a man moving in a train would not know if he were moving or not if the windows of his train car were covered. The laws of physics are independent of the state of motion of the observer.

The square of the length of a vector may be written using a *metric tensor* $g_{\mu\nu}$, whose components can be displayed as a matrix

$$[g_{\mu\nu}] = \begin{bmatrix} 1 & 0 & 0 & 0 \\ 0 & -1 & 0 & 0 \\ 0 & 0 & -1 & 0 \\ 0 & 0 & 0 & -1 \end{bmatrix}. \tag{12.53}$$

The g_{00} element of this matrix is equal to one, and the other diagonal elements of the matrix are equal to minus one. Using the metric tensor, the square of the length of x^μ can be written

$$|x|^2 = g_{00}\, x^0 x^0 + g_{11}\, x^1 x^1 + g_{22}\, x^2 x^2 + g_{33}\, x^3 x^3 \, .$$

This last equation, which expresses $|x|^2$ as a sum over the time and space components of x_μ, may be written

$$|x|^2 = \sum_{\mu=0}^{3} \sum_{\nu=0}^{3} g_{\mu\nu} x^\mu x^\nu,$$

or with the help of the Einstein summation convention

$$|x|^2 = g_{\mu\nu} x^\mu x^\nu. \tag{12.54}$$

The metric tensor thus enables us to combine products of the components of a four-vector to form an invariant length squared. Combining products of the components x'^μ in this way, we obtain

$$|x'|^2 = g_{\mu\nu} x'^\mu x'^\nu.$$

We may substitute Eq. (12.49) into this last equation to obtain the following expression for $|x'|^2$ in terms of the components in the reference frame S

$$|x'|^2 = g_{\mu\nu} \Lambda^\mu{}_\rho \Lambda^\nu{}_\sigma x^\rho x^\sigma.$$

Using the above equation to evaluate $|x'|^2$ and Eq. (12.54) to evaluate $|x|^2$, Eq. (12.52) may be written

$$g_{\mu\nu} \Lambda^\mu{}_\rho \Lambda^\nu{}_\sigma x^\rho x^\sigma = g_{\rho\sigma} x^\rho x^\sigma \, ,$$

where the indices ρ and σ on the right-hand side serve the role of μ and ν in Eq. (12.54). We may now identify the coefficients of $x^\rho x^\sigma$ on both sides of this last equation to get

$$g_{\mu\nu} \Lambda^\mu{}_\rho \Lambda^\nu{}_\sigma = g_{\rho\sigma}. \tag{12.55}$$

This is a condition imposed on the coefficients $\Lambda^\mu{}_\nu$ by the requirement that the norm of the position vector is invariant with respect to Lorentz transformations.

The concept of invariance has been used in the previous section to describe the invariant interval between two events in space-time and in the present section to describe the length of a four-vector. We shall use the concept of Lorentz invariance in the next chapter when describing the Dirac equation, which is the relativistic wave equation for the electron, and we will use Lorentz invariance in describing the interaction of elementary particles in following chapters. All possible Lorentz invariants can be described using two different kinds of four-vectors, which are called *contravariant vectors* and *covariant vectors*. The position-time four vector (12.43) is an example of a contravariant vector, while the derivatives of the time and the space coordinates transform as a covariant vector. Since the letter "v" is used to denote velocity, we shall use instead the letter "t" to stand for a general four-vector. Four numerical values t^μ will be referred to as a contravariant vector if the components transform with respect to Lorentz transformations according to the equation

$$t'^\mu = \Lambda^\mu{}_\nu t^\nu. \tag{12.56}$$

Comparing this equation with Eq. (12.49), we see that the position-time vector x^μ satisfies the defining condition for a contravariant vector.

We define the *covariant* four-vector t_μ (with index down) by the equation

$$t_\mu = g_{\mu\nu} t^\nu. \tag{12.57}$$

Since the g_{00} component of the metric tensor (12.53) is one and the diagonal position components are minus one, the 0-component of the covariant vector will be equal to the 0-component of the original contravariant vector, while the position components will differ in sign. An example is provided by the position-time vector which has covariant components

$$x_\mu = (ct, -x, -y, -z).$$

We shall denote the inverse of the metric tensor by $g^{\mu\nu}$. The inverse has the same components as the metric tensor: it is diagonal and has components $g^{00} = +1$, $g^{11} = g^{22} = g^{33} = -1$. The product of the metric tensor and its inverse satisfy the equation

$$g^{\mu\nu} g_{\nu\sigma} = \delta^\mu_\sigma, \tag{12.58}$$

where as before the summation over the repeated ν index is implied. The Kronecker delta δ^μ_σ is equal to one if μ is equal to σ and is zero otherwise. Multiplying Eq. (12.57) with $g^{\sigma\mu}$ and using Eq. (12.58) gives

$$g^{\sigma\mu} t_\mu = g^{\sigma\mu} g_{\mu\nu} t^\nu = \delta^\sigma_\nu t^\nu = t^\sigma. \tag{12.59}$$

According to Eqs. (12.57) and (12.59), $g_{\mu\nu}$ and $g^{\mu\nu}$ may be used to lower or raise the indices of a vector.

The square of the length of a four-vector can be written simply using both the contravariant and covariant components of the vector. For instance, using Eq. (12.54), the square of the length of the position-time vector can be written

$$|x|^2 = g_{\mu\nu} x^\mu x^\nu = x_\nu x^\nu.$$

The square of the length of a vector, which is a Lorentz invariant, can be formed by carrying out a sum of products where one index is down and the other index is up. As we have discussed before in another context, the square of the length of a four-vector t^μ need not be positive. If $|t|^2$ is positive we say that t^μ is timelike, while if $|t|^2$ is negative we say that t^μ is spacelike. If $|t|^2 = 0$, we say that t^μ is lightlike.

In order to derive a transformation equation involving the covariant components of the vector t_μ, we multiply the transformation equation (12.56) from the left with $g_{\sigma\mu} \Lambda^\sigma{}_\tau$ to obtain

$$g_{\sigma\mu} \Lambda^\sigma{}_\tau t'^\mu = g_{\sigma\mu} \Lambda^\sigma{}_\tau \Lambda^\mu{}_\nu t^\nu.$$

Using Eqs. (12.55) and (12.57), this equation can be written

$$\Lambda^\sigma{}_\tau t'_\sigma = g_{\tau\nu} t^\nu = t_\tau.$$

The covariant components of the position vector t_ν thus transform according to the equation

$$t_\nu = \Lambda^\mu{}_\nu t'_\mu. \tag{12.60}$$

In order to derive a transformation equation for derivatives with respect to the components of the position-time vector, we make use of the chain rule to write

$$\frac{\partial \psi}{\partial x^\nu} = \frac{\partial x'^\mu}{\partial x^\nu} \frac{\partial \psi}{\partial x'^\mu}, \tag{12.61}$$

where we have used the summation convention of Einstein summing over the μ index. The partial derivatives of x'^μ with respect to x^ν may be evaluated using the transformation equation (12.49) to obtain

$$\frac{\partial x'^\mu}{\partial x^\nu} = \Lambda^\mu{}_\nu.$$

Substituting this last equation into Eq. (12.61) gives

$$\frac{\partial \psi}{\partial x^\nu} = \Lambda^\mu{}_\nu \frac{\partial \psi}{\partial x'^\mu}.$$

Comparing this equation with Eq. (12.60), we see that the components of the derivative operator transform as the covariant components of a vector. For this reason, we denote the derivative operator by ∂_μ. This operator has components

$$\partial_\mu = \frac{\partial}{\partial x^\mu} = (\frac{\partial}{\partial x^0}, \frac{\partial}{\partial x^1}, \frac{\partial}{\partial x^2}, \frac{\partial}{\partial x^3}) = (\partial^0, \nabla).$$

This notation can be readily used to express common relations in classical physics. The *Lorentz condition* in the classical theory of electromagnetism,

$$\nabla \cdot \mathbf{A} + \frac{1}{c} \frac{\partial \phi}{\partial t} = 0,$$

can be written simply

$$\partial_\mu A^\mu = 0,$$

where the four-vector A^μ is defined

$$A^\mu = (\phi, \mathbf{A}).$$

The Lorentz transformations can thus be cast into a form where ct appears on the same footing as the space coordinates. This has led us to define the four-vector x^μ with components, $x^0 = ct$, x^1, x^2, and x^3, and to obtain a Lorentz invariant by forming the square of the length of this vector, $|x|^2 = (x^0)^2 - (x^1)^2 - (x^2)^2 - (x^3)^2$. The concepts of four-vectors and Lorentz invariants will play an important role in coming chapters where we consider the relativistic Dirac equation and the interactions between elementary particles.

Suggestions for further reading

P.G. Bergmann, *Introduction to the Theory of Relativity* (Englewood Cliffs, New Jersey: Prentice-Hall, 1942).
E.F. Taylor and J.A. Wheeler, *Spacetime Physics*, Second Edition (New York: Freeman, 1992).
A. Pais, *Subtle Is the Lord: The Science and the Life of Albert Einstein* (Oxford: Oxford University Press, 1982).
W. Isaacson, *Einstein: His Life and Universe* (New York: Simon and Schuster, 2007).
D. Bohm *The Special Theory of Relativity* (New York: Benjamin, 1965).
J.H. Smith, *Introduction to Special Relativity* (New York: Benjamin, 1967).

Basic equations

Galilean transformations

Coordinate transformations

$$x' = x - ut$$
$$y' = y$$
$$z' = z$$

Velocity transformations

$$v'_x = v_x - u$$
$$v'_y = v_y$$
$$v'_z = v_z$$

The relativistic transformations

Lorentz transformations

$$x' = \gamma(x - u\,t)$$
$$y' = y$$
$$z' = z$$
$$t' = \gamma(t - u\,x/c^2),$$

where

$$\gamma = \frac{1}{\sqrt{1 - u^2/c^2}}$$

Inverse transformations

$$x = \gamma(x' + u\,t')$$
$$y = y'$$
$$z = z'$$
$$t = \gamma(t' + u\,x'/c^2)$$

Lorentz contraction

$$L_M = \sqrt{1 - u^2/c^2}\,L_R$$

Time dilation

$$\Delta t_M = \frac{\Delta t_R}{\sqrt{1 - u^2/c^2}}$$

The invariant space-time interval

$$(\text{invariant interval})^2 = (c \times \text{time separation})^2 - (\text{space separation})^2$$

Velocity transformation

$$v' = \frac{v - u}{1 - uv/c^2}$$

Doppler effect

$$f = \sqrt{\frac{1 \pm \beta}{1 \mp \beta}}\,f_0,$$

where the upper signs apply for an approaching light source, and the lower signs apply for a receding light source.

Four vectors

The position-time vector

$$x^\mu = (ct, x, y, z)$$

Lorentz transformation

$$x'^{\mu} = \Lambda^{\mu}{}_{\nu}x^{\nu},$$

where a summation is implied over the repeated index ν.

Transformation matrix

$$[\Lambda^{\mu}{}_{\nu}] = \begin{bmatrix} \gamma & -\beta\gamma & 0 & 0 \\ -\beta\gamma & \gamma & 0 & 0 \\ 0 & 0 & 1 & 0 \\ 0 & 0 & 0 & 1 \end{bmatrix},$$

where $\beta = u/c$ and $\gamma = 1/\sqrt{1 - u^2/c^2}$.

Square of the length of the four-vector x^{μ}

$$|x|^2 = (x^0)^2 - (x^1)^2 - (x^2)^2 - (x^3)^2$$

Metric tensor

$$[g_{\mu\nu}] = \begin{bmatrix} 1 & 0 & 0 & 0 \\ 0 & -1 & 0 & 0 \\ 0 & 0 & -1 & 0 \\ 0 & 0 & 0 & -1 \end{bmatrix}$$

Square of the length of the four-vector x^{μ}

$$|x|^2 = g_{\mu\nu}x^{\mu}x^{\nu},$$

where a summation is implied over the repeated indices μ and ν.

Summary

The requirement of relativity that the speed of light should be the same in all frames of reference moving at a constant velocity with respect to each other leads to the Lorentz transformations, which relate the space coordinates and the time of events in different frames of reference. The Lorentz transformations enable one to derive formulas describing Lorentz contraction and time dilation relating the length and time separation of events with respect to different reference frames.

An invariant space-time interval can be defined by the equation

$$(\text{interval})^2 = (c \times \text{time separation})^2 - (\text{space separation})^2.$$

For two events occurring at the same spatial location, the invariant interval is equal to c times the time separation of the events. The time interval in a reference frame in which events happen at the same point is called the proper time interval.

The Lorentz transformations can be cast into a form where ct appears on the same footing as the space coordinates. This leads to the definition the position-time vector x^{μ} with components, $x^0 = ct$, x^1, x^2, and x^3, and to the general definition of a four-vector t^{μ} as a set of four quantities that transform as the components of x^{μ}. Using the concept of a four-vector, one can construct Lorentz invariants, which are independent of the state of motion of the observer.

Questions

1. Write down the Galilean transformation relating the x'- and x-coordinates of two reference frames moving with respect to each other.
2. Suppose that a train is moving with speed u and that a ball is thrown with speed v in the same direction the train is moving. According to the Galilean transformations, what would be the speed of the ball with respect to an observer on the train?
3. How did the description of electromagnetic radiation provided by Maxwell cause conceptual problems for physicists of his day?
4. What are the fundamental postulates of the theory of relativity?

5. How could one determine whether or not two events occurred at the same time?
6. How is the length of a moving object defined?
7. Write down the transformation equation relating the x and x' coordinates for the Galilean and Lorentz transformations. How do these equations differ?
8. Which relativistic transformation should be used to derive the length contraction formula?
9. Which relativistic transformation should be used to derive the time dilation formula?
10. One event occurs at the origin when $t = 0$, and a second event occurs at the point $x = 3$ m along the x-axis when $ct = 5$ m. What would be the velocity of a particle that travels between these two events?
11. What is the space-time interval between the two events described in the preceding question?
12. How would the frequency of radiation we receive from a distant galaxy be affected if the galaxy were moving away from us?
13. Draw the worldline of a particle that moves a distance of 2 m in 4 m of time.
14. What is the slope of the worldline of the particle described in the preceding question?
15. Draw the worldline of a particle that travels from the origin to the point with $x = 6$ m in 10 m of time and then returns to the origin in the same amount of time.
16. What is the proper time for the trip described in the preceding question? Is your result consistent with the time dilation formula?
17. Sketch the worldline of a particle that is accelerating in the positive x-direction.
18. Is the separation of the two events described in Question 10 lightlike, timelike, or spacelike?
19. One event occurs at the origin at time t equal to zero, and a second event occurs at the point $x = 5$ m along the x-axis at time with $ct = 4$ m. Is the separation of these two events lightlike, timelike, or spacelike?
20. Could the first event described in the preceding problem cause the second event? Explain your answer.
21. Give the four components of the position-time vector x^{μ}.
22. What important property does the square of the length of x^{μ} have?
23. What is the square of the length of the four-vector (t^0, t^1, t^2, t^3)?
24. What are diagonal components of the metric tensor $g^{\mu\nu}$?
25. How do the derivatives with respect to the time and the spatial coordinates transform with respect to Lorentz transformations?

Problems

1. Suppose that a spaceship which is traveling at half the speed of light ($c/2$) passes a light source. What is the velocity of light from the source in the reference frame of the spaceship?
2. Suppose that an event occurs in the reference frame S with coordinates $x = 2$ m, $y = 0$ m and $z = 0$ m at $t = 4$ m/c. The frame S' moves in the positive x-direction with velocity $u = 0.2c$. The origins of S and S' coincide at $t = t' = 0$. (a) Use the Lorentz transformations (12.15) to find the coordinates of the event in S'. (b) Use the inverse transformation on the results of (a) to obtain the original coordinates and time.
3. Suppose that the frame S' moves along the positive x-axis of the frame S with velocity $u = 0.25c$. The origins of S and S' coincide at $t = t' = 0$. An event occurs in the frame S' with coordinates $x' = 2$ m, $y' = 0$ m, and $z' = 0$ m at $t' = 4$ m/c. (a) Use the inverse Lorentz transformations (12.17) to find the coordinates of the event in S. (b) Use the Lorentz transformation on the results of (a) to obtain the original coordinates and time.
4. How fast would an object have to be moving for its length to be contracted by one half?
5. A meter stick moves parallel to its length with speed $v = 0.5c$ relative to you. (a) How long would you measure the length of the meter stick to be? (b) How much time would it take the stick to pass by you?
6. A meter stick in frame S' makes an angle of 30° with the x' axis. If the frame moves parallel to the x-axis with speed $0.80c$ relative to frame S, what would an observer in S measure the length of the meter stick to be?
7. Supersonic jets can now attain speeds of about $3 \times 10^{-6} c$. By what percentage would a modern jet be contracted in length? How many seconds would a pilot's clock loose during one year compared to an observer on the ground?
8. Rocket A leaves a space station with a speed of $0.8c$. Later, rocket B leaves the space station traveling in the same direction with speed $0.6c$. How fast would the space traveler on rocket B observe rocket A to be moving?
9. In an asymmetric colliding beam experiment, a beam of electrons e^- strike a beam of positrons e^+ traveling in the opposite direction. Suppose that the electrons are moving in the positive x-direction and the positrons are moving in the negative x-direction. Denoting the position of an electron by x_e, the position of a positron by x_p, the position of the center of mass of an electron positron pair by X, and the mass of an electron or a positron by m, the position of

the center of mass is given by the equation

$$2mX = mx_e + mx_p .$$

(a) Suppose the speed of the electrons is $0.95c$ and the speed of the positrons is $0.2c$. Calculate the velocity of the center of mass of an electron-positron pair. (b) Suppose that the collision between an electron and a positron produces a particle at rest in the center of mass system, which decays in 2.0×10^{-8} s. What will the lifetime of the particle be in the laboratory frame of reference? The asymmetric colliding beam technique has been used in an experiment performed by scientists at Cornell University. The time-dilation effect makes the experiment possible.

10. A space traveler takes off from Earth and moves at speed $0.99c$ toward the star Sirius, which is 8.6 light years away. How long does it take to get there (a) as measured on Earth (b) as measured by a traveler on the spaceship?

11. Calculate the Doppler shift of the Sodium D_2 line with $\lambda = 589.0$ nm if the source is moving with speed $0.3c$ (a) toward the observer, (b) away from the observer, (c) in a transverse direction.

12. Suppose that a distant galaxy is moving away from us with speed $120,000$ km/s. What would be the relative shift $(\lambda - \lambda_0)/\lambda_0$ of its Hydrogen α line with $\lambda = 656.5$ nm?

13. A spaceship moving away from Earth with a speed of $0.80c$ reports back to us by sending a radio wave with a frequency of 100 MHz. To what frequency must we tune our receivers to get the report?

14. How fast must yellow ($\lambda = 590$ nm), green ($\lambda = 525$ nm), and blue ($\lambda = 460$ nm) lights be moving away from us to appear red ($\lambda = 650$ nm)?

15. Using Eqs. (12.38) and Eq. (12.16), derive the relativistic velocity transformation (12.39).

16. One event occurs at the origin at $t = 0$, and a second event has space and time coordinates, $x = 3$ m and $ct = 8$ m. Draw the worldline of an observer who travels between the two events with a constant velocity. Calculate the proper time measured by that observer.

17. Another observer travels between the two events given in the preceding problem. The second observer remains stationary for one meter of time, then travels with a constant velocity to $x = 4$ m in five meters of time, and finally returns to the point $x = 3$ m in two meters of time. Draw the worldline of the second observer and calculate the proper time interval between the two events measured by the second observer.

18. Draw a space-time diagram of S showing S' moving at the speed $u = 0.2c$. Find the coordinates in S' of an event in S with coordinates $x = 2$ m and $ct = 4$ m by reading the x' and ct' coordinates directly from your diagram. Compare your answer with the solution of Problem 2.

19. Draw a space-time diagram of S showing S' moving at the speed $u = 0.25c$. Find the coordinates in S of an event in S' with coordinates $x' = 2$ m and $ct' = 4$ m by reading the x and ct coordinates directly from your diagram. Compare your answer with the solution of Problem 3.

20. Obtain the solution of part (a) of Problem 2 by using the form (12.44) of the Lorentz transformations.

21. Obtain the inverse Lorentz transformations (12.17) by solving Eqs. (12.15) for the unprimed variables.

22. Using the transformation equation,

$$v'^{\mu} = \Lambda^{\mu}{}_{\nu} v^{\nu},$$

and Eqs. (12.57) and (12.59), which relate the covariant and contravariant components of a vector, show that the Lorentz transformation for the covariant components of the vector may be written

$$v'_{\mu} = \Lambda_{\mu}{}^{\nu} v_{\nu},$$

where

$$\Lambda_{\mu}{}^{\nu} = g_{\mu\rho} \Lambda^{\rho}{}_{\sigma} g^{\sigma\nu}.$$

23. Using the result of the preceding problem and the explicit form of the transformation matrix $\Lambda^{\mu}{}_{\nu}$, show that the corresponding transformation matrix $\Lambda_{\mu}{}^{\nu}$ is

$$[\Lambda_{\mu}{}^{\nu}] = \begin{bmatrix} \gamma & \gamma\beta & 0 & 0 \\ \gamma\beta & \gamma & 0 & 0 \\ 0 & 0 & 1 & 0 \\ 0 & 0 & 0 & 1 \end{bmatrix} .$$

24. The condition (12.52) can be expressed in terms of the covariant components of the position-time vector

$$|x|^2 = g^{\mu\nu} x_\mu x_\nu = |x'|^2 = g^{\rho\sigma} x'_\rho x'_\sigma \ .$$

Using Eq. (12.60) show that the coefficients defining the Lorentz transformations satisfy the additional condition

$$g^{\mu\nu} \Lambda^\rho_{\ \mu} \Lambda^\sigma_{\ \nu} = g^{\rho\sigma} \ .$$

Chapter 13

The relativistic wave equations and general relativity

Contents

13.1 Momentum and energy   311
13.2 Conservation of energy and momentum   314
13.3 * The Dirac theory of the electron   318
13.4 * Field quantization   327
13.5 The general theory of relativity   329

Suggestions for further reading   336
Basic equations   336
Summary   338
Questions   338
Problems   339

Quantum field theory is … with certain qualifications … the only way to reconcile quantum mechanics with special relativity.

Steven Weinberg

Relativity theory has provided us a means of seeing how the space and time coordinates of particles change as one goes from one coordinate system to another. In this chapter, we shall be concerned with finding definitions of the momentum and energy which are appropriate for particles traveling at high speeds, and we shall be concerned with formulating conservation laws for the momentum and energy that are independent of the relative motion of the observer.

Conservation laws that are not dependent upon any particular coordinate system can be obtained by expressing these laws in terms of vector quantities. As an example, consider the conservation of momentum, which for two colliding particles has the form

$$\mathbf{p_1}^i + \mathbf{p_2}^i = \mathbf{p_1}^f + \mathbf{p_2}^f.$$

This equation states that the sum of the momentum of the two particles before the collision (initially) is equal to the sum of the momentum of the two particles after the collision (finally). If one were to rotate the coordinate axes, the components of the vectors on each side of the equation would change, but the vector equation would remain valid. Expressing the basic quantities in terms of vectors enables us to ensure that the equations involving them are independent of our choice of coordinate axes.

13.1 Momentum and energy

These ideas can be applied to relativity theory. We can ensure that physical laws are independent of the relative motion of the observer by expressing physical relations in terms of four-vectors. The velocity of a particle in relativity theory is defined

$$v^\mu = \frac{dx^\mu}{d\tau}, \tag{13.1}$$

where dx^μ being a differential of the position-time vector x^μ has components (cdt, dx^1, dx^2, dx^3) and $d\tau$ is an increment of the proper time. This equation may be written out explicitly in terms of its components

$$v^0 = \frac{d(ct)}{d\tau}$$

$$v^1 = \frac{dx^1}{d\tau}$$

Modern Physics with Modern Computational Methods. https://doi.org/10.1016/B978-0-12-817790-7.00020-2

$$v^2 = \frac{dx^2}{d\tau}$$

$$v^3 = \frac{dx^3}{d\tau}.$$

The point to notice here is that the differential $dx^\mu = (cdt, dx^1, dx^2, dx^3)$ is the difference of two time-position vectors and is hence itself a four-vector. As we have seen, the proper time interval $d\tau$ is an interval of time in the rest frame of the particle. Since dx^μ transforms like a vector and $d\tau$ is invariant with respect to Lorentz transformations, v^μ transforms like a four-vector.

We suppose that the particle is at rest in the reference frame S', which is moving with a velocity v with respect to the reference frame S. Since the particle is moving with respect to the frame S and at rest with respect to the frame S', the time interval dt in the frame S is related to the time interval $d\tau$ in the frame S' by Eq. (12.26). We have

$$dt = \frac{d\tau}{\sqrt{1 - v^2/c^2}}.$$

Using this result, the equation for the relativistic velocity can be written

$$v^\mu = \gamma \frac{dx^\mu}{dt} \tag{13.2}$$

where

$$\gamma = \frac{1}{\sqrt{1 - v^2/c^2}}. \tag{13.3}$$

As for the original definition of the four-velocity, Eq. (13.2) can be written out in component form

$$v^0 = \gamma \frac{d(ct)}{dt} = \gamma c$$

$$v^1 = \gamma \frac{dx^1}{dt}$$

$$v^2 = \gamma \frac{dx^2}{dt}$$

$$v^3 = \gamma \frac{dx^3}{dt}.$$

The components of v^μ may thus be written

$$v^\mu = \gamma(c, v_x, v_y, v_z) = \gamma(c, \mathbf{v}). \tag{13.4}$$

The v^0 component is equal to γc, while the spatial components of v^μ are equal to γ times the corresponding components of the ordinary velocity $\mathbf{v}$. The square of the length of the relativistic velocity is

$$v_\mu v^\mu = \gamma^2(c^2 - v_x{}^2 - v_y{}^2 - v_z{}^2) = \gamma^2 c^2(1 - v^2/c^2) = c^2, \tag{13.5}$$

which is clearly invariant with respect to Lorentz transformations. As the velocity of the particle $\mathbf{v}$ approaches zero, γ approaches one. For small velocities, the spatial components of the relativistic velocity thus approach the components of the ordinary velocity of classical mechanics.

We can construct the four-momentum p^μ of a particle with mass m by multiplying the four-velocity v^μ by the mass of the particle

$$p^\mu = mv^\mu. \tag{13.6}$$

Since v^μ is a four-vector and m is a constant, p^μ transforms as a four-vector. This implies that relations expressed in terms of p^μ will be valid in any inertial reference system. Using Eqs. (13.3) and (13.4), the spatial part of the relativistic momentum can be written

$$\mathbf{p} = \gamma m \mathbf{v} = \frac{m\mathbf{v}}{\sqrt{1 - v^2/c^2}}, \tag{13.7}$$

while the time component of the momentum is

$$p^0 = \gamma mc. \tag{13.8}$$

We define the relativistic energy of a particle as

$$E = \gamma mc^2 = \frac{mc^2}{\sqrt{1 - v^2/c^2}}. \tag{13.9}$$

The validity of this expression can be appreciated by considering the limiting form of the equation as the velocity of the particle approaches zero. For small velocities, the denominator in Eq. (13.9) can be expanded in a Taylor series to give

$$\frac{1}{\sqrt{1 - v^2/c^2}} = 1 + \frac{1}{2}\frac{v^2}{c^2} + \frac{3}{8}\frac{v^4}{c^4} + \cdots. \tag{13.10}$$

(See Problem 5.) Substituting this expansion into the expression for the energy (13.9) gives

$$E = mc^2 + \frac{1}{2}mv^2 + \frac{3}{8}m\frac{v^4}{c^2} + \cdots. \tag{13.11}$$

Notice that the leading term on the right-hand side (mc^2) is a constant, while the second term is the nonrelativistic expression for the kinetic energy. In classical mechanics, *changes* in the energy of a particle affect its state of motion, but a constant additive term has no effect. The relativistic expression for the energy (13.9) applies to velocities in the range $0 \leq v < c$ and reduces to a constant plus the nonrelativistic expression for the kinetic energy when the velocity is small and the higher order terms in the expansion are negligible. The constant term in Eq. (13.11) is called the *rest energy*

$$R = mc^2. \tag{13.12}$$

The rest energy is the energy that a particle has when it is at rest. The *relativistic kinetic energy*, which is defined to be the difference between the energy and the rest energy of the particle

$$KE = E - R = (\gamma - 1)mc^2, \tag{13.13}$$

is the contribution to the energy that is attributable to the motion of the particle.

Comparing Eqs. (13.8) and (13.9), we see that the zeroth component of p^μ is E/c. The relativistic momentum p^μ thus has components

$$p^\mu = (E/c, p_x, p_y, p_z) = (E/c, \mathbf{p}).$$

Since the time component of p^μ is equal to E/c and the spatial components are equal to $\mathbf{p}$, p^μ is often referred to as the *energy-momentum four-vector*. The square of the length of p^μ is

$$p_\mu p^\mu = \frac{E^2}{c^2} - \mathbf{p}^2.$$

Using Eqs. (13.5) and (13.6), $p_\mu p^\mu$ may also be written

$$p_\mu p^\mu = m^2 c^2.$$

Equating the right-hand sides of these last two equations, we get

$$\frac{E^2}{c^2} - \mathbf{p}^2 = m^2 c^2.$$

This last equation, which relates the relativistic energy and momentum, is called the *energy-momentum relation*. It can be written

$$E^2 = \mathbf{p}^2 c^2 + m^2 c^4. \tag{13.14}$$

In the following, we shall refer to **p** simply as the *momentum* and reserve the more cumbersome expression *four-momentum* for the four-vector p^μ. For a particle with zero mass, Eq. (13.14) becomes simply

$$E = |\mathbf{p}|c. \tag{13.15}$$

The third term in the expansion for the energy (13.11) may be used to estimate the error that would be caused by using the classical expression for the kinetic energy, $\frac{1}{2}mv^2$. One may show that the velocity of a particle must be 11.5% the speed of light for the third term in the expansion for the energy to be one percent of the second term in the expansion. The relativistic formula for the kinetic energy (13.13) must thus be used for particles having a speed about ten percent the speed of light. For slowly moving particles, one may use the classical expression for the kinetic energy.

Example 13.1

Find the kinetic energy of an electron moving with velocity (*a*) $v = 1.00 \times 10^{-4}c$, (*b*) $v = 0.9c$.

Solution

(*a*) Since the velocity $v = 1.00 \times 10^{-4}c$ is far less than 11.5 percent the speed of light, we use the classical expression for the kinetic energy. We may write

$$KE = \frac{1}{2}mv^2 = \frac{1}{2}mc^2 \left(\frac{v}{c}\right)^2 = \frac{1}{2}mc^2(1.00 \times 10^{-4})^2.$$

Here the term mc^2 is the rest energy of the electron. Using the value 0.511 MeV given in Appendix A, we obtain

$$KE = 2.56 \times 10^{-3} \text{ eV}.$$

(*b*) Since the velocity $v = 0.9c$ is close to the speed of light, we use Eq. (13.13) in this case to obtain

$$KE = \left(\frac{1}{\sqrt{1-(0.9)^2}} - 1.0\right)(0.511 \text{ MeV}) = 0.661 \text{ MeV}.$$

The kinetic energy of the electron is now greater than its rest energy.

13.2 Conservation of energy and momentum

The importance of the energy and momentum depends upon the fact that these quantities are *conserved* in scattering processes. The scattering processes that occur in particle physics provide many clear examples of the use of these conservation laws. Consider, for example, a collision process in which particles A and B collide to produce particles C and D

$$A + B \rightarrow C + D.$$

The conservation of energy and momentum for this collision process may be stated succinctly by requiring that the sum of the energy-momentum four-vectors before and after the collision be equal

$$p^\mu{}_A + p^\mu{}_B = p^\mu{}_C + p^\mu{}_D.$$

If the kinetic energy is conserved in a collision, the collision is said to be *elastic*; while if the kinetic energy is not conserved, the collision is said to be *inelastic*. Since the energy is the sum of the rest and kinetic energies, and the energy is always conserved, any change in the sum of the masses of the particles will have to be made up by a corresponding change in their kinetic energies. As an example, we consider the process

$$e^- + p \rightarrow n + \nu,$$

in which an electron collides with a proton to produce a neutron and a neutrino. Since the sum of the masses of the neutron and the neutrino is more than the sum of the masses of the electron and the proton, the kinetic energy of the neutron and neutrino is less than the kinetic energy of the incident particles.

We would like to do apply the laws of the conservation of energy and momentum to a number of scattering problems.

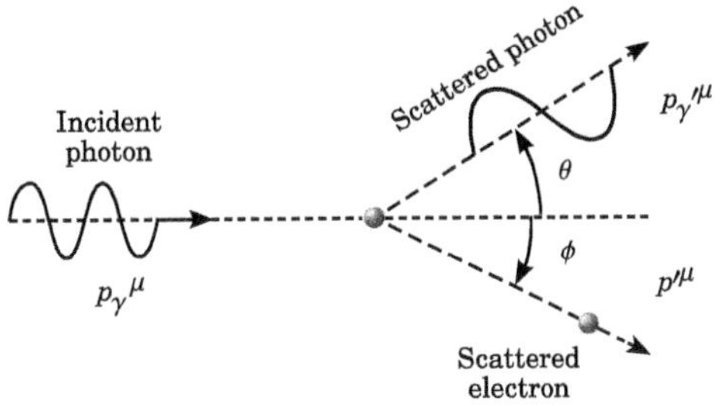

FIGURE 13.1 Compton scattering in which a photon scatters off an electron that is initially at rest. The energy-momentum four-vectors of the initial and final photons are denoted by p_γ^μ and $p_\gamma'^\mu$, and the four-vector of the final electron is denoted p'^μ.

Example 13.2

The Compton effect depicted in Fig. 13.1 is a collision event in which a photon scatters off an electron that is initially at rest

$$\gamma + e \to \gamma + e.$$

Here the photon is denoted by γ. Find the frequency of the scattered photon as a function of the scattering angle.

Solution

We shall denote the momentum four-vectors of the initial and final photons by $p_\gamma{}^\mu$ and $p_\gamma{}'^\mu$, and momentum four-vectors of the initial and final electrons by p^μ and p'^μ. The condition that the energy and momentum be conserved is equivalent to the requirement that the sum of the momentum four-vectors before and after the collision be equal

$$p_\gamma{}^\mu + p^\mu = p_\gamma{}'^\mu + p'^\mu.$$

Taking the electron four-vector p'^μ over to the left-hand side of the equation and the photon four-vector $p_\gamma{}^\mu$ over to the right-hand side and forming the square of the sum of four-vectors on the left and the right, we obtain

$$|p^\mu - p'^\mu|^2 = |p_\gamma{}'^\mu - p_\gamma{}^\mu|^2. \qquad (13.16)$$

The electron four-momenta are now on the left and the photon four-momenta are on the right.

If the electron is initially at rest, the initial energy of the electron, which we denote by E, is equal to the rest energy mc^2 and the spatial part of the initial momentum of the electron $\mathbf{p}$ is zero. We denote the energy and the spatial part of the momentum of the outgoing electron by E' and $\mathbf{p}'$. The left-hand side of Eq. (13.16) can be evaluated by taking the difference of the square of the zeroth and spatial components of the four-vector $p^\mu - p'^\mu$. We get

$$|p^\mu - p'^\mu|^2 = \frac{1}{c^2}(mc^2 - E')^2 - \mathbf{p}'^2.$$

Using the energy-momentum relation (13.14), the right-hand side of this last equation can be simplified giving

$$|p^\mu - p'^\mu|^2 = 2m^2c^2 - 2mE'. \qquad (13.17)$$

We shall denote the initial and final energies of the photon by hf and hf', respectively. According to Eq. (13.15), the magnitude of the momentum of a photon is its energy divided by c. We shall thus denote the initial and final momenta of the photon by $(hf/c)\hat{\mathbf{p}}$ and $(hf'/c)\hat{\mathbf{p}}'$, where $\hat{\mathbf{p}}$ and $\hat{\mathbf{p}}'$ are unit vectors pointing in the incoming and outgoing directions of the photon. With this notation, the right-hand side of Eq. (13.16) can be written

$$|p_\gamma{}'^\mu - p_\gamma{}^\mu|^2 = \frac{h^2}{c^2}(f' - f)^2 - \frac{h^2}{c^2}(f'\,\hat{\mathbf{p}}' - f\,\hat{\mathbf{p}})^2$$

We note that the terms on the right-hand side of this last equation which involve the squares of the frequencies cancel out, and we get

$$|p_\gamma{}'^\mu - p_\gamma{}^\mu|^2 = -\frac{2h^2 f f'}{c^2}(1 - \cos\theta), \qquad (13.18)$$

where θ is the angle between the outgoing and incoming directions of the photon. Recall that Eqs. (13.17) and (13.18) give the left- and right-hand sides of Eq. (13.16). Equating these two results, we obtain

$$2m^2c^2 - 2mE' = -\frac{2h^2 ff'}{c^2}(1 - \cos\theta) . \tag{13.19}$$

The energy E' of the outgoing electron, which appears on the left-hand side of this equation, can be removed by using the conservation of the energy. The condition that energy is conserved in the process is

$$hf + mc^2 = hf' + E' .$$

Solving this equation for E' and substituting this value into Eq. (13.19), we get

$$-2mh(f - f') = -\frac{2h^2 ff'}{c^2}(1 - \cos\theta) .$$

Finally, multiplying this last equation through by $-c^2/2h$, we obtain

$$mc^2(f - f') = hff'(1 - \cos\theta) . \tag{13.20}$$

In 1923, Arthur Compton performed an experiment in which light scattered off electrons in graphite obtaining results consistent with the above formula. His experiment played a very important role in the development of physics because it showed that photons could scatter like ordinary particles.

Example 13.3

The intermediate vector boson Z_0, which has a rest mass energy of 91.187 GeV, is produced in collisions of positrons and electrons

$$e^+ + e^- \rightarrow Z_0.$$

How much energy must the positrons and electrons in symmetric colliding beams have to produce the Z_0?

Solution

According to Eq. (13.14), the energy and momentum of the Z_0 satisfy the equation

$$E_Z^2 - \mathbf{p}_Z^2 c^2 = M_Z^2 c^4. \tag{13.21}$$

Denoting the energy and momentum of the positron by E_p and $\mathbf{p}_p$ and the energy and momentum of the electron by E_e and $\mathbf{p}_e$, the requirements that the energy and momentum be conserved in the collision are

$$E_p + E_e = E_Z$$
$$\mathbf{p}_p + \mathbf{p}_e = \mathbf{p}_Z.$$

Substituting these conservation laws into Eq. (13.21) gives

$$(E_p + E_e)^2 - (\mathbf{p}_p + \mathbf{p}_e)^2 c^2 = M_Z^2 c^4. \tag{13.22}$$

In a colliding beam experiment involving particles of equal mass, the center of mass of two colliding particles is stationary, and the momenta of the particles is equal and opposite

$$\mathbf{p}_p = -\mathbf{p}_e.$$

The energies of the two particles is equal

$$E_p = c\sqrt{\mathbf{p}_p^2 + m_p^2 c^2} = E_e.$$

Eq. (13.22) can thus be written simply

$$2E_p = M_Z c^2,$$

or

$$E_p = M_Z c^2/2.$$

The kinetic energy of the positron is equal to the energy E_p minus its rest energy $m_p c^2 = 0.511$ MeV. So, the kinetic energy of the positrons – and the electrons – is

$$KE = M_Z c^2/2 - m_p c^2 = 45.08 \text{ GeV}.$$

Example 13.4

Suppose that a beam of positrons strike electrons that are at rest. What must the kinetic energy of the positrons be to produce the Z_0?

Solution

As in the previous example, the energy and momentum of the Z_0 satisfy Eq. (13.21). However, the conditions that the energy and momentum of the particles be conserved in the laboratory frame of reference are

$$E_p + m_e c^2 = E_Z$$
$$\mathbf{p}_p = \mathbf{p}_Z.$$

Substituting these values into Eq. (13.21) gives

$$(E_p + m_e c^2)^2 - \mathbf{p}_p^2 c^2 = M_Z^2 c^4. \tag{13.23}$$

Using Eq. (13.14), the left-hand side of this equation may be written

$$E_p^2 + 2m_e c^2 E_p + m_e^2 c^4 - \mathbf{p}_p^2 c^2 = 2m_e c^2 (E_p + m_e c^2),$$

and Eq. (13.23) becomes

$$2m_e c^2 (E_p + m_e c^2) = M_Z^2 c^4.$$

Solving this equation for E_p, we get

$$E_p = \frac{M_Z^2 c^4}{2m_e c^2} - m_e c^2,$$

and the kinetic energy of a positron in the beam is

$$KE = \frac{M_Z^2 c^4}{2m_e c^2} - 2m_e c^2 = 8.136 \times 10^6 \text{ GeV}.$$

For an experiment in which the electrons are at rest in the laboratory, the kinetic energy of the positrons must be $180,000$ times larger than they would have to be in a colliding beam experiment. A few calculations of this kind are sufficient for one to understand the popularity of collider experiments in recent years. Much of the energy in traditional scattering experiments goes into increasing the velocity of the center of mass of the colliding particles rather than increasing the velocity with which the particles approach each other in the center of mass frame.

Example 13.5

A rho meson (ρ^+) with a rest mass energy of 775.5 MeV decays into a pion (π^+) with a rest mass energy of 139.6 MeV and a gamma ray according to the reaction formula

$$\rho^+ \rightarrow \pi^+ + \gamma$$

Find the velocity of the π^+ produced by the reaction.

Solution

We shall suppose the rho meson is at rest when it decays. Denoting the energy and the momentum of the rho meson by E_ρ and $\mathbf{p}_\rho$, the energy and momentum of the pion by E_π and $\mathbf{p}_\pi$, and the energy and momentum of the gamma ray by E_γ and $\mathbf{p}_\gamma$, the requirement that the energy and the momentum be conserved in the rest frame of the rho leads to the equations

$$m_\rho c^2 = E_\pi + E_\gamma$$
$$0 = \mathbf{p}_\pi + \mathbf{p}_\gamma.$$

Since a gamma ray is a photon with zero mass, the energy of the gamma is equal to $|\mathbf{p}_\gamma|c$ and the second equation for the conservation of momentum implies that $\mathbf{p}_\pi = -\mathbf{p}_\gamma$. Using this information, the first equation for the conservation of energy may thus be written

$$m_\rho c^2 = E_\pi + c|\mathbf{p}_\pi|$$

We may now use Eq. (13.14) to rewrite this last equation

$$m_\rho c^2 = E_\pi + \sqrt{E_\pi^2 - m_\pi^2 c^4}$$

To solve this equation we first bring E_π over to the left-hand side and square both sides of the resulting equation to obtain

$$(m_\rho c^2 - E_\pi)^2 = E_\pi^2 - m_\pi^2 c^4$$

Explicitly squaring the term on the left-hand side, we obtain

$$m_\rho^2 c^4 - 2m_\rho c^2 E_\pi + E_\pi^2 = E_\pi^2 - m_\pi^2 c^4$$

The two E_π^2 terms cancel, and we may solve for E_π to obtain

$$E_\pi = \frac{m_\rho^2 c^4 + m_\pi^2 c^4}{2m_\rho c^2}$$

Substituting the rest mass energies of the rho meson and pi meson into this last equation, we get

$$E_\pi = 400.3 \text{ MeV}.$$

We may then solve the equation, $E_\pi = \gamma(m_\pi c^2)$, to find γ has the value 2.867 and use Eq. (13.47) to get

$$\frac{v}{c} = \sqrt{1 - 1/\gamma^2} = 0.937.$$

13.3 * The Dirac theory of the electron

We now look into the possibility of finding a relativistic analogue of the time-dependent Schrödinger equation considered in earlier chapters.

13.3.1 Review of the Schrödinger theory

In three dimensions, the Schrödinger equation can be written

$$i\hbar\frac{\partial \psi(\mathbf{r}, t)}{\partial t} = -\frac{\hbar^2}{2m}\nabla^2 \psi(\mathbf{r}, t) + V(\mathbf{r})\psi(\mathbf{r}, t), \tag{13.24}$$

where ∇ is the gradient operator and ∇^2 is the Laplacian operator. The properties of the gradient and Laplacian operators that are relevant to the following discussion are summarized in Appendix AA. Eq. (13.24) can be obtained from the nonrelativistic expression for the energy

$$E = \frac{\mathbf{p}^2}{2m} + V$$

by replacing the energy and the momentum with the operators

$$\hat{E} = i\hbar\frac{\partial}{\partial t}, \quad \hat{\mathbf{p}} = -i\hbar\nabla \tag{13.25}$$

and allowing the resulting expression to act on the wave function $\psi(\mathbf{r}, t)$.

The Schrödinger equation has a number of properties that are consistent with a naive understanding of the motion of the electron and the probabilistic nature of the theory. Since the operator $(-\hbar^2/2m)\nabla^2$, which corresponds to the kinetic energy, is invariant with respect to rotations and translations, the Schrödinger equation of a free particle,

$$i\hbar\frac{\partial \psi(\mathbf{r},t)}{\partial t} = -\frac{\hbar^2}{2m}\nabla^2\psi(\mathbf{r},t), \tag{13.26}$$

also has this property. The invariance of the theory is consistent with intuitive ideas about the homogeneity of space.

As we have seen in Chapter 2, the Schrödinger equation has free-particle solutions. In three dimensions, a free-particle is described by a wave function of the form

$$\psi_\mathbf{k}(\mathbf{r}) = A_\mathbf{k}e^{i(\mathbf{k}\cdot\mathbf{r}-\omega t)}. \tag{13.27}$$

This function is an eigenfunction of the energy and momentum operators (13.25) corresponding to the eigenvalues

$$E = \hbar\omega, \quad \mathbf{p} = \hbar k. \tag{13.28}$$

Substituting the wave function (13.27) into the Schrödinger equation for a free-particle (13.26) gives

$$\hbar\omega A_\mathbf{k}e^{i(\mathbf{k}\cdot\mathbf{r}-\omega t)} = \frac{\hbar^2 k^2}{2m}A_\mathbf{k}e^{i(\mathbf{k}\cdot\mathbf{r}-\omega t)}.$$

Using the equations for the eigenvalues (13.28), this equation can be written

$$(E - \frac{\mathbf{p}^2}{2m})A_\mathbf{k}e^{i(\mathbf{k}\cdot\mathbf{r}-\omega t)} = 0.$$

The plane wave solutions of the Schrödinger equation thus maintain the correct energy-momentum relation.

The Schrödinger equation may be used to derive a continuity equation for the probability density $\psi^*(\mathbf{r},t)\psi(\mathbf{r},t)$. Multiplying the Schrödinger equation (13.24) from the left with ψ^*, gives

$$i\hbar\psi^*\frac{\partial \psi}{\partial t} = -\frac{\hbar^2}{2m}\psi^*\nabla^2\psi + V(\mathbf{r})\psi^*\psi. \tag{13.29}$$

Similarly, multiplying the complex conjugate of the Schrödinger equation,

$$-i\hbar\frac{\partial \psi^*}{\partial t} = -\frac{\hbar^2}{2m}\nabla^2\psi^* + V(\mathbf{r})\psi^*,$$

from the right with ψ, gives

$$-i\hbar\frac{\partial \psi^*}{\partial t}\psi = -\frac{\hbar^2}{2m}\psi\nabla^2\psi^* + V(\mathbf{r})\psi^*\psi. \tag{13.30}$$

The continuity equation is obtained by subtracting Eq. (13.30) from Eq. (13.29) giving

$$i\hbar(\psi^*\frac{\partial \psi}{\partial t} + \frac{\partial \psi^*}{\partial t}\psi) = -\frac{\hbar^2}{2m}(\psi^*\nabla^2\psi - \psi\nabla^2\psi^*). \tag{13.31}$$

Eq. (13.31) may be expressed concisely in terms of the probability density and the probability current. As before, we define the probability density ρ to be

$$\rho = \psi^*\psi \tag{13.32}$$

In electromagnetic theory, the current density is defined to be the amount of charge per area which passes through a cross section of a conducting wire every second. The probability current is the probability per area that crosses a perpendicular surface each second. If the probability current $\mathbf{j}$ is defined

$$\mathbf{j} = \frac{\hbar}{2mi}(\psi^*\nabla\psi - \psi\nabla\psi^*), \tag{13.33}$$

the continuity equation (13.31) can be written

$$\frac{\partial \rho}{\partial t} + \nabla \cdot \mathbf{j} = 0, \tag{13.34}$$

where we have used the equality

$$\nabla \cdot \left(\psi^* \nabla \psi - \psi \nabla \psi^* \right) = \psi^* \nabla^2 \psi - \psi \nabla^2 \psi^* .$$

The continuity equation is important from a conceptual point of view because it insures that the increase of probability within an infinitesimal volume is equal to the flow of probability into the volume. Unless such a condition is fulfilled, the probability density defined by Eq. (13.32) is meaningless.

13.3.2 The Klein-Gordon equation

The most natural way to obtain a relativistic wave equation is to start with the relativistic expression for the energy

$$E^2 = \mathbf{p}^2 c^2 + m^2 c^4 \tag{13.35}$$

and adopt the same line of argument used to derive the Schrödinger equation. Replacing the energy and momentum with the operators (13.25) and letting the resulting expression act on the wave function $\psi(\mathbf{r}, t)$ leads to the relativistic wave equation

$$-\hbar^2 \frac{\partial^2 \psi}{\partial t^2} = (-\hbar^2 c^2 \nabla^2 + m^2 c^4) \psi, \tag{13.36}$$

which is known as the *Klein-Gordon equation*.

The free-electron functions (13.27) are solutions of the Klein-Gordon equation (13.36) just as they are solutions of the Schrödinger equation. As in the nonrelativistic theory, these functions are eigenfunctions of the energy and momentum operators corresponding to the eigenvalues (13.28). Substituting Eq. (13.27) into (13.36), one may show that the energy and momentum eigenvalues in the Klein-Gordon theory are related by the relativistic expression for the energy (13.35). Using Eq. (13.35), the energy eigenvalues can be written

$$E = \pm (\mathbf{p}^2 c^2 + m^2 c^4)^{1/2}.$$

There are thus positive and negative energy solutions of the Klein-Gordon equations. The negative energy solutions do not appear to be physically reasonable since transitions could occur to more and more negative energies. A particle could spiral down into oblivion.

In order to identify the probability density corresponding to the Klein-Gordon equation, we multiply the equation from the left with $-i\psi^*$ and subtract the equation obtained by multiplying the complex conjugate equation from the right with $-i\psi$ as we have in deriving Eq. (13.31). This gives

$$i\hbar^2 \left(\psi^* \frac{\partial^2 \psi}{\partial t^2} - \frac{\partial^2 \psi^*}{\partial t^2} \psi \right) = i\hbar^2 c^2 \left(\psi^* \nabla^2 \psi - \psi \nabla^2 \psi^* \right).$$

This equation may be expressed as a continuity condition identical to Eq. (13.34) with the probability density

$$\rho = \frac{i\hbar}{2mc^2} \left(\psi^* \frac{\partial \psi}{\partial t} - \frac{\partial \psi^*}{\partial t} \psi \right), \tag{13.37}$$

and with the current $\mathbf{j}$ defined as before. The probability density of free-electrons in the Klein-Gordon theory is obtained by substituting the free-electron wave function (13.27) into Eq. (13.37) giving

$$\rho = \frac{\hbar \omega}{mc^2} |A_{\mathbf{k}}|^2 = \frac{E}{R} |A_{\mathbf{k}}|^2, \tag{13.38}$$

where we have used Eq. (13.28) and the definition of the rest energy $R = mc^2$. For the positive energy states, Eq. (13.38) gives reasonable results, however, for the negative energy states the probability density given by the equation is negative which is unreasonable. Schrödinger originally derived the Klein-Gordon equation to describe the energy levels of hydrogen.

He found that the equation gave a fine structure for hydrogen in disagreement with the accurate measurements of F. Paschen and that the equation led to an expression for the probability density that could be negative. For these reasons, Schrödinger discarded the Klein-Gordon equation and derived the nonrelativistic equation that has come to bear his name. We now know that the Klein-Gordon equation applies to bosons with spin equal to zero.

13.3.3 The Dirac equation

Paul Dirac sought to derive a relativistic equation that did not have the negative features of the Klein-Gordon theory. He wanted in some way to avoid negative energy states and negative probability densities. Since the negative energy eigenvalues of the Klein-Gordon equation are an unavoidable consequence of the quadratic nature of the relation between the energy and momentum operators, Dirac sought to find a linear equation of the form

$$i\hbar\frac{\partial\psi}{\partial t} = -i\hbar c\left(\alpha^1\frac{\partial\psi}{\partial x^1} + \alpha^2\frac{\partial\psi}{\partial x^2} + \alpha^3\frac{\partial\psi}{\partial x^3}\right) + \beta mc^2\psi, \tag{13.39}$$

where α^1, α^2, α^3, and β, are coefficients that are as yet unknown. We shall find that for this equation to satisfy all of the conditions we would naturally impose upon it, the coefficients, α^1, α^2, α^3, and β, must be matrices. A review of the elementary properties of matrices is given in Appendix CC.

Eq. (13.39), which is known as the *Dirac equation* may be written more conveniently using vector notation

$$i\hbar\frac{\partial\psi}{\partial t} = c\alpha\cdot(-i\hbar\nabla\psi) + mc^2\beta\psi, \tag{13.40}$$

where the dot product in the first term on the right-hand side indicates a summation over products of the α matrices and the corresponding partial derivatives. We note that the time and the spatial coordinates in Eq. (13.40) appear on the same footing, which make it easy to satisfy the condition of Lorentz covariance. The Dirac equation is the appropriate relativistic wave equation for particle with spin one half such as electrons.

To provide a correct description of relativistic particles, the Dirac equation must have free-particle solutions that satisfy the correct energy-momentum relation (13.35), and it must be possible to define a probability density and a current satisfying a continuity equation. As shown in the books by Rose and Gross cited at the end of this chapter, the solutions of the Dirac equation (13.40) will also be solutions of the Klein-Gordon equation (13.36) and thus have plane wave solutions that satisfy the correct energy-momentum relation (13.35) if the Dirac matrices, α^i and β, satisfy the following conditions

$$\begin{aligned} \alpha^i\alpha^j + \alpha^j\alpha^i &= 2\delta_{ij}I \\ \alpha^i\beta + \alpha^j\beta &= 0 \\ \beta^2 &= I, \end{aligned} \tag{13.41}$$

where δ_{ij} is the Kronecker delta and I is the unit matrix. These equations can only be satisfied if the "coefficients", α^1, α^2, α^3, and β, are matrices.

The first two equations in Eq. (13.41) are similar to the commutation relations considered previously except that the two terms on the left-hand side are added together rather than subtracted. If i is not equal to j, the first equation can be written

$$\alpha^i\alpha^j = -\alpha^j\alpha^i.$$

The two matrices, α^i and α^j, can be interchanged provided that their overall sign is changed. Relations of this kind are called *anticommutation* relations.

The smallest matrices satisfying the conditions (13.41) are 4×4 matrices. The choice of the α and β matrices is not unique. The Dirac-Pauli representation is most commonly used

$$\alpha^i = \begin{bmatrix} 0 & \sigma^i \\ \sigma^i & 0 \end{bmatrix}, \quad \beta = \begin{bmatrix} \mathbf{I} & 0 \\ 0 & -\mathbf{I} \end{bmatrix}, \tag{13.42}$$

where **I** denotes the unit 2×2 matrix and σ^i are the Pauli matrices

$$\sigma^1 = \begin{bmatrix} 0 & 1 \\ 1 & 0 \end{bmatrix}, \quad \sigma^2 = \begin{bmatrix} 0 & -i \\ i & 0 \end{bmatrix}, \quad \sigma^3 = \begin{bmatrix} 1 & 0 \\ 0 & -1 \end{bmatrix}. \tag{13.43}$$

This notation of expressing α- and β-matrices in terms of 2×2 matrices is very useful. The 4×4 matrices built up of σ's and I's may be manipulated as ordinary matrices. For example, we have

$$
\begin{aligned}
\alpha^i \beta + \beta \alpha^i &= \begin{bmatrix} 0 & \sigma^i \\ \sigma^i & 0 \end{bmatrix} \begin{bmatrix} 1 & 0 \\ 0 & -1 \end{bmatrix} + \begin{bmatrix} 1 & 0 \\ 0 & -1 \end{bmatrix} \begin{bmatrix} 0 & \sigma^i \\ \sigma^i & 0 \end{bmatrix} \\
&= \begin{bmatrix} 0 & -\sigma^i \\ \sigma^i & 0 \end{bmatrix} + \begin{bmatrix} 0 & \sigma^i \\ -\sigma^i & 0 \end{bmatrix} \\
&= \begin{bmatrix} 0 & 0 \\ 0 & 0 \end{bmatrix}.
\end{aligned}
$$

One may readily show that the α- and β-matrices given by Eq. (13.42) satisfy the anticommutation relations (13.41).

Each solution ψ of the Dirac equation is a four-component column vector and is thus of the general form

$$
\psi = \begin{bmatrix} a \\ b \\ c \\ d \end{bmatrix}.
$$

The row vector whose components are the complex conjugate of the components of ψ is called the *adjoint* wave function and is denoted by $\psi^\dagger$. We write

$$
\psi^\dagger = \begin{bmatrix} a^* & b^* & c^* & d^* \end{bmatrix}.
$$

More generally the adjoint of a matrix is the matrix obtained by transposing the matrix and taking the complex conjugate of each of its elements. The conjugate equation of the Dirac equation (13.40) is obtained by replacing each matrix in the equation by its adjoint and replacing i with $-i$. We have

$$
-i\hbar \frac{\partial \psi^\dagger}{\partial t} = i\hbar c (\nabla \psi^\dagger) \cdot \alpha^\dagger + mc^2 \psi^\dagger \beta^\dagger. \tag{13.44}
$$

Notice the order of the adjoint function $\psi^\dagger$ and the $\alpha^\dagger$ matrix on the right-hand side of this equation. The order of these two terms is due to the fact that the adjoint of the product of two matrices is equal to the product of the two adjoint matrices in reverse order.

We can obtain a continuity equation of the form (13.34) by multiplying the Dirac equation (13.40) from the left with the adjoint function $\psi^\dagger$ and subtracting from this equation the result of multiplying the adjoint equation (13.44) from the right with ψ. We obtain

$$
i\hbar \left(\psi^\dagger \frac{\partial \psi}{\partial t} + \frac{\partial \psi^\dagger}{\partial t} \psi \right) = -i\hbar c \left(\psi^\dagger \alpha \cdot \nabla \psi + \nabla \psi^\dagger \cdot \alpha^\dagger \psi \right) + mc^2 \left(\psi^\dagger \beta \psi - \psi^\dagger \beta^\dagger \psi \right). \tag{13.45}
$$

This equation can be written in a form identical to Eq. (13.34) with the following definitions of the probability density and the components of the current

$$
\begin{aligned}
\rho &= \psi^\dagger \psi \\
j^k &= c \psi^\dagger \alpha^k \psi
\end{aligned} \tag{13.46}
$$

provided that the α- and β-matrices satisfy the equations

$$
\begin{aligned}
\alpha^\dagger &= \alpha \\
\beta^\dagger &= \beta.
\end{aligned}
$$

A matrix which is equal to its adjoint is said to be *self-adjoint* or *Hermitian*. The α- and β-matrices given by Eq. (13.42) have this property. Since the Dirac equation leads to a continuity equation, the solution of the equation may be interpreted in terms of a probability density just as we have for the nonrelativistic Schrödinger equation. This was an important part of Dirac's motivation for deriving the equation.

In the previous chapter, we found that the theory of relativity grew out of an attempt to make the laws of physics independent of the state of motion of the observer. The principle of relativity, which expresses this idea is held as firmly by physicists today as in the time of Einstein. We require that the form of fundamental equations be invariant with respect to Lorentz transformations. Equations having this property are said to be *covariant*. We can cast the Dirac equation into a form in which its transformation properties with respect to the Lorentz transformations are more apparent by multiplying the Dirac equation (13.40) from the left with $\beta/\hbar c$ and using the last of Eqs. (13.41) to obtain

$$i\beta\alpha^i \frac{\partial \psi}{\partial x^i} + i\beta \frac{\partial \psi}{\partial x^0} - \frac{mc}{\hbar}\psi = 0,$$

where the repeated index i is summed over in keeping with the Einstein convention introduced in the previous chapter. Defining

$$\gamma^i = \beta\alpha^i, \quad \gamma^0 = \beta, \tag{13.47}$$

the Dirac equation may be written

$$\left(i\gamma^\mu \frac{\partial}{\partial x^\mu} - k_0\right)\psi = 0, \tag{13.48}$$

where

$$k_0 = \frac{mc}{\hbar}, \tag{13.49}$$

and the summation index μ runs over the values $(0, 1, 2, 3)$. Eq. (13.48) is called the covariant form of the Dirac equation.

The continuity equation, which is satisfied by the Dirac current, may be cast in a covariant form by defining the current four-vector

$$j^\mu = (c\rho, \mathbf{j}). \tag{13.50}$$

Multiplying the first term of the continuity equation (13.34) by c/c, the equation can be written

$$\frac{\partial(c\rho)}{\partial(ct)} + \nabla \cdot \mathbf{j} = 0.$$

Using the component $x^0 = ct$ in place of time and the definition of the four-current (13.50), the equation becomes simply

$$\partial_\mu j^\mu = 0.$$

The Dirac current, which we have considered previously, can readily be expressed using the γ-matrices. The charge density ρ and the current $\mathbf{j}$ of a Dirac particle are given by Eq. (13.46). The components of the four-current (13.50) may be written

$$j^0 = c\psi^\dagger\psi$$
$$j^k = c\psi^\dagger\alpha^k\psi.$$

Using the equation $\beta^2 = I$, we may insert $\beta\beta$ following $\psi^\dagger$ in each of these equations and use the definition of the γ-matrices (13.47) to obtain

$$j^0 = c\psi^\dagger\gamma^0\gamma^0\psi$$
$$j^k = c\psi^\dagger\gamma^0\gamma^k\psi.$$

These last equations may be written more conveniently by defining the adjoint function

$$\overline{\psi} = \psi^\dagger\gamma^0. \tag{13.51}$$

The Dirac current may then be written

$$j^\mu = c\overline{\psi}\gamma^\mu\psi, \tag{13.52}$$

and the Hamiltonian for the interaction of an electron with an electromagnetic field expressed simply

$$H = eA_\mu j^\mu. \tag{13.53}$$

The Dirac equation, which provides a quantum mechanical description of particles of spin one half, is essential for describing physics problems for which the velocity of the particle is comparable to the speed of light.

13.3.4 Plane wave solutions of the Dirac equation

We now consider the free-particle solutions of the Dirac equation. For this purpose, we use a system of units which is referred to in particle physics as *natural units*. In this system of units, the velocity of light c and $\hbar$ are both equal to one. The system of units is then completely defined by specifying the unit of energy. In particle physics, one usually gives the energy in GeV (1 GeV $= 10^9$ eV). With c equal to one, the energy-momentum four-vector is

$$p^\mu = (E, \mathbf{p}).$$

The condition that $\hbar$ be equal to one implies that the wave vector $\mathbf{k}$ is equal to the momentum $\mathbf{p}$ and the angular frequency ω is equal to the energy E.

In natural units, the Dirac equation (13.40) is

$$(-i\alpha \cdot \nabla + m\beta) \psi = i \frac{\partial \psi}{\partial t}. \tag{13.54}$$

The free-particle solutions of the Dirac equation have the same dependence upon the space and time coordinates as the free particle solutions of the Schrödinger equation (13.27); however, one must take into account the fact that the coefficients α and β in the Dirac equation are 4×4 matrices. We shall suppose that the free particle solutions of the Dirac equation are of the form

$$\psi_\mathbf{p}(x) = u(\mathbf{p}) e^{i(\mathbf{p} \cdot \mathbf{r} - Et)}, \tag{13.55}$$

where $u(\mathbf{p})$ is a four-component vector independent of the space and time coordinates. Substituting Eq. (13.55) into Eq. (13.54) gives

$$(\alpha \cdot \mathbf{p} + m\beta) u(\mathbf{p}) = E u(\mathbf{p}). \tag{13.56}$$

This equation serves to define the Dirac spinors $u(\mathbf{p})$ of the free particle solutions of the Dirac equation.

Example 13.6

Construct the matrix $\alpha \cdot \mathbf{p} + m\beta$ that occurs in the above equation.

Solution

In the Dirac-Pauli representation of the α and β matrices (13.42), the matrix α^i is

$$\alpha^i = \begin{bmatrix} 0 & \sigma^i \\ \sigma^i & 0 \end{bmatrix}, \tag{13.57}$$

where σ^i is a two-dimensional Pauli matrix defined by Eq. (13.43). Multiplying the above α-matrix with p^i gives

$$\alpha^i p^i = \begin{bmatrix} 0 & \sigma^i p^i \\ \sigma^i p^i & 0 \end{bmatrix}.$$

The dot product is accomplished by summing over the i indices to obtain

$$\alpha \cdot \mathbf{p} = \begin{bmatrix} 0 & \sigma \cdot \mathbf{p} \\ \sigma \cdot \mathbf{p} & 0 \end{bmatrix}. \tag{13.58}$$

Similarly, the β matrix is

$$\beta = \begin{bmatrix} \mathbf{I} & 0 \\ 0 & -\mathbf{I} \end{bmatrix},$$

where **I** denotes the unit 2×2 matrix. Multiplying β by m gives

$$m\beta = \begin{bmatrix} m\mathbf{I} & 0 \\ 0 & -m\mathbf{I} \end{bmatrix}.$$

The 2×2 unit matrices **I** that occur in this last equation can be understood writing simply

$$m\beta = \begin{bmatrix} m & 0 \\ 0 & -m \end{bmatrix}. \tag{13.59}$$

Combining Eqs. (13.58) and (13.59), we obtain

$$\alpha \cdot \mathbf{p} + m\beta = \begin{bmatrix} m & \sigma \cdot \mathbf{p} \\ \sigma \cdot \mathbf{p} & -m \end{bmatrix}. \tag{13.60}$$

We would like now to construct four-component vectors $u(\mathbf{p})$ satisfying Eq. (13.56). For this purpose, we divide $u(\mathbf{p})$ into two-component vectors, u_A and u_B

$$u(\mathbf{p}) = \begin{bmatrix} u_A \\ u_B \end{bmatrix}. \tag{13.61}$$

Substituting Eqs. (13.61) and (13.60) into Eq. (13.56) gives

$$\begin{bmatrix} m & \sigma \cdot \mathbf{p} \\ \sigma \cdot \mathbf{p} & -m \end{bmatrix} \begin{bmatrix} u_A \\ u_B \end{bmatrix} = E \begin{bmatrix} u_A \\ u_B \end{bmatrix}.$$

Evaluating the matrix times vector multiplication on the left and equating corresponding components of the vectors on the two sides of the equation, we obtain the equations

$$mu_A + \sigma \cdot \mathbf{p}u_B = Eu_A$$
$$\sigma \cdot \mathbf{p}u_A - mu_B = Eu_B,$$

which may be written

$$\sigma \cdot \mathbf{p}u_B = (E - m)u_A \tag{13.62}$$
$$\sigma \cdot \mathbf{p}u_A = (E + m)u_B. \tag{13.63}$$

There are both positive and negative energy solutions of the Dirac equation for a free particle. For the positive energy solutions, the term $E + m$ that occurs in Eq. (13.63) is nonzero, while the term $E - m$ in Eq. (13.62) approaches zero for small $\mathbf{p}$. We thus use Eq. (13.63) to solve for u_B in terms of u_A

$$u_B = \frac{\sigma \cdot \mathbf{p}}{E + m} u_A .$$

Two independent positive-energy solutions can be obtained by expressing the vector u_A in terms of the two vectors

$$\chi^{(1)} = \begin{bmatrix} 1 \\ 0 \end{bmatrix}, \quad \chi^{(2)} = \begin{bmatrix} 0 \\ 1 \end{bmatrix}. \tag{13.64}$$

The four-component vectors for positive energy can then be written

$$u^{(r)}(\mathbf{p}) = C_N \begin{bmatrix} \chi^{(r)} \\ \dfrac{\sigma \cdot \mathbf{p}}{E + m} \chi^{(r)} \end{bmatrix}, \quad E > 0, \tag{13.65}$$

where C_N is a constant, and r has the values 1 or 2.

For the negative energy solutions, the term $E - m$ in Eq. (13.62) is nonzero, and Eq. (13.62) can be solved to obtain

$$u_A = \frac{-\sigma \cdot \mathbf{p}}{|E| + m} u_B .$$

The vectors for the negative energy solutions of the Dirac equation can thus be written

$$u^{(r+2)}(\mathbf{p}) = D_N \begin{bmatrix} \dfrac{-\sigma \cdot \mathbf{p}}{|E| + m} \chi^{(r)} \\ \chi^{(r)} \end{bmatrix}, \quad E < 0, \tag{13.66}$$

where the two-component vectors $\chi^{(r)}$ are given as before by Eq. (13.64). The constant C_N for the positive energy solutions and the constant D_N for the negative energy solutions may be evaluated using the normalization condition for the free-particle states. A discussion of the normalization condition can be found in the book by Griffiths cited at the end of this chapter.

The two-component vectors $\chi^{(r)}$ in Eqs. (13.65) and (13.66) may be identified as the "spin-up" and "spin-down" states in the nonrelativistic theory of a particle with spin 1/2. Using the Pauli matrix σ^3, which is the third matrix given in Eq. (13.43), we may define a spin matrix by the equation

$$s_z = (1/2)\sigma^3 .$$

The vectors $\chi^{(1)}$ and $\chi^{(2)}$ are eigenvectors of the matrix s_z corresponding to the eigenvalues, $\pm 1/2$. We have

$$s_z \chi^{(1)} = (1/2) \begin{bmatrix} 1 & 0 \\ 0 & -1 \end{bmatrix} \begin{bmatrix} 1 \\ 0 \end{bmatrix} = (1/2) \begin{bmatrix} 1 \\ 0 \end{bmatrix} = (1/2)\chi^{(1)}$$

and

$$s_z \chi^{(2)} = (1/2) \begin{bmatrix} 1 & 0 \\ 0 & -1 \end{bmatrix} \begin{bmatrix} 0 \\ 1 \end{bmatrix} = (-1/2) \begin{bmatrix} 0 \\ 1 \end{bmatrix} = (-1/2)\chi^{(2)} .$$

The two independent states that occur for positive and negative energy can thus be associated with the two spin orientations of the electron. We note that the spin of the electron appears naturally in the solutions of the relativistic Dirac equation, while the spin had to be added to the nonrelativistic Schrödinger theory.

The negative energy solutions to the Dirac equation, which do not have an obvious physical counterpart, posed very serious problems for Dirac when he originally proposed the theory. Of course, even in classical physics, the relativistic equation for the energy

$$E^2 = \mathbf{p}^2 c^2 + m^2 c^4$$

has two solutions

$$E = \pm (\mathbf{p}^2 c^2 + m^2 c^4)^{1/2}.$$

However, in classical physics one can simply assume that physical particles have positive energy. The problem of negative energies is much more troublesome in relativistic quantum mechanics. There is nothing to prohibit an electron from making a transition down to a negative energy state.

To give some historical background, we shall first say how Dirac interpreted the negative energy states and then give the view particle physicists have today. Dirac suggested that all the negative energy states were occupied and that the Pauli exclusion principle prevented an electron from making a transition to a state of negative energy. As illustrated in Fig. 13.2, a *hole* in the sea of negative energy states could be created by the excitation of an electron from a negative energy state to a positive energy state. The absence of an electron of charge $-e$ and energy $-E$ could be interpreted as the presence of an *antiparticle* of charge $+e$ and energy $+E$. Thus, the net effect of the excitation is to produce a particle/antiparticle pair

$$e^-(E') + e^+(E).$$

Dirac originally associated the holes with the proton, which was the only particle having a positive charge known at the time. We now know that the antiparticle of the electron is the positron. The positron has the same rest energy as the

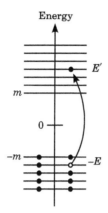

FIGURE 13.2 A hole in the sea of negative energy states created by the excitation of an electron from a negative energy state to a positive energy state.

electron, but it has a positive charge. We also know now that all charged particles have antiparticles – even particles called bosons that have an integral value of the spin and do not satisfy the Pauli principle.

The view held by physicists today is that the solutions of the Dirac equation with negative energy correspond to antiparticles. The negative-energy solutions of the Dirac equation with momentum $\mathbf{p}$ and energy E correspond to positrons with momentum $-\mathbf{p}$ and energy $-E$. The wave function of a positron may thus be obtained by making the replacements, $\mathbf{p} \rightarrow -\mathbf{p}$ and $E \rightarrow -E$, in Eq. (13.55) to obtain

$$\psi_{\mathbf{p}}(x) = v^{(r)}(\mathbf{p})e^{-i(\mathbf{p}\cdot\mathbf{r}-Et)} ,$$

where the positron spinor is denoted $v^{(r)}(\mathbf{p})$. By making similar replacements in Eq. (13.66), we obtain the following expression for the positron spinors

$$v^{(r)}(\mathbf{p}) = D_N \left[\begin{array}{c} \dfrac{\sigma \cdot \mathbf{p}}{|E| + m} \chi^{(r)} \\[2mm] \chi^{(r)} \end{array} \right] .$$

The free electron and positron states discussed in this section play an important role in understanding high-energy scattering events.

13.4   * Field quantization

In this section, we would like to overcome a particular asymmetry in our description of particles and fields. We have described the state of a particle by a wave function and have depicted particles making transitions from one quantum state to another under the influence of an external field. We have described the fields, however, entirely in classical terms. Recall our account of the absorption of light by the hydrogen atom in Chapter 4. In the oscillating electromagnetic field produced by a light source, the atom makes a transition from one energy level to another. The field is depicted as the cause and the backdrop for the dynamic process. A more balanced way of describing the absorption process is to treat both the material object and the electromagnetic field as quantum mechanical systems. In an absorption process, the atom goes from one state to another, and the electromagnetic field makes a transition from its initial state to a state with one fewer photon.

A scattering process in which two electrons collide is depicted in classical terms in Fig. 13.3(A). The two electrons in the figure repel each other, and, for this reason, the paths of the electrons are altered. Fig. 13.3(B) illustrates the semiclassical treatment of scattering processes that we have given thus far in this book. In the semiclassical description, the state of an incoming particle corresponds to a de Broglie wave, which is represented in Fig. 13.3(B) by three parallel lines perpendicular to the horizontal axis. The potential field of the interacting particle distorts the incoming waves causing the particle to make a transition to a plane wave moving in one of various directions. A single scattered wave is shown in Fig. 13.3(B) leaving the scattering center in the upper right. A fully quantum mechanical description of the scattering process is depicted in Fig. 13.3(C). In this figure, electrons approach each other, exchange a quantum of the interacting field, and depart in an altered state. Diagrams of this kind called *Feynman diagrams* were introduced by Richard Feynman in the early 1940's. Interacting particles are represented in Feynman diagrams by lines with arrows pointing toward the right. A particle exchanged in an interaction corresponds to an internal line passing from one vertex to another.

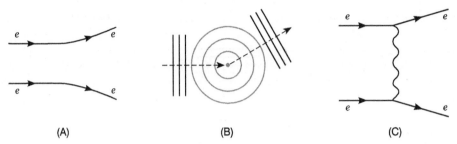

FIGURE 13.3 (A) A scattering process in which two electrons interact classically. (B) A semiclassical description of the scattering process. (C) A fully quantum mechanical description of the scattering of two electrons.

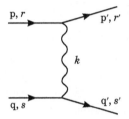

FIGURE 13.4 Feynman diagram for the scattering of two electrons. The momentum and spin orientation of the two incoming electrons are denoted by $\mathbf{p}, r$ and $\mathbf{q}, s$, and the momentum and spin orientation of the two outgoing electrons are denoted by $\mathbf{p}', r'$ and $\mathbf{q}', s'$.

A good model for understanding the quantum mechanical properties of fields is the harmonic oscillator, which was considered in Chapter 2 and is considered in more abstract terms in Appendix KK. The simple harmonic oscillator can be solved classically and quantum mechanically. An essential feature of the classical oscillator is that it vibrates with a single frequency. The spectrum of energy levels of the quantum oscillator is obtained in Appendix KK using purely algebraic methods. The lowest quantum mechanical state of the oscillator is the analogue of the zero point energy considered in Chapter 7 on statistical physics. For excited states of the oscillating field, one and more quanta of the field are present. Particles are thus the quanta of fields.

Quantum field theory is the area of physics which apples quantum mechanics to oscillating fields. Fields extend over the entire space in which scattering events occur and these quantum fields are the fundamental quantities. Material particles can be thought of as quanta of the field to which they correspond. Just as photons are the quanta of the electro-magnetic field, electrons and positrons are the quanta of a Dirac field. In this way, quantum field theory resolves the wave-particle paradox we have considered in the early chapters of this book. As Steven Weinberg has said, "quantum fields are the basic ingredients of the universe and particles are just bundles of energy and momentum of the fields."

The Feynman diagram in Fig. 13.3(C) is shown in more detail in Fig. 13.4 where the momentum and spin orientation of the two incoming electrons are denoted by $\mathbf{p}, r$ and $\mathbf{q}, s$, and the momentum and the spin orientation of the two outgoing electrons are denoted by $\mathbf{p}', r'$ and $\mathbf{q}', s'$. The mathematical expression for the Feynman diagram can be obtained by using a set of rules called *Feynman rules*, which, like the diagrams themselves, are due to Feynman. The free lines in a Feynman diagram correspond to the plane states of particles, the vertices correspond to interactions, and an interior line going from one vertex to another is called a *propagator*. For the electron-scattering event depicted in Fig. 13.4, the incoming- and outgoing lines correspond to the plane-wave solutions of the Dirac equation given by Eq. (13.65). Since the interaction involves the electro-magnetic field and a fermion field, the vertices of the diagram correspond to the term, $e\gamma^{\mu}$, appearing in Eqs. (13.52) and (13.53), and the propagator for the electromagnetic field is the Fourier transform of the Coulomb interaction between the two electrons. Just as Fig. 13.4 corresponds to electron-electron scattering, other Feynman diagrams can be drawn that correspond to positron-positron scattering and Compton scattering. A discussion of Feynman diagrams and calculations involving quantum fields can be found in the books by Mandl and Shaw and by Halzen and Martin cited at the end of this chapter. A more substantial description of quantum field theory can be found in the book by Ryder, which is also cited.

The scattering process depicted in Fig. 13.4 can be understood in physical terms. At a vertex of the Feynman diagram, an incoming electron excites the electromagnetic field promoting it to an excited state with an additional photon. The photon created by one electron propagates to the location of the other electron where the photon is absorbed. Feynman diagrams, which give a clear schematic description of scattering processes, will be used in the next chapter to describe the interaction of elementary particles.

Although the idea that the de Broglie wave associated with a particle should be distorted by an interacting field provides a framework for understanding collision processes at low energies. This idea is not particularly useful for describing collision processes at high energies where particles are typically created and destroyed. Scattering events at high energies can be understood using the concepts of quantum field theory. The incoming particles in a modern accelerator experiment create field quanta which carry the interaction between elementary particles. Photons, which are quanta of an electromagnetic field, and electrons and positrons, which are quanta of a Dirac field, both serve as carriers of the electromagnetic interaction.

13.5   The general theory of relativity

We consider now the general theory of relativity, which like the special theory of relativity considered in the previous chapter, owes its origins to Einstein's work in the early part of the twentieth century. Einstein would consider very simple problems and use these simple problems to reach far reaching conclusions. The simple problem that lead Einstein to formulate the *Principle of Equivalence*, which forms the basis of the general theory of relativity, concerns a man standing in an elevator. From the weight the man feels in his feet, he would not be able to decide whether the elevator was standing still in a gravitational field or the elevator was accelerating upward with there being no gravity.

13.5.1   The principle of equivalence

The *principle of equivalence* states that it is not possible to distinguish between being in a gravitational field and being accelerated. A person in a freely falling elevator would not feel the effects of gravity because the bodies within the elevator and the elevator itself would respond to the gravitational field in the same way. This can easily be proved. Suppose that the coordinate frame with axes x_1, x_2, and x_3 and the frame with axes ξ_1, ξ_2, and ξ_3 coincided at $t = 0$ and the frame axes ξ_1, ξ_2, and ξ_3 is attached to a freely falling body. The relation between the x_3 and ξ_3 coordinates at a later time would be

$$\xi_3 = x_3 - \frac{1}{2}gt^2,$$

where g is the acceleration of gravity. The acceleration of the body measured by the ξ_3 coordinate would be

$$\frac{d^2\xi_3}{dt^2} = 0. \tag{13.67}$$

Whether or not the person experiences a normal force on his feet depends on whether or not he is falling freely with the elevator.

The equivalence of gravity and acceleration has far-reaching consequences. If a particle or a beam of light is following a straight-line trajectory in a reference frame $\xi_1, \ldots \xi_3$, the particle or the beam of light would follow a curved path in another frame x_1, x_2, and x_3 with respect to which the first frame is the accelerating. Since acceleration and gravity are equivalent, one would expect a beam of light to follow a curved path in a gravitational field.

These ideas provide an alternate way of describing the motion of the Earth around the Sun. According to Issac Newton, the Earth is attracted to the Sun by a gravitational force which causes the Earth to follow a curved path in its motion around the Sun. According to general relativity, the mass of the Sun curves the space around the Sun. The Earth follows a curved path because space itself is curved.

13.5.2   The path of a freely-falling body in curvilinear coordinates

Consider now a particle falling under the influence of purely gravitational forces. According to the principal of equivalence, there is freely falling coordinate frame in which the particle moves in a straight line. As before we denote the coordinates in this reference frame by ξ_0, ξ_1, ξ_2, and ξ_3. The square of the length of a vector may be written using a *metric tensor* which we denote by $\eta_{\mu\nu}$. The elements of the metric tensor in this coordinate system attached to the particle can be displayed as a matrix

$$[\eta_{\mu\nu}] = \begin{bmatrix} 1 & 0 & 0 & 0 \\ 0 & -1 & 0 & 0 \\ 0 & 0 & -1 & 0 \\ 0 & 0 & 0 & -1 \end{bmatrix} \tag{13.68}$$

with the η_{00} element of this matrix equal to one, and the other diagonal elements of the matrix are equal to minus one. Using the summation convention of Einstein that repeated indices are summed over, the proper time element is

$$d\tau^2 = \eta_{\alpha\beta} d\xi^\alpha d\xi^\beta$$

Now suppose we view the motion of the particle from another frame of reference that may be curvilinear. The freely-falling coordinates ξ^α are functions of the new coordinates x^μ with the metric tensor $g_{\mu\nu}$ in the new coordinates defined by

$$g_{\mu\nu} \equiv \frac{\partial \xi^\alpha}{\partial x^\mu} \frac{\partial \xi^\beta}{\partial x^\nu} \eta_{\alpha\beta}. \tag{13.69}$$

The equation of motion in the new reference frame with coordinates ξ^α is

$$0 = \frac{d}{d\tau}\left(\frac{\partial \xi^\alpha}{\partial x^\mu} \frac{dx^\mu}{d\tau} \right) = \frac{\partial \xi^\alpha}{\partial x^\mu} \frac{d^2 x^\mu}{d\tau^2} + \frac{\partial^2 \xi^\alpha}{\partial x^\mu \partial x^\nu} \frac{dx^\nu}{d\tau} \frac{dx^\mu}{d\tau}. \tag{13.70}$$

Multiplying this last equation by $\partial x^\lambda / \partial \xi^\alpha$ and using the familiar chain rule

$$\frac{\partial x^\lambda}{\partial \xi^\alpha} \frac{\partial \xi^\alpha}{\partial x^\mu} = \delta^\lambda_\mu \tag{13.71}$$

to simplify the first term on the right-hand side of the resulting expression, we obtain

$$0 = \frac{d^2 x^\lambda}{d\tau^2} + \frac{\partial x^\lambda}{\partial \xi^\alpha} \frac{\partial^2 \xi^\alpha}{\partial x^\mu \partial x^\nu} \frac{dx^\nu}{d\tau} \frac{dx^\mu}{d\tau}.$$

This last equation can be written more compactly by introducing a *Christoffel symbol* defined by the following equation

$$\Gamma^\lambda_{\mu\nu} \equiv \frac{\partial x^\lambda}{\partial \xi^\alpha} \frac{\partial^2 \xi^\alpha}{\partial x^\mu \partial x^\nu}. \tag{13.72}$$

The Christoffel symbol is also known as the *Affine connection*. With this notation, the equation of motion of a particle falling freely in a gravitational field is

$$0 = \frac{d^2 x^\lambda}{d\tau^2} + \Gamma^\lambda_{\mu\nu} \frac{dx^\mu}{d\tau} \frac{dx^\nu}{d\tau}. \tag{13.73}$$

13.5.3 Relations between partial derivatives of $g_{\mu\nu}$ and $\Gamma^\lambda_{\mu\nu}$

The appearance of the Christoffel symbol in Eq. (13.73) indicates that an accelerated frame of reference has been used while the appearance of the metric tensor determines the proper time interval between two events. We shall now show that the Christoffel symbol $\Gamma^\lambda_{\mu\nu}$ is related to the rate of change of the metric tensor $g_{\mu\nu}$. Differentiating Eq. (13.69) with respect to x^λ gives

$$\frac{\partial g_{\mu\nu}}{\partial x^\lambda} = \frac{\partial^2 \xi^\alpha}{\partial x^\lambda \partial x^\mu} \frac{\partial \xi^\beta}{\partial x^\nu} \eta_{\alpha\beta} + \frac{\partial \xi^\alpha}{\partial \xi^\mu} \frac{\partial^2 \xi^\beta}{\partial x^\lambda \partial x^\nu} \eta_{\alpha\beta}. \tag{13.74}$$

To simplify this last equation, we first multiply the Christoffel symbol (13.72) by $\partial \xi^\beta / \partial x^\lambda$ and use the chain rule

$$\frac{\partial \xi^\beta}{\partial x^\lambda} \frac{\partial x^\lambda}{\partial \xi^\alpha} = \delta^\beta_\alpha$$

to obtain

$$\frac{\partial^2 \xi^\beta}{\partial x^\mu \partial x^\nu} = \Gamma^\lambda_{\mu\nu} \frac{\partial \xi^\beta}{\partial x^\lambda}$$

Using this last equation to simplify Eq. (13.74), we obtain

$$\frac{\partial g_{\mu\nu}}{\partial x^\lambda} = \Gamma^\rho_{\lambda\mu} \frac{\partial \xi^\alpha}{\partial x^\rho} \frac{\partial \xi^\beta}{\partial x^\nu} \eta_{\alpha\beta} + \Gamma^\rho_{\lambda\nu} \frac{\partial \xi^\beta}{\partial x^\rho} \frac{\partial \xi^\alpha}{\partial x^\mu} \eta_{\alpha\beta}. \tag{13.75}$$

Finally, using Eq. (13.69), we get

$$\frac{\partial g_{\mu\nu}}{\partial x^\lambda} = \Gamma^\rho_{\lambda\mu} g_{\rho\nu} + \Gamma^\rho_{\lambda\nu} g_{\rho\mu}. \tag{13.76}$$

This last equation gives a derivative of a metric tensor in terms of Christoffel symbols. To get an expression for a Christoffel symbol in terms of derivatives of the metric tensors, we add to Eq. (13.76) the same equation with μ and λ interchanged and subtract the same equation with ν and λ interchanged. Recalling that $\Gamma^\kappa_{\mu\nu}$ and $g_{\mu\nu}$ are symmetric with respect to an interchange of μ and ν, we obtain

$$\frac{\partial g_{\mu\nu}}{\partial x^\lambda} + \frac{\partial g_{\lambda\nu}}{\partial x^\mu} - \frac{\partial g_{\mu\lambda}}{\partial x^\nu} = 2\Gamma^\rho_{\lambda\mu} g_{\rho\nu}.$$

We now define an inverse $g^{\rho\nu}$ of the metric tensor by the equation

$$g^{\nu\rho} g_{\sigma\nu} = \delta^\rho_\sigma \tag{13.77}$$

and multiply by $g^{\nu\kappa}$ to obtain

$$\Gamma^\kappa_{\lambda\mu} = \frac{1}{2} g^{\nu\kappa} \left[\frac{\partial g_{\mu\nu}}{\partial x^\lambda} + \frac{\partial g_{\lambda\nu}}{\partial x^\mu} - \frac{\partial g_{\mu\lambda}}{\partial x^\nu} \right]. \tag{13.78}$$

13.5.4 A slow moving particle in a weak gravitational field

To make contact with the theory of Newton, we consider a particle moving slowly in a weak stationary gravitational field. If the particle is moving slowly, we may neglect the spatial components of its velocity and write Eq. (13.73) as

$$\frac{d^2 x^\lambda}{d\tau^2} + \Gamma^\lambda_{00} \left(\frac{dt}{d\tau} \right)^2 = 0.$$

Since the field is stationary, all of the time derivatives of $g_{\mu\nu}$ vanish and Eq. (13.78) becomes

$$\Gamma^\mu_{00} = -\frac{1}{2} g^{\mu\nu} \frac{\partial g_{00}}{\partial x^\nu}. \tag{13.79}$$

Finally, since the field is weak, we may adopt a nearly Cartesian coordinate system in which

$$g_{\alpha\beta} = \eta_{\alpha\beta} + h_{\alpha\beta} \qquad |h_{\alpha\beta}| \ll 1 \tag{13.80}$$

or to first order in $h_{\alpha\beta}$,

$$\Gamma^\alpha_{00} = -\frac{1}{2} \eta^{\alpha\beta} \frac{\partial h_{00}}{\partial x^\beta}. \tag{13.81}$$

Using the Christoffel symbol in the equation of motion then gives

$$\frac{d^2 \mathbf{x}}{d\tau^2} = \frac{1}{2} \left(\frac{dt}{d\tau} \right)^2 \nabla h_{00}$$

$$\frac{d^2 t}{d\tau^2} = 0.$$

The solution of the second equation is that $dt/d\tau$ is a constant. So, dividing the first equation by $(dt/d\tau)^2$ gives

$$\frac{d^2 \mathbf{x}}{d\tau^2} = \frac{1}{2} \nabla h_{00}$$

The corresponding Newtonian result is

$$\frac{d^2\mathbf{x}}{d\tau^2} = -\nabla\phi.$$

Comparing these last two equations we see that

$$h_{00} = -2\phi + constant.$$

Since the coordinate system must become Minskowskian for great distances, the constant in the last equation must be equal to zero. Returning to the metric defined by Eq. (13.80), we find that

$$g_{00} = -(1 + 2\phi). \tag{13.82}$$

13.5.5 Vectors and tensors

In the previous chapter, we found conservation laws that did not depend upon a particular reference frame by expressing these laws in terms of vector quantities. We shall adopt such an approach here defining properties that have particularly simple transformation properties. A *contravariant vector*, V^μ, transforms according to the transformation law

$$V'^\mu = \frac{\partial x'^\mu}{\partial x^\nu} V^\nu \tag{13.83}$$

Since the rules for partial differentiation give

$$dx'^\mu = \frac{\partial x'^\mu}{\partial x^\nu} dx^\nu,$$

the differentials (dx^μ) form the components of a contravariant vector. Another related quantity is a *covariant vector*, U_μ, which transforms according to the transformation law

$$U'_\mu = \frac{\partial x^\nu}{\partial x'^\mu} V_\nu. \tag{13.84}$$

For instance, if ϕ is a scalar field, then $\partial\phi/\partial x^\mu$ transforms as a covariant vector because due to the chain rule

$$\frac{\partial\phi}{\partial x'^\mu} = \frac{\partial x^\nu}{\partial x'^\mu} \frac{\partial\phi}{\partial x^\nu}$$

Our definitions of contravariant and covariant vectors can easily be generalized to tensors of higher rank. For instance, under a coordinate transformation the second-order contravariant tensor, $T^{\mu\nu}$ transforms according to the equation

$$T'^{\mu\nu} = \frac{\partial x'^\mu}{\partial x^\rho} \frac{\partial x'^\nu}{\partial x^\sigma} T^{\rho\sigma}$$

and the second-order covariant tensor, $T_{\mu\nu}$ transforms according to the equation

$$T'_{\mu\nu} = \frac{\partial x'^\rho}{\partial x^\mu} \frac{\partial x'^\sigma}{\partial x^\nu} T_{\rho\sigma}.$$

An important example of a covariant tensor is the metric tensor $g_{\mu\nu}$ defined by Eq. (13.69)

$$g_{\mu\nu} = \frac{\partial\xi^\alpha}{\partial x^\mu} \frac{\partial\xi^\beta}{\partial x^\nu} \eta_{\alpha\beta}.$$

In a different coordinates system x'^μ, the metric tensor is

$$g'_{\mu\nu} = \frac{\partial\xi^\alpha}{\partial x'^\mu} \frac{\partial\xi^\beta}{\partial x'^\nu} \eta_{\alpha\beta} = \frac{\partial\xi^\alpha}{\partial x^\rho} \frac{\partial x^\rho}{\partial x'^\mu} \frac{\partial\xi^\beta}{\partial x^\sigma} \frac{\partial x^\sigma}{\partial x'^\nu} \eta_{\alpha\beta}, \tag{13.85}$$

and therefore

$$g'_{\mu\nu} = \frac{\partial x^\rho}{\partial x'^\mu} \frac{\partial x^\sigma}{\partial x'^\nu} g_{\rho\sigma}.$$

13.5.6 Transformation of the affine connection

We would now like to consider how the affine connection transforms with respect to a coordinate transformation. The affine connection or Christoffel symbol is given by Eq. (13.72), which we rewrite

$$\Gamma^{\lambda}_{\mu\nu} = \frac{\partial x^{\lambda}}{\partial \xi^{\alpha}} \frac{\partial^2 \xi^{\alpha}}{\partial x^{\mu} \partial x^{\nu}}. \tag{13.86}$$

Passing from the coordinate system with coordinate x^{μ} to a different coordinate system with coordinate x'^{μ}, we obtain

$$\begin{aligned}
\Gamma'^{\lambda}_{\mu\nu} &= \frac{\partial x'^{\lambda}}{\partial \xi^{\alpha}} \frac{\partial^2 \xi^{\alpha}}{\partial x'^{\mu} \partial x'^{\nu}} \\
&= \frac{\partial x'^{\lambda}}{\partial x^{\rho}} \frac{\partial x^{\rho}}{\partial \xi^{\alpha}} \frac{\partial}{\partial x'^{\mu}} \left(\frac{\partial x^{\sigma}}{\partial x'^{\nu}} \frac{\partial \xi^{\alpha}}{\partial x^{\sigma}} \right) \\
&= \frac{\partial x'^{\lambda}}{\partial x^{\rho}} \frac{\partial x^{\rho}}{\partial \xi^{\alpha}} \left[\frac{\partial x^{\sigma}}{\partial x'^{\nu}} \frac{\partial x^{\tau}}{\partial x'^{\mu}} \frac{\partial^2 \xi^{\alpha}}{\partial x^{\tau} \partial x^{\sigma}} + \frac{\partial^2 \xi^{\alpha}}{\partial x'^{\mu} \partial x'^{\nu}} \frac{\partial \xi^{\alpha}}{\partial x'^{\sigma}} \right].
\end{aligned}$$

Using Eq. (13.86), this last equation can be written

$$\Gamma^{\lambda}_{\mu\nu} = \frac{\partial x'^{\lambda}}{\partial x^{\rho}} \frac{\partial x^{\tau}}{\partial x'^{\mu}} \frac{\partial x^{\sigma}}{\partial x'^{\nu}} \Gamma^{\rho}_{\tau\sigma} + \frac{\partial x'^{\lambda}}{\partial x^{\rho}} \frac{\partial^2 x^{\rho}}{\partial x'^{\mu} \partial x'^{\nu}} \tag{13.87}$$

For future developments, it is useful to produce another representation of the second derivative term in this last expression for the Christoffel symbol. We consider the following identity

$$\frac{\partial x'^{\lambda}}{\partial x^{\rho}} \frac{\partial x^{\rho}}{\partial x'^{\nu}} = \delta^{\lambda}_{\nu}.$$

Differentiating the identity with respect of x'^{μ} gives

$$\frac{\partial x'^{\lambda}}{\partial x^{\rho}} \frac{\partial^2 x^{\rho}}{\partial x'^{\mu} \partial x'^{\nu}} + \frac{\partial x^{\rho}}{\partial x'^{\nu}} \frac{\partial^2 x'^{\lambda}}{\partial x'^{\mu} \partial x^{\rho}} = 0. \tag{13.88}$$

The second term in this last equation may be brought over to the right-hand side and rewritten using the chain rule to obtain

$$\frac{\partial x'^{\lambda}}{\partial x^{\rho}} \frac{\partial^2 x^{\rho}}{\partial x'^{\mu} \partial x'^{\nu}} = - \frac{\partial x^{\rho}}{\partial x'^{\nu}} \frac{\partial x^{\sigma}}{\partial x'^{\mu}} \frac{\partial^2 x'^{\lambda}}{\partial x^{\rho} \partial x^{\sigma}}$$

Using this last equation to evaluate the second derivative in Eq. (13.87), we obtain finally

$$\Gamma^{\lambda}_{\mu\nu} = \frac{\partial x'^{\lambda}}{\partial x^{\rho}} \frac{\partial x^{\tau}}{\partial x'^{\mu}} \frac{\partial x^{\sigma}}{\partial x'^{\nu}} \Gamma^{\rho}_{\tau\sigma} - \frac{\partial x^{\rho}}{\partial x'^{\nu}} \frac{\partial x^{\sigma}}{\partial x'^{\mu}} \frac{\partial^2 x'^{\lambda}}{\partial x^{\rho} \partial x^{\sigma}} \tag{13.89}$$

13.5.7 Covariant differentiation

The differentiation of a tensor does not generally produce a tensor. For instance, consider a contravariant vector V^{μ} that transforms according to the equation

$$V'^{\mu} = \frac{\partial x'^{\mu}}{\partial x^{\nu}} V^{\nu}. \tag{13.90}$$

Differentiating with respect to x'^{λ} gives

$$\frac{\partial V'^{\mu}}{\partial x'^{\lambda}} = \frac{\partial x'^{\mu}}{\partial x^{\nu}} \frac{\partial x^{\rho}}{\partial x'^{\lambda}} \frac{\partial V'^{\nu}}{\partial x^{\rho}} + \frac{\partial^2 x'^{\mu}}{\partial x^{\rho} \partial x^{\nu}} \frac{\partial x^{\rho}}{\partial x'^{\lambda}} V^{\nu} \tag{13.91}$$

The first term on the right-hand side of this last equation is what we would expect if $\partial V^{\mu}/\partial x^{\lambda}$ were a tensor, but the second term in the equation destroys the tensor character of the derivative.

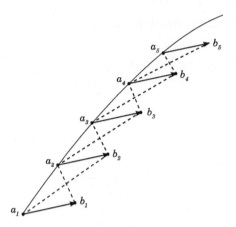

FIGURE 13.5 Schild's Ladder.

Although $\partial V^\mu / \partial x^\lambda$ is not a tensor, we can use it to construct a tensor. Using Eq. (13.89) to represent $\Gamma^\mu_{\lambda\kappa}$ and Eq. (13.90) to represent V'^κ, we see that

$$
\begin{aligned}
\Gamma^\mu_{\lambda\kappa} V'^\kappa &= \frac{\partial x'^\mu}{\partial x^\rho} \frac{\partial x^\tau}{\partial x'^\lambda} \frac{\partial x^\sigma}{\partial x'^\kappa} \Gamma^\rho_{\tau\sigma} \frac{\partial x'^\kappa}{\partial x^\eta} V^\eta - \frac{\partial x^\rho}{\partial x'^\kappa} \frac{\partial x^\sigma}{\partial x'^\lambda} \frac{\partial^2 x'^\mu}{\partial x^\rho \partial x^\sigma} \frac{\partial x'^\kappa}{\partial x^\eta} V^\eta \\
&= \frac{\partial x'^\mu}{\partial x^\rho} \frac{\partial x^\tau}{\partial x'^\lambda} \Gamma^\rho_{\tau\sigma} V^\sigma - \frac{\partial^2 x'^\mu}{\partial x^\rho \partial x^\sigma} \frac{\partial x^\sigma}{\partial x'^\lambda} V^\rho
\end{aligned}
\tag{13.92}
$$

Apart from the sign, the last term on the right hand side of Eq. (13.91) differs from the term containing a second derivative in this last equation only by dummy summation indices. Notice that replacing the dummy index σ with ρ and replacing the dummy index ρ with ν in the term involving a second derivative in this last equation makes this term identical to the corresponding term in Eq. (13.91). Adding Eqs. (13.91) and (13.89), the terms involving the second derivatives thus cancel and we obtain

$$
\frac{\partial V'^\mu}{\partial x'^\lambda} + \Gamma^\mu_{\lambda\kappa} V'^\kappa = \frac{\partial x'^\mu}{\partial x^\rho} \frac{\partial x^\tau}{\partial x'^\lambda} \left(\frac{\partial V^\rho}{\partial x^\tau} + \Gamma^\rho_{\tau\sigma} V^\sigma \right)
\tag{13.93}
$$

We are this led to define a *covariant derivative*

$$
V^\mu_{;\lambda} = \frac{\partial V^\mu}{\partial x^\lambda} + \Gamma^\mu_{\lambda\kappa} V^\kappa.
\tag{13.94}
$$

Eq. (13.93) then tells us that $V^\mu_{;\lambda}$ is a tensor that transforms by the rule

$$
V'^\mu_{;\lambda} = \frac{\partial x'^\mu}{\partial x^\rho} \frac{\partial x^\tau}{\partial x'^\lambda} V'^\rho_{;\tau}.
$$

13.5.8 The parallel transport of a vector along a curve

We now imagine that a particle is falling in a gravitational field. A clock falling with the particle records the proper time τ associated with the particle at each point along its path in space-time. We also imagine that a vector such as the spin $S_\mu(\tau)$ or momentum $P_\mu(\tau)$ is associated with the particle. As shown in Fig. 13.5, the vector associated with the particle is transported along its trajectory in space-time. We imagine that the clock traveling with the particle emits a flash of light at regular time intervals and those flashes are represented by the solid dots along the path associated with the points, $a_1, a_2, \ldots$, and the tip of the vector is directed to the points, $b_1, b_2, \ldots$. Knowing the vector pointing from a_1 to the point b_1, the point b_2 can be located by drawing a line from the point a_1 through the midpoint of the dotted line between a_2 and b_1 and continuing twice the distance beyond the dotted line. Continuing in this way, we obtain the illustration in Fig. 13.5 which shows how a vector is transported along the path of the particle. The drawing shown in Fig. 13.5 is known as Schild's ladder.

Thus far, we have been concerned with transformations in all of space but are now wanted to describe derivatives with respect to the parameter τ that measures the position of a point along the trajectory of the particle. We consider the contravariant vector, $A^\mu(\tau)$ that transforms according to the equation

$$A'^\mu(\tau) = \frac{\partial x'^\mu}{\partial x^\nu} A^\nu(\tau).$$ (13.95)

Differentiating this last equation with respect to τ and keeping in mind the fact that x^ν depends upon τ, we obtain

$$A'^\mu(\tau) = \frac{\partial x'^\mu}{\partial x^\nu} \frac{A^\nu(\tau)}{d\tau} + \frac{\partial^2 x'^\mu}{\partial x^\nu \partial x^\lambda} \frac{dx'^\lambda}{d\tau} A^\nu(\tau).$$

The term with second derivative that appears in this last equation is similar to the term with the second derivative that appears in Eq. (13.89) for the affine connection. Eliminating this term as we have before, we obtain the following equation for the covariant derivative along the curve

$$\frac{DA^\mu}{D\tau} \equiv \frac{dA^\mu}{d\tau} + \Gamma^\mu_{\nu\lambda} \frac{dx^\lambda}{d\tau} A^\nu,$$ (13.96)

which transforms as a contravariant vector

$$\frac{DA'^\mu}{D\tau} = \frac{\partial x'^\mu}{\partial x^\nu} \frac{DA^\nu}{D\tau}.$$

13.5.9 The curvature tensor

There are a number of naive questions that can be raised at this point. Suppose we were to do a parallel transport of a vector around a closed loop returning to the same point. Would the vector then point in the same direction it did before? There is a straight-forward way of answering this question. We may define a curvature tensor by the equation

$$R^\lambda_{\mu\nu\kappa} \equiv \frac{\partial \Gamma^\lambda_{\mu\nu}}{\partial x^\kappa} - \frac{\partial \Gamma^\lambda_{\mu\kappa}}{\partial x^\nu} + \Gamma^\eta_{\mu\nu} \Gamma^\lambda_{\kappa\eta} - \Gamma^\eta_{\mu\kappa} \Gamma^\lambda_{\nu\eta}$$ (13.97)

that transforms according to the rule

$$R'^\tau_{\rho\sigma\eta} = \frac{\partial x'^\tau}{\partial x^\lambda} \frac{\partial x^\mu}{\partial x'^\rho} \frac{\partial x^\nu}{\partial x'^\sigma} \frac{\partial x^\kappa}{\partial x'^\eta} R^\lambda_{\mu\nu\kappa}.$$ (13.98)

Consider now a closed curve C that bounds an area A. If the curvature tensor $R^\lambda_{\mu\nu\kappa}$ vanishes for all points on A, an arbitrary vector will not change when parallelly-transported around C. The region within C is then said to be flat. However, if the curvature tensor is nonzero for some points in the area A, parallel transport around a loop will generally cause a change of the vector.

The algebraic properties of the curvature tensor are improved by forming the fully covariant tensor

$$R_{\lambda\mu\nu\kappa} \equiv g_{\lambda\sigma} R^\sigma_{\mu\nu\kappa}.$$

The tensor is then antisymmetric with respect to an interchange of any two of its indices and symmetric with respect to a cyclic permutation of the last three indices. The important *Ricci tensor* is formed by the contraction

$$R_{\mu\kappa} \equiv R^\lambda_{\mu\lambda\kappa}$$ (13.99)

and the *curvature scalar* is

$$R = g^{\mu\nu} R_{\mu\nu}.$$ (13.100)

13.5.10 Einstein's field equations

At the heart of general relativity are equations relating the mass and energy to the curvature of space. One knows from the start that the gravitational field equation will be more complicated than the equations for electromagnetism. Maxwell's equations are linear because the electromagnetic fields do not carry an electric charge. By contrast gravitational fields do carry energy and momentum and must thus contribute as a source of the gravitational fields. This suggests that the gravitational field equations must be nonlinear partial differential equations.

We can begin our attempt to motivate the field equation by considering a weak gravitational field for which the Newtonian gravitational potential ϕ satisfies Poisson's equation

$$\nabla^2 \phi = 4\pi G \rho, \tag{13.101}$$

where G is Newton's gravitational constant. According to Eq. (13.82), the metric tensor g_{00} for a particle moving slowly in a weak electromagnetic field is equal to $-(1 + 2\phi)$. Since the mass density ρ is equal to T_{00}, Poisson's equation for a weak gravitational field can be written

$$\nabla^2 g_{00} = -8\pi G T_{00}.$$

We are thus led to *guess* that the equations for weakly interacting fields with a general distribution $T_{\mu\nu}$ of energy and momentum are of the form

$$G_{\mu\nu} = -8\pi G T_{\mu\nu},$$

where $G_{\mu\nu}$ is a linear combination of the metric tensor and its first and second derivatives.

For fields of any strength, the *Einstein Field Equations* can be written

$$R_{\mu\nu} - \frac{1}{2} g_{\mu\nu} R - \lambda g_{\mu\nu} = -8\pi G T_{\mu\nu}, \tag{13.102}$$

where $R_{\mu\nu}$ is the Ricci tensor defined by Eq. (13.99), R is the curvature scalar defined by Eq. (13.100), and $T_{\mu\nu}$ is the energy momentum tensor. When Einstein first proposed this equation, he thought the universe was static and added the term $-\lambda g_{\mu\nu}$ to the equation to hinder the expansion of the universe. Later, when Edwin Hubble encourage Einstein to look through his telescope at Mount Wilson to convince Einstein that the universe was indeed expanding, Einstein described his adding of the term $-\lambda g_{\mu\nu}$ to his field equations as "the greatest blunder of my professional career." This term, which is now called the *cosmological constant*, has recently been found to be important to describe the effect of *dark energy*.

Knowing the energy momentum tensor, one can solve Einstein's field equation for the curvature of the universe. However, in today's world with precise measurements of the location of distant galaxies and uncertainties about the nature of dark matter and energy, the calculations generally go the other way. One uses one's knowledge of the distribution of matter in the visible universe to calculate the curvature tensor and then uses Einstein's field equation to calculate the energy-momentum tensor.

Suggestions for further reading

P. Bergmann, *Introduction to the Theory of Relativity* (Englewood Cliffs, New Jersey: Prentice-Hall, 1942).
E. Taylor and J. Wheeler, *Spacetime Physics*, Second Edition (New York: Freeman, 1992).
R. Hagedorn, *Relativistic Kinematics* (New York: Benjamin, 1964).
F. Mandl and G. Shaw, *Quantum Field Theory*, Second Edition (New York: Wiley, 2010).
F. Halzen and A. Martin, *Quarks and Leptons* (New York: Wiley, 1984).
L. Ryder, *Quantum Field Theory*, Second Edition (Cambridge: Cambridge University Press, 1996).

Basic equations

Definitions

Four-velocity

$$v^\mu = \frac{dx^\mu}{d\tau} = \gamma \frac{dx^\mu}{dt},$$

where

$$\gamma = \frac{1}{\sqrt{1 - v^2/c^2}}$$

The velocity four-vector can be written

$$v^\mu = \gamma(c, v_x, v_y, v_z) = \gamma(c, \mathbf{v})$$

Four-momentum

$$p^\mu = mv^\mu$$

$$\mathbf{p} = \gamma m \mathbf{v} = \frac{m\mathbf{v}}{\sqrt{1 - v^2/c^2}} \, ,$$

$$p^0 = \gamma mc$$

Energy

$$E = \gamma mc^2 = \frac{mc^2}{\sqrt{1 - v^2/c^2}}$$

The energy-momentum four-vector can be written

$$p^\mu = (E/c, p_x, p_y, p_z) = (E/c, \mathbf{p})$$

Energy-momentum relation

$$E^2 = \mathbf{p}^2 c^2 + m^2 c^4$$

Rest energy

$$R = mc^2$$

Kinetic energy

$$K = E - R = (\gamma - 1)mc^2$$

The Dirac theory of the electron

Dirac equation

$$i\hbar \frac{\partial \psi}{\partial t} = c\alpha \cdot (-i\hbar \nabla \psi) + mc^2 \beta \psi,$$

where α and β are four by four matrices.

Conditions satisfied by the Dirac matrices

$$\alpha^i \alpha^j + \alpha^j \alpha^i = 2\delta_{ij} I$$
$$\alpha^i \beta + \alpha^j \beta = 0$$
$$\beta^2 = I$$

Dirac-Pauli representation of the Dirac matrices

$$\alpha^i = \begin{bmatrix} 0 & \sigma^i \\ \sigma^i & 0 \end{bmatrix}, \quad \beta = \begin{bmatrix} \mathbf{I} & 0 \\ 0 & -\mathbf{I} \end{bmatrix},$$

where $\mathbf{I}$ denotes the unit 2×2 matrix and σ^i are the Pauli matrices

$$\sigma^1 = \begin{bmatrix} 0 & 1 \\ 1 & 0 \end{bmatrix}, \quad \sigma^2 = \begin{bmatrix} 0 & -i \\ i & 0 \end{bmatrix}, \quad \sigma^3 = \begin{bmatrix} 1 & 0 \\ 0 & -1 \end{bmatrix}$$

Covariant form of the Dirac equation

$$\left(i\gamma^\mu \frac{\partial}{\partial x^\mu} - k_0 \right) \psi = 0,$$

where $\gamma^i = \beta \alpha^i$, $\gamma^0 = \beta$ and $k_0 = mc/\hbar$.

Plane wave solutions of the Dirac equation

$$\psi_{\mathbf{p}}(x) = u(\mathbf{p})e^{i(\mathbf{p}\cdot\mathbf{r}-Et)},$$

where the four-component vector $u(\mathbf{p})$ for positive energy is

$$u^{(r)}(\mathbf{p}) = C_N \begin{bmatrix} \chi^{(r)} \\ \dfrac{\sigma\cdot\mathbf{p}}{E+m}\chi^{(r)} \end{bmatrix}, \quad E > 0 \ \ r = 1, 2$$

and for negative energy is

$$u^{(r+2)}(\mathbf{p}) = D_N \begin{bmatrix} \dfrac{-\sigma\cdot\mathbf{p}}{|E|+m}\chi^{(r)} \\ \chi^{(r)} \end{bmatrix}, \quad E < 0 \ \ r = 1, 2$$

Plane wave states of an antiparticle

$$\psi_{\mathbf{p}}(x) = v^{(r)}(\mathbf{p})e^{-i(\mathbf{p}\cdot\mathbf{r}-Et)},$$

where the four-component vector $v^{(r)}(\mathbf{p})$ is

$$v^{(r)}(\mathbf{p}) = D_N \begin{bmatrix} \dfrac{\sigma\cdot\mathbf{p}}{|E|+m}\chi^{(r)} \\ \chi^{(r)} \end{bmatrix}, \quad r = 1, 2$$

Summary

Four-vectors are defined representing the velocity and momentum of a particle. The energy is defined and found to be related to the momentum of a particle by the equation

$$E^2 = \mathbf{p}^2 c^2 + m^2 c^4.$$

The same kind of arguments that lead to the Schrödinger equation is found to lead to relativistic wave equations. The relativistic wave equation for an electron, which is called the Dirac equation, is necessary for describing scattering processes at high energy. For high-energy scattering processes in which particles are created and destroyed, the field describing the interaction of particles must be treated quantum mechanically. This leads to a new kind of theory called quantum field theory which is based on the ideas of relativity and quantum mechanics.

The equivalence principle led Einstein to the idea that the mass of an object curves the space around it. The Earth orbits the Sun because the mass of the Sun curves the space around it. In the section on general relativity, the differential structure of Minkowski space is studied by considering the trajectory of a particle falling in a gravitational field. The curvature tensor is defined, and it is found that a vector points in its original direction if it is parallelly transported around a closed curve if the curvature tensor is equal to zero. The curvature tensor is the solution of the Einstein field equation with the energy momentum tensor being the driving term in the equation

Knowing the energy momentum tensor, one can solve Einstein's field equation for the curvature of the universe. However, in today's world with precise measurements of the location of distant galaxies and uncertainties about the nature of dark matter and energy, the calculations generally go the other way. One uses one's knowledge of the distribution of matter in the visible universe to calculate the curvature tensor and solves Einstein's field equation to calculate the energy-momentum tensor.

Questions

1. What advantage can one hope to achieve by expressing basic physical quantities in terms of vectors.
2. Write down the defining equation of the velocity in relativity theory.
3. Give the four components of the four-velocity.
4. What is the square of the length of the velocity vector v^μ?
5. How are the relativistic momentum and velocity vectors related?

6. Write down the defining equation of the energy in relativity theory.
7. How is the rest energy defined?
8. How is the kinetic energy defined in relativity theory?
9. Give the four components of the energy-momentum vector.
10. What is the square of the length of the momentum vector p^μ?
11. Write down an equation relating the energy E and the momentum **p** in relativity theory.
12. How are the energy and momentum of a particle with zero mass related?
13. About how fast must a particle be traveling for there to be a significant difference between the relativistic and nonrelativistic expressions for the kinetic energy?
14. Suppose that positrons and electrons collide in a symmetric colliding beam experiment. How are the momenta of the positron and the electron related?
15. Suppose that a positron strikes an electron at rest to produce a Z particle. Write down the condition that the energy of the particles is conserved.
16. What condition does the continuity equation impose upon the current and probability of a particle?
17. Write down the Klein-Gordon equation.
18. Why did Schrödinger discard the Klein-Gordon equation?
19. Write down the Dirac equation.
20. What conditions are satisfied by the Dirac matrices, α^i and β?
21. Write down the covariant form of the Dirac equation.
22. Give the four components of the Dirac current.
23. What is the general form of the free-particle solutions of the Dirac equation?
24. How did Dirac account for the fact that particles in positive energy states do not make transitions into negative energy states?
25. How did Dirac describe antiparticles?
26. What significance do we give to the negative energy solutions of the Dirac equation today?
27. Compare the classical description of the electromagnetic interaction between two charged particles with the modern description in which the interacting field is treated quantum mechanically.
28. To what do the free lines and the vertices of a Feynman diagram correspond?
29. Give the names of two particles which are the quanta of a Dirac field.

Problems

1. An electron is traveling with a speed of $v = 0.2\,c$ in the x-direction. (a) Determine all four components of its four-velocity. (b) Determine all four components of its four-momentum.
2. What is the kinetic energy of an electron with velocity $v = 0.2\,c$?
3. How fast must a particle be traveling for its kinetic energy to be equal to its rest energy?
4. What is the speed of a particle whose total energy is twice its rest energy?
5. Expressing the function, $1/\sqrt{1 - v^2/c^2}$, as $f(x) = (1 - x)^{-1/2}$, where $x = v^2/c^2$, and using the Taylor series expansion about the origin

$$f(x) = f(0) + f'(0)\,x + \frac{1}{2}f''(0)\,x^2 + \dots,$$

show that

$$\frac{1}{\sqrt{1 - v^2/c^2}} = 1 + \frac{1}{2}\frac{v^2}{c^2} + \frac{3}{8}\frac{v^4}{c^4} + \dots.$$

6. Electrons are accelerated to high speeds by a two-stage machine. The first stage accelerates the electrons from rest to $0.99\,c$, and the second stage accelerates the electrons from $0.99\,c$ to $0.999\,c$. By how much does each stage increase the kinetic energy of an electron?
7. What is the speed of an electron whose kinetic energy is 100 MeV?
8. A proton with rest energy of 938 MeV has a total energy of 1500 MeV. (*a*) What is the speed of the proton? (*b*) What is the magnitude of the momentum **p**?
9. For a two-body scattering event

$$A + B \rightarrow C + D,$$

it is convenient to introduce the Mandelstam variables

$$s = \frac{|p_A + p_B|^2}{c^2}$$

$$t = \frac{|p_A - p_C|^2}{c^2}$$

$$u = \frac{|p_A - p_D|^2}{c^2}.$$

The laboratory frame in a scattering experiment is defined by the equations

$$\mathbf{p}_B = 0, \quad E_B = m_B c^2,$$

while the center of mass frame is defined by the equation

$$\mathbf{p}_A + \mathbf{p}_B = 0.$$

(a) Derive explicit expressions for s in the laboratory and center of mass frames which depend only upon the masses, m_A and m_B, the energies, E_A and E_B, and momenta, $\mathbf{p}_A$ and $\mathbf{p}_B$.

(b) Show that

$$s + t + u = m_A^2 + m_B^2 + m_C^2 + m_D^2.$$

10. A ρ^0 ($mc^2 = 775.5$ MeV) decays by the emission of two π^0 mesons ($mc^2 = 135.0$ MeV) according to the formula

$$\rho^+ \rightarrow 2\pi^0$$

What is the velocity of the two π^0's?

11. A Λ^0 ($mc^2 = 1115.7$ MeV) decays by the emission of a proton (p) with rest energy ($mc^2 = 938.3$ MeV) and a π^- with rest energy ($mc^2 = 139.6$ MeV) according to the formula

$$\Lambda \rightarrow p + \pi^-$$

What is the velocity of the proton and the π^-?

12. Two particles, each having a mass m and a speed $(2/3)c$, collide head-on and stick together. Using the laws of the conservation of momentum and energy, find the mass of the resulting particle.

13. An electron having a kinetic energy of 10 GeV makes a head-on collision with a positron having the same energy. The collision produces two muons ($mc^2 = 105.7$ MeV) moving in opposite directions. Find the kinetic energy and velocity of each muon.

14. An experiment is designed in which a proton and an antiproton collide producing a particle with a mass of 9700 MeV. What must be the incident kinetic energies and velocities of the colliding particles?

15. A pion (mass $= 139.6$ MeV/c^2) decays into a muon (mass $= 105.7$ MeV/c^2, mean lifetime $= 2.2 \times 10^{-6}$ s) and a neutrino (mass ≈ 0). (a) Find the speed of the muon? (b) Find the mean distance traveled by the muon before it decays.

16. Using the Dirac-Pauli representation, show that the α- and β-matrices satisfy the anticommutation relations (13.41).

17. Using the Dirac-Pauli representation of the α and β matrices, construct the matrices for $\gamma^\mu p_\mu + m$ and $\gamma^\mu p_\mu - m$.

18. Draw the Feynman diagram for the electromagnetic scattering of an electron (e^-) and a muon (μ^-) showing the momentum and the spin indices of the incoming and outgoing particles. (The muon is a fermion.)

Chapter 14

Particle physics

Contents

14.1	Leptons and quarks	341	14.9 Supersymmetry	385
14.2	Conservation laws	348	Suggestions for further reading	388
14.3	Spatial symmetries	354	Basic equations	388
14.4	Isospin and color	357	Summary	389
14.5	Feynman diagrams	364	Questions	389
14.6	The $R(3)$ and $SU(3)$ symmetry groups	369	Problems	390
14.7	* Gauge invariance and the electroweak theory	380		
14.8	Spontaneous symmetry breaking and the discovery of the Higgs	382		

The goal of particle physics is to understand the tiniest objects of which the universe is made and the forces that govern them.

14.1 Leptons and quarks

All matter is composed of leptons, quarks, and elementary particles called bosons which serve as the carriers of the force between particles. The lepton family includes the electron e^- which has an electric charge and interacts with other charged particles by means of the *electromagnetic force*. The electron also interacts by means of a force called the *weak force*, which is considerably weaker than the electromagnetic force. Associated with the electron is an illusive particle called the electron neutrino ν_e, which only interacts by means of the weak force. The other members of the lepton family are the muon μ^- with its neutrino ν_μ, and the tau τ^- with its neutrino ν_τ. The leptons are divided into distinct doublets or *generations* as follows

$$
\begin{bmatrix} \nu_e \\ e^- \end{bmatrix}, \quad \begin{bmatrix} \nu_\mu \\ \mu^- \end{bmatrix}, \quad \begin{bmatrix} \nu_\tau \\ \tau^- \end{bmatrix}. \tag{14.1}
$$

Strongly interacting particles are composed of quarks. The particles composed of quarks and interacting by the strong interaction are known collectively as *hadrons* from the Greek word, hadros, meaning robust. Among the hadrons, the proton and neutron are members of a family of particles called *baryons*, which are made up of three quarks. Another family of strongly interacting particles is the *mesons*, which are made up of a quark/antiquark pair.

Quarks come in six types, called *flavors* denoted by *up* (u), *down* (d), *strange* (s), *charmed* (c), *bottom* (b), and *top* (t) quarks. The b and t quarks are also referred to by the more appealing names of *beauty* and *truth*. Like the leptons, the quarks are divided into three generations

$$
\begin{bmatrix} u \\ d \end{bmatrix}, \quad \begin{bmatrix} c \\ s \end{bmatrix}, \quad \begin{bmatrix} t \\ b \end{bmatrix}. \tag{14.2}
$$

An unusual property of quarks is that they have fractional charges (Q). For each quark doublet, the upper member (u, c, t) has electric charge $Q = +2/3$ times the charge of a proton and the lower member (d, s, b) has charge $Q = -1/3$ times the charge of a proton. The proton is made up of two up-quarks and one down-quark (uud), while the neutron is made up of one up-quark and two down-quarks (udd). Using this information, one can easily confirm that the proton has electric charge equal to one, while the neutron has an electric charge equal to zero.

The properties of the leptons are summarized in Table 14.1.

The electron (e^-), muon (μ^-), and tau (τ^-) have negative charges. All of the particles shown in Table 14.1 have antiparticles. The antiparticle of the electron is the positron (e^+), and the antiparticles of muon and tau are the positive

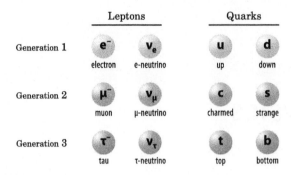

FIGURE 14.1 The three generations of quarks and leptons.

TABLE 14.1 The family of particles called Leptons. Leptons participate in the electromagnetic and weak interaction, but do not participate in the strong interaction. The masses and life times are those given by the Particle Data Group in 2008.

Particle	Mc^2	Lifetime	Decay mode
e^-	0.510998 MeV	$> 4.6 \times 10^{26}$ yr	
μ^-	105.658 MeV	2.197×10^{-6} s	$\mu^- \to e^- + \overline{\nu_e} + \nu_\mu$
τ^-	1776.99 MeV	290.6×10^{-15} s	$\tau^- \to \mu^- + \overline{\nu_\mu} + \nu_\tau$ $\to e^- + \overline{\nu_e} + \nu_\tau$
ν_1	< 200 meV	stable	
ν_2	< 200 meV	stable	
ν_3	< 200 meV	stable	

muon (μ^+) and tau (τ^+). While the electron is stable, the muon decays in 2.197 micro-seconds and the tau decays in 290.6 femtoseconds. A few of the decay modes of the muon and tau are shown in Table 14.1.

The determination of the mass of neutrinos is currently a subject of active research. Experiments designed to measure the flux of solar neutrinos upon the surface of the Earth have consistently measured a flux below the value predicted by the accepted model of the Sun. This discrepancy between measurements conducted in particle physics and the standard solar model can be understood if some of the electron neutrinos emitted by the Sun are converted into muon and tau neutrinos. Neutrinos are emitted and absorbed in states with a definite flavor but the flavor states are not mass eigenstates. The flavor of a neutrino can thus change as the neutrino travels through free space. However, neutrinos are detected by absorption and emission in states of definite flavor. The situation is analogous to the Stern-Gerlach experiment discussed in Chapter 4. While an electron can be in a state which is a superposition of states having a spin-up and spin-down character, the spin of the electron is always measured to be up or down. In the same way, the mass eigenstates of the neutrino are linear combinations of states with a definite flavor, but the neutrino always interacts as a particle with a definite flavor. The neutrino mass corresponding to the stationary neutrino states are generally denoted by m_1, m_2, and m_3. In Table 14.1, we have given only upper bounds to these three masses. While the current values of the neutrino masses are not very accurate, the experimental data does rule out the possibility that neutrinos have zero mass.

The charge and mass of the three generations of quarks are given in Table 14.2. An illustration of the three generations of leptons and quarks is given in Fig. 14.1.

Convincing evidence for the quark model can be obtained from scattering experiments in which high-energy electrons collide with protons. At energies of a few hundred MeV, the proton target behaves as a particle with a continuous distribution of matter. However, the collisions that occur at much higher energies, for which the electron has more than 20 GeV of energy, can only be explained by supposing that protons are composed of three spin 1/2 particles. The short wavelength associated with such high-energy electrons probes the finer details of the structure of protons. High-energy scattering experiments showing that the proton has an internal structure are analogous to the scattering experiments of Rutherford, which showed that atoms have nuclei.

The forces between elementary particles are transmitted by means of a class of particles called bosons. The carrier of the electromagnetic force is the *photon*, while the *gluon* is the carrier of the strong force and the *graviton* has been postulated as the carrier of the gravitational force. All of these bosons have zero mass, and the force associated with each of these

TABLE 14.2 Properties of the quarks. The masses of the quarks are those given by the Particle Data Group in 2008.

Flavor	Symbol	Q	Mc^2 (MeV)
down	d	$-1/3$	3.5–6.0
up	u	$+2/3$	1.5–3.3
strange	s	$-1/3$	104
charmed	c	$+2/3$	1,270
bottom	b	$-1/3$	4,200
top	t	$+2/3$	171,200

particles has infinite range. In contrast, the carriers of the weak force are the massive W^+, W^-, and Z bosons. The weak force has a very short range.

The quark compositions and several of the properties of a few of the lightest mesons are given in Table 14.3.

TABLE 14.3 Properties of a few of the lightest mesons. Data is that given by the Particle Data Group in 2008.

Name	Symbol	Composition	Mc^2 (MeV)	Decay Mode	Lifetime/Width
pion	π^-	$d\bar{u}$	139.6	$\pi^- \to \mu^- + \bar{\nu}_\mu$	2.603×10^{-8} s
	π^+	$u\bar{d}$	139.6	$\pi^+ \to \mu^+ + \nu_\mu$	2.603×10^{-8} s
	π^0	$u\bar{u}, d\bar{d}$	135.0	$\pi^0 \to 2\gamma$	8.4×10^{-17} s
kaon	K^+	$u\bar{s}$	493.7	$K^+ \to \mu^+ + \nu_\mu$	1.238×10^{-8} s
	K^-	$s\bar{u}$	493.7	$K^- \to \mu^- + \bar{\nu}_\mu$	1.238×10^{-8} s
	K^0	$d\bar{s}$	497.6		
	$\overline{K}^0$	$\bar{d}s$	497.6		
eta	η	$u\bar{u}, d\bar{d}, s\bar{s}$	547.8	$\eta \to 2\gamma$	1.30 keV
rho	ρ^+	$u\bar{d}$	775.5	$\rho^+ \to \pi^+ + \pi^0$	149.4 MeV
	ρ^-	$d\bar{u}$	775.5	$\rho^- \to \pi^- + \pi^0$	149.4 MeV
	ρ^0	$u\bar{u}, d\bar{d}$	775.5	$\rho^0 \to 2\pi^0$	149.4 MeV
omega	ω	$u\bar{u}, d\bar{d}$	782.7	$\omega \to \pi^+ + \pi^-$	8.49 MeV

In Table 14.3, the name of each particle and the symbol used to denote the particle are given in the first two columns. Columns three and four give the quark composition and rest-mass energy of the particles. All mesons are composed of a quark/antiquark pair. The π^-, for instance, is composed of a d quark and a $\bar{u}$ antiquark, while the π^+ is composed of a u quark and a $\bar{d}$ antiquark. The π^0 and ρ^0 are each linear combinations of $u\bar{u}$ and $d\bar{d}$. From the quark compositions in the table, one can see that the π^+ is the antiparticle of the π^-, the ρ^+ is the antiparticle of the ρ^-, the K^+ is the antiparticle of the K^-, and the $\overline{K}^0$ is the antiparticle of the K^0. All of the particles described in Table 14.3 are unstable. Column five of the table gives one of the possible decay processes of the particle, while column six gives information about the decay time.

In the introduction, we said that processes involving the strong force take place within 10^{-22} seconds, while processes involving the electromagnetic force typically take place in 10^{-14} to 10^{-20} seconds. Processes due to the weak interaction occur generally within 10^{-8} to 10^{-13} seconds. Using this information, one can see that the decay of the π^+, π^-, K^+, and K^- occur by the weak interaction. This possibility is confirmed by the fact that the particles produced by the decay of these particles include the neutrino, which only participates in the weak interaction. Similarly, the decay of π^0, producing two photons, occurs by the electromagnetic interaction.

While the K-mesons are produced by the strong interactions, they decay by means of the weak interaction. One can see that the decay modes and the decay times of K^0 and $\overline{K}^0$ are not given in Table 14.3. These particles are not approximate eigenstates of the weak interaction and do not have decay modes and times associated with them. The neutral kaons, which are approximate eigenstates of the weak interaction, are generally denoted by K_S^0 (for K^0 short) and K_L^0 (for K^0 long). The states of K_S^0 and K_L^0 are linear combinations of the states of K^0 and $\overline{K}^0$. The K_S^0 meson generally decays in 8.958×10^{-11} s

FIGURE 14.2 Big European Bubble Chamber used at CERN in Switzerland. (From http://commons.wikimedia.org/wiki.)

into $\pi^0 + \pi^0$ or into $\pi^+ + \pi^-$, while the K_L^0 decays in 5.116×10^{-8} s most often into $\pi^+ + e^- + \overline{\nu}_e$ or $\pi^+ + \mu^- + \overline{\nu}_\mu$. The properties of neutral kaons are similar to the properties of neutrinos, which, as we have seen earlier, are emitted or absorbed in states of definite flavor, which are not eigenstates of the mass.

We saw earlier in connection with the Heisenberg uncertainty principle that states that exist for a long time have well-defined values of the energy, while the energy of shorter-lived resonance states is more poorly defined. The rho meson decays by the strong interaction within 10^{-22} seconds and thus corresponds to defuse states having a poorly-defined value of the energy. The lifetime of particles that decay by the strong interaction is most commonly reported by giving the width of the resonance state. As can be seen in Table 14.3, the width of the rho resonance is 149.4 MeV.

A number of mesons and baryons were discovered using the bubble chamber technology invented by Donald Glaser in 1952 and developed as a scientific instrument by Luis Averez and his coworkers at Berkeley. The bubble chamber used a superheated fluid, which boiled into tiny bubbles of vapor along the tracks of particles. A chamber filled with liquid hydrogen provided a dense concentration of hydrogen nuclei which served as the targets for nuclear scattering events, while a chamber filled with deuterium was used to study collision processes involving neutrons. Fig. 14.2 shows a picture of the Big European Bubble Chamber at CERN in Switzerland. When the large piston on the bottom of the chamber was lowered, the liquid within the chamber became superheated and bubbles would form along the tracks of charged particles. The tracks of the particles curved due to the presence of a strong magnetic field with the tracks of positively charged particles curved in one way and the tracks of negatively charged particles curved in the other. After being used for many years in high-energy scattering experiments, the bubble chamber has been replaced by quicker electronic counters. The Big European Bubble Chamber now stands on a lawn outside the laboratory in CERN as a scientific exhibit.

Example 14.1

Find the length of the tracks produced in a bubble chamber by a particle traveling with a speed equal to $0.96c$ that decays by the weak interaction in 10^{-10} s. What would the length of the track be if the particle were to decay by the electromagnetic interaction in 10^{-16} s, or the strong interaction in 10^{-24} s?

Solution

Using Eq. (12.26) of Chapter 12, the lifetime of the particle in the laboratory frame of reference would be

$$\Delta t = \frac{10^{-10} \text{ s}}{\sqrt{1 - (0.96)^2}} = 3.57 \times 10^{-10} \text{ s}.$$

Hence, the length of the track in the bubble chamber would be

$$\Delta x = 0.96c \times 3.57 \times 10^{-10} \text{ s}.$$

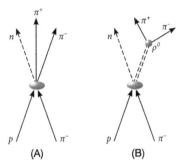

FIGURE 14.3 The reaction $\pi^- + p \rightarrow n + \pi^+ + \pi^-$ can proceed in the following two ways. (A) The three particles in the final state are produced all at once. (B) Two particles, n and ρ^0, are produced, and the ρ^0 subsequently decays into π^- and π^+.

Using the value for the velocity of light given in Appendix A, we obtain

$$\Delta x = 0.96 \times 2.998 \times 10^8 \text{ m/s} \times 3.57 \times 10^{-10} \text{ s} = 0.1027 \text{ m}.$$

The length of the track is thus about 10.3 cm long and could be easily observed.

Using the same approach, the length of the track left by a particle that decayed by the electromagnetic interaction would be 0.1027×10^{-6} m and the length of the track of a particle decaying by the strong interaction would be 0.1027×10^{-14} m $= 1.027$ fm. While it might be possible to observe the track of a rapidly moving particle that decays electromagnetically, the length of the path of a particle that decays by the strong interaction would be about equal to the radius of an atomic nucleus and would not be observable.

An example of a collision between two particles is illustrated in Figs. 14.3(A) and (B). The figures depict a scattering event in which a negative pion collides with a proton producing a neutron and positive and negative pions as follows

$$\pi^- + p \rightarrow n + \pi^+ + \pi^- .$$

The process can occur as shown in Fig. 14.3(A) with the three particles in the final state being created independently, or it can occur as shown in Fig. 14.3(B) with a neutron and a neutral rho being produced. The neutral rho then decays into two pions. Since the neutron and the rho are neutral particles, they will not leave tracks in a bubble chamber. In order to decide which of these two processes actually occurs, we define the *invariant mass* of the two outgoing pions by the following equation

$$m_{12} = \frac{1}{c^2} \left[(E_1 + E_2)^2 - (\mathbf{p}_1 + \mathbf{p}_2)^2 c^2 \right]^{1/2} , \tag{14.3}$$

where E_1 and $\mathbf{p}_1$ being the energy and momentum of the π^+ and E_2 and $\mathbf{p}_2$ being the energy and the momentum of the π^-.

The momenta of the two pions can be determined by measuring the curvature of their tracks in a magnetic field and the energy of the pions can be determined by the amount of ionization they produce. Eq. (14.3) can then be used to calculate the invariant mass for every observed pair. If the reaction proceeds as shown in Fig. 14.3(A) with no correlation occurring between the two pions, the pions will share energy and momentum statistically. Plotting the number of events with a particular value of the invariant mass versus the invariant mass will then lead to the distribution having the form shown in Fig. 14.4(A). If, on the other hand, the reaction proceeds as shown in Fig. 14.3(B) with the production of a ρ, energy and momentum conservation lead to the following equations

$$E_\rho = E_1 + E_2, \quad \mathbf{p}_\rho = \mathbf{p}_1 + \mathbf{p}_2 .$$

According to Eq. (14.3), the invariant mass is then given by the equation

$$m_{12} = \frac{1}{c^2} \left[E_\rho^2 - \mathbf{p}_\rho^2 c^2 \right]^{1/2} .$$

Using the relation Eq. (13.14) of Chapter 13, we may then identify the right-hand side of this last equation as being equal to the mass of the ρ. We thus have

$$m_{12} = m_\rho .$$

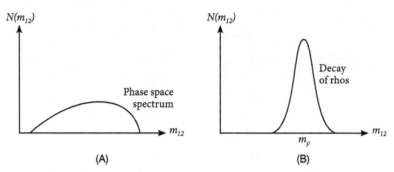

FIGURE 14.4 The number of events with a particular invariant mass versus the invariant mass (A) when no correlation occurs between the two pions, (B) when a rho is produced which decays into two pions.

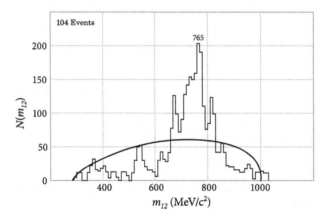

FIGURE 14.5 The invariant mass spectrum for $\pi^- + p \rightarrow n + \pi^+ + \pi^-$ in an early experiment.

If the reaction proceeds as shown in Fig. 14.3(B), a plot of the number of events versus the invariant mass will thus lead to the distribution shown in Fig. 14.4(B). The invariant mass of the two pions in this case will be equal to the mass of the decaying particle.

Distribution curves such as those shown in Figs. 14.4(A) and 14.4(B) are known as *phase-space spectra*. Since the scattering of a π^- and a proton can occur in either of the two ways described, the distribution curve for an actual experiment will be some combination of these two distribution curves. Fig. 14.5 shows the invariant mass spectrum obtained in an early experiment. A broad peak at an invariant mass of 765 MeV is clearly visible. Even though the rho lives for only about 6×10^{-24} s, its existence is well established and its mass has been determined.

Notice that we have given a single mass for the rho even though it has three distinct charge states (ρ^+, ρ^-, and ρ^0). The rho resonance is so broad that it is not possible to resolve experimentally the three charge states. This is consistent with the uncertainty principle. Since the rho decays in such a short time, its energy and, hence, its mass is more poorly determined than it is for particles that decay by the weak and electromagnetic interactions.

The quark compositions and several of the properties of a few of the lightest baryons are given in Table 14.4.

The mesons shown in Table 14.3 and the baryons in Table 14.4 occur in groups with the same generic name. According to Table 14.3, the mass of π^- and its antiparticle π^+ are equal to 139.6 MeV, while the mass of π^0 is equal to 135.0 MeV. These particles all have about the same mass and may be regarded as members of a charge multiplet. The eta in Table 14.3 is a charge singlet. The masses of the baryons shown in Table 14.4 also cluster about certain common values. The proton and neutron have very similar masses and together form the first baryon multiplet. Other multiplets correspond to the lambda, sigma, delta, xi, and omega. Notice that the rho mesons given in Table 14.2 have the same quark composition as the pi mesons, and the Δ^+ and Δ^0 given in Table 14.4 have the same quark composition as the proton and neutron. The rho mesons and the delta baryons can be thought of as resonances or excited states of lower lying pi and nucleon states.

Another example of an early bubble-chamber experiment is illustrated in Fig. 14.6(A). Here again a negative pion having a few GeV of energy collides with a proton in a bubble chamber experiment. In Fig. 14.6(A), the negative pion disappears at the point where it hits the proton and farther downstream two V-like events appear. By measuring the energy and momentum of the four particles forming the Vs, one V was shown to consist of a proton and a pion and the other V shown to consist of two pions. As indicated in Fig. 14.6(B), the experimental data is consistent with two neutral particles

TABLE 14.4 Properties of a few of the lightest baryons. Data is that given by the Particle Data Group in 2008.

Name	Symbol	Composition	Mc^2 (MeV)	Decay Mode	Lifetime/Width
Nucleon	p	uud	938.3		$> 10^{31}$ yr
	n	udd	939.6	$n \to p + e^- + \overline{\nu}_e$	885.7 s
Lambda	Λ	uds	1115.7	$\Lambda \to p + \pi^-$	2.63×10^{-10} s
Sigma	Σ^+	uus	1189.4	$\Sigma^+ \to p + \pi^0$	8.02×10^{-11} s
	Σ^0	uds	1192.6	$\Sigma^0 \to \Lambda + \gamma$	7.4×10^{-20} s
	Σ^-	dds	1197.5	$\Sigma^- \to n + \pi^-$	1.48×10^{-10} s
Delta	Δ^{++}	uuu	1232	$\Delta^{++} \to p + \pi^+$	118 MeV
	Δ^+	uud	1232	$\Delta^+ \to n + \pi^+$	118 MeV
	Δ^0	udd	1232	$\Delta^0 \to p + \pi^-$	118 MeV
	Δ^-	ddd	1232	$\Delta^- \to n + \pi^-$	118 MeV
Xi	Ξ^0	uss	1314.9	$\Xi^0 \to \Lambda + \pi^0$	2.90×10^{-10} s
	Ξ^-	dss	1321.7	$\Xi^- \to \Lambda + \pi^-$	1.639×10^{-10} s
Omega	Ω^-	sss	1672.5	$\Omega^- \to \Lambda + K^-$	8.21×10^{-11} s

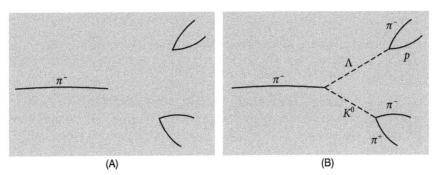

(A)   (B)

FIGURE 14.6 For an early experiment in which a π^- collides with a proton, (A) an illustration of the bubble chamber tracks, (B) an illustration of the tracks with the various particles identified.

being produced in the collision of the negative pion and the proton, and these neutral particles decaying to produce the two Vs. The identity of the neutral particles can be found by finding the invariant mass associated with the two Vs. Using Eq. (14.3), the invariant mass of the proton-pion pair was found to be 1116 MeV, while the invariant mass of the pion pair was found to be about 500 MeV. Scattering experiments of the kind we have just described enabled physicists to identify the neutral lambda having a mass of 1115.7 MeV and the neutral kaon having a mass of 497.7 MeV. The initial collision of the negative pion and the proton is described by the reaction formula

$$\pi^- + p \to \Lambda + K^0 \,, \tag{14.4}$$

while the decays of the particles forming the Vs are described by the formulas

$$\Lambda \to p + \pi^- \tag{14.5}$$

and

$$K^0 \to \pi^+ + \pi^- \,. \tag{14.6}$$

The momentum and energy of the particles forming the Vs could be determined from their tracks, and the momentum and energy of the neutral particles could be calculated using the conservation of momentum and energy. The lifetime of each neutral particles could then be determined using the distance it traveled in the bubble chamber.

14.2 Conservation laws

Particles accelerated to high-energy in modern accelerators collide to produce an astounding variety of new particles. Conservation laws provide a means of characterizing the possible outcomes of scattering events and describing what can and cannot occur.

14.2.1 Energy, momentum, and charge

We have already found that energy and momentum are conserved in scattering processes. As an example of the conservation of energy, we consider again the process in which a negative pion collides with a proton to produce a lambda and a neutral kaon. This process is described by Eq. (14.4) and illustrated in Fig. 14.6(B). Using the data given in Tables 14.3 and 14.4, we find that the rest energy of the incoming particles is 1077.9 MeV, and the rest energy of the outgoing particles is 1613.3 MeV. The reaction can only occur if the incoming particles have sufficient kinetic energy to make up this mass difference.

The total electric charge of colliding particles is also conserved in collision processes. Since the charge of an assembly of particles is the sum of the charges of the individual particles and is always a multiple of the basic unit e, the charge is referred to as an *additive quantum number*. We shall soon define other quantum numbers of this kind. Additive quantum numbers always have opposite values for the members of a particle-antiparticle pair. To see this, we consider the pair creation process in which an incoming photon produces a particle-antiparticle pair

$$\gamma \rightarrow e^- + e^+, \tag{14.7}$$

and the corresponding annihilation process in which an electron and a positron collide to produce a pair of photons

$$e^- + e^+ \rightarrow \gamma + \gamma. \tag{14.8}$$

Additive quantum numbers always have the value zero for photons, which are entirely characterized by their energy and polarization. Since the additive quantum numbers are conserved in pair creation and annihilation processes, the additive quantum numbers of the antiparticle must cancel the quantum numbers of the particle in each case.

14.2.2 Lepton number

We now turn our attention to scattering processes involving leptons. We begin by considering how neutrinos and antineutrinos appear in reaction formulas. While the π^+ decays into a positron and a neutrino,

$$\pi^+ \rightarrow e^+ + \nu_e, \tag{14.9}$$

the π^- decays into an electron and an antineutrino,

$$\pi^- \rightarrow e^- + \overline{\nu_e}. \tag{14.10}$$

Notice that a positron appears together with a neutrino on the right-hand side of Eq. (14.9), while an electron appears together with an antineutrino on the right-hand side of Eq. (14.10).

Further information about neutrinos can be obtained from neutrino reactions. We consider the following process in which an electron neutrino is captured by a neutron

$$\nu_e + n \rightarrow e^- + p. \tag{14.11}$$

The corresponding capture process for an antineutrino is

$$\overline{\nu_e} + p \rightarrow e^+ + n. \tag{14.12}$$

A neutrino and an electron appear on opposite sides of Eq. (14.11), while an antineutrino and a positron appear on opposite sides of Eq. (14.12).

In order to explain which processes can and cannot occur, E. Konopinski and H. Mahmoud introduced the idea of lepton number L and lepton conservation. They assigned the value $L = 1$ to e^-, μ^-, ν_e, and ν_μ and the value $L = -1$ to the antileptons e^+, μ^+, $\overline{\nu_e}$, and $\overline{\nu_\mu}$. We may readily confirm that lepton number is conserved in the reactions (14.9) and (14.10) and in the reactions (14.11) and (14.12). Taking the lepton numbers of the π^+ and π^- to be zero, the lepton numbers of the left-hand sides of Eqs. (14.9) and (14.10) are equal to zero. The lepton numbers of the right-hand sides of these equations are also zero since the right-hand side of each equation contains one lepton with $L = 1$ and another lepton with $L = -1$. Each side of Eq. (14.11) has a lepton with $L = 1$, while each side of Eq. (14.12) has a lepton with $L = -1$.

The rule of lepton conservation allows the reactions that we have found to occur and it prohibits some reactions that have been found not to occur. For instance, the reaction (14.12) is consistent with lepton conservation and has been found to occur, while the reaction

$$\overline{v_e} + n \rightarrow e^- + p \tag{14.13}$$

is inconsistent with lepton conservation and has been found not to occur. (The lepton number of the left-hand side of this last equation is -1, while the lepton number of the right-hand side is $+1$.) However, some reactions are consistent with all of the conservation laws discussed so far and still do not occur. An example of such a forbidden reaction is

$$v_\mu + n \rightarrow e^- + p. \tag{14.14}$$

The collision of a muon neutrino with a neutron leads to a muon and a proton, but not to an electron and a proton. The law of lepton conservation does not distinguish between electrons and muons or their corresponding neutrinos. Another example of the weakness of the lepton conservation law involves the decay of the muon. The μ^- decays according to the following equation

$$\mu^- \rightarrow e^- + \overline{v_e} + v_\mu . \tag{14.15}$$

Another possible decay of the muon is

$$\mu^- \rightarrow e^- + \gamma . \tag{14.16}$$

This last reaction, which is allowed by lepton conservation, has been found not to occur. All attempts to find the gamma decay of the muon have been unsuccessful.

The simplest way to explain why the reactions (14.14) and (14.16) do not occur is to assign to the muon and its neutrino a muon lepton number (L_μ). L_μ has the value $+1$ for the μ^- and v_μ, is equal to -1 for μ^+ and $\overline{v_\mu}$, and is zero for all other particles. Similarly, we define an electron lepton number L_e which is equal to $+1$ for the electron and the electron neutrino, -1 for the positron and the electron antineutrino, and is zero for all other particles. One can then see that the processes we have considered, which do occur, conserve both electron and muon lepton numbers, while a number of processes that do not conserve electron and muon numbers do not occur. To see whether Eq. (14.11) conserves electron and muon numbers, we write the equation with the appropriate value of the electron lepton number (L_e) and the muon lepton number (L_μ) under each term

$$\begin{array}{ccccccc}
 & v_e & + & n \rightarrow & e^- & + & p \\
L_e: & 1 & & 0 & 1 & & 0 \\
L_\mu: & 0 & & 0 & 0 & & 0 .
\end{array} \tag{14.17}$$

The sum of L_e for the left- and right-hand sides of Eq. (14.17) is equal to one, while the corresponding sums for L_μ are equal to zero. One may also see that Eq. (14.14) does not conserve muon and electron lepton numbers. Writing L_e and L_μ under each term of Eq. (14.14) as before, we obtain

$$\begin{array}{ccccccc}
 & v_\mu & + & n \rightarrow & e^- & + & p \\
L_e: & 0 & & 0 & 1 & & 0 \\
L_\mu: & 1 & & 0 & 0 & & 0 .
\end{array} \tag{14.18}$$

The lepton numbers, L_e and L_μ, are clearly not conserved in this reaction. One may also show that L_e and L_μ are not conserved for the process (14.16).

The discovery of the tau lepton has led to the introduction of yet another quantum number, the tau lepton number (L_τ). We assign the value $L_\tau = +1$ for τ^- and v_τ and the value $L_\tau = -1$ for τ^+ and $\overline{v_\tau}$. L_τ is zero for all other particles. The tau decays in a number of different ways including

$$\begin{aligned}
\tau^- &\rightarrow e^- + \overline{v_e} + v_\tau \\
&\rightarrow \mu^- + \overline{v_\mu} + v_\tau \\
&\rightarrow \pi^- + v_\tau .
\end{aligned} \tag{14.19}$$

Since L_τ has the value $+1$ for both τ^- and ν_τ and zero for the other particles involved in Eq. (14.19), one can readily see that the tau lepton number is conserved in all of these decay processes.

The values of L_e, L_μ, and L_τ for the leptons are shown in Table 14.5. Since the lepton numbers for an assembly of particles are equal to the algebraic sums of the lepton numbers for the individual particles, the lepton numbers are additive quantum numbers. As for the electric charge, the values of the lepton numbers for antiparticles are the negative of the values for the corresponding particles.

TABLE 14.5 The values of the electron lepton number (L_e), muon lepton number (L_μ), and tau lepton number (L_τ) for the leptons. These quantum numbers are equal to zero for all other particles.

Particle	L_e	L_μ	L_τ
electron (e^-)	$+1$	0	0
positron (e^+)	-1	0	0
electron neutrino (ν_e)	$+1$	0	0
electron antineutrino ($\overline{\nu}_e$)	-1	0	0
negative muon (μ^-)	0	$+1$	0
positive muon (μ^+)	0	-1	0
muon neutrino (ν_μ)	0	$+1$	0
muon antineutrino ($\overline{\nu}_\mu$)	0	-1	0
negative tau (τ^-)	0	0	$+1$
positive tau (τ^+)	0	0	-1
tau neutrino (ν_τ)	0	0	$+1$
tau antineutrino ($\overline{\nu}_\tau$)	0	0	-1

14.2.3 Baryon number

Another quantity that is conserved in particle reactions is the baryon number, which can be expressed in terms of the number of quarks $N(q)$ and the number of antiquarks $N(\overline{q})$ by the following formula

$$B = \frac{1}{3}\left[N(q) - N(\overline{q})\right]. \tag{14.20}$$

Baryons are composed of three quarks and thus have baryon number B equal to $+1$, while antibaryons are composed of three antiquarks and have baryon number equal to -1. Mesons, which are made up of a quark/antiquark pair, have baryon number equal to zero.

Baryon number is conserved in strong and electromagnetic interactions because in these interactions quarks and antiquarks are only created or destroyed in particle/antiparticle pairs. Consider, for example, the strong interaction process

$$p + p \rightarrow p + n + \pi^+.$$

The quark description of the particle involved in this interaction is

$$(uud) + (uud) \rightarrow (uud) + (udd) + (u\overline{d}).$$

Comparing the quarks of each flavor in the initial and final states, we see that the final state contains the same number of quarks of each flavor as the initial state plus an additional $d\overline{d}$ pair. The baryon number is equal to two for each side of the equation. Similarly, the π^0, which is a linear combination of $u\overline{u}$ and $d\overline{d}$, decays electromagnetically

$$\pi^0 \rightarrow \gamma + \gamma.$$

The baryon number is zero for each side of the equation. For each of these two interaction processes, the number of quarks minus the number of antiquarks remains the same.

A quark/antiquark pair can also be created in a weak interaction; however, a weak interaction may also change the flavor of a quark. Consider the following decay of the neutron

$$n \rightarrow p + e^- + \bar{\nu}_e,$$

having the quark description

$$(udd) \rightarrow (uud) + e^- + \bar{\nu}_e.$$

A d quark is changed into an u quark in this interaction. While the number of quarks and antiquarks of any flavor is not conserved by the weak interaction, the number of baryons is still conserved since quarks are not changed into antiquarks or vice versa.

For a particle reaction to occur, the sum of the baryon numbers of the incoming particles must be equal to the sum of the baryon numbers of the outgoing particles. Consider, for example, the following scattering process

$$\begin{array}{ccccc} \pi^- & + & p \rightarrow & K^+ & + & \Sigma^- \\ B: \quad 0 & & 1 & 0 & & 1. \end{array} \tag{14.21}$$

The proton and the sigma minus have baryon number one, while the pi minus and the K plus are mesons with baryon number zero. The sum of the baryon numbers for the incoming particles and the sum of the baryon numbers for the outgoing are both equal to one. The process thus conserves baryon number and is allowed. As a second example, we consider the decay process

$$\begin{array}{cccc} n \rightarrow & \pi^+ & + & \pi^- \\ B: \quad 1 & 0 & & 0. \end{array} \tag{14.22}$$

The baryon number of the initial state in this reaction is equal to one, while the baryon number of the final state is equal to zero. Hence, baryon number is not conserved and this process does not occur.

The decay modes of baryons provide many examples of baryon conservation. Consider, for example, the following observed decay modes of Σ^+

$$\begin{aligned} \Sigma^+ &\rightarrow p + \pi^0 \\ &\rightarrow n + \pi^+ \\ &\rightarrow \Lambda + e^+ + \nu_e. \end{aligned} \tag{14.23}$$

Baryon number is conserved in each of these processes. Since baryon number is conserved in the decay processes of baryons and since the proton is the only stable baryon, all baryons decay ultimately into a proton.

14.2.4 Strangeness

Soon after the discovery of the pion, other mesons and baryons were observed which were produced in the strong interaction but decayed by the weak interaction. These particles had the unlikely or strange property that they were produced in 10^{-22} seconds and yet lived long enough to produce considerable tracks in a bubble chamber. Fig. 14.7 shows one of the first observed weak decays of a strongly interacting particle, in which the K^+ meson decays into a μ^+ and a ν_μ. Another example of the weak decay of a strongly interacting particle has been discussed previously in conjunction with Figs. 14.6(A) and (B). The neutral Λ and K^0 particles depicted in Fig. 14.6(B) are produced by strong interactions but decay over a longer span of time producing characteristic V-like patterns. In 1952, A. Pais made the first step in explaining this paradox by observing that strange particles are always produced in pairs. The solution to this problem came the following year when M. Gell-Mann and K. Nishijima both introduced a new quantum number. Gell-Mann called the quantum number *strangeness*, and this name has been adopted. The strangeness quantum number S is conserved by the strong and electromagnetic interactions, but may be violated by weak interactions.

Using the idea of strangeness, the production of strange particles can be easily explained. One of the two strange particles produced by the strong interaction has a positive value of the strangeness quantum number and the other has a negative value. The total amount of strangeness produced by the strong interaction is equal to zero. Following its production, each strange particle decays by the weak interaction, which may involve a change of the strangeness quantum number. Consider the reaction

$$\pi^- + p \rightarrow \Lambda + K^0. \tag{14.24}$$

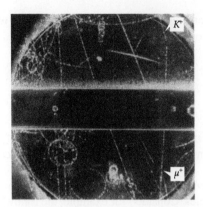

FIGURE 14.7 An early observed weak decays of a strongly interacting particle, in which the K^+ meson decays into a μ^+ and a ν_μ. (From http://commons.wikimedia.org/wiki.)

The strangeness quantum number S is taken to be zero for pions and nucleons. Since strangeness is conserved by the strong interaction, the total strangeness for both sides of Eq. (14.24) must be equal to zero. The strangeness of the K^0 must therefore be the negative of the strangeness of the Λ. This explains the rule of Pais. If the incoming particles all have $S = 0$, strange particles must occur in conjunction with other particles. Also, the strange particles produced in the reaction can only decay to nonstrange particles by the weak interaction. This explains the rapid creation and slow decay of the Λ and K^0.

The assignment of strangeness to hadrons is based on reactions that are observed to occur by the strong interaction. If the K^+ is assigned a strangeness $S = +1$, the reaction

$$\pi^- + p \rightarrow n + K^+ + K^- \,, \tag{14.25}$$

which is observed to proceed by the strong interaction, may then be used to assign a strangeness $S = -1$ to K^-. We have seen previously that particle/antiparticle pairs have opposite values of additive quantum numbers. Since the positive and negative kaons have opposite strangeness, it is natural to suppose that K^- is the antiparticle of K^+.

The strangeness quantum number associated with the particles involved in a reaction can be understood in terms of the quark model. Consider for example the reaction

$$\Lambda \rightarrow \pi^- + p.$$

The quark description of the particles appearing in this reaction is

$$
\begin{array}{cccc}
(uds) & \rightarrow & (d\overline{u})+ & (uud) \\
S: \quad -1 & & 0 & 0 \,.
\end{array}
$$

In the reaction, a s quark changes into an u quark and an additional $d\overline{u}$ pair of quarks is produced. Only the weak interaction can change the flavor of a quark.

A strange particle can either be described as a particle with a nonzero value of the strangeness quantum number S or as a particle which contains one or more strange or antistrange quarks. The strangeness quantum number is related to the number of strange and antistrange quarks by the equation

$$S = -[N(s) - N(\overline{s})], \tag{14.26}$$

where $N(s)$ is the number of strange quarks and $N(\overline{s})$ is the number of antistrange quarks. The lightest strange mesons and baryons have a single strange quark or a single antistrange quark. Since only the weak interaction can change the flavor of a quark, the lightest strange mesons and baryons can only decay by the weak interaction. Of the mesons shown in Table 14.3 and the baryons shown in Table 14.4, the K-mesons and the Lambda- and Σ-baryons all have a single strange quark or a single antistrange quark.

Example 14.2

State whether each of the following processes can occur. If the reaction cannot occur or if it can only occur by the weak interaction, state which conservation law is violated.

$$(a) \qquad \pi^- + p \rightarrow \Lambda + \overline{\Sigma}^0$$
$$(b) \qquad p + p \rightarrow \Sigma^+ + n + K^0 + \pi^+ + \pi^0$$
$$(c) \qquad \mu^- \rightarrow e^- + \nu_e + \nu_\mu$$
$$(d) \qquad \Lambda \rightarrow p + e^- + \overline{\nu_e}$$
$$(e) \qquad \Xi^- \rightarrow \Sigma^0 + K^-$$

Solution

(a) This reaction does not occur because baryon number is not conserved. The proton (p) and lambda (Λ) are baryons having baryon number equal to $+1$, while the $\overline{\Sigma}^0$ is the antiparticle of a baryon and has baryon number equal to -1. Thus, the total baryon number of the initial state is $+1$, while the total baryon number of the final state is 0.

(b) This process is allowed. Since the baryon number of p, Σ^+, and n are all equal to $+1$, the total baryon number of the initial and final states are equal to $+2$. Since the strangeness quantum number S is equal to -1 for the Σ^+ and $+1$ for K^0, the total strangeness is equal to zero for both the initial and final states. No leptons are involved in this reaction.

(c) This process is not allowed because electron lepton number L_e is not conserved. The total electron lepton number of the initial state is zero; however, the electron lepton number of both e^- and ν_e are equal to $+1$, and, hence, the total electron number of the final state is equal to $+2$.

(d) Since Λ has a strangeness quantum number S equal to -1 and all the particles in the final state have $S = 0$, strangeness is not conserved in this reaction, and the reaction can only occur by the weak interaction.

(e) While the lepton numbers, baryon number, and strangeness are all conserved for this process, the process will not occur because energy is not conserved. According to the data given in Tables 14.4 and 14.3, the rest energies of Ξ^-, Σ^0, and K^- are 1321.3 MeV, 1192.6 MeV, and 493.7 MeV, respectively. Hence, the energy of the initial state is 1321.7 MeV, while the minimum energy of the final state is 1686.3 MeV.

14.2.5 Charm, beauty, and truth

All hadrons discovered during the early years of particle physics can be described as bound states of the u, d, and s quarks. Then in 1974, a heavy particle was discovered at the Brookhaven National Laboratory and at the Stanford Linear Accelerator Center. The new particle, which the Brookhaven group named J and the Stanford group named ψ, has come to be known as J/ψ. The properties of this particle show that it is the lightest of a family of particles which are bound states of the charmed quark and anticharmed quark

$$J/\psi\,(3097) = c\overline{c}$$

Just as the strangeness quantum number S can be defined in terms of the number of strange quarks and the antiparticle of the strange quark, the charm quantum number can be defined in terms of the number of charmed quarks and the antiparticle of the charmed quark

$$C = N(c) - N(\overline{c}). \tag{14.27}$$

The J/ψ particle is made up of a $c/\overline{c}$ pair of particles. Other particles have since been detected, which have a single c-quark or antiquark. The lightest charmed mesons are the D-mesons with the quark structure

$$D^+(1869) = c\overline{d}, \quad D^0(1865) = c\overline{u} \ \ (C = +1),$$
$$D^-(1869) = d\overline{c}, \quad \overline{D}^0(1865) = u\overline{c} \ \ (C = -1),$$

and D_s mesons with quark structures

$$D_s^+(1969) = c\bar{s}, \quad (C = +1, S = +1),$$
$$D_s^-(1969) = s\bar{c}, \quad (C = -1, S = -1).$$

The lightest charmed baryon is

$$\Lambda_c^+(2285) = udc, \quad (C = +1).$$

The mesons and baryons with a single charmed quark or antiquark all decay in about 10^{-13} s, which is to be expected of particles decaying by means of the weak interaction. Just as for the strange hadrons, charmed hadrons are produced together with other charmed hadrons.

The prevailing theory of elementary particles requires that the number of leptons and quarks be the same, implying that there be six quarks to match the six known leptons. Evidence for the fifth quark – the bottom quark b with the associated quantum number beauty $\tilde{B}$ – came with the discovery in 1977 of one of the lightest particles consisting of a $b/\bar{b}$ pair

$$\Upsilon(9460) = b\bar{b}, \quad (\tilde{B} = 0).$$

The B-mesons with a single b-quark or antiquark have the quark structure

$$B^+(5279) = u\bar{b}, \quad D^0(5279) = d\bar{b} \quad (\tilde{B} = +1),$$
$$B^-(5279) = b\bar{u}, \quad \overline{D}^0(5279) = b\bar{d} \quad (\tilde{B} = -1).$$

The lightest baryon with the b-quark is

$$\Lambda_b^0(5461) = udb, \quad (\tilde{B} = -1).$$

The hadrons containing the top quark have a much higher rest mass energy.

Like the strangeness quantum number S, the charm, bottom (beauty), and top (truth) quantum numbers, which we denote by C, $\tilde{B}$, T, are conserved by the electromagnetic and strong interactions but violated by the weak interaction. The additive quantum numbers of all six quarks are summarized in Table 14.6.

TABLE 14.6 Additive quantum numbers of the quarks.

Quark	Q	B	S	C	$\tilde{B}$	T
d	−1/3	1/3	0	0	0	0
u	2/3	1/3	0	0	0	0
s	−1/3	1/3	−1	0	0	0
c	2/3	1/3	0	1	0	0
b	−1/3	1/3	0	0	−1	0
t	2/3	1/3	0	0	0	1

14.3 Spatial symmetries

We shall now consider the effect of spatial symmetries upon the states of leptons and upon the states of composite systems made up of particles having an intrinsic spin. Issues involving spatial symmetry were discussed before when we considered the hydrogen atom which is made up of a proton and a single electron. Because the electrostatic potential of the proton is spherically symmetric, the wave functions of the electron can have well-defined values of the orbital angular momentum. The importance of angular momentum for many-electron atoms depends upon the fact that to a good approximation atomic electrons move independently of each other in a spherically symmetric field due to the nucleus and other electrons.

14.3.1 Angular momentum of composite systems

For elementary particles with no internal structure and for composite particles made up of quarks, the spin of the particle is defined as the angular momentum of the particle in its own rest frame. The angular momentum of a composite particle

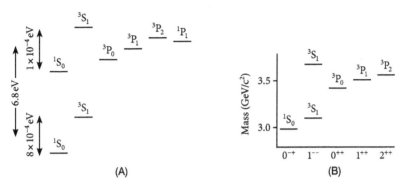

FIGURE 14.8 (A) The lowest states of positronium with principal quantum numbers $n = 1$ and $n = 2$, (B) the lowest states of charmonium.

can be determined using the rule given in Section 4.3.4 for combining two angular momenta. For given values j_1 and j_2 of the angular momenta of the two parts of a composite system, the quantum number J of the total angular momentum can have the values $J = j_1 + j_2, j_1 + j_2 - 1, \ldots, |j_1 - j_2|$, and for each value of J, the azimuthal quantum number can be $M = -J, -J + 1, \ldots, J$. We used this rule in Chapter 4 to combine the spin and orbital angular momentum of a single electron and in Chapter 5 to combine the spins and orbital angular momenta of electrons to form the total spin S, the total orbital angular momentum L, and the total angular momentum J.

The first example of a composite system we will consider here is positronium, which is a hydrogen-like bound system consisting of an electron and a positron. An important difference between the hydrogen atom and positronium is that the nucleus of hydrogen is the proton, which is very much more massive than the electron, while the positron and electron have the same mass. Both systems rotate about their center of mass. However, while the center of mass of hydrogen is very close to the center of the proton, the center of mass of positronium is at the mid-point of a line from the positron to the electron. One can show that the level spacings for positronium are approximately one half of the level spacings for hydrogen. When a hydrogen atom makes the transition $2p \rightarrow 1s$, a photon with energy 10.2 eV is emitted. The corresponding transition for positronium leads to the emission of a 5.1 eV photon.

The states of positronium with the principal quantum numbers $n = 1$ and $n = 2$ are illustrated in Fig. 14.8(A). Each state is denoted using the spectroscopic notation introduced in Chapter 5 with the upper-case letters $S, P, D, \ldots$ standing for $L = 0, 1, 2, \ldots$. The superscript in each case gives the value of $2S + 1$ and the subscript gives the total angular momentum J for the state. The two lowest states have principal quantum number $n = 1$ with the lowest state ${}^1S_0$ having spin $S = 0$, orbital angular momentum $L = 0$, and total angular momentum $J = 0$, and the second state ${}^3S_1$ having $S = 1$, $L = 0$, and $J = 1$. A point to notice here is that the electron and the positron are different particles, and so they can be in a triplet S state while two electrons with $n = 1$ cannot. All of the higher-lying states shown in the figure have principal quantum number $n = 2$.

Mesons are bound states of a quark/antiquark pair. In the rest frame of the quark/antiquark system, there is a single orbital angular momentum and two spins. While the orbital angular momentum quantum number L can have several values, we expect the lightest mesons to have orbital angular momentum $L = 0$. As was shown in Example 4.3, the possible values of the total spin of two spin one-half particles are $S = 0$ and $S = 1$. Using the spectrographic notation as before, we thus expect the two least massive mesons to have the spectroscopic designation

$$ {}^1S_0, {}^3S_1. $$

The total orbital angular momentum L is equal to zero for both states with $S = J = 0$ for the first state and $S = J = 1$ for the second state. The lowest observed states of charmonium, which consists of a $c/\bar{c}$ pair, are shown in Fig. 14.8(B). The energy level structure of charmonium is very similar to the energy level structure of positronium shown in Fig. 14.8(A).

Baryons are bound states of three quarks. As for mesons, the possible values of the spin quantum number of two quarks are $S = 0$ and $S = 1$. Combining the third quark to these angular momenta gives $S = 1/2$ and $S = 3/2$. Expecting the lowest states again to have $L = 0$, the states of the lightest baryons should be

$$ {}^2S_{1/2}, {}^4S_{3/2}, $$

where the state with $S = 1/2$ has $2S + 1 = 2$ and the state with $S = 3/2$ has $2S + 1 = 4$. Of the baryons we have considered thus far, the proton, neutron, lambda, and sigma have spin J equal to one half, while the delta baryons have spin J equal to three halves.

As we have said, the *spin* of a composite particle made up of quarks is the total angular momentum J in the rest frame of the particle.

14.3.2 Parity

We now consider the parity transformation in which the coordinates of particles are inverted through the origin

$$\mathbf{x} \to -\mathbf{x}. \tag{14.28}$$

Under a parity transformation, the velocity $\mathbf{v}$ and the momentum $\mathbf{p}$ change sign. The orbital angular momentum $\mathbf{l} = \mathbf{r} \times \mathbf{p}$ and the spin of a particle are unaffected by a parity transformation.

A parity transformation can be achieved by a mirror reflection followed by a rotation of 180° about an axis perpendicular to the mirror. Since the laws of nature are invariant under rotations, the question of whether parity is conserved depends upon whether an event and its mirror image occur with the same probability. While the strong and electromagnetic interactions are invariant with respect to parity transformations, we shall find that the weak interactions do not always have this property.

To study the effect of spatial inversions upon single-particle states, we introduce a parity operator $\hat{P}$ that acts on the wave function describing a single particle

$$\hat{P}\psi(\mathbf{x}, t) = P_a \psi(-\mathbf{x}, t), \tag{14.29}$$

where a identifies a particular type of particle such as an electron e^- or a quark. We thus suppose that each particle has an intrinsic parity in addition to the parity of its wave function. Since two successive parity operations leave the system unchanged, we require that

$$\hat{P}^2\psi(\mathbf{x}, t) = \psi(\mathbf{x}, t),$$

implying that the intrinsic parity P_a be equal to $+1$ or -1.

In addition to its intrinsic parity, the wave function of a particle has a parity associated with its orbital angular momentum. We found while studying selection rules for atomic transitions in Chapter 4 that a wave function is even or odd with respect to spatial inversion depending upon whether the orbital angular momentum quantum number is an even or an odd number. The parity associated with orbital angular momentum of a state is thus $(-1)^L$.

The free-particle states of electrons and positrons are represented by four-component wave functions in the relativistic Dirac theory. A careful study of the electron and positron shows that these particles have opposite parity. This is shown, for instance, in the book by Perkins which is cited at the end of this chapter. The parity of electrons and positrons cannot be determined in an absolute sense because they are always created or destroyed in pairs. We shall follow the ordinary conventions and assign to the electron, muon, and tau a positive parity

$$P_{e^-} = P_{\mu^-} = P_{\tau^-} = 1$$

and assign to the corresponding antiparticles a negative parity

$$P_{e^+} = P_{\mu^+} = P_{\tau^+} = -1.$$

Quarks like electrons are only created and destroyed in pairs. The usual convention for quarks is

$$P_u = P_d = P_s = P_c = P_b = P_t = 1,$$

and for antiquarks

$$P_{\bar{u}} = P_{\bar{d}} = P_{\bar{s}} = P_{\bar{c}} = P_{\bar{b}} = P_{\bar{t}} = -1.$$

The (intrinsic) parity of mesons and baryons can be predicted from the parity of the quarks. We recall that the rest frame of a meson corresponds to the center-of mass frame of a quark/antiquark pair. The spin states of the quark and antiquark are unaffected by an inversion of the spatial coordinates. Denoting the quark by a and the antiquark by $\bar{b}$ and denoting the orbital angular momentum of the quark/antiquark pair by L, the parity of a meson M is

$$P_M = P_a P_{\bar{b}} (-1)^L = (-1)^{L+1}. \tag{14.30}$$

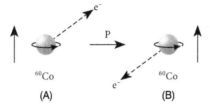

FIGURE 14.9 (A) An electron being emitted from a ^{60}Co nucleus, (B) the result of a parity transformation upon the ^{60}Co nucleus.

The quark labels a and b can each be u, d, s, c, b, or t. The least massive mesons with $L = 0$ are expected to have negative parity. This is consistent with the observed parities of pi- and K-mesons which have been found to have spin zero and negative parity.

Baryons are composed of three quarks. The orbital angular momentum of two quarks can be combined to form total orbital angular momentum L_{12} and this angular momentum can then be combined with the orbital angular momentum of the third quark, which we denote by L_3. Denoting the quarks by a, b, and c, the parity of a baryon is

$$P_B = P_a P_b P_c (-1)^{L_{12}} (-1)^{L_3} = (-1)^{L_{12}+L_3}, \tag{14.31}$$

and the corresponding antibaryon has parity

$$P_{\overline{B}} = P_{\overline{a}} P_{\overline{b}} P_{\overline{c}} (-1)^{L_{12}} (-1)^{L_3} = -(-1)^{L_{12}+L_3}. \tag{14.32}$$

Low-lying baryons with $L_{12} = L_3 = 0$ are predicted to have positive parity, while the low-lying antibaryons are predicted to have negative parity. These predictions are consistent with the observed parities of the p, n, and Λ.

A thorough study of the experimental evidence for parity conservation was conducted by C.N. Yang and T.D. Lee in 1956. They showed that while there was strong evidence for parity conservation in electromagnetic and strong interactions, there was no evidence for parity conservation in the weak interactions. In 1957, following suggestions by Yang and Lee, C.S. Wu and her coworkers at Columbia University placed a sample of cobalt-60 inside a solenoid and cooled it to a temperature of 0.01 K. At such low temperatures, the cobalt nuclei align parallel to the direction of the magnetic field. Polarized cobalt-60 nuclei decay to an excited state of nickel-60 by the process

$$^{60}Co \rightarrow ^{60}Ni^* + e^- + \overline{\nu_e}.$$

Parity violation was established by the observation that more electrons were emitted in the direction of the nuclear spins than the backward directions.

An illustration of an electron being emitted from a ^{60}Co nucleus is shown in Fig. 14.9(A). The spin of the cobalt nucleus is illustrated by an arrow indicating the rotational motion of the nucleus, and by an arrow beside the nucleus pointing upward because a right-hand screw would move up if it were to rotate in the way the cobalt nucleus is spinning. As shown in Fig. 14.9(B), a parity transformation reverses the direction of the emitted electron but leaves the direction of the nuclear spin unchanged. Parity is violated since a beta decay in the forward direction of the spin of the cobalt nucleus such as that shown in Fig. 14.9(A) occurs more often than a beta decay in the backward direction as shown in Fig. 14.9(B).

Another example of parity violation is provided by the dominant decay mode of π^+ described by the formula,

$$\pi^+ \rightarrow \mu^+ + \nu_\mu. \tag{14.33}$$

In this decay process illustrated in Fig. 14.10(A), the spin of the μ^+ is indicated by the downward arrow next to the particle. The emitted muon has negative *helicity* which means that its spin points in the direction opposite to its motion. In the parity transformed process shown in Fig. 14.10(B), the μ^+ is rotating as before and the spin still points down. The process shown in Fig. 14.10(B), for which the muon has positive helicity, does not occur.

A number of interesting examples of parity violation is provided by weak decays of mesons.

14.4 Isospin and color

The purpose of this section is to describe two fundamental properties of strongly interacting particles, *isospin* and *color*. We shall find that isospin is an approximate symmetry that provides a framework for understanding a wealth of experimental data, while color is related to an exact symmetry of the strong interactions.

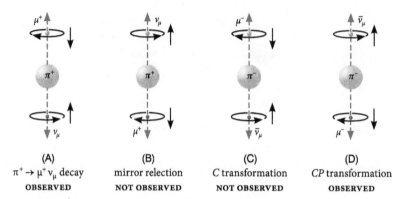

(A)   (B)   (C)   (D)

$\pi^+ \to \mu^+ \nu_\mu$ decay   mirror relection C transformation CP transformation

OBSERVED   **NOT OBSERVED**   **NOT OBSERVED**   **OBSERVED**

FIGURE 14.10 (A) A μ^+ emitted from π^+ in the decay process (14.33), (B) the result of a parity transformation upon the π^+ decay, (C) the result of a charge conjugation transformation upon the π^+ decay, (D) the result of both a parity and charge conjugation transformation upon the π^+ decay.

14.4.1 Isospin

One of the most striking properties of hadrons is that they occur in families of particles with approximately equal masses. Within a given family, all particles have the same spin, parity, baryon number, strangeness, charm, bottom, and top quantum numbers but differ in their electric charge. Of the mesons shown in Table 14.3, the π^- with quark composition $d\overline{u}$, the π^+ with quark composition $u\overline{d}$, and the π^0, which is a combination of $u\overline{u}$ and $d\overline{d}$, have approximately the same mass. This is also true of the K^+ with quark composition $u\overline{s}$ and the K^0 with quark composition $d\overline{s}$. Of the baryons shown in Table 14.4, the proton with quark composition uud and the neutron with composition udd have about the same mass. For all of these examples, the members of a charge multiplet can be distinguished by the varying number of u and d quarks. The similarity of the masses within a charge multiplet is due to the fact that the u and d quarks have approximately the same mass and interact by means of the same strong interaction.

To describe the isospin symmetry, we introduce three quantum numbers that are conserved by the strong interactions. Two of these quantum numbers are combinations of quantum numbers introduced previously. The first new quantum number is the *hypercharge* defined by the following equation

$$Y = B + S + C + \tilde{B} + T, \tag{14.34}$$

where B, S, C, $\tilde{B}$, and T are the baryon number, strangeness, charm, bottom, and top quantum numbers, respectively. Since the quantum numbers appearing on the right-hand side of this last equation have the same values for all members of an isospin multiplet, so does the hypercharge. The second combination of quantum numbers is the azimuthal isospin quantum number I_3 defined by the equation

$$I_3 = Q - Y/2, \tag{14.35}$$

where Q is the charge. The different members of an isospin multiplet have different charges and hence different values of I_3. We define the isospin I to be the maximum value of I_3 within a multiplet.

14.4.1.1 Quarks

In keeping with the fact that baryons are made of three quarks, the Baryon number of all six quarks is equal to one third. Of the six quarks, the d, s, and b quarks, and the u, c, and t quarks have similar properties. The d, s, and b quarks all have charge equal to $-1//3$, while the u, c, and t quarks have charge equal to $4/3$.

As we have seen earlier the strange particles are produced in pairs with one of the particles produced having positive strangeness and the other particle having negative strangeness. The K^+ meson is traditionally assigned strangeness $S = +1$ and the K^- having a single strange quark is assigned strangeness $S = -1$. The strange quark is thus assigned a strangeness $S = -1$, while the antiquark $\overline{s}$ has strangeness $S = +1$. Similarly, the bottom and antibottom quarks have bottom quantum number $\tilde{B}$ equal to -1 and $+1$ respectively. The charmed quark c and the top quark t have quantum numbers $C = 1$ and $T = 1$ respectively, while the anticharmed quark and the antitop quarks have quantum number $\overline{C} = -1$ and $\overline{T} = -1$. These basic quark conventions can be used to calculate values of the hypercharge Y and the isospin quantum number I_3 for each of the quarks. The hypercharge quantum number Y can be assigned to the quarks using Eq. (14.34), and the isospin quantum number I_3 can be assigned using (14.35). These quantum numbers together with the baryon quantum number B, the charge Q, and the isospin I are given in Table 14.7.

TABLE 14.7 Values of the baryon number B, hypercharge Y, charge Q, and isospin quantum numbers I_3 and I for quarks.

Quark	B	Y	Q	I_3	I
d	1/3	1/3	$-1/3$	$-1/2$	1/2
u	1/3	1/3	2/3	1/2	1/2
s	1/3	$-2/3$	$-1/3$	0	0
c	1/3	4/3	2/3	0	0
b	1/3	$-2/3$	$-1/3$	0	0
t	1/3	4/3	2/3	0	0

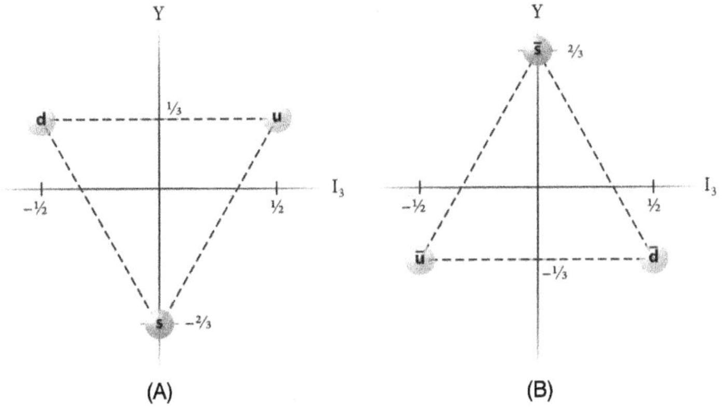

FIGURE 14.11 The hypercharge and isospin quantum numbers for (A) u, d, and s quarks and for (B) $\bar{u}$, $\bar{d}$, and $\bar{s}$ antiquarks.

Only the u and d quarks have isospin quantum numbers, I_3 and I, different from zero. The isospin quantum number I is equal to 1/2 for both the up and down quarks, while I_3 is equal to $+1/2$ and $-1/2$, for u and d quarks, respectively. The hypercharge and isospin quantum numbers for the u, d, and s quarks and for $\bar{u}$, $\bar{d}$, and $\bar{s}$ antiquarks are shown in Fig. 14.11.

14.4.1.2 The light mesons

The states of the lightest mesons have $L = 0$ and according to Eq. (14.30) have negative parity. The quark and antiquark, of which each meson is composed, have an intrinsic spin equal to one half. Hence, mesons must have total spin S equal to 0 or 1. Since the total orbital angular momentum L is equal to zero, the total angular momentum J must like the spin S be equal to 0 or 1. The lightest mesons are observed experimentally to consist of a family of nine mesons with spin-parity $J = 0^-$ and a family of nine mesons with spin-parity $J = 1^-$. Using the Greek word *nonet* to describe nine objects, the states are said to belong to scalar and vector nonets. These two families of mesons with their quark assignments are given in Table 14.8.

TABLE 14.8 Light Mesons with spin equal to zero and one.

Quarks	0^- meson	1^- meson	I_3	I	Y
$u\bar{s}$	$K^+(494)$	$K^{*+}(892)$	1/2	1/2	1
$d\bar{s}$	$K^0(498)$	$K^{*0}(896)$	$-1/2$	1/2	1
$u\bar{d}$	$\pi^+(140)$	$\rho^+(776)$	1	1	0
$u\bar{u}, d\bar{d}$	$\pi^0(135)$	$\rho^0(776)$	0	1	0
$d\bar{u}$	$\pi^-(140)$	$\rho^-(776)$	-1	1	0
$s\bar{d}$	$\bar{K}^0(498)$	$\bar{K}^{*0}(896)$	1/2	1/2	-1
$s\bar{u}$	$K^-(494)$	$K^{*-}(892)$	$-1/2$	1/2	-1
$u\bar{u}, d\bar{d}, s\bar{s}$	$\eta(548)$	$\omega(783)$	0	0	0
$u\bar{u}, d\bar{d}, s\bar{s}$	$\eta'(958)$	$\phi(1019)$	0	0	0

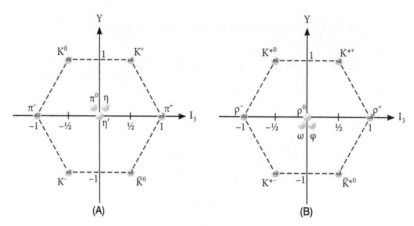

FIGURE 14.12 (A) The scalar meson nonet, and (B) the vector meson nonet.

While all the charged particles shown in Table 14.8 correspond to a specific quark/antiquark pair, the neutral particles correspond to a linear combination of quark states. The π^0 and ρ^0 are a linear combination of $u\bar{u}$ and $d\bar{d}$, while the η, η', ω, and ϕ correspond to linear combinations of $u\bar{u}$, $d\bar{d}$, and $s\bar{s}$. The scalar and vector meson nonets are illustrated in Fig. 14.12.

14.4.1.3 The light baryons

Like the mesons, baryons occur in families of particles with the same baryon number, spin, and parity. These families are called *supermultiplets*. While the supermultiplets of mesons have nine members and are called nonets, the supermultiplets of baryons can have one, eight, or ten members and are called *singlets*, *octets*, and *decuplets*. The lightest baryons observed experimentally are the octet with $J^P = \frac{1}{2}^+$ shown in Table 14.9 and the decuplet with $J^P = \frac{3}{2}^+$ shown in Table 14.10.

TABLE 14.9 States of the $\frac{1}{2}^+$ octet of light baryons.

Quarks	Baryon	I_3	I	Y
uud	p(938)	1/2	1/2	1
udd	n(940)	−1/2	1/2	1
uds	Λ(1116)	0	0	0
uus	Σ^+(1189)	1	1	0
uds	Σ^0(1193)	0	1	0
dds	Σ^-(1197)	−1	1	0
uss	Ξ^0(1315)	1/2	1/2	−1
dss	Ξ^-(1322)	−1/2	1/2	−1

One can see that Δ^+ and Δ^0 shown in Table 14.10 have the same quark composition as the proton and neutron shown in Table 14.9. The Δ's in Table 14.10 may be thought of as excited or resonance states of nucleons. The baryon octet and decuplet are illustrated in Fig. 14.13.

14.4.1.4 Pion-nucleon scattering

We now consider the scattering processes that occur when a beam of pions is incident upon a proton or neutron. The isospin symmetry, which we have just discussed, provides information about which scattering processes can occur and about the relative amplitude of different scattering processes.

The isospin quantum numbers of pions are given in Table 14.8, and the isospin quantum numbers of protons and neutrons are given in Table 14.9. The quantum number I is equal to one for the pions with $I_3 = 1, 0, -1$ for π^+, π^0, and π^-, respectively. For the proton, the isospin quantum numbers are $I = 1/2$ and $I_3 = +1/2$, while the isospin quantum numbers are $I = 1/2$ and $I_3 = -1/2$ for the neutron. Using the rule for addition of angular momentum, we immediately see that a pion-nucleon system can have a total isospin $I = 3/2$ or $I = 1/2$. The different possible combinations of pions

TABLE 14.10 States of the $\frac{3}{2}^+$ decuplet of light baryons.

Quarks	Baryon	I_3	I	Y
uuu	$\Delta^{++}(1232)$	3/2	3/2	1
uud	$\Delta^{+}(1232)$	1/2	3/2	1
udd	$\Delta^{0}(1232)$	−1/2	3/2	1
ddd	$\Delta^{-}(1232)$	−3/2	3/2	1
uus	$\Sigma^{+}(1383)$	1	1	0
uds	$\Sigma^{0}(1384)$	0	1	0
dds	$\Sigma^{-}(1387)$	−1	1	0
uss	$\Xi^{0}(1532)$	1/2	1/2	−1
dss	$\Xi^{-}(1535)$	−1/2	1/2	−1
sss	$\Omega^{-}(1672)$	0	0	−2

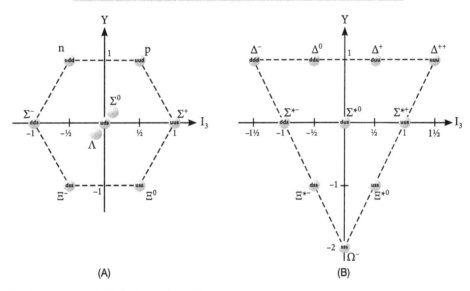

(A) (B)

FIGURE 14.13 (A) The baryon octet, and (B) the baryon decuplet.

and nucleons that can occur in a pion-nucleon scattering experiment and the isospin quantum number I_3, to which they correspond, are given in Table 14.11.

TABLE 14.11 Possible combinations of pions and nucleons. The I_3 values shown are equal to the sum of the I_3 quantum numbers of the pion and nucleon.

Pion and nucleon	I_3
$\pi^+ p$	3/2
$\pi^+ n, \pi^0 p$	1/2
$\pi^0 n, \pi^- p$	−1/2
$\pi^- n$	−3/2

To a good approximation, the isospin quantum numbers, I and I_3, are conserved in a collision process. Since the combination of a π^+ and a p is the only combination of a pion and a nucleon having $I_3 = 3/2$, collisions of a π^+ and a p are elastic scattering processes described by the formula

$$\pi^+ + p \rightarrow \pi^+ + p. \tag{14.36}$$

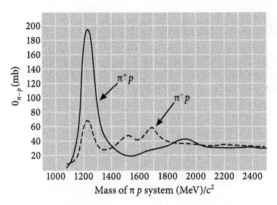

FIGURE 14.14 The cross sections for $\pi^+ p$ and $\pi^- p$ scattering.

Similarly, the combination of a π^- and a n is the only combination of a pion and a nucleon having $I_3 = -3/2$. Hence, the collisions of a π^- and a n are elastic scattering processes described by the formula

$$\pi^- + n \rightarrow \pi^- + n. \tag{14.37}$$

The two processes, (14.36) and (14.37), differ only in the sign of the azimuthal quantum number I_3. The cross-sections for these two scattering processes have approximately the same dependence upon the energy of the incoming pion.

As we saw in the first section of this chapter when we considered the collision process depicted in Fig. 14.3, the scattering of a pion and a nucleon can occur independently or by forming a resonance state. Since the isospin quantum number is conserved, a baryon resonance corresponding to either $\pi^+ p$ or $\pi^- n$ scattering should have isospin $I = 3/2$. The cross sections for $\pi^+ p$ and $\pi^- p$ scattering are shown in Fig. 14.14. We note that a sharp peak occurs for $\pi^+ p$ scattering for an effective mass around 1232 MeV, which is the mass of the Δ^{++} resonance. The Δ^{++} resonance has isospin equal to $3/2$. Closer inspection of the scattering data shows that a shoulder corresponding to a smaller peak occurs in the range between 1600 and 1700 MeV. A more well-defined peak in the cross section for $\pi^+ p$ scattering occurs at 1900 MeV. A review of the data at the web site of the Particle Data Group in 2008 shows that Δ resonances occur at 1600, 1700, and 1905 MeV.

The collision of a π^- and a p can lead to the elastic scattering process

$$\pi^- + p \rightarrow \pi^- + p \tag{14.38}$$

or to the exchange process

$$\pi^- + p \rightarrow \pi^0 + n. \tag{14.39}$$

The $\pi^- p$ and $\pi^0 n$ states are included in Table 14.11. These two states both have $I_3 = -1/2$ and are the only possible outcome of a collision of π^- and a proton. The state $\pi^- p$ state having $I_3 = -1/2$ is a linear combination of isospin states with $I = 3/2$ and $I = 1/2$. A careful analysis of the role of isospin in pion-nucleon scattering can be found in the book by Perkins, which is cited at the end of this chapter. We shall use only a few features of this analysis here to describe the scattering curve of $\pi^- p$ scattering shown in Fig. 14.14. While the states of $\pi^+ p$ have $I = 3/2$, one can show that one third of the states of the $\pi^- p$ system have a $I = 3/2$ character and two thirds have a $I = 1/2$ character. This is the reason the peak of the $\pi^- p$ cross section corresponding to the Δ^{++} resonance is approximately one third of the corresponding peak for $\pi^+ p$ scattering.

Reviewing the cross section for $\pi^- p$ scattering shown in Fig. 14.14, one can see that the $\pi^- p$ cross section has two additional peaks. One of these peaks occurs in the range between 1600 and 1700 MeV lying above the shoulder of the $\pi^+ p$ cross section we have referred to previously. This peak may be due to the Δ^{++} resonance considered earlier. Another peak occurs slightly above 1500 MeV where there is no corresponding peak in the $\pi^+ p$ cross section. A review of the data of the Particle Data Group in 2008 shows that a nucleon resonance with $J = 1/2$ occurs at 1535 MeV. The isolated peak of the cross section for $\pi^- p$ scattering occurring above 1500 MeV may well correspond to this resonance.

14.4.2 Color

The spatial wave function of the lowest-lying state of physical systems are generally symmetric with respect to the interchange of like particle. We shall find that the number of hadron states in Tables 14.9 and 14.10 can be successfully accounted for if one also supposes that the spin wave functions of these baryons are symmetric with respect to the interchange of two quarks. This simple hypothesis shall allow us to account for the number of states that can be formed with up, down, and strange quarks.

We consider first the following six combinations of three quarks, which consist of two identical quarks and an additional quark of another kind

$$uud, uus, ddu, dds, ssu, ssd. \tag{14.40}$$

It can be shown quite generally that the spin-wave function of two fermions with spin $S = 1$ is symmetric with respect to an interchange of the two spins while the spin wave function with $S = 0$ is antisymmetric. The requirement that the spin wave functions of the particles (14.40) be symmetric with respect to the interchange of two spins then implies that the two identical quarks are in an $S = 1$ state. Coupling the additional quark with spin one half to the spin-one pair then results in spin one half or spin three halves. Since orbital angular momentum L is zero for these low-lying states, the three quarks will have total angular momentum J equal to one half and three halves. The quark states (14.40) thus contribute six states to the baryon octet shown in Fig. 14.13(A) and six states to the baryon decuplet shown in Fig. 14.13(B).

We next consider the states of three identical quarks,

$$uuu, ddd, sss. \tag{14.41}$$

The states of three identical quarks for which each quark is in a spin-up state will have a total value of the azimuthal quantum number M_S equal to $3/2$. For such a state, the spin quantum number S must also have the value $3/2$ since only the $S = 3/2$ state can have $M_S = 3/2$. Other states with $S = 3/2$ and with $M_S = 1/2, -1/2, -3/2$ can be generated from the $M = 3/2$ states by replacing the spin-up functions of one of the three quarks with spin-down function. The states (14.41) contribute three states to the baryon decuplet.

The only remaining state of a particle consisting of three quarks is

$$uds, \tag{14.42}$$

in which all the quarks are different. The spin angular momenta of the ud pair can be combined to form spin states with total spin angular momentum S equal to 0 and 1. The spin-zero state of the ud-pair can be combined with the spin of the s quark giving S equal to $1/2$, while the spin-one state of the ud pair can be combined with the spin of the s quark to give S equal to $1/2$ or $3/2$. The states (14.42) thus contribute two baryons to the octet and one baryon to the decuplet.

The assumption that the spatial and spin wave functions of baryons is symmetric with respect to an interchange of identical quarks allows one to explain the mass spectra of the light baryons, and yet this assumption appears to contradict the basic assumption of quantum mechanics that the wave function of fermions be antisymmetric with an exchange of two particles. The requirement that the wave function of many-fermion systems be antisymmetric insures that fermion states automatically satisfy the Pauli exclusion principle. One can see immediately that the state of the Δ^{++} resonance with all three u quarks having spin up appears to violate the Pauli exclusion principle.

The apparent contradiction between the quark model and the Pauli principle was resolved in 1964 when Oscar W. Greenberg suggested quarks possess another attribute, which he called color. The combined space and spin wave function can then be symmetric with respect to the interchange of two quarks of the same flavor – as required by experiment – provided that the color part of the wave function is antisymmetric. The basic assumption of the color theory proposed by Greenberg is that the quarks of any flavor can exist in three different color states, *red*, *green*, and *blue*, denoted by r, g, b.

Just as the electromagnetic and weak interactions depend upon the hypercharge Y and isospin I_3 of the particles, the strong interaction depends on the two color charges called *color hypercharge* Y^C and *color isospin* I_3^C. The values of these new quantum numbers for the color states r, g, and b are given in Table 14.12. The color charges Y^C and I_3^C are conserved by the strong interaction.

The quantum numbers I_3^C and Y^C are additive quantum numbers with values for antiparticles being the negative of the values for particles.

Notice that the sum of the columns of Table 14.12 giving the total values of the color charges for red, green, and blue quarks and for red, green, and blue antiquarks is equal to zero. Hadrons only exist in color singlet states with zero values of the color charges. All baryons are made of three quarks of different colors, while mesons consist of a quark and an antiquark

TABLE 14.12 Values of the color charges I_3^C and Y^C for the color states of quarks and antiquarks.

	Quarks			Antiquarks	
	I_3^C	Y^C		I_3^C	Y^C
r	1/2	1/3	$\bar{r}$	−1/2	−1/3
g	−1/2	1/3	$\bar{g}$	1/2	−1/3
b	0	−2/3	$\bar{b}$	0	2/3

of the same color. For such states, the total value of the additive quantum numbers, I_3^C and Y^C, are equal to zero. This is called *color confinement*. The idea that quarks are always found in nature in color-neutral states was part of the motivation for using the word "color". Just as white light can be obtained by combining the three primary colors, baryons combine red, green, and blue quarks into a color neutral state.

The modern theory of the strong interactions, which is called *quantum chromodynamics* has been successful in describing a broad range of experimental data. Here, we only give one success of the theory. The differential cross section for the electromagnetic interaction

$$e^+ + e^- \rightarrow q + \bar{q},$$

in which a positron/electron pair annihilate to produce a quark/antiquark pair, can be obtained from the cross section for the electromagnetic interaction producing a μ^+/μ^- pair,

$$e^+ + e^- \rightarrow \mu^+ + \mu^-,$$

by replacing the square of the charge of the quark, $(e\,e_q)^2$ with the square of the charge of a muon, e^2, times a factor of three to take into account the fact that there are three colors.

14.5 Feynman diagrams

We would now like to look more closely at the scattering processes that occur when beams of particles are directed upon a target. An incident beam of particles can be characterized by its flux, which is the total number of particles that would pass through a surface perpendicular to the beam per area and per time. If the density of particles in the beam is denoted by n_i and the velocity of the beam is denoted by v_i, then the flux of the beam is

$$\phi_i = n_i v_i. \tag{14.43}$$

One can easily see this is true by considering the number of particles that would pass through an area A, which is perpendicular to the beam, in an infinitesimal time dt. In a time dt, all of the particles within a box with volume $v_i dt A$ would pass through the area A. The number of particles within this box would be $n_i v_i dt A$, and the number of particles passing through A per area and per time would be $n_i v_i$.

The transition rate W is the total number of particles deflected from the beam by an interaction with a particle in the target, and the cross section σ is the transition rate divided by the incident flux. The relation between the transition rate and the cross section can be written

$$W = \phi_i \sigma. \tag{14.44}$$

The transition rate itself is given by an equation which can be derived by quantum mechanics. This equation, which Enrico Fermi called the *Golden Rule*, is

$$W = \frac{2\pi}{\hbar} |M_{if}|^2 \rho_f. \tag{14.45}$$

The quantity ρ_f, which appears on the right-hand side of this equation, is the density of final states, and M_{if} is the matrix element of the interacting potential V between the initial and final states

$$M_{ij} = \int \psi_f^* V \psi_i dV. \tag{14.46}$$

We found in Chapter 13 that scattering matrix elements M_{if} can be obtained using Feynman diagrams.

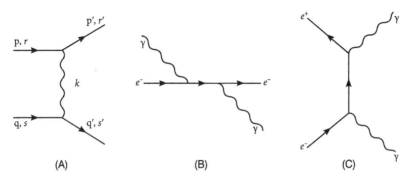

FIGURE 14.15 Feynman diagrams for (A) electron-electron scattering, (B) Compton scattering, and (C) positron-electron annihilation.

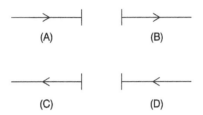

FIGURE 14.16 (A) Particle absorption, (B) Particle creation, (C) Hole absorption, (D) Hole creation.

14.5.1 Electromagnetic interactions

The Feynman diagram corresponding to two electrons scattering by means of the electromagnetic interaction is shown in Fig. 13.4 and reproduced in this chapter as Fig. 14.15(A). The diagram corresponds to the mathematical expression

$$\mathcal{M} = (-e^2)\delta_{p+q,p'+q'}\overline{u}^{(r')}(\mathbf{p}')\overline{u}^{(s')}(\mathbf{q}')\gamma^\mu \frac{i}{k^2}\gamma^\nu u^{(r)}(\mathbf{p})u^{(s)}(\mathbf{q}). \tag{14.47}$$

As discussed in Chapter 13, each term in the expression on the right-hand side of Eq. (14.47) corresponds to a part of the Feynman diagram. The in-coming and out-going lines of the diagram correspond to free-particle solutions of the Dirac equation, the factors, $ie\gamma^\mu$ and $ie\gamma^\nu$, correspond to the two vertices of the diagram, and the wavy line corresponds to the *propagator*. The propagator of the electromagnetic interaction is represented by the term i/k^2 appearing in Eq. (14.47). The delta function $\delta_{p+q,p'+q'}$ ensures that the total four-momentum of the electrons is conserved in the interaction.

The Feynman diagrams we will consider include both particles and antiparticles with the lines corresponding to antiparticles obeying certain conventions due to Feynman and Dirac. While the lines corresponding to particles are always directed to the right, the lines corresponding to antiparticle are directed to the left. This is due to Feynman's suggestion that antiparticles move backward in time. Fig. 14.16 shows lines that can appear in Feynman diagrams. Notice that the lines in (A) and (B) are directed to the right and correspond to particles, while the lines in (C) and (D) are directed to the left and correspond to antiparticles. The line in (A), which approaches and ends at a vertical line corresponds to particle absorption while the line in (B), which emanates from a vertical line, corresponds to particle creation. In contrast, the line in (D) that approaches and ends with a vertical line corresponds to antiparticle creation. This is consistent with Dirac's idea that antiparticle are holes in a sea of negative energy states. Creating an antiparticle corresponds to absorbing a particle and thereby creating a hole. Similarly, the line in (C) that emanates from a vertical line and moves to the left corresponds to antiparticle absorption. Creating a particle corresponds to filling a hole.

The Feynman diagram shown in Fig. 14.15(B) describes Compton scattering in which a photon scatters off a free electron producing a photon and an electron with different momentum and energy. The in-coming and out-going photons in this scattering process are represented by free wavy lines, while the solid line joining the two vertices is referred to as the *Fermion propagator*. Fig. 14.15(C) describes a process in which an electron and positron annihilate producing two photons. Arrows directed to the right in Feynman diagrams correspond to particles, while lines directed to the left correspond to antiparticles. Notice that the incoming positron is represented by a free solid line with an arrow directed away from the vertex. This is consistent with our description or particle and hole lines in Fig. 14.16. Absorbing a positron corresponds to filling a hole and is hence represented by an arrow directed away from the vertex. In a similar fashion, the creation of a positron correspond to an arrow directed toward a vertex.

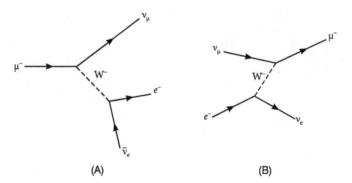

FIGURE 14.17 Feynman diagram for (A) muon decay and (B) inverse muon decay.

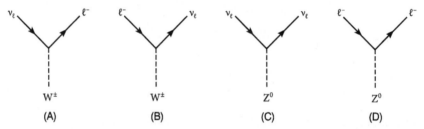

FIGURE 14.18 Vertices in which (A) a neutrino is absorbed and a lepton of the same generation is created due to an interaction mediated by $W^\pm$, (B) a lepton is absorbed and a neutrino of the same generation is created due to an interaction mediated by $W^\pm$, (C) a neutrino is absorbed and another created due to an interaction mediated by Z^0, and (D) a lepton is absorbed and another created due to an interaction mediated by Z^0.

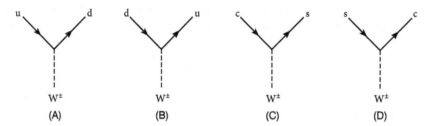

FIGURE 14.19 Vertices in which due to an interaction mediated by $W^\pm$ (A) an u quark is absorbed and a d quark is created, (B) a d quark is absorbed and a u quark is created, (C) a c quark is absorbed and a s quark is created, or (D) a s quark is absorbed and a c quark is created.

14.5.2 Weak interactions

The Feynman diagrams for the electromagnetic processes we have considered have vertices with two lines for charged particles and one line for a photon. We would now like to consider weak interactions of leptons mediated by the vector bosons. As an example of weak interaction processes of this kind, we consider muon decay,

$$\mu^- \to e^- + \overline{\nu}_e + \nu_\mu,$$

which is represented by the Feynman diagram shown in Fig. 14.17(A), and inverse muon decay,

$$\nu_\mu + e^- \to \mu^- + \nu_e,$$

which is represented by the Feynman diagram shown in Fig. 14.17(B).

Processes mediated by W^+ or W^- have vertices of the kind shown in Figs. 14.18(A) and (B), while processes mediated by the neutral Z^0 have vertices of the kind shown in Figs. 14.18(C) and (D). In Fig. 14.18(A), a neutrino (ν_l) of one generation is absorbed at the vertex with a lepton of the same generation (l^-) being created. The interaction line corresponds to a $W^\pm$ boson. Fig. 14.18(B) shows the corresponding process with a lepton being absorbed and a neutrino being produced. The fact that the leptons and neutrinos involved in these interactions belong to the same generation corresponds to conservation laws we have seen earlier which require that electron lepton number, muon lepton number, and tau lepton number are individually conserved. Weak interaction vertices involving quarks are shown in Fig. 14.19.

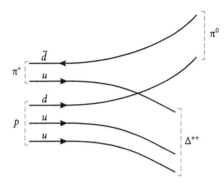

FIGURE 14.20 Feynman diagram for the reaction $\pi^+ + p \to \Delta^{++} + \pi^0$.

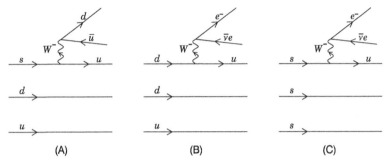

FIGURE 14.21 Feynman diagrams for (A) $\Lambda \to \pi^- + p$, (B) $n \to p + e^- + \bar{\nu}_e$, (C) $\Omega^- \to \Xi^0 + e^- + \bar{\nu}_e$.

The vertices shown in Figs. 14.18(C) and (D) for the neutral Z^0 are similar to the photon vertices seen earlier. At energies small compared to the Z^0 mass, the Z^0 exchange interactions can be neglected in comparison to the corresponding photon exchange interactions. However, at very high energies, Z^0 exchange interactions become comparable to photon exchange interactions.

Example 14.3

Give the quark composition of the particles in the following processes and draw the Feynman diagram

$$(a) \qquad \pi^+ + p \to \Delta^{++} + \pi^0$$
$$(b) \qquad \Lambda \to \pi^- + p$$
$$(c) \qquad n \to p + e^- + \bar{\nu}_e$$
$$(d) \qquad \Omega^- \to \Xi^0 + e^- + \bar{\nu}_e$$

Solution

We shall first write each reaction formula showing the quark composition of each particle and then say how the quarks have changed and draw the appropriate Feynman diagram.

$$(a) \quad \pi^+ + \quad p \quad \to \quad \Delta^{++} + \quad \pi^0.$$
$$\qquad u\bar{d} \quad uud \qquad uuu \qquad d\bar{d}$$
$$(b) \qquad\qquad \Lambda \quad \to \quad \pi^- + \quad p$$
$$\qquad\qquad uds \qquad d\bar{u} \qquad uud$$
$$(c) \qquad\qquad n \quad \to \quad p + \quad e^- + \bar{\nu}_e$$
$$\qquad . \qquad udd \qquad uud \qquad e^- \quad \bar{\nu}_e$$
$$(d). \qquad\qquad \Omega^- \quad \to \quad \Xi^0 + \quad e^- + \bar{\nu}_e.$$

According to the Feynman diagram shown in Fig. 14.20, the quarks in part (a) are simply rearranged. For the (b) part, an s quark is changed into an u quark and an additional $d\bar{u}$ pair is created. This is shown in Fig. 14.21(A). The (B) and (C) figures of Fig. 14.21, describe the reactions given in the (c) and (d) parts of Example 14.3. For the (c) part a d quark is changed into an u

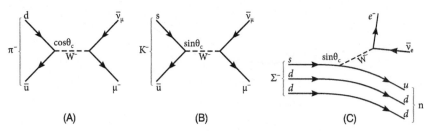

FIGURE 14.22 The decays (A) $\pi^- \to \mu^- + \bar{\nu}_\mu$, (B) $K^- \to \mu^- + \bar{\nu}_\mu$, and (C) $\Sigma^- \to n + e^- + \bar{\nu}_e$.

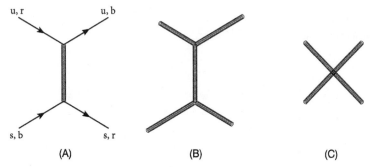

FIGURE 14.23 (A) Interaction of two quarks by gluon exchange, (B) interaction of two gluons by gluon exchange, (C) a zero range "contact" interaction between two gluons.

---

quark and an additional $e/\bar{\nu}_e$ pair is formed. For the (d) part an s quark is changed to a u quark and an additional $e/\bar{\nu}_e$ pair is formed.

We have found earlier that a mixing occurs between the generations of neutrinos that substantially affects the number of solar neutrinos reaching the Earth. A similar mixing between the first- and second-generations of quarks is described by the equations

$$d' = d\cos\theta_c + s\sin\theta_c$$
$$s' = -d\sin\theta_c + s\cos\theta_c,$$

(14.48)

where the parameter θ_c is called the *Cabibbo angle*. The experimental value of the Cabibbo angle is about 12.7°. Because the Cabibbo angle is small, the d' quark is mainly a d and only a small portion of the s quark mixes into the d; however this small mixing between generations of quarks influences both the magnitude and the scope of weak interactions.

We now give a number of examples of weak decays in which Cabibbo mixing plays a role. The decay of the π^- by the reaction

$$\pi^- \to \mu^- + \bar{\nu}_\mu$$

is described by the diagram shown in Fig. 14.22(A). The decay of π^- would occur without Cabibbo mixing. Still, as shown in Fig. 14.22(A), the Cabibbo mixing contributes a factor $\cos\theta_c$ to the quark vertex.

The decay of the K^- is described by the Feynman diagram in Fig. 14.22(B) occurs because of Cabibbo mixing, which contributes a factor of $\sin\theta_c$ to the quark vertex. Another example of a weak interaction process in which Cabibbo mixing plays a role is the decay of the Σ^- shown in Fig. 14.22(C).

14.5.3 Strong interactions

The electromagnetic and strong interactions are both mediated by massless spin-1 bosons. However, there is an important difference between these two interactions which affect the kinds of processes that can occur. Unlike photons that are electrically, gluon interact among themselves. Fig. 14.23(A) shows the interaction of two quarks by gluon exchange. The gluon is represented in this diagram by a "quarkscrew" line to distinguish it from the electromagnetic and weak interactions. The two types of gluon-gluon interaction are illustrated in Figs. 14.23(B) and (C). Fig. 14.23(B) illustrates the exchange of a gluon between two other gluons, while Fig. 14.23(C) illustrates a zero range "contact" interaction between two gluons.

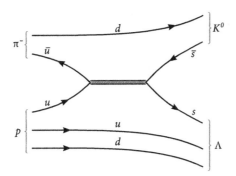

FIGURE 14.24 Feynman diagram for the process $\pi^- + p \to K^0 + \Lambda$.

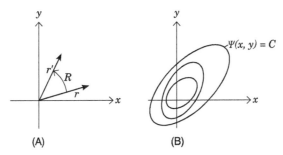

FIGURE 14.25 (A) The rotation of a vector in two dimensions, (B) Representation of a function by means of contour curves on which the function has a constant value.

Of the many scattering processes involving quarks that occur by means of the strong interaction, we consider only the following example

$$\pi^- + p \to K^0 + \Lambda,$$

which corresponds to the diagram shown in Fig. 14.24. In this scattering process, a $\bar{u}/u$ pair is absorbed giving a gluon which produces a $\bar{s}/s$ pair.

The strong interaction has the distinctive properties of color confinement and asymptotic freedom. The property of color confinement means that observed particles all have zero color charge. Mesons are made up of a quark/antiquark pair with the two constituents having opposite values of the color charge, while baryons are made up of three quarks of different colors which are color neutral. Asymptotic freedom means the interaction gets weaker at short distances. At distances less than about 0.1 fm the lowest order Feynman diagrams dominate.

14.6 The $R(3)$ and $SU(3)$ symmetry groups

We shall begin now to study the role of higher symmetries in particle physics. To orient ourselves we first consider the rotation group in three dimensions and then consider the $SU(3)$ group which consists of unitary 3×3 matrices with determinant equal to one. Murry Gell-Mann and Yuval Ne'eman were the first to suggest that the $SU(3)$ group, which consists of unitary 3×3 matrices with determinant equal to one, was the appropriate generalization of the $SU(2)$ symmetry group. Gell-Mann showed that the lightest mesons and baryons could be described in a unified way using the $SU(3)$ symmetry, and he was able to make predictions of the possible decay modes of these particles.

14.6.1 The rotation group in three dimensions

We first consider functions in two-dimensional space. Fig. 14.25(A) shows the effect of a rotation transforming a vector **r** into another vector **r**$'$ and we would like to introduce rotation operators that act upon functions. As shown in Fig. 14.25(B), a function $\psi(x, y)$ can be represented graphically by means a family of contour curves along which the function has a constant value. As illustrated in Fig. 14.26, we can define a rotation operator that rotates the contours of a function.

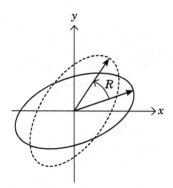

FIGURE 14.26 The rotation of a function can be illustrated by rotating its contour curves.

In a polar coordinate system, we can represent an angle defining rotations about the z-axis by ϕ_z. Using this polar coordinate frame, we define an operator $p(d\phi_z)$ that rotates the function $\psi(r, \theta_z, \phi_z)$ by an angle $d\phi_z$ as

$$\psi'(r, \theta_z, \phi_z) = p_z(d\phi_z)\psi(r, \theta_z, \phi_z).$$

Then the new function $\psi'(r, \theta_z, \phi_z)$ has the same value at the angle ϕ_z as the original function $\psi(r, \theta_z, \phi_z)$ has at an angle $\phi_z - d\phi_z$

$$p_z(d\phi_z)\psi(r, \theta_z, \phi_z) = \psi(r, \theta_z, \phi_z - d\phi_z). \tag{14.49}$$

We are now in a position to give a definition of the angular momentum operators in terms of rotation operators. In a system of units in which $\hbar = 1$, the operator corresponding to the z component of the angular momentum is

$$l_z = -i\frac{\partial}{d\phi_z}. \tag{14.50}$$

In order to express the z-component of the angular momentum in terms of a rotation operator, we first write the derivative explicitly as a limit of finite difference ratio

$$\frac{\partial}{\partial\phi_z}\psi(r, \theta, \phi_z) = \frac{\psi(r, \theta, \phi_z) - \psi(r, \theta, \phi_z - \delta\phi_z)}{\delta\phi_z},$$

where we have allowed the limit $\delta\phi_z \to 0$ to be understood. Using Eq. (14.49), this last equation can be written

$$\frac{\partial}{\partial\phi_z}\psi(r, \theta, \phi_z) = \left[\frac{1 - p_z(\delta\phi_z)}{\delta\phi_z}\right]\psi(r, \theta_z, \phi_z).$$

The differential operator ∂ may thus be written

$$\frac{\partial}{\partial\phi_z} = \left[\frac{1 - p_z(\delta\phi_z)}{\delta\phi_z}\right],$$

which leads to the following expression for the z component of the angular momentum operator

$$l_z = -i\left[\frac{1 - p_z(\delta\phi_z)}{\delta\phi_z}\right], \tag{14.51}$$

or

$$p_z(\delta\phi_z) = 1 - i\delta\phi_z l_z. \tag{14.52}$$

The operator l_z serves as the generator of rotations about the z-axis. Similarly, the operators l_x and l_y serves as generators of rotations about the x- and y-axes. We have

$$p_x(\delta\phi_x) = 1 - i\delta\phi_x l_x, \tag{14.53}$$

and

$$p_y(\delta\phi_y) = 1 - i\delta\phi_y l_y. \tag{14.54}$$

The effect of rotations upon the functions Y_{lm_l} for particular values of l and m_l depends upon the effect of the angular momentum operators, l_x, l_y, and l_z. When the operators, l_z and $p_z(\delta\phi_z)$, act on the function Y_{lm_l}, they produce a number times Y_{lm_l}. They do not mix the spherical harmonic with different values of m_l. To understand how l_x and $p_x(\delta\phi_x)$ and l_y and $p_y(\delta\phi_y)$ effect the functions Y_{lm_l}, we must recall some of the elementary properties of the angular momentum operators.

The operators, l_x, l_y, and l_z, satisfy the commutation relations

$$[l_x, l_y] = l_x l_y - l_y l_x = i l_z. \tag{14.55}$$

Using these commutation relations, one can show that the operators,

$$l_+ = l_x + i l_y \tag{14.56}$$

and

$$l_- = l_x - i l_y \tag{14.57}$$

step up and step down the values of m_l of the spherical harmonics. We have

$$l_+ Y_{lm_l} = \sqrt{l(l+1) - m_l(m_l+1)} Y_{lm_l+1} \tag{14.58}$$

and

$$l_- Y_{lm_l} = \sqrt{l(l+1) - m_l(m_l-1)} Y_{lm_l-1} \tag{14.59}$$

An equation for l_x can be obtained by adding Eqs. (14.56) and (14.57) to obtain

$$l_x = \frac{1}{2}(l_+ + l_-). \tag{14.60}$$

Similarly, an equation for l_y can be obtained by subtracting Eq. (14.57) from Eq. (14.56) to obtain

$$l_y = -\frac{i}{2}(l_+ - l_-). \tag{14.61}$$

Example 14.4

Show that the states of a p electron, Y_{11}, Y_{10}, and Y_{1-1}, transform according to a three-dimensional representation of the rotation group $R(3)$.

Solution
According to Eq. (14.52), the product of the rotation operator $p\delta(\phi_z)$ and a state Y_{lm_l} is

$$p_z(\delta\phi_z)Y_{lm_l} = (1 - i\delta\phi_z l_z)Y_{lm_l} = (1 - i\delta\phi_z m_l)Y_{lm_l}$$

The product of $p_z(\delta\phi_z)$ or l_z times a spherical harmonic gives a constant times a spherical harmonic with the same value l and the same value of m_l.

According to Eq. (14.53), the product of $p_x(\delta\phi_x)$ times a spherical harmonic Y_{lm_l} is $(1 - i\delta\phi_x l_x)Y_{lm_l}$. Then according to Eq. (14.60) and Eqs. (14.58) and (14.59), we get spherical harmonics with the same value of l and one value greater or one value less of the azimuthal quantum number m. The same can be said of the product of $p_y(\delta\phi_y)$ times a spherical harmonic.

In each case, an infinitesimal rotation of the spherical harmonics of a p-electron gives linear combinations of p-electron states

---

We shall now consider a configuration of two p-electrons. The states of two p-electrons form the basis of a nine dimensional representation of the rotation group.

Just as l_z is a generator for the rotation group of a single electron.

$$L_z = l_z(1) + l_z(2)$$

is a generator of the rotation group for simultaneous rotations of the two electrons. 'The operators of the total angular moment, L_x, L_y, and L_z satisfy the commutation relations

$$[L_x, L_y] = iL_z$$

and serve to label the states of two electrons just as the angular momentum operators of a single electron may be used to label the states of a single electron. We found in Chapter 5 that the total orbital angular momentum of two p-electrons can have the values $L = 2, 1, 0$. For each of those values of L, M_L can have the values $M_L = L, L - 1, \ldots, 0$. By making a transformation to the states of the total angular momentum, the nine-dimension matrices of the $SU(3)$ np^2 configuration are reduced to matrices with $5 \times 5, 3 \times 3$ a a 1×1 blocks running down the diagonal. The nine dimensional representation of the np^2 configuration reduce to a 5-dimensional, a 3-dimensional, and a 1-dimensional representation of the rotation group in three dimensions.

In the next section, we shall consider the $SU(3)$ symmetry group. We shall find that the u, d, and s quarks transform according to the fundamental representation of $SU(3)$, while the $\bar{u}, \bar{d}$, and $\bar{s}$ transform according to the complex conjugate representation of $SU(3)$. Mesons which are composed of a quark/antiquark pair transform according to a nine-dimensional representation of $SU(3)$ which reduces to an octet with eight mesons and a single meson state. Two of the light mesons shown in Fig. 14.12(A) with $I_3 = 0$ and $Y = 0$ transform according to the singlet representation of $SU(3)$ while the other transform according to the eight-dimensional octet. Baryons which are composed of three quarks transform according to a twenty seven dimensional representation of $SU(3)$ which reduces to a decuplet with ten states two octets and one singlet; however, only the $J = 3/2$ decuplet and the $J = 1/2$ octet correspond to states whose spatial wave functions are antisymmetric with respect to an interchange of two quarks, and only these representations occur in nature.

14.6.2 The $SU(3)$ symmetry group

As for the $R(3)$ symmetry group, the transformations of the symmetry group $SU(3)$ can be built up from infinitesimal transformations

$$\mathbf{W}_{inf} = \mathbf{I}_3 + i\chi, \tag{14.62}$$

where $\mathbf{I}_3$ is the 3×3 identity matrix and the Hermitian matrix χ has infinitesimal elements. The χ matrix, which involves *eight* independent parameters, can be written

$$\chi = \eta \cdot \mathbf{T} = \eta_1 \mathbf{T}_1 + \eta_2 \mathbf{T}_2 + \cdots + \eta_8 \mathbf{T}_8, \tag{14.63}$$

where $\eta = (\eta_1, \eta_2, \ldots, \eta_8)$ and the matrices $\mathbf{T}_i$ are defined by the following equations

$$
\mathbf{T}_1 = \frac{1}{2}\begin{bmatrix} 0 & 1 & 0 \\ 1 & 0 & 0 \\ 0 & 0 & 0 \end{bmatrix} \quad
\mathbf{T}_2 = \frac{1}{2}\begin{bmatrix} 0 & -i & 0 \\ i & 0 & 0 \\ 0 & 0 & 0 \end{bmatrix} \quad
\mathbf{T}_3 = \frac{1}{2}\begin{bmatrix} 1 & 0 & 0 \\ 0 & -1 & 0 \\ 0 & 0 & 0 \end{bmatrix}
$$

$$
\mathbf{T}_4 = \frac{1}{2}\begin{bmatrix} 0 & 0 & 1 \\ 0 & 0 & 0 \\ 1 & 0 & 0 \end{bmatrix} \quad
\mathbf{T}_5 = \frac{1}{2}\begin{bmatrix} 0 & 0 & -i \\ 0 & 0 & 0 \\ i & 0 & 0 \end{bmatrix} \quad
\mathbf{T}_6 = \frac{1}{2}\begin{bmatrix} 0 & 0 & 0 \\ 0 & 0 & 1 \\ 0 & 1 & 0 \end{bmatrix} \tag{14.64}
$$

$$
\mathbf{T}_7 = \frac{1}{2}\begin{bmatrix} 0 & 0 & 0 \\ 0 & 0 & -i \\ 0 & i & 0 \end{bmatrix} \quad
\mathbf{T}_8 = \frac{1}{2\sqrt{3}}\begin{bmatrix} 1 & 0 & 0 \\ 0 & 1 & 0 \\ 0 & 0 & -2 \end{bmatrix}.
$$

These matrices, which are called the Gell-Mann matrices, can be constructed by adding an additional row and column of zeroes to the Pauli matrices given by Eq. (13.43) of Chapter 13 and multiplying the resulting matrix by an additional factor of one half. We shall begin our study of implications of the $SU(3)$ symmetry, by considering the commutation relations of the Gell-Mann matrices. We recall that the commutator of two matrices, $\mathbf{A}$ and $\mathbf{B}$, is

$$[\mathbf{A}, \mathbf{B}] = \mathbf{AB} - \mathbf{BA}$$

and the matrices are said to commute if

$$\mathbf{AB} - \mathbf{BA} = 0$$

or equivalently if $\mathbf{AB} = \mathbf{BA}$. The matrices $\mathbf{T}_i$ satisfy commutation relations

$$[\mathbf{T}_i, \mathbf{T}_j] = i f_{ijk} \mathbf{T}_k, \tag{14.65}$$

where the constants f_{ijk} are real.

Of the matrices defined by Eqs. (14.64), the matrices $\mathbf{T}_3$ and $\mathbf{T}_8$ will play a special role in our discussion because they commute with each other. We shall make the following associations with these matrices

$$\mathbf{H}_1 = \mathbf{T}_3 \qquad \mathbf{H}_2 = \mathbf{T}_8 . \tag{14.66}$$

One can easily find linear combinations of the other $\mathbf{T}$-matrices which have single nonzero elements. For instance, the combination $\mathbf{T}_4 + i\mathbf{T}_5$ gives

$$\mathbf{T}_4 + i\mathbf{T}_5 = \frac{1}{2} \begin{bmatrix} 0 & 0 & 1 \\ 0 & 0 & 0 \\ 1 & 0 & 0 \end{bmatrix} + \frac{i}{2} \begin{bmatrix} 0 & 0 & -i \\ 0 & 0 & 0 \\ i & 0 & 0 \end{bmatrix} = \begin{bmatrix} 0 & 0 & 1 \\ 0 & 0 & 0 \\ 0 & 0 & 0 \end{bmatrix}.$$

The matrix $\mathbf{T}_4 + i\mathbf{T}_5$ satisfies the following commutation relations with $\mathbf{H}_1$ and $\mathbf{H}_2$

$$[\mathbf{H}_1, (\mathbf{T}_4 + i\mathbf{T}_5)] = \frac{1}{2}(\mathbf{T}_4 + i\mathbf{T}_5),$$

$$[\mathbf{H}_2, (\mathbf{T}_4 + i\mathbf{T}_5)] = \frac{\sqrt{3}}{2}(\mathbf{T}_4 + i\mathbf{T}_5).$$

The commutation relations involving $\mathbf{T}_4$ and $\mathbf{T}_5$ can be written more concisely by defining a vector

$$\alpha_1 = \begin{bmatrix} 1/2 \\ \sqrt{3}/2 \end{bmatrix} \tag{14.67}$$

and the matrix

$$\mathbf{E}_{\alpha_1} = \mathbf{T}_4 + i\mathbf{T}_5 = \begin{bmatrix} 0 & 0 & 1 \\ 0 & 0 & 0 \\ 0 & 0 & 0 \end{bmatrix}. \tag{14.68}$$

With this notation, the matrix $\mathbf{E}_{\alpha_1}$ satisfies the equations

$$[\mathbf{H}_1, \mathbf{E}_{\alpha_1}] = [\alpha_1]_1 \, \mathbf{E}_{\alpha_1},$$

$$[\mathbf{H}_2, \mathbf{E}_{\alpha_1}] = [\alpha_1]_2 \, \mathbf{E}_{\alpha_1}.$$

The vector α_1, which determines the commutation relations of the generators $\mathbf{H}_1$ and $\mathbf{H}_2$ with the matrix $\mathbf{E}_{\alpha_1}$, is called a *root vector* and the matrix $\mathbf{E}_{\alpha_1}$ is regarded as the generator corresponding to that root.

One may easily identify other roots. The matrix $\mathbf{E}_{-\alpha_1}$, defined by the equation

$$\mathbf{E}_{-\alpha_1} = \mathbf{T}_4 - i\mathbf{T}_5 = \begin{bmatrix} 0 & 0 & 0 \\ 0 & 0 & 0 \\ 1 & 0 & 0 \end{bmatrix}, \tag{14.69}$$

satisfies the equations

$$[\mathbf{H}_1, \mathbf{E}_{-\alpha_1}] = -1/2 \, \mathbf{E}_{-\alpha_1}$$

$$[\mathbf{H}_2, \mathbf{E}_{-\alpha_1}] = -\sqrt{3}/2 \, \mathbf{E}_{-\alpha_1},$$

and is thus associated with the root vector

$$-\alpha_1 = \begin{bmatrix} -1/2 \\ -\sqrt{3}/2 \end{bmatrix}. \tag{14.70}$$

Similarly, the matrix $\mathbf{E}_{\alpha_2}$, defined by the equation

$$\mathbf{E}_{\alpha_2} = \mathbf{T}_6 - i\mathbf{T}_7 = \begin{bmatrix} 0 & 0 & 0 \\ 0 & 0 & 0 \\ 0 & 1 & 0 \end{bmatrix}, \tag{14.71}$$

corresponds to the root

$$\alpha_2 = \begin{bmatrix} 1/2 \\ -\sqrt{3}/2 \end{bmatrix}, \tag{14.72}$$

and the matrix $\mathbf{E}_{-\alpha_2}$, defined by the equation

$$\mathbf{E}_{-\alpha_2} = \mathbf{T}_6 + i\mathbf{T}_7 = \begin{bmatrix} 0 & 0 & 0 \\ 0 & 0 & 1 \\ 0 & 0 & 0 \end{bmatrix}, \tag{14.73}$$

corresponds to the root $-\alpha_2$. The four matrices, $\mathbf{E}_{\pm\alpha_1}$ and $\mathbf{E}_{\pm\alpha_2}$, correspond to the roots $\pm\alpha_1$ and $\pm\alpha_2$. These matrices may be used to replace the matrices $\mathbf{T}_4$, $\mathbf{T}_5$, $\mathbf{T}_6$, and $\mathbf{T}_7$.

One may also show that the matrix $\mathbf{E}_{[1,0]}$, defined by the equation

$$\mathbf{E}_{[1,0]} = \mathbf{T}_1 + i\mathbf{T}_2 = \begin{bmatrix} 0 & 1 & 0 \\ 0 & 0 & 0 \\ 0 & 0 & 0 \end{bmatrix}, \tag{14.74}$$

corresponds to the root

$$\begin{bmatrix} 1 \\ 0 \end{bmatrix},$$

and the matrix $\mathbf{E}_{[-1,0]}$, defined by the equation

$$\mathbf{E}_{[-1,0]} = \mathbf{T}_1 - i\mathbf{T}_2 = \begin{bmatrix} 0 & 0 & 0 \\ 1 & 0 & 0 \\ 0 & 0 & 0 \end{bmatrix}, \tag{14.75}$$

corresponds to the root

$$\begin{bmatrix} -1 \\ 0 \end{bmatrix}.$$

The properties of the generators of $SU(3)$ are important for our further discussion. While the operators $\mathbf{H}_1$ and $\mathbf{H}_2$ commute with each other, the matrices $\mathbf{E}_\alpha$ satisfy the commutation relations

$$[\mathbf{H}_i, \mathbf{E}_\alpha] = \alpha_i \mathbf{E}_\alpha. \tag{14.76}$$

The matrices $\mathbf{H}_1$ and $\mathbf{H}_2$ are like the angular momentum operator l_z whose eigenvalues can be used to label physical states, while the matrices $\mathbf{E}_\alpha$ are like the step-up and step-down operators, l_+ and l_-, which act upon angular momentum states to produce states with larger and smaller values of the quantum number m.

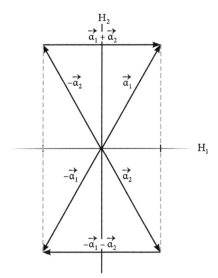

FIGURE 14.27 The roots of $SU(3)$.

All of the roots of $SU(3)$ may be written in terms of the roots α_1 and α_2, which are given by Eqs. (14.67) and (14.72). Before further developing the theory, we summarize the equations defining the roots

$$\alpha_1 = \begin{bmatrix} 1/2 \\ \sqrt{3}/2 \end{bmatrix}, \quad -\alpha_1 = \begin{bmatrix} -1/2 \\ -\sqrt{3}/2 \end{bmatrix},$$

$$\alpha_2 = \begin{bmatrix} 1/2 \\ -\sqrt{3}/2 \end{bmatrix}, \quad -\alpha_2 = \begin{bmatrix} -1/2 \\ \sqrt{3}/2 \end{bmatrix}, \tag{14.77}$$

$$\begin{bmatrix} 1 \\ 0 \end{bmatrix} = \alpha_1 + \alpha_2, \quad \begin{bmatrix} -1 \\ 0 \end{bmatrix} = -\alpha_1 - \alpha_2.$$

The roots of $SU(3)$, which are given by these equations, are all illustrated in Fig. 14.27. A root is said to be *positive* if the first nonzero component is positive. In this sense, the two roots, α_1 and α_2, are positive. Since the first nonzero component of the vector $\alpha_1 - \alpha_2$ is positive, α_1 is called the highest root.

14.6.3 The representations of $SU(3)$

A representation of a group consists of a set of operators or matrices acting upon a vector space with a unique operator or matrix of the representation corresponding to each member of the group. The simplest representation of the group $SU(3)$ is formed by the matrices of the group acting on the column vectors

$$\mathbf{v}_1 = \begin{bmatrix} 1 \\ 0 \\ 0 \end{bmatrix}, \quad \mathbf{v}_2 = \begin{bmatrix} 0 \\ 1 \\ 0 \end{bmatrix}, \quad \mathbf{v}_3 = \begin{bmatrix} 0 \\ 0 \\ 1 \end{bmatrix}. \tag{14.78}$$

The representation of $SU(3)$ formed by the matrices of the group acting on the column vectors (14.78) is known as a *fundamental representation*.

We may assign weights to each column vector, $\mathbf{v}_1$, $\mathbf{v}_2$, and $\mathbf{v}_3$, by seeing how it is affected by the diagonal generators of the group. The diagonal generators, $\mathbf{H}_1$ and $\mathbf{H}_2$, are defined by Eqs. (14.66) and (14.64). For instance, we may derive the following equations

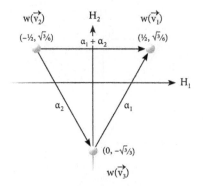

FIGURE 14.28 The three weights of the fundamental representation together with the roots α_1, α_2, and $\alpha_1 + \alpha_2$.

$$\mathbf{H}_1\mathbf{v}_1 = \mathbf{T}_3\mathbf{v}_1$$

$$= (1/2)\begin{bmatrix} 1 & 0 & 0 \\ 0 & -1 & 0 \\ 0 & 0 & 0 \end{bmatrix}\begin{bmatrix} 1 \\ 0 \\ 0 \end{bmatrix} = (1/2)\begin{bmatrix} 1 \\ 0 \\ 0 \end{bmatrix} = (1/2)\mathbf{v}_1 , \tag{14.79}$$

$$\mathbf{H}_2\mathbf{v}_1 = \mathbf{T}_8\mathbf{v}_1$$

$$= (\sqrt{3}/6)\begin{bmatrix} 1 & 0 & 0 \\ 0 & 1 & 0 \\ 0 & 0 & -2 \end{bmatrix}\begin{bmatrix} 1 \\ 0 \\ 0 \end{bmatrix} = (\sqrt{3}/6)\begin{bmatrix} 1 \\ 0 \\ 0 \end{bmatrix} = (\sqrt{3}/6)\mathbf{v}_1 . \tag{14.80}$$

We may thus assign to the vector $\mathbf{v}_1$ a weight vector with components $1/2$ and $\sqrt{3}/6$. The components of the weight determine the result of operating with the generators $\mathbf{H}_1$ and $\mathbf{H}_2$ upon the corresponding vector. Constructing the weights for the vectors $\mathbf{v}_2$ and $\mathbf{v}_3$ in a similar manner, we obtain the following three weight vectors

$$w(\mathbf{v}_1) = \begin{bmatrix} 1/2 \\ \sqrt{3}/6 \end{bmatrix}, \quad w(\mathbf{v}_2) = \begin{bmatrix} -1/2 \\ \sqrt{3}/6 \end{bmatrix}, \quad w(\mathbf{v}_3) = \begin{bmatrix} 0 \\ -\sqrt{3}/3 \end{bmatrix}. \tag{14.81}$$

Each weight corresponds to a particular column vector. The roots translate one weight into another. Using Eqs. (14.77) and (14.81), one can derive, for example, the following equation

$$\alpha_1 + w(\mathbf{v}_3) = w(\mathbf{v}_1).$$

The three weights of the fundamental representation together, with the roots α_1, α_2, and $\alpha_1 + \alpha_2$, are illustrated in Fig. 14.28. Each root translates one weight into another. The root α_1 translates the weight, $w(\mathbf{v}_3)$, into the weight, $w(\mathbf{v}_1)$. Similarly, the root α_2 translates the weight, $w(\mathbf{v}_2)$, into the weight, $w(\mathbf{v}_3)$, while the root $\alpha_1 + \alpha_2$ translates the weight, $w(\mathbf{v}_2)$, into the weight, $w(\mathbf{v}_1)$. The root diagram can be thought of as a map which shows us how to get from one weight to another.

Another representation of $SU(3)$ may be obtained by taking the negative of the complex conjugate of the matrices of the fundamental representation. The new representation obtained in this way is called the *complex conjugate representation*. Taking the complex conjugate of Eq. (14.65) and using the fact that the constants f_{ijk} are real, we obtain

$$[\mathbf{T}_i{}^*, \mathbf{T}_j{}^*] = -if_{ijk}\mathbf{T}_k{}^*.$$

This equation can be written

$$[-\mathbf{T}_i{}^*, -\mathbf{T}_j{}^*] = if_{ijk}(-\mathbf{T}_k{}^*),$$

which shows that the matrices $-\mathbf{T}_i{}^*$ satisfy the same algebraic equations as the $\mathbf{T}_i$ matrices themselves.

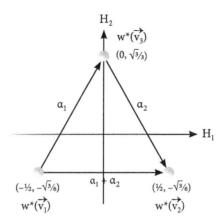

FIGURE 14.29 The weights of the complex conjugate representation of $SU(3)$.

We assign weights to each vector of the complex conjugate representation by allowing the diagonal generators, $-\mathbf{H}_1^*$ and $-\mathbf{H}_2^*$, to act upon the vector. For the vector $\mathbf{v}_1$, we obtain

$$-\mathbf{H}_1{}^*\mathbf{v}_1 = \frac{1}{2}\begin{bmatrix} -1 & 0 & 0 \\ 0 & 1 & 0 \\ 0 & 0 & 0 \end{bmatrix}\begin{bmatrix} 1 \\ 0 \\ 0 \end{bmatrix} = -\frac{1}{2}\mathbf{v}_1,$$

$$-\mathbf{H}_2{}^*\mathbf{v}_1 = \frac{\sqrt{3}}{6}\begin{bmatrix} -1 & 0 & 0 \\ 0 & -1 & 0 \\ 0 & 0 & 2 \end{bmatrix}\begin{bmatrix} 1 \\ 0 \\ 0 \end{bmatrix} = -\frac{\sqrt{3}}{6}\mathbf{v}_1 .$$

We may thus assign to the vector $\mathbf{v}_1$ a weight vector with components $-1/2$ and $-\sqrt{3}/6$. Since the matrices $\mathbf{H}_1^\star$ and $\mathbf{H}_2^\star$, are the negatives of $\mathbf{H}_1$ and $\mathbf{H}_2$, the weights of the complex conjugate representation can be obtained by reflecting the weights of the fundamental representation through the origin. Using the weights of the fundamental representation, which are given by Eq. (14.81), the weights of the complex conjugate representation are seen to be

$$w^*(\mathbf{v}_1) = \begin{bmatrix} -1/2 \\ -\sqrt{3}/6 \end{bmatrix}, \quad w^*(\mathbf{v}_2) = \begin{bmatrix} 1/2 \\ -\sqrt{3}/6 \end{bmatrix}, \quad w^*(\mathbf{v}_3) = \begin{bmatrix} 0 \\ \sqrt{3}/3 \end{bmatrix}. \tag{14.82}$$

The weights of the complex conjugate representation are shown in Fig. 14.29.

We recall that a root is said to be positive if the first nonzero component is positive and the highest root has the property that the difference between this root and any other root is positive. Applying these definitions to the weights, the highest weight of the fundamental representation is

$$\mu_1 = \begin{bmatrix} 1/2 \\ \sqrt{3}/6 \end{bmatrix}, \tag{14.83}$$

while the highest weight of the complex conjugate representation is

$$\mu_2 = \begin{bmatrix} 1/2 \\ -\sqrt{3}/6 \end{bmatrix}. \tag{14.84}$$

The representations of $SU(3)$ may be characterized by their highest weights.

As we have said, a representation of a group consists of a set of operators or matrices acting upon a vector space. An important representation of $SU(3)$ called the *adjoint representation* is obtained by allowing the matrices corresponding to the generators of the group to play the role of *both* the operators and the states of the vector space. In the adjoint representation, the product of two generators is equal to the commutator of the generators. Since the commutation relations are used to define the roots and the product of the generators $\mathbf{H}_i$ with the states of the representation, the weights of the

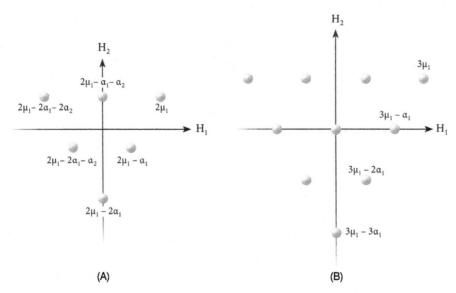

FIGURE 14.30 (A) The weights of the $(2, 0)$ representation and (B) the weights of the $(3, 0)$ representation.

adjoint representation are equal to the roots of the group. The root diagram shown in Fig. 14.27 thus serves also as the weight diagram of the adjoint representation.

Various notations are used by particle physicists to denote the different representations of $SU(3)$. The fundamental and the complex conjugate representations of $SU(3)$, which have the highest weights μ_1 and μ_2, respectively, are denoted by $(1, 0)$ and $(0, 1)$, respectively. Representations of $SU(3)$ can also be denoted by their dimension with the complex conjugate representation being distinguished by a bar. The two fundamental representations can thus be denoted

$$(1, 0) \equiv 3 \quad (0, 1) \equiv \overline{3}. \tag{14.85}$$

The representation $(2, 0)$ has the highest weight

$$2\mu_1 = \begin{bmatrix} 1 \\ \sqrt{3}/3 \end{bmatrix}.$$

The other weights of this representation may be obtained by shifting the highest weight by the root vectors, α_1 and α_2. The weights of the $(2, 0)$ representation are shown in Fig. 14.30(A), and the weights of the $(3, 0)$ representation with highest weight $3\mu_1$ are shown in Fig. 14.30(B). In his book, *Lie Algebras in Particle Physics*, Howard Georgi gives a beautiful pictorial scheme for finding all the weights of representations of $SU(3)$.

14.6.3.1 The flavor $SU(3)$ symmetry

One may notice that the figures of the weights of the representations of $SU(3)$ in the preceding section are very similar to the figures in Section 14.4.1 giving the isospins I_3 and hypercharge Y of supermultiplets of particles. We consider first the weights of the fundamental representation of $SU(3)$, which are given by Eq. (14.81) and shown in Fig. 14.28. Using Fig. 14.11, the vectors with components I_3 and Y, associated with the three quarks are

$$u = \begin{bmatrix} 1/2 \\ 1/3 \end{bmatrix}, \quad d = \begin{bmatrix} -1/2 \\ 1/3 \end{bmatrix}, \quad s = \begin{bmatrix} 0 \\ -2/3 \end{bmatrix}. \tag{14.86}$$

The vectors appearing in Eq. (14.86) become identical to the vectors in Eq. (14.81) if we multiply the second component of each vector in Eq. (14.86) by $\sqrt{3}/2$. This amounts to labeling the second axis of Fig. 14.11(A) by $(\sqrt{3}/2Y)$ rather than Y, while still labeling the first axis by I_3. A similar argument may be used for the complex conjugate representation. Multiplying the values of hypercharge Y by $\sqrt{3}/2$ in Fig. 14.11(B) give the weights of the complex conjugate representation shown in Fig. 14.28.

Quarks thus transform according to the fundamental representation of $SU(3)$ and antiquarks transform according to the complex conjugate representation. Mesons, which are composed of a quark and an antiquark, transform according to the direct-product of these two representations of $SU(3)$. This nine-dimensional direct-product representation is reducible, which means that it is possible to find linear combinations of the quark/antiquark states that transform among themselves according to the $SU(3)$ symmetry. We have already encountered this effect in atomic physics. We have seen that angular momenta with quantum numbers j_1 and j_2 can be combined to form states with total angular momentum $J = j_1 + j_2, \ldots, |j_1 - j_2|$. Two p-electrons can combine to form states with $L = 2, 1, 0$. This result can be stated in the language of group theory. The three states of a p-electron with $l = 1$ and $m_l = 1, 0, -1$ transform among themselves under rotations and form the basis of a three-dimensional representation of the rotation group. Two p-electrons thus transform according to the direct product of two three-dimensional representations of the rotation group $R(3)$. However, this representation is reducible. It is possible to show that linear combinations of the np^2 functions transform among themselves with respect to rotations. These linear combinations correspond to states for which the electrons have total angular momentum L equal to 2, 1, or 0. The nine-dimensional representation of the rotation group due to two p-electrons decomposes into a five-dimensional, a three-dimensional, and a one-dimensional representation. In spectroscopic notation these states of p^2 are denoted by D, P, and S. In the same way, the nine-dimensional representation of $SU(3)$ corresponding to mesons decomposes into an octet corresponding to eight mesons and a singlet corresponding to a single mesonic state.

The lightest mesons are described in Table 14.8 and Fig. 14.12. Of the spin-zero mesons in Table 14.8, the π^0, η, and η' have I_3 and Y equal to zero. The π^0 and η are members of the octet representation of $SU(3)$, while the η' with wave function

$$\psi = \frac{1}{3}\left[d\bar{d} + u\bar{u} + s\bar{s}\right] \tag{14.87}$$

is an $SU(3)$ singlet. For the spin-one mesons shown in Fig. 14.12(B), octet-singlet mixing occurs with the ω and ϕ both being mixtures of octet and singlet states.

The lightest baryons described in Tables 14.9 and 14.10 and illustrated in Fig. 14.13 all consist of three quarks and hence correspond to a reducible twenty-seven dimensional representation of $SU(3)$. The twenty-seven dimensional representation of $SU(3)$ corresponds to a decuplet with $J^P = \frac{3}{2}^+$, an octet with $J^P = \frac{1}{2}^+$, and one other octet and singlet; however, only the $J = 3/2$ decuplet and the $J = 1/2$ octet correspond to states whose spatial wave functions are antisymmetric with respect to an interchange of two quarks, and only these representations occur in nature.

14.6.3.2 The color $SU(3)$ symmetry

We would like now to consider the color component of the wave functions of quarks and gluons. The color component of wave functions is conveniently represented by the three spinors

$$\mathbf{r} = \begin{bmatrix} 1 \\ 0 \\ 0 \end{bmatrix}, \quad \mathbf{g} = \begin{bmatrix} 0 \\ 1 \\ 0 \end{bmatrix}, \quad \mathbf{b} = \begin{bmatrix} 0 \\ 0 \\ 1 \end{bmatrix}, \tag{14.88}$$

which transform according to the fundamental representation of $SU(3)$. As for the antiquarks, the anticolor states, $\bar{r}$, $\bar{g}$, and $\bar{b}$, transform according to the complex conjugate representation of $SU(3)$.

We have found that mesons, which are composed of quark/antiquark pairs, transform according to the nine-dimensional representation of $SU(3)$ which is the product of the fundamental and complex conjugate representations. This representation is reducible, being composed of an octet and a singlet representation of $SU(3)$. The singlet state of mesons is described by the wave function (14.87). Similarly, the states of gluons are made up of color/anticolor states that transform according to the octet representation of $SU(3)$. There are eight independent color states of the gluon. The linear combination of color/anticolor states which transforms according to the singlet representation of $SU(3)$ is

$$\psi = \frac{1}{3}\left[r\bar{r} + g\bar{g} + b\bar{b}\right]. \tag{14.89}$$

This state is analogous to the singlet state (14.87) of mesons.

In Section 14.4.2, we saw that the color component of baryon wave functions is antisymmetric with respect to an exchange of two quarks. An appropriate color wave function of this kind can be written

$$\psi_B^C = \frac{1}{\sqrt{6}}\left[r_1 g_2 b_3 - g_1 r_2 b_3 + b_1 r_2 g_3 - b_1 g_2 r_3 + g_1 b_2 r_3 - r_1 b_2 g_3\right]. \tag{14.90}$$

Since the space-spin part of baryon wave functions is symmetric with respect to the interchange of two quarks and the color part of the wave function is antisymmetric, the total wave function of baryons is antisymmetric and thus satisfies the Pauli exclusion principle.

There is now a good deal of experimental evidence that the color $SU(3)$ symmetry is exact and differs in this way from the approximate flavor $SU(3)$ symmetry, which was found first.

14.7 * Gauge invariance and the electroweak theory

The most fundamental and enduring question in particle physics concerns the existence of higher symmetries. We have already considered the parity symmetry and the isospin and $SU(3)$ symmetries. As we shall now show, gauge symmetries provide a general framework for understanding the interactions between elementary particles.

The concept of a gauge symmetry first arose in classical electromagnetic theory. The electric field and the magnetic field can be expressed in terms of a scalar potential V and a vector potential $\mathbf{A}$. Together the scalar potential and the vector potential form the components of a four-vector $A_\mu = (V, \mathbf{A})$.

The electric and magnetic fields can be shown to be unaffected by gauge transformations of the kind

$$A'_\mu(x) = A_\mu(x) - \frac{1}{e}\partial\alpha(x), \tag{14.91}$$

where $\alpha(x)$ is a function of the spatial coordinates and ∂_μ has components

$$\partial_\mu = \frac{\partial}{\partial^\mu} = (\frac{\partial}{\partial x^0}, \frac{\partial}{\partial x^1}, \frac{\partial}{\partial x^2}, \frac{\partial}{\partial x^3})$$

Maxwell's equations are invariant with respect to gauge transformations, and we might expect this symmetry to carry over to quantum mechanics.

The Dirac equation of a fermion with charge e moving in an electromagnetic field is

$$\left[i\gamma^\mu(\partial_\mu + ieA_\mu) - k_0\right]\psi = 0, \tag{14.92}$$

where

$$k_0 = \frac{mc}{\hbar}. \tag{14.93}$$

In so-called natural units with $\hbar$ and c equal to one, k_0 is equal to the mass of the particle. Eq. (14.92) for a particle moving in an electromagnetic field can be obtained from the Dirac equation of a free particle given in Chapter 13 by making the replacement $\partial_\mu \to \partial_\mu + ieA_\mu$. The Dirac equation (14.92) is not invariant with respect to a gauge transformation because the wave function ψ cannot satisfy Eq. (14.92) and the analogous equation obtained by replacing A_μ by A'_μ. We must keep in mind, however, that the phase of the wave function ψ is not an observable quantity. Any transformation of the kind

$$\psi'(x) = e^{i\alpha(x)}\psi(x) \tag{14.94}$$

will not affect the probability density of a particle. For the Dirac equation to be gauge invariant, the gauge transformations (14.91) must be accompanied by a transformation of the wave functions described by Eq. (14.94) so that the transformed wave function ψ' satisfies the equation

$$\left[i\gamma^\mu(\partial_\mu + ieA'_\mu) - k_0\right]\psi' = 0, \tag{14.95}$$

To obtain this result, we first define the *covariant derivative*

$$D_\mu = \partial_\mu + ieA_\mu, \tag{14.96}$$

and see how the product $D_\mu\psi(x)$ transforms with respect to a gauge transformation. Using Eqs. (14.91) and (14.94), we obtain

$$D'_\mu\psi' = (\partial_\mu + ieA'_\mu)e^{i\alpha(x)}\psi(x) = i\frac{\partial\alpha}{\partial x^\mu}\psi + e^{i\alpha}\partial_\mu\psi + ieA_\mu e^{i\alpha}\psi - i\frac{\partial\alpha}{\partial x^\mu}\psi. \tag{14.97}$$

The first and last terms on the right-hand side of this last equation cancel, and we obtain

$$D'_\mu\psi' = e^{i\alpha}D_\mu\psi. \tag{14.98}$$

Thus, although the space-time function, α, feels the effect of the derivatives, the terms due to the derivatives of the phase function cancel and the phase function passes through the operator D_μ allowing $D'_\mu \psi'$ to transform in the same way as the wave function ψ. The Dirac equation of a fermion with charge e moving in an electromagnetic field, which can be written

$$(i\gamma^\mu D_\mu - k_0)\psi = 0,$$

is thus invariant with respect to a gauge transformation.

We have thus far supposed that the wave function of a charged fermion in an electromagnetic field satisfies the Dirac equation (14.92) and then showed that the Dirac equation is invariant with respect to the gauge transformation defined by Eqs. (14.91) and (14.94). This line of argument can be reversed. One can start by demanding that the theory be invariant with respect to the phase transformation

$$\psi(x) \rightarrow \psi'(x) = e^{i\alpha(x)}\psi(x)$$

Such a gauge invariance is not possible for a free particle, but rather requires the existence of a vector field A_μ, which we have seen in Chapter 13 is coupled to the Dirac current j_μ. The demand of phase invariance has thus lead to the introduction of a vector field that interacts with a conserved current. Because, the gauge transformation of the Dirac equation corresponds to a unitary transformation depending upon a single parameter α, this gauge symmetry is referred to as a $U(1)$ symmetry.

The standard model of the electroweak interactions is based on the more complicated symmetry group $SU(2) \otimes U(1)$. The model contains fields, W^1_μ, W^2_μ, and W^3_μ, associated with the gauge group $SU(2)$ and a field B_μ associated with the group $U(1)$. As we shall find in the next section, the electromagnetic field, A_μ, and the field associated with the neutral boson, Z_0, are linear combinations of the fields, W^3 and B_μ. Just as isospin and hypercharge can be used to describe multiplet of particles that interact by the strong interaction, weak isospin and weak hypercharge can be used to describe the quanta of the weak interaction. We shall denote the weak isospin quantum number by I_W, the azimuthal quantum number of the weak isospin by I^3_W, and weak hypercharge by Y_W. The charge of the bosons that are the carriers of the weak force are related to I^3_W and Y_W by the equation

$$Q = I^3_W + \frac{Y_W}{2}. \tag{14.99}$$

This last equation is entirely analogous to the equation we have seen earlier giving I_3 in terms of the hypercharge.

The quantum fields associated with the charged W^+ and W^- bosons are

$$W^+_\mu = \sqrt{\frac{1}{2}}(W^1_\mu - W^2_\mu)$$

$$W^-_\mu = \sqrt{\frac{1}{2}}(W^1_\mu + W^2_\mu)$$

The isospin quantum numbers of W^+_μ are $I_W = 1$ and $I^3_W = 1$, while the isospin quantum numbers of W^-_μ are $I_W = 1$ and $I^3_W = -1$. The field W^3_μ has $I_W = 1$ and $I^3_W = 0$ and the neutral field B_μ has $I_W = 0$ and $I^3_W = 0$.

Just as we have achieved gauge invariance for the Dirac theory, we introduce the following covariant derivative for the electro-weak theory

$$D_\mu = \partial_\mu + ig_W \mathbf{W}_\mu \cdot \mathbf{T}_\mu + \frac{1}{2}iY_W g'_W B_\mu, \tag{14.100}$$

where g_W is the constant associated with the weak $SU(2)$ coupling and g'_W is the coupling constant associated with the $U(1)$ coupling. The inner product $\mathbf{W}_\mu \cdot \mathbf{T}_\mu$ that occurs in this last equation can be expanded in terms of the individual isospin matrices

$$\mathbf{W}_\mu \cdot \mathbf{T}_\mu = W^3_\mu T^3_\mu + \frac{1}{\sqrt{2}}W^+_\mu T^+ + \frac{1}{\sqrt{2}}W^-_\mu T^-. \tag{14.101}$$

The form of the isospin $\mathbf{T}$ matrices depends upon the spin of the states on which they act. For example, in the doublet representation of $SU(2)$ we have

$$T^3 = \begin{bmatrix} \frac{1}{2} & 0 \\ 0 & -\frac{1}{2} \end{bmatrix}, T^+ = \begin{bmatrix} 0 & 1 \\ 0 & 0 \end{bmatrix}, T^- = \begin{bmatrix} 0 & 0 \\ 1 & 0 \end{bmatrix} \tag{14.102}$$

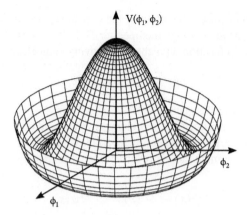

FIGURE 14.31 The potential energy function associated with the Higgs boson.

While the electro-weak theory with the covariant derivative in Eq. (14.100) describes the electromagnetic and weak interactions in a unified way, the theory can be shown to be gauge invariant only if the vector bosons have zero mass. The theory cannot account for the observed experimental data because the photon is the only carrier of the electro-weak force with zero mass. This difficulty will be overcome in the next section in which we describe the Higgs mechanism of spontaneous symmetry breaking which maintains the gauge invariance of the theory while giving mass to all but one of the vector bosons.

14.8 Spontaneous symmetry breaking and the discovery of the Higgs

We begin this section by considering a possibility. Suppose that a field has a gauge symmetry but the ground state of the system – the vacuum state – does not have the symmetry of the field. Like a pencil standing on its end, the state of the system will eventually fall into another state for which the potential energy of the system has a local minimum. To be concrete, we consider the potential energy function, $V(\phi_1, \phi_2)$, illustrated in Fig. 14.31. This function – popularly described as a "Mexican Sombrero" – has a maximum in the center and a circular trough around the center. A particle moving in this potential could fall arbitrarily in any direction and find itself in this circular trough – a distance v from the origin. Since a motion of the particle around the trough would cost no energy, this direction would be associated with a quantum of the field with zero mass, while motion in the other directions would be associated with massive field quanta.

One can imagine that in the first instant after the big bang – the first 10^{-35} or 10^{-45} seconds – the Universe was in a pure state in which the vector bosons which are the carriers of the weak interaction had zero mass. And then the Universe fell from grace. The symmetry of the Universe was spontaneously broken. The photon and the Higgs boson were born and the W^+, W^-, and Z^0 bosons, which are the carriers of the weak force, acquired a mass. We now want to find the mass of the vector bosons after this primordial symmetry was broken.

In the previous section, we found that the equations of the quantum fields associated with the weak interaction could be made gauge invariant if we replaced ordinary partial derivatives in the field equations by the covariant derivative (14.100). In the following, we shall write the expression for the appropriate covariant derivative

$$D_\mu = \partial + G, \tag{14.103}$$

where G is equal to

$$G = i g_W \mathbf{W}_\mu \cdot \mathbf{T} + (i/2) Y_W g'_W B_\mu. \tag{14.104}$$

After the gauge symmetry is spontaneously broken, the state of the Higgs field ϕ will have fallen into a local minimum in the trough surrounding the central bulge in Fig. 14.31. Which direction the state vector falls to is entirely arbitrary. The state of the Higgs field can then be represented by the vector

$$\phi_0 = \frac{1}{\sqrt{2}} \begin{bmatrix} 0 \\ v \end{bmatrix}. \tag{14.105}$$

The zero that occurs in the first component of this vector may be identified as the charge of the vacuum state while v in the second component is the distance from the local minimum to the origin shown in Fig. 14.31.

The kinetic energy in quantum field theory is associated with products of partial derivatives just as in nonrelativistic quantum mechanics, while the mass of particles in quantum field theory is associated with quadratic expressions of the fields. We shall now form the product of G and ϕ_0. The masses of the vector bosons can be obtained by taking the square of the norm of $G\phi_0$ and identify the coefficients of squares of quantum fields in the resulting expression as the masses of the corresponding particles. The masses of the vector bosons being nonzero depends upon the fact that the vacuum expectation value of the Higgs field is not equal to zero.

We first use Eqs. (14.104) and (14.105) to form the product of G and the Higgs field

$$G\phi_0 = \frac{1}{\sqrt{2}}(ig_W \mathbf{W}_\mu \cdot \mathbf{T}_\mu + ig'_W B_\mu)\begin{bmatrix} 0 \\ v \end{bmatrix} \tag{14.106}$$

where we have taken the weak hypercharge $Y_W = 1$. The right-hand side of this last equation can be simplified using Eqs. (14.101) and (14.102) to obtain

$$G\phi_0 = -\frac{ig_W W_\mu^3}{2\sqrt{2}}\begin{bmatrix} 0 \\ v \end{bmatrix} + \frac{ig_W W_\mu^+}{2}\begin{bmatrix} v \\ 0 \end{bmatrix} + \frac{ig'_W B_\mu}{2\sqrt{2}}\begin{bmatrix} 0 \\ v \end{bmatrix} \tag{14.107}$$

The square of the norm of a vector can be obtained by taking the adjoint of the vector and then multiplying the adjoint times the vector itself. The adjoint of $G\phi_0$ is

$$(G\phi_0)^\dagger = \frac{ig_W W_\mu^3}{2\sqrt{2}}\begin{bmatrix} 0 & v \end{bmatrix} - \frac{ig_W W_\mu^-}{2}\begin{bmatrix} v & 0 \end{bmatrix} - \frac{ig'_W B_\mu}{2\sqrt{2}}\begin{bmatrix} 0 & v \end{bmatrix} \tag{14.108}$$

Multiplying Eqs. (14.108) and (14.107), we obtain

$$||G\phi_0||^2 = \frac{v^2}{8}\left[(g_W W_\mu^3 - g'_W B_\mu)[(g_W W_\mu^3 - g'_W B_\mu) + 2g_W^2 W_\mu^- W_\mu^+\right]. \tag{14.109}$$

The above expression for the square of the norm of $G\phi_0$ contains products of different fields; however, the quadratic form may be transformed into a quadratic expression containing the sum of squares of individual fields by making the transformations

$$W_\mu^3 = \cos\theta_W Z_\mu + \sin\theta_W A_\mu \tag{14.110}$$

$$B_\mu = -\sin\theta_W Z_\mu + \cos\theta_W A_\mu \tag{14.111}$$

where the angle θ_W (the Weinberg or electroweak mixing angle) is defined by the relative strengths of the coupling constants

$$\sin^2\theta_W = \frac{g_W'^2}{g_W^2 + g_W'^2} \tag{14.112}$$

The transformation defined by Eqs. (14.110) and (14.111) expresses the fields, W_μ^3 and B_μ, having definite values of the isospin quantum numbers in terms of the physical Z_μ and A_μ fields. After these transformations, the quadratic terms in the expression for $||G\phi_0||^2$ becomes

$$||G\phi_0||^2 = \frac{g_W^2 v^2}{4} W_\mu^+ W_\mu^- + \frac{(g_W^2 + g_W'^2)v^2}{4} Z_\mu Z_\mu \tag{14.113}$$

We therefore find that the W and Z bosons have acquired masses given by the formulas

$$m_W = \frac{1}{2} v g_W \tag{14.114}$$

$$m_Z = \frac{1}{2} v\sqrt{G_W^2 + g_W'^2} = \frac{m_W}{\cos\theta_W} \tag{14.115}$$

The photon remains massless because there are no terms quadratic in the field A_μ in Eq. (14.109).

The Higgs field ϕ_0 thus acquired a nonzero vacuum expectation value at a particular point on a circle of minima away from ϕ_0 and the gauge symmetry was broken. As a result, the three vector mesons, W^+, W^-, and Z, acquired a mass but the

FIGURE 14.32 Photo credit: "CERN LHC Tunnel" by Julian Herzog. Wikimedia Commons.

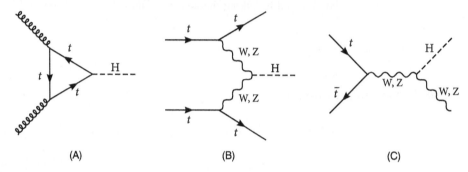

FIGURE 14.33 The most processes for the production of the Higgs.

photon remained massless. The Higgs boson itself acquired a mass but the gauge theory does not provide any information about the mass of the Higgs. The interaction of the Higgs boson with other particles depends upon the mass of the particle with which the Higgs interacts. Its interaction is greatest with the massive top quark and with the W^+, W^-, and Z^0 bosons.

Robert Brout, Peter Higgs, and Francois Englert presented their theory of spontaneous symmetry breaking in 1964 and the particle physics community has devoted much of its resources in the intervening years trying to discover the Higgs boson. The existence of the Higgs would confirm the theory of spontaneous symmetry breaking that accounts for the masses of the W and Z. Other theories have been formulated to describe the masses of the vector bosons, but none of the other theories are as simple or as beautiful as the Higgs mechanism.

The first attempts to find the Higgs boson were carried out at the large electron-positron collider (LEP). As described in the first section of this chapter, a short-lived resonance state such as the Higgs can be identified as a local maxima of an invariant mass distribution curve. While early experiments at LEP and LEP II were not able to identify a Higgs resonance, these experiments did establish a lower limit of 114 GeV for the Higgs mass. A very massive Higgs boson should not produce a narrow, easily identifiable resonance state.

The opening of the Large Hadron Collider (LHC) at the CERN laboratory near Geneva in Switzerland in 2010 offered the possibility of performing experiments at very high energies. The CERN collider, which is shown in Fig. 14.32, produces a colliding beam of protons. The collider should be able to identify a Higgs between the lower limit of 114 GeV provided by the LEP II experiments extending up to around 800 MeV.

A central difficulty in the search for the Higgs boson has been that the possible mechanisms for producing the Higgs and the possible decay mechanisms all depend upon the mass of the Higgs which was entirely unknown. At an energy of a few hundred GeV, the most important production mechanism is so-called gluon-gluon fusion, which is described by the Feynman diagram in Fig. 14.33(A). In this process, two gluons produced by the proton beams interact with top quarks which then produce a Higgs. The next most important processes are the boson fusion processes depicted in Fig. 14.33(B) and 14.33(C). In both of these processes, top quarks produced by the interacting protons produce W or Z bosons which then produce the Higgs. On July 4, 2012, the CERN laboratory announce that it had identified a resonance state at 125 GeV

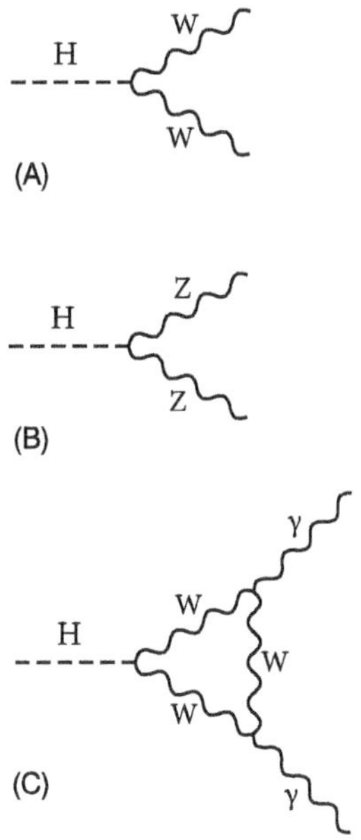

(A)

(B)

(C)

FIGURE 14.34 The most important processes for the decay of the Higgs.

which could be the Higgs. Further analysis since performed at CERN has verified the presence of a resonance at this energy and have studied the decay channels of the resonance.

As shown by the Feyman diagrams in Fig. 14.34(A) and (B), the Higgs can decay into W^+ and W^- bosons or into two Z bosons. The decay into W^+ and W^- should occur more often; however, all of the decay modes of W^+ and W^- bosons involve neutrinos which are difficult to detect. The decay of the Higgs by producing two Z bosons is more promising because Z bosons decay into two leptons that can be easily detected. Another important decay process is described by the Feynman diagram shown in Fig. 14.33(C). In higher-order processes of this kind, which can include the top quark, the decay of the Higgs produces a pair of gamma rays.

Since the discovery of a Higgs-like resonance was announced by CERN in July of 2012, the ATLAS and CMS collaborations at CERN have succeeded in establishing that the resonance has a spin zero and even parity – as the Higgs should. They have also found that the resonance decays producing four leptons and two gamma rays. There can be little doubt that the resonance found at CERN is the particle postulated by Higgs and his collaborators so long ago. Peter Higgs and Francois Englert received the Nobel Prize in 2013 for conceiving of spontaneous symmetry breaking and the Higgs boson. We should note though that the breaking of the gauge symmetry by a single scalar field is the simplest possible way the gauge symmetry could be broken. It is unlikely that the discovery of the Higgs resonance on July 4, 2012 will be the last act in the Higgs drama.

14.9 Supersymmetry

The standard model of particle physics has been successful in describing a range of events up to 235 GeV, and yet this theory is entirely fragmented. On a fundamental level, leptons and quarks are spin one half particles that obey Fermi-Dirac statistics while the carriers of the force between particles have spin zero or one and obey Bose-Einstein statistics.

14.9.1 Symmetries in physics

Supersymmetry is a general theory that maps particles and fields with spin one half into particles and fields of integer spin and vice versa. It is a general theory of matter and radiation. The generators of supersymmetry Q act to transform Fermions into Bosons according to the equation

$$Q|\text{Fermions} >= |\text{Boson} >$$

and viceversa. From this basic equation, two facts should immediately become clear

- Q changes the spin of a particle and thus its space-time properties. Supersymmetry is a space-time symmetry – not an internal symmetry.
- Each one-particle state has at least one superpartner. One has to deal with (super)multiplets of particles.

Before we begin studying this new symmetry, we recall how the $R(3)$ and the $SU(3)$ symmetries were treated. We found that the operator $p_z(\delta\phi_z)$ that rotate wave functions by an infinitesimal angle $\delta\phi_z$ about the z-axis was related to the angular momentum operator l_z which generate such rotations by the equation

$$p_z(\delta\phi_z) = 1 - i\delta\phi_z l_z.$$

Similarly, the operators l_x and l_y serves as generators of rotations about the x- and y-axes. We found that the angular momentum operators satisfied the commutation relations

$$[l_x, l_y] = l_x l_y - l_y l_x = i \tag{14.116}$$

The states $|lm>$ with $m = l, l-1, \ldots, -l$ serve as a basis for a matrix representation of the angular momentum operators which are the generators of spacial rotations. The algebra of the root operators of the $SU(3)$ symmetry group is somewhat more complex, but it is again true that physical states form the basis of a representation of the generators of the $SU(3)$ symmetry group and the commutation relations satisfied by the generators of $SU(3)$.

14.9.2 The Poincaré algebra

We begin our study of supersymmetry by trying to find the algebraic relations satisfied by the generators of the transformations of supersymmetry. Any symmetry group that governs the interaction between elementary particle must include the generators of the Lorentz group which forms the basis of relativity theory. The Lorentz group has three generators, J_i, corresponding to rotations about the three coordinate axes and three generators, K_i, corresponding to boosts in the three coordinate directions. The six generators of the Lorentz group satisfy the commutation relations

$$[J_i, J_j] = i\epsilon_{ijk} J_k, \qquad [J_i, K_j] = i\epsilon_{ijk} K_k, \qquad [K_i, K_j] = -i\epsilon_{ijk} J_k \tag{14.117}$$

In order to construct representations of the Lorentz algebra, it is useful to introduce the following linear combinations of the generators, J_i and K_i

$$J_i^\pm = \frac{1}{2}(J_i \pm K_i), \tag{14.118}$$

where the $J_i^\pm$ are hermitian. In terms of $J_i^\pm$ the Lorentz algebra (14.117) becomes

$$[J_i^\pm, J_j^\pm] = i\epsilon_{ijk} J_k^\pm \quad \text{and} \quad [J_i^\pm, J_j^\mp] = 0. \tag{14.119}$$

This shows that the Lorentz algebra is equivalent to two $SU(2)$ algebras.

Because of the isomorphism between the Lorentz group and two $SU(2)$ groups, the states ψ_a and $\overline{\psi}_{\dot{a}}$ that are four vectors that transform according to the Lorentz group can be labeled in terms of the representations of $SU(2)$ as

$$\psi_a = \left(\frac{1}{2}, 0\right) \tag{14.120}$$

$$\overline{\psi}_{\dot{a}} = \left(0, \frac{1}{2}\right), \tag{14.121}$$

where the representation of the group $(SU(2), 0)$ is labeled by a and the representation of the group $(0, SU(2))$ is labeled by $\dot{a}$.

The generators of the Lorentz group can be expressed in terms of the antisymmetric tensor $M_{\mu\nu}$ defined by the equations

$$M_{\mu\nu} = -M_{\nu\mu} \quad \text{with} \quad M_{0i} = K_i \quad \text{and} \quad M_{ij} = \epsilon_{ijk} J_k, \tag{14.122}$$

where $\mu = 0, 1, 2, 3$ and $i, j = 1, 2, 3$. In terms of the $M_{\mu\nu}$ matrices, the Lorentz algebra is

$$[M_{\mu\nu}, M_{\rho\sigma}] = -i\eta_{\mu\rho}M_{\nu\sigma} - i\eta_{\nu\sigma}M_{\mu\rho} + i\eta_{\mu\sigma}M_{\nu\rho} + i\eta_{\nu\rho}M_{\mu\sigma}, \tag{14.123}$$

where η is the flat Minkowski metric that has diagonal elements equal to $+, -, -, -$.

The fundamental group that must be included in any higher symmetry group is the Poincaré group which is the Lorentz group augmented by the space-time translations generated by P_μ. In terms of the generators P_μ and $M_{\mu\nu}$, the Poincaré algebra is

$$[P_\mu, P_\nu] = 0$$
$$[M_{\mu\nu}, M_{\rho\tau}] = -i\eta_{\mu\rho}M_{\nu\sigma} - i\eta_{\nu\sigma}M_{\mu\rho} + i\eta_{\mu\sigma}M_{\nu\rho} + i\eta_{\nu\rho}M_{\mu\sigma} \tag{14.124}$$
$$[M_{\mu\nu}, P_\rho] = -i\eta_{\rho\mu}P_\nu + -i\eta_{\rho\nu}P_\mu$$

Because of the isomorphism between the Lorentz group and two $SU(2)$ groups, the states that transform according to the Lorentz group can be labeled in terms of the representations of $SU(2)$. The Casimir operators of the two $SU(2)$ groups have eigenvalues n(n+1) and m(m+1) with n, m = 0,1/2,.... Hence, we can label the representations of the Lorentz group by pairs (n, m) and identify the spin of the representation as n + m.

14.9.3 The supersymmetry algebra

A considerable effort has been made to include known symmetries of the scattering S-matrix such as charge conjugation (C), parity (P), and time-reversal (T) into the Poincaré group. A no-go theorem proved by Coleman and Mandula shows that under certain reasonable assumptions like locality, causality, and the energy being positive, the only possible symmetries of the S matrix other than C, P, and T are

- Poincaré symmetries with generators P_μ and $M_{\mu\nu}$
- Some internal symmetry group with generators B_l, which are Lorentz scalars and which are typically related to conserved quantum numbers like electric charge or spin

The supersymmetry algebra includes the Poincaré algebra (14.124) together with the commutation relations

$$[P_\mu, Q_a^I] = 0$$
$$[P_\mu, \overline{Q}_{\dot{a}}^I] = 0$$
$$[M_{\mu\nu}, Q_a^I] = i(\sigma_{\mu\nu})_a^b Q_b^I$$
$$[M_{\mu\nu}, \overline{Q}^{I\dot{a}}] = i(\sigma_{\mu\nu})_{\dot{b}}^{\dot{a}}\overline{Q}^{I\dot{b}}$$

and the anticommutation relations

$$\{Q_a^I, \overline{Q}_{\dot{b}}^J\} = 2(\sigma^\mu)_{a\dot{b}}P_\mu\delta^{IJ}$$
$$\{Q_a^I, Q_b^J\} = \epsilon_{ab}Z^{IJ}, \quad Z^{IJ} = -Z^{JI}$$
$$\{\overline{Q}_{\dot{a}'}^I, \overline{Q}_{\dot{b}}^J\} = \epsilon_{\dot{a}\dot{b}}(Z^{IJ})^*$$

Here Q_a^I and $\overline{Q}_{\dot{a}}^I$ are the two spinors appearing previously in Eqs. (14.120) and (14.121). The index I goes from 1 to N where N is the number of supersymmetries. There are different supersymmetry algebra (smaller or larger) depending on the value of N. The $N = 1$ case, which is the only supersymmetry to be considered here, is called minimal supersymmetry. Supersymmetries with $N > 1$ are called extended supersymmetry.

The fact that the generators Q^I of supersymmetry satisfy anticommutation relations in addition to commutation relations allows for their not being excluded by the Coleman-Mandula theorem.

The representations of the supersymmetry algebra are nothing more than the irreducible representations of the Poincare group connected by the Q^I's. Supersymmetry does not give us new physical states but rather a new, more-general way to understand how these states are interconnected. We should also note that the applications of supersymmetry reach beyond the Poincaré group and can be applied to general relativity as well – with the theories being referred to generically as *supergravity*.

Suggestions for further reading

B.R. Martin and G. Shaw, *Particle Physics*, Second Edition (Chichester, England: Wiley, 1997).

Donald H. Perkins, *Introduction to High Energy Physics*, Fourth Edition (Menlo Park, California: Benjamin/Cummings, 1995).

I.J.R. Aitchison and A.J.G. Hey, *Gauge Theories in Particle Physics*, Volumes I and II, Third Edition (Bristol and Philadelphia: Institute of Physics, 2004).

Howard Georgi, *Lie Algebras in Particle Physics: From Isospin to Unified Theories*, Second Edition (Boulder, Colorado: Westview Press, 1999).

David Griffiths, *Introduction to Quantum Mechanics*, Second Edition (New Jersey: Benjamin Cummings, 1994).

Kurt Weisskopf and Victor Gottfried, *Concepts of Particle Physics*, Volume 1 (New York: Oxford University Press, 1986).

Gordan Kane, *Modern Elementary Particle Physics: The Fundamental Particles and Forces* (Menlo Park, California: Addison-Wesley, 1993).

Matteo Bertolinni, *Lectures on Supersymmetry* (Trieste, Italy: Sissa (International School of Advanced Studies)).

Basic equations

Leptons and quarks

Generations of leptons

$$\begin{bmatrix} \nu_e \\ e^- \end{bmatrix}, \qquad \begin{bmatrix} \nu_\mu \\ \mu^- \end{bmatrix}, \qquad \begin{bmatrix} \nu_\tau \\ \tau^- \end{bmatrix}$$

Generations of quarks

$$\begin{bmatrix} u \\ d \end{bmatrix}, \qquad \begin{bmatrix} c \\ s \end{bmatrix}, \qquad \begin{bmatrix} t \\ b \end{bmatrix}$$

Definition of hypercharge and isospin

Hypercharge

$$Y = B + S + C + \tilde{B} + T,$$

where B, S, C, $\tilde{B}$, and T are the baryon number, strangeness, charm, beauty, and truth, respectively.

Isospin

$$I_3 = Q - Y/2$$

For every isospin multiplet, I is the maximum value of I_3.

Feynman diagrams

Definition of flux

$$\phi_i = n_i v_i$$

Golden rule

$$W = \frac{2\pi}{\hbar} |M_{if}|^2 \rho_f$$

SU(3) symmetry

Diagonal generators

$$[H_1, H_2] = 0$$

Equations defining roots

$$[\mathbf{H}_1, \mathbf{E}_{\alpha_1}] = [\alpha_1]_1 \, \mathbf{E}_{\alpha_1},$$
$$[\mathbf{H}_2, \mathbf{E}_{\alpha_1}] = [\alpha_1]_2 \, \mathbf{E}_{\alpha_1}$$

Equations defining weights

$$H_i \mathbf{v}_i = w(v_i) \mathbf{v}_i$$

Summary

All matter is composed of leptons, quarks, and elementary particles called bosons, which serve as the carriers of the force between particles. The lepton family includes the electron, muon, tau, and their neutrinos. Quarks come in six types, called flavors, denoted by up (u), down (d), strange (s), charmed (c), bottom (b), and top (t) quarks. All interactions conserve the lepton numbers and the number of baryons (B). The weak interaction can change the flavor of a quark and thus violate the conservation laws associated with the quark quantum numbers (strangeness, charm, bottom, and top). The strongly interacting particles called hadrons occur as mesons, which are composed of quark/antiquark pairs, and baryons, which consist of three quarks.

The strongly interacting particles occur in isospin multiplets with all members of a multiplet having approximately the same mass. The members of an isospin multiplet all have the same value of the hypercharge Y, which is the sum of B, S, C, $\tilde{B}$, and T. The members of each multiplet are distinguished by the charge Q and I_3. Particles can also be grouped in larger families of particles called supermultiplets. The lightest mesons consist of two nonets with each consisting of nine mesons, while the lightest baryons consist of an octet with eight baryons and a decuplet with ten baryons.

The interactions between leptons and quarks are described by Feynman diagrams. These diagrams describe processes by which particles interact by exchanging quanta of the interaction fields. Recent work on gauge symmetries provide a general framework for understanding the interactions between elementary particles. There are three gauge symmetries associated with the electromagnetic, weak, and strong interactions. Associated with each gauge symmetry is a gauge field and gauge bosons which serve as carriers of the interactions. The photon is the gauge particle associated with the electromagnetic interaction, while the W^+, W^-, and Z^0 are the gauge particles associated with the weak interaction, and gluons serve as the gauge particles of the strong interaction. It has been conjectured that the states of Fermions with one half spin and the states of bosons with integral spin together form the basis for a more general group of symmetry operations called supersymmetry that transforms Fermions into bosons and vice versa.

Questions

1. Give the names of the particles corresponding to the first and second generations of leptons.
2. Give the names of the particles corresponding to the first and second generations of quarks.
3. Which family of particles interacts by means of the weak and electromagnetic forces but not by means of the strong force?
4. What evidence is there for the quark model?
5. Which particles serve as the carriers of the weak force?
6. Suppose that two particles produced in a scattering experiment have energy and momentum, E_1, $\mathbf{p}_1$ and E_2, $\mathbf{p}_2$, respectively. Write down a formula for the invariant mass of the two particles.
7. Which of the mesons we have encountered have a single strange quark or a single strange antiquark?
8. Which of the baryons we have encountered have two strange quarks?
9. The μ^- decays into e^- and two other particles. Using the conservation of electron and muon lepton numbers determine the identity of the other particles.
10. The μ^+ decays into e^+ and two other particles. Using the conservation of electron and muon lepton numbers determine the identity of the other particles.
11. Which conservation law is violated by the decay process $n \rightarrow \pi^+ + \pi^-$?
12. What quantum number can be assigned to particles that are produced by the strong interaction but decay by the weak interaction?
13. How is the spin of a composite particle defined?
14. What property of the decay of cobalt-60 nuclei by the reaction $^{60}Co \rightarrow \,^{60}Ni^* + e^- + \overline{\nu_e}$ showed that parity was violated?

15. What does it mean to say that the μ^+ has negative helicity?
16. How is the hypercharge Y related to the baryon number (B), strangeness (S), charm (C), beauty ($\tilde{B}$), and truth (T)?
17. Write down an equation relating I_3 to Q and Y.
18. What are the possible values of the isospin I of the baryon resonances produced in pion-nucleon scattering?
19. What are the possible values of the isospin I of the meson resonances produced in pion-nucleon scattering?
20. Draw a Feynman diagram for Compton scattering.
21. Draw a Feynman diagram for μ^- decay.
22. Draw an acceptable Feynman vertex involving a lepton l^- and the Z^0.
23. Draw an acceptable Feynman vertex involving quarks of the second generation and the W^-.
24. Draw a Feynman diagram in which a $\bar{u}/u$ pair is absorbed giving a gluon that produces a $\bar{s}/s$ pair.
25. Draw the root diagram for $SU(3)$.
26. Draw a diagram showing the weights of the fundamental representation of $SU(3)$.
27. Make a drawing showing how the root α_2 and the weights $w(\mathbf{v_2})$ and $w(\mathbf{v_3})$ of the fundamental representation of $SU(3)$ are related.
28. What causes the gauge symmetry of the interaction of the Higgs to be spontaneously broken?

Problems

1. State whether each of the following processes can occur. If the reaction cannot occur or if it can only occur by the weak interaction, state which conservation law is violated.

   (a) $\quad p \rightarrow e^+ + \gamma$
   (b) $\quad \Sigma^0 \rightarrow \Lambda + \pi^0$
   (c) $\quad \Delta^+ \rightarrow p + \pi^0$
   (d) $\quad \pi^+ \rightarrow \mu^+ + \gamma$
   (e) $\quad K^- \rightarrow \pi^- + \pi^0$
   (f) $\quad \rho^0 \rightarrow \pi^0 + \pi^0$
   (g) $\quad K^+ \rightarrow \pi^+ + \gamma$

2. State whether each of the following processes can occur. If the reaction cannot occur or if it can only occur by the weak interaction, state which conservation law is violated.

   (a) $\quad p + K^- \rightarrow \Sigma^+ + \pi^- + \pi^+ + \pi^- + \pi^0$
   (b) $\quad e^+ + e^- \rightarrow v + \bar{v}$ (give neutrino type)
   (c) $\quad v_\mu + p \rightarrow \mu^+ + n$
   (d) $\quad v_\mu + p \rightarrow \mu^- + n + \pi^+$
   (e) $\quad v_\mu + e^- \rightarrow \mu^- + v_e$
   (f) $\quad v_e + p \rightarrow e^+ + \Sigma^0 + K^0$
   (g) $\quad K^0 + p \rightarrow \Lambda + \pi^+ + K^- + \pi^+$

3. Find the total rest mass energy of the initial and final states of each of the following reactions. For which of the following reactions must the kinetic energy of the initial state be greater than the kinetic energy of the final state?

   (a) $\quad \pi^- + p \rightarrow n + \pi^0$
   (b) $\quad \Lambda \rightarrow p + \pi^0$
   (c) $\quad \pi^+ + p \rightarrow K^+ + \Sigma^+$

4. For which of the following reactions is the conservation law of electron lepton number violated? Explain any violation of the laws that occur.

$$(a) \qquad \mu^- \to e^- + \overline{\nu_e} + \nu_\mu$$
$$(b) \qquad \nu_e + p \to n + e^+$$
$$(c) \qquad n \to p + e^- + \nu_e$$
$$(d) \qquad p \to n + e^+ + \nu_e$$

5. Determine the change in strangeness for each of the following reactions, and state whether the reaction can occur by the strong, electromagnetic, or weak interactions. Give reasons for your answers.

$$(a) \qquad \Omega^- \to \Lambda + K^-$$
$$(b) \qquad \Sigma^0 \to \Lambda + \gamma$$
$$(c) \qquad \Lambda \to p + \pi^-$$

6. Determine the change in strangeness for each of the following reactions, and state whether the reaction can occur by the strong, electromagnetic, or weak interactions. Give reasons for your answers.

$$(a) \qquad \Xi^0 \to \Lambda + \pi^0$$
$$(b) \qquad \Sigma^- \to \Lambda + e^- + \overline{\nu_e}$$
$$(c) \qquad K^- \to \pi^- + \pi^0$$

7. For each of the following decay processes, state which are strictly forbidden, which are weak, electromagnetic, or strong. Give reasons for your answers.

$$(a) \qquad p + \overline{p} \to \pi^+ + \pi^- + \pi^0 + \pi^+ + \pi^-$$
$$(b) \qquad \pi^- + p \to p + K^-$$
$$(c) \qquad \pi^- + p \to \Lambda + \overline{\Sigma^0}$$
$$(d) \qquad \overline{\nu_\mu} + p \to e^+ + n$$
$$(e) \qquad \nu_e + p \to e^+ + \Lambda + K^0$$

8. Give the quark composition of the particles in the following processes and describe how the quark composition of the final state differs from the quark composition of the initial state.

$$(a) \qquad \Lambda \to p + \pi^0$$
$$(b) \qquad \pi^- + p \to n + \pi^0$$
$$(c) \qquad \pi^+ + p \to K^+ + \Sigma^+$$

9. Give the quark composition of the particles in the following processes and describe how the quark composition of the final state differs from the quark composition of the initial state.

$$(a) \qquad \Omega^- \to \Lambda + K^-$$
$$(b) \qquad \pi^- + p \to \Lambda + K^0$$
$$(c) \qquad p + K^- \to \Xi^- + K^+$$

10. For each of the following processes, give the isospin quantum numbers of the particles involved in the collision and the quantum number I_3 of the total isospin. Give the possible values of I and I_3 for the final state and give an example

of what the final state might be.

$$(a) \quad \pi^- + p$$
$$(b) \quad \pi^- + n$$
$$(c) \quad \pi^+ + p$$
$$(d) \quad n + n$$
$$(e) \quad n + p$$

11. Give the quark composition of the particles in the following processes and draw the Feynman diagram giving the cosine or sine of the Cabibbo angle where appropriate.

$$(a) \quad \Lambda \rightarrow p + p^0$$
$$(b) \quad \pi^+ + p \rightarrow K^+ + e^- + \sigma^+$$
$$(c) \quad \Xi^0 - \rightarrow \Sigma^+ + e^+ + \bar{\nu}_e$$
$$(d) \quad \Omega^- \rightarrow \Xi^0 + e^- + \bar{\nu}_e$$

12. Give the weights of the $(3, 0)$ representation of $SU(3)$ showing how each of the weights can be obtained from the highest weight by a translation by one or more of the root vectors.

Chapter 15

Nuclear physics

Contents

15.1 Properties of nuclei	393	Basic equations	416
15.2 Decay processes	401	Summary	417
15.3 The nuclear shell model	408	Questions	417
15.4 Excited states of nuclei	412	Problems	417
Suggestions for further reading	416		

The people in particle physics have been preoccupied lately with the electroweak theory. The work of understanding in detail the nature of the strong force has fallen to nuclear physicists.

Malcolm MacFarlane

The histories of elementary particle physics and nuclear physics are intertwined. The rays emitted by radioactive substances were studied by many physicists in the early decades of the twentieth century. Using the first three letters of the Greek alphabet, Rutherford classified the different forms of radiation by the letters, alpha (α), beta (β), and gamma (γ). We now know that α-radiation consists of the nuclei of the most common isotope of helium with two protons and two neutrons. β-radiation consists of electrons and positrons, while γ-rays are a very penetrating form of electromagnetic radiation having a very short wavelength.

Scattering experiments designed by Rutherford in 1911 showed that each atom has a small nucleus at its center containing most of the atomic mass. Rutherford's discovery led Bohr, who worked in Rutherford's laboratory, to formulate his model of the atom. Important discoveries made since the pioneering work of Rutherford and Bohr include the discovery of the neutron by J. Chadwick, the discovery of the positron by C. Anderson, and the first scattering experiments using accelerated beams of protons by J. Cockcoft and E. Watson. These three developments, all of which occurred in 1932, were important milestones in the effort to understand the atomic nucleus.

We begin this chapter by describing the composition of the atomic nucleus and the physical properties of nuclei such as their size and binding energy. Ensuing sections will be devoted to radioactive decay processes and to the nuclear shell model.

15.1 Properties of nuclei

As discussed in the Introduction to this book, the number of protons in a nucleus is referred to as the *atomic number* (Z), while the total number of protons and neutrons is referred to as the *atomic mass number* (A). We shall use the generic term *nucleon* to refer to a particle which may be either a proton or a neutron. For a particular element, the different atomic species having differing numbers of neutrons in their nucleus are referred to as isotopes of the element. The atomic mass number of an isotope is indicated by a superscript, while the atomic number is indicated by a subscript. Consider as an example the carbon isotope, $_6^{13}C$. The atomic number of this isotope is six indicating that it has six protons, while the atomic mass number is thirteen indicating that the nucleus has thirteen nucleons and, hence, seven neutrons. In this notation, the α-particle being a helium nucleus with two protons and two neutrons is denoted by $_2^4He$. Table 15.1 gives some properties of a number of naturally occurring isotopes.

This table gives the atomic number (Z), atomic mass number (A), atomic mass in atomic mass units (u), and nuclear spin (I) of each isotope. The atomic mass of the most common occurring isotope of carbon ($_6^{12}C$) is by definition equal to 12.0. The natural abundance of each isotope is given in the last column of Table 15.1.

Modern Physics with Modern Computational Methods. https://doi.org/10.1016/B978-0-12-817790-7.00022-6

TABLE 15.1 Some of the properties of a number of light isotopes.

Name	Z	A	Atomic mass (u)	I	Natural abundance
H	1	1	1.007825	1/2	99.989%
		2	2.014102	1	0.011%
He	2	3	3.016029	1/2	1.37×10^{-4}%
		4	4.002603	0	99.99986%
Li	3	6	6.015122	1	7.59%
		7	7.016004	3/2	92.41%
Be	4	9	9.012182	3/2	100%
B	5	10	10.012937	3	19.9%
		11	11.009306	3/2	80.1%
C	6	12	12.000000	0	98.93%
		13	13.003355	1/2	1.07%
N	7	14	14.003074	1	99.632%
		15	15.000109	1/2	0.368%
O	8	16	15.994915	0	99.757%
		17	16.999132	5/2	0.038%
		18	17.999160	0	0.205%

15.1.1 Nuclear sizes

Precise information about the structure of nuclei can be obtained from high-energy electron scattering experiments. There is an obvious advantage in using charged leptons to probe nuclei since leptons interact mainly by the electromagnetic force, which is well understood. The wavelength associated with an electron is related to the momentum of the electron by the de Broglie relation, $\lambda = h/p$. For electrons to probe the inner structure of nuclei, the wavelength of the incident electrons must be comparable to the size of the nucleus.

Example 15.1

Find the kinetic energy of an electron with a de Broglie wavelength of 10 fm.

Solution

Using Eq. (13.14) of Chapter 13 and the de Broglie relation, the energy and the wavelength of an electron are seen to be related by the equation

$$E^2 = \left(\frac{hc}{\lambda}\right)^2 + m^2 c^4.$$

As we have discussed in the Introduction, the product of constants hc is equal to 1240 MeV · fm. Substituting this value of hc and $\lambda = 10$ fm into the above equation, we obtain

$$E = \sqrt{(124 \text{ MeV})^2 + (0.511 \text{ MeV})^2} \approx 124 \text{ MeV}.$$

The kinetic energy of an electron is equal to the difference between its energy and its rest energy of the electron

$$K.E. = 124 \text{ MeV} - 0.511 \text{ MeV} = 123.5 \text{ MeV}.$$

Incident electrons must thus have kinetic energies of a few hundred million electron volts to provide detailed information about the structure of nuclei.

The results of electron scattering experiments are usually described by giving the differential cross-section as a function of the scattering angle. The differential cross section is the number of particles scattered in a particular angle divided by the flux of the incident beam. The differential cross-section for the elastic scattering of electrons from gold $^{197}_{79}Au$ is shown in Fig. 15.1. In this figure, the scattering to be expected if the gold nucleus had a point structure is represented by a dashed line.

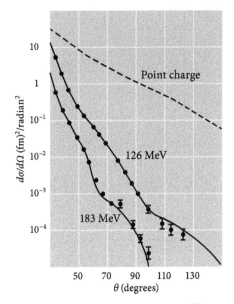

FIGURE 15.1 The differential cross-section for the elastic scattering of electrons from gold $^{197}_{79}Au$.

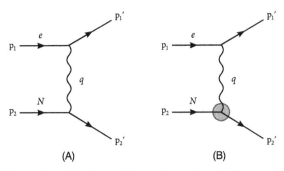

FIGURE 15.2 (A) Feynman diagram for the scattering of an electron from a point nucleus, (B) Feynman diagram with the form factor of the nucleus represented by a circle around the nuclear vertex.

As one would expect, the point nucleus, which is *harder* than an extended charge distribution, produces a larger scattering cross section for all energies. The differences between the scattering cross-sections for the extended charge distribution and a point charge becomes larger as the scattering angle increases.

The scattering of an electron from a point nucleus is described by the Feynman diagram shown in Fig. 15.2(A). In this diagram, the lower external line corresponds to the nucleus, and the internal line denoted by **q** represents the momentum transferred from the incident electron to the nucleus. The effect of the charge distribution of the nucleus can be included in the Feynman diagram by adding a function $F(\mathbf{q})$ of the momentum transfer to the nuclear vertex. This function, which is called the *nuclear form factor*, is the Fourier transform of the nuclear charge distribution. In Fig. 15.2(B), the form factor of the nucleus is represented by a circle around the nuclear vertex. Since the scattering amplitude is proportional to the square of the Feynman amplitude, the scattering cross section for electrons scattered from an extended nucleus will be equal to $|F(\mathbf{q})|^2$ times the cross section of a point nucleus. The absolute value squared of the nuclear form factor is thus equal to the ratio of the scattering cross sections of the extended and point nuclei. Once the form factor of the nucleus is obtained from the scattering data, the charge distribution of the nucleus $\rho_{ch}(\mathbf{r})$ can be obtained by taking the inverse Fourier transformation.

Nuclei having only one or two nucleons or one or two holes outside closed shells have been found to be approximately spherically symmetric. The charge density, which for such nuclei only depends upon the distance from the center of the nucleus, is generally constant for some distance and then falls off very quickly. A charge distribution of this kind can be described by the function

$$\rho_{ch}(r) = \frac{\rho_{ch}^0}{1 + e^{-(r-R)/a}} , \tag{15.1}$$

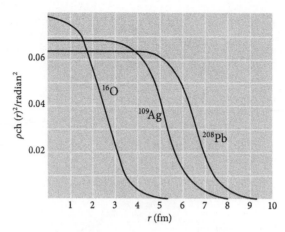

FIGURE 15.3 The charge distributions of the light (^{16}O), medium (^{109}Ag), and heavy (^{208}Pb) nuclei. (Data from Hofstadter, R., 1963.)

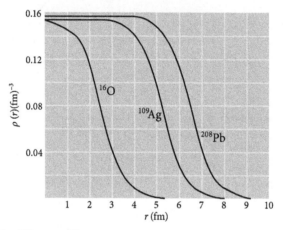

FIGURE 15.4 The nucleon densities of ^{16}O, ^{109}Ag, and ^{208}Pb.

where R and a may be regarded as adjustable parameters, and ρ_{ch}^0 is a normalization constant determined by the condition

$$\int \rho_{ch}(r)dV = 4\pi \int \rho_{ch}(r)r^2 dr = Z.$$

By fitting the values of R and a in the empirical formula (15.1) to electron scattering data, R. Barrett and R. Jackson obtained the charge distributions of the light (^{16}O), medium (^{109}Ag), and heavy (^{208}Pb) nuclei shown in Fig. 15.3. The charge densities of these three nuclei all have the same general form with an approximately level inner portion and a thin shell region where the charge falls off exponentially to zero. The inner region of the charge distributions of oxygen and silver are higher than the central portion of the charge distribution of lead.

Some indication of how protons are distributed within a complex nucleus can be inferred from the charge distribution. If the protons within a nucleus were point particles, the density of protons ρ_p would be related to the charge density ρ_{ch} by the formula, $\rho_{ch} = e\rho_p$. To the extent that the isospin symmetry holds, and protons and neutrons are equivalent, the density of nucleons within a nucleus would be related to the charge density by the formula

$$\rho(r) = (A/eZ)\rho_{ch}(r). \tag{15.2}$$

We note that the factor, A/eZ in this formula has the effect of lowering the inner region of the density of light nuclei in relation to the density of heavy nuclei which have relatively higher numbers of neutrons. The nuclear densities of ^{16}O, ^{109}Ag, and ^{208}Pb are shown in Fig. 15.4. These curves show that at the center of a nucleus the density of nuclear matter is roughly the same for all nuclei. It increases with A, but appears to approach a limiting value ρ_0 of about 0.17 nucleons per fm^3 for large A.

The existence of a limiting value of the nuclear density ρ_0 for large A is an important result. Using this idea, we can obtain an approximate relationship between the atomic mass number A and the nuclear radius R. We set the product of the volume of a sphere of radius R and the nuclear density ρ_0 equal to the atomic mass number A to obtain

$$\left(\frac{4\pi}{3}\right) R^3 \rho_0 = A.$$

Solving this equation for R and using the fact that ρ_0 is equal to 0.17 fm^{-3}, we obtain

$$R = 1.12 A^{1/3} \text{ fm}. \tag{15.3}$$

This formula will be used in later sections to estimate the radius of nuclei having particular values of A.

15.1.2 Binding energies

The protons and neutrons in the nucleus can only be separated by working against the strong attractive forces holding them together. The binding energy is the amount of work that would be needed to pull the protons and neutrons in the nucleus entirely apart. We can calculate the binding energy $B(N, Z)$ of a nucleus $^A_Z X$ with Z protons and N neutrons by finding the difference between the total rest energy of the constituent protons and neutrons and the nucleus itself

$$B(N, Z) = \left[Zm_p + Nm_n - m_{\text{nuc}}(N, Z)\right]c^2. \tag{15.4}$$

The quantity within square brackets in this equation is the mass which would be lost if the nucleus were to be assembled from its constituents. In Eq. (15.4), the mass loss is converted into a binding energy by multiplying it by c^2 according to Einstein's formula, $E = mc^2$. The amount of energy corresponding to a single atomic mass unit is 931.5 MeV. Notice that the proton and the neutron, which have masses slightly larger than one atomic mass unit, have rest energies mc^2 equal to 938.3 MeV and 939.6 MeV, respectively.

The masses of atoms are measured experimentally rather than the masses of the bare nuclei. Since the binding energy of the electrons in an atom are very much smaller than the binding energy of the nucleus, we can find the nuclear binding energy by calculating the difference in mass of the atomic constituents. In place of Eq. (15.4), we thus write

$$B(N, Z) = \left[Zm(^1_1H) + Nm_n - m(N, Z)\right]c^2, \tag{15.5}$$

where $m(^1_1H)$ is the mass of a hydrogen atom, m_n is the mass of a neutron, and $m(N, Z)$ is the mass of the atomic isotope $^A_Z X$. Since the atom $^A_Z X$ has Z electrons, the rest energy of the electrons in Z hydrogen atoms is equal to the electronic contribution to $m(N, Z)c^2$.

Example 15.2

Using the atomic masses of the ^1_1H and ^4_2He isotopes given in Table 15.1 and the mass of the neutron given in Appendix A, calculate the binding energy of the ^4_2He nucleus.

Solution

The reaction in which two hydrogen atoms combine with two neutrons to form the ^4_2He atom is

$$2^1_1H + 2n \rightarrow ^4_2He.$$

The terms on each side of this equation contain two protons, two neutrons, and two electrons. To be concrete, we write the mass under each term in the above reaction to obtain

$$2^1_1H \quad + \quad 2n \quad \rightarrow \quad ^4_2He$$

$$2(1.007825)\,u \qquad 2(1.008665)\,u \qquad 4.002602\,u$$

The mass lost when two ^1_1H atoms combine with two neutrons to form the ^4_2He isotope is

$$\Delta m = 2(1.007825)\,u + 2(1.008665)\,u - 4.002602\,u = 0.030378\,u.$$

The binding energy of the helium nucleus is obtained by multiplying this mass loss by 931.5 MeV. We obtain

$$B(2, 2) = 28.297 \text{ MeV.}$$

The binding energies of some light nuclei are given in Table 15.2.

TABLE 15.2 Binding energies of some light nuclei.

Nucleus	Binding energy (MeV)	Binding energy of last nucleon (MeV)	Binding energy per nucleon (MeV)
^2_1H	2.22	2.2	1.1
^3_1H	8.48	6.3	2.8
^4_2He	28.30	19.8	7.1
^5_2He	27.34	−1.0	5.5
^6_3Li	31.99	4.7	5.3
^7_3Li	39.25	7.3	5.6
^8_4Be	56.50	17.3	7.1
^9_4Be	58.16	1.7	6.5
$^{10}_5B$	64.75	6.6	6.5
$^{11}_5B$	76.21	11.5	6.9
$^{12}_6C$	92.16	16.0	7.7
$^{13}_6C$	97.11	5.0	7.5
$^{14}_7N$	104.66	7.6	7.5
$^{15}_7N$	115.49	10.8	7.7
$^{16}_8O$	127.62	12.1	8.0
$^{17}_8O$	131.76	4.1	7.8

The binding energies of all of the nuclei in Table 15.2 can be calculated using the method we have employed in Example 15.2, and the binding energy per nucleon can then be calculated by dividing the total binding energy of the nucleus by the number of nucleons. We note that the binding energy per nucleon of some nuclei is larger than for the nuclei around them. This is true of the ^4_2He, $^{12}_6C$, and $^{16}_8O$ nuclei. As we shall find in a later section, the great stability of these nuclei can be understood by considering their shell structure.

One can see from Table 15.2 that the binding energy of the ^8_4Be nuclei is 0.1 MeV less than the binding energy of two ^4_2He nuclei. So, the ^8_4Be nuclei is unstable and eventually decays into two ^4_2He nuclei.

The binding energy of the last nucleon, which is given in the third column of Table 15.2, is calculated by taking the difference of the binding energy of a particular nucleus with an isotope having one fewer nucleon. We note that the binding energy of the last nucleon of the ^5_2He nucleus, which is obtained by taking the difference of the binding energies of ^5_2He and ^4_2He, is negative. The unstable ^5_2He nucleus decays into a neutron and an α-particle (^4_2He). Except for ^8_4Be mentioned before, all of the other nuclei in Table 15.2 are stable.

15.1.3 The semiempirical mass formula

All nuclei have a shell structure, which contributes to their stability and influences the kinds of decay processes that can occur. The effects of the shell structure are superimposed on a slowly varying binding energy per nucleon. When we studied the spatial distribution of nuclear matter within the nucleus, we found that nuclei all have an inner region where the density is approximately uniform and a thin surface region where the distribution of nuclear matter falls off exponentially to zero. The same could be said of the drops of a liquid. The validity of thinking of nuclei as drops of a liquid is made more precise by giving an empirical formula for the binding energy of nuclei. With a few parameters, this formula fits the binding energy of all but the lightest nuclei to a high degree of accuracy. Following W.N. Cottingham and D.A. Greenwood whose book is cited at the end of this chapter, we give the following version of the formula for the binding energy

$$B(N, Z) = aA - bA^{2/3} - \frac{dZ^2}{A^{1/3}} - s\frac{(N-Z)^2}{A} - \frac{\delta}{A^{1/2}}, \tag{15.6}$$

where A is the number of nucleons, N is the number of neutrons, and Z is the number of protons. The parameters a, b, d, s, and δ can be obtained by fitting the formula to the measured binding energies. The values of these parameters given by *Handbuch der Physik*, XXXVIII/1 are

$$\begin{aligned} a &= 15.835 \text{ MeV} \\ b &= 18.33 \text{ MeV} \\ d &= 0.714 \text{ MeV} \\ s &= 23.20 \text{ MeV} \end{aligned} \tag{15.7}$$

and

$$\delta = \begin{cases} +11.2 \text{ MeV}, & \text{odd-odd nuclei } (Z \text{ odd, } N \text{ odd}) \\ 0, & \text{even-odd nuclei } (Z \text{ even, } N \text{ odd or } N \text{ even, } Z \text{ odd}) \\ -11.2 \text{ MeV}, & \text{for even-even nuclei } (Z \text{ even, } N \text{ even}). \end{cases} \tag{15.8}$$

If nuclear matter were entirely homogeneous, the number of nucleons in a nucleus would be proportional to the volume of the nucleus. In the analogy between nuclei and liquid drops, it is the atomic mass number (A) which is the analogue of the volume of the liquid. The term aA in Eq. (15.6) depends upon A in the same way that the cohesive energy of a fluid depends upon the volume of the fluid. Since the surface area of a sphere depends upon the radius squared and the volume of a sphere depends upon the radius raised to the third power, the surface area of a sphere depends upon the volume of the sphere raised to the two-thirds power. Hence, the term $bA^{2/3}$ is analogous to the surface energy of a liquid sphere. The surface tension of a liquid keeps drops spherical when – as for the nucleus – there is no long-range attractive force like gravity. Protons and neutrons on the surface of a nucleus are bound by fewer particles to the collective than protons and neutrons in the interior of the nucleus. This effect and the fact that the surface tension plays an important role in the breakup of a drop when it is distorted helps us understand why the surface term is negative.

The term, $-dZ^2/A^{1/3}$, also has a simple explanation. It represents the electrostatic Coulomb repulsion between the positive charged protons in the nucleus. We have already mentioned that heavy nuclei tend to have many more neutrons than protons. The reason for this is that the cumulative effect of the Coulomb repulsion among protons makes heavy nuclei having comparable numbers of protons and neutrons unstable. We can easily estimate the effect of Coulomb repulsion by using the expression for the nuclear radius obtained previously. The electrostatic energy of a uniform sphere of charge eZ and radius R is

$$E_{\text{Coul}} = \frac{3}{5}\frac{(eZ)^2}{(4\pi\epsilon_0)R}.$$

Substituting into this equation the approximate expression for the nuclear radius R given by Eq. (15.3), we find that E_{Coul} is of the form, $dZ^2/A^{1/3}$, and we obtain the approximate value of the constant $d = 0.77$ MeV, which is fairly close to the empirical value, 0.714 MeV, given above.

While the Coulomb term, $-dZ^2/A^{1/3}$, discourages the formation of states with high numbers of protons, the term, $-s(N-Z)^2/A$ discourages the formation of states having unequal numbers of protons and neutrons. This can be described as a statistical effect that depends upon the properties of identical particles. Due to the Pauli exclusion principle, which prohibits two nucleons from being in the same quantum state, the average neutron-proton attraction in a nucleus is greater than the average proton-proton and the neutron-neutron attractions. States having nearly equal values of Z and N have comparatively higher numbers of proton-neutron pairs and are more stable. The factor of A in the denominator of the term, $-s(N-Z)^2/A$, ensures that this term depends linearly with upon A for a fixed ratio of neutrons to protons.

The final term in the semiempirical formula (15.6) describes a pairing effect that can be important for light nuclei. The pairing term makes even-even nuclei more stable than their odd-odd counterparts with the same A.

Fig. 15.5 shows the binding energy per nucleon for stable nuclei. In this figure, the smooth curve gives the value of B/A obtained from the semiempirical formula (15.6), while the dots represent experimental data. As we shall discuss in the next section, nuclei can decay in a number of different ways. They can emit a compact α-particle, which is a $^4_2 He$

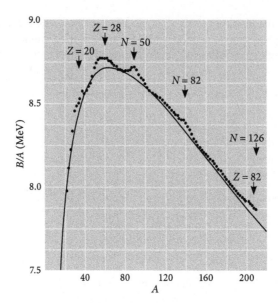

FIGURE 15.5 The binding energy per nucleon for stable nuclei. (From W.N. Cottingham and D.A. Greenwood, 2001.)

nucleus. They can emit a beta particle, which is an electron or a positron, or they can emit a gamma ray. In drawing Fig. 15.5, we have chosen, for each value of Z, the nucleus which is the most stable with respect to β-decay. Notice that there are some discrepancies between the mass formula and the experimental data. These differences are primarily due to the nuclear shell structure which will be discussed in a later section. More sophisticated versions of the formula (15.6) include terms describing shell effects; however, these additional terms are relatively unimportant for nuclei heavier than neon with $Z = 10$.

Of the various contributions to the binding energy, only the bulk term is positive with the negative surface and Coulomb contributions giving the largest reductions in the bulk value. The initial rise of B/A with A may be attributed to the fact the magnitude of the surface contribution to the binding energy decreases relative to the bulk contribution as the size of the nucleus increases. As A and therefore Z increase further, though, the Coulomb term becomes more important producing a maximum on the curve.

A formula for the mass of an atom can be obtained by substituting the expression for the binding energy $B(N, Z)$ given by Eq. (15.6) into Eq. (15.5) and then solving for $m(Z, N)$. We thus obtain the following empirical formula for the mass $m(N, Z)$ of a neutral atom with atomic mass number A, Z protons, and N neutrons

$$m(N, Z)c^2 = [Zm(_1^1 H) + Nm_n]c^2 - aA + bA^{2/3} + \frac{dZ^2}{A^{1/3}} + s\frac{(N - Z)^2}{A} + \frac{\delta}{A^{1/2}} . \tag{15.9}$$

This formula is called the *semiempirical mass formula*.

Example 15.3

Using the semiempirical formulas, calculate the binding energy per nucleon and mass of the isotopes $_{26}^{56}Fe$ and $_{82}^{208}Pb$.

Solution

For $_{26}^{56}Fe$, $A = 56$, $Z = 26$, and $N = 30$. Substituting these values into Eq. (15.6) with the values of the parameters given by Eq. (15.7), we obtain the binding energy $B(N, Z) = 487.1696$; MeV. The binding energy per nucleon is $B(N, Z)/56 = 8.699$ MeV.

The mass of $_{26}^{56}Fe$ can then be calculated using Eq. (15.5). We obtain

$$m(N, Z) = Zm(_1^1 H) + Nm_n - B(N, Z)/c^2 = 26 \times 1.007825 + 30 \times 1.008665 - \frac{487.1696 \text{ MeV}}{931.5 \text{ MeV}/u} = 55.9404 \, u. \tag{15.10}$$

This value agrees fairly well with the experimental value, 55.9349 MeV.

In exactly the same way, the binding energy of $_{82}^{208}Pb$ is found to be 1624.7481 MeV or 7.811 MeV per nucleon. The mass of the $_{82}^{208}Pb$ isotope obtained by using the calculated value of the binding energy together with Eq. (15.5) is 207.9892123, which agrees well with the experimental value of 207.976627. Notice that the binding energy per nucleon we have obtained for $^{208}_{82}Pb$ is *less* than the binding energy per nucleon of $^{56}_{26}Fe$. The isotope of iron with $A = 56$ is the most stable isotope.

The importance of the empirical formulas for binding energy and mass is not that they enable us to predict new or exotic nuclear phenomena, but rather that they enable us to understand the properties of nuclei in simple physical terms. The empirical models give us some insight as to which nuclear species should be stable and what decay processes are likely to occur.

15.2   Decay processes

In addition to the stable nuclei existing in nature, there are numerous other unstable nuclei which decay emitting radiation. Radioactive isotopes can be found in metal ores, and many other radioactive isotopes have been produced in modern accelerators. Like the cell mutations, which lead to cancer in our own bodies, the process by which the unstable nuclei of radioactive isotopes decay is random in nature. The number of nuclei decaying in a time interval is proportional to the time interval dt and to the number of nuclei present N. Denoting the number of nuclei that decay by dN, we may thus write

$$dN = -\lambda N \, dt, \tag{15.11}$$

where the proportionality constant, λ is called the *decay constant*. The negative sign that occurs in Eq. (15.11) is due to the fact that the number of unstable nuclei decreases with time. The significance of the proportionality constant λ can be found by rewriting Eq. (15.11) in the following form

$$\lambda = -\frac{dN/N}{dt} .$$

We thus see that λ is equal to the fractional number or probability that a nucleus decays per time.

The solution of Eq. (15.11) is

$$N(t) = N_0 e^{-\lambda t} , \tag{15.12}$$

where N_0 is the number of nuclei at time $t = 0$. The decay rate can be obtained by taking the derivative of this last equation giving

$$R = -\frac{dN}{dt} = \lambda N_0 e^{-\lambda t} .$$

Notice that both the number of nuclei and the decay rate decay exponentially.

The decay of a radioactive isotope is often described in terms of its *half-life* ($t_{1/2}$), which is defined to be the length of time required for the number of radioactive nuclei to decrease to half its original value. Using Eq. (15.12), we obtain the following equation for the half-life

$$\frac{1}{2} = \frac{N}{N_0} = e^{-\lambda t_{1/2}} .$$

This last equation can be written

$$e^{\lambda t_{1/2}} = 2.$$

Taking the natural logarithm of both sides of the above equation and solving for $t_{1/2}$, we obtain

$$t_{1/2} = \frac{\ln 2}{\lambda} . \tag{15.13}$$

Example 15.4

The isotope $^{239}_{94}Pu$ has a half-life of 24, 100 years. What percentage of an original plutonium sample will be left after one thousand years?

Solution

Using Eq. (15.13), the decay constant λ of $^{239}_{94}Pu$ is found to be 2.876×10^{-5} years^{-1}. Eq. (15.12) can be written

$$\frac{N}{N_0} = e^{-\lambda t}.$$

Substituting the value of λ we have obtained and $t = 1000$ years into this last equation, we find that N/N_0 is equal to 0.972 or 97.2%.

One of the reasons that debates concerning the use of nuclear energy lead to controversy is that experts in different fields have different ideas of the length of time over which they can make predictions with a reasonable amount of confidence. Nuclear physicists and geologists feel comfortable talking about a thousand years. However, the makeup of the human society surrounding a waste disposal site will itself change as the buried isotopes decay. No sociologist with any sense wants to talk about a thousand years!

The conservation laws considered in the previous chapter apply to the decay of unstable nuclei just as they apply to elementary particles. We consider now in turn the most important quantities that are conserved in nuclear reactions, and how the conservation laws help us understand nuclear decay processes

1) *Conservation of energy.* The total energy of the nuclei involved in a nuclear reaction must be equal to the total energy of the products of the nuclear reaction. For decay processes, this implies that the rest energy of the decaying nucleus must be greater than the total rest energy of the products of the decay process. The energy liberated in a decay, which is called the *Q-value* of the reaction, can be calculated just as we have calculated the binding energy of nuclei by finding the mass loss in the process and converting the mass loss into an energy.

2) *Conservation of momentum.* The condition that the momentum be conserved implies that the total momentum of the decay products must be equal to zero in the rest frame of the decaying nucleus.

3) *Conservation of electric charge.* The total electric charge before a nuclear reaction must be equal to the total charge after the reaction.

4) *Conservation of atomic mass number.* For nuclear reactions in which the only baryons involved are protons and neutrons, the conservation of baryon number implies that the total number of nucleons must remain the same. This in turn means that the atomic mass number is conserved in the reaction.

15.2.1   Alpha decay

In α-decay, an unstable nucleus disintegrates into a lighter nucleus and an α-particle (^4_2He nucleus) according to the formula

$$^A_Z X_N \rightarrow ^{A-4}_{Z-2} X' + ^4_2 He.$$

The decaying nucleus X is called the *parent* nucleus. The nucleus X', which is produced in the decay, is called the *daughter* nucleus. In alpha decay, X' has an atomic mass number that is four less than the atomic mass number of the nucleus X that decayed, and X' has an atomic number two less than the atomic number of X.

The fact that α-decay occurs so commonly for the heavy elements can be understood by considering Fig. 15.5, which shows how the binding energy per nucleon varies with atomic number. Since the ratio B/A gradually decreases for heavy nuclei, a heavy nucleus can split into two smaller nuclei with a greater total binding energy. The breakup of a nucleus into two or more smaller nuclei is called *fission* with α-decay being the most common fission process. As we found in Example 15.2, an α-particle, which is a ^4_2He nucleus, has a comparatively large binding energy of 28.2 MeV. The condition for a nucleus (A, Z) to decay by α-emission to a nucleus $(A-4, Z-2)$ is

$$B(A, Z) < B(A - 4, Z - 2) + 28.3 \text{ MeV}.$$

This condition is always satisfied for A sufficiently large. Of the nuclei beyond $^{209}_{83}Bi$ in the periodic table only a few isotopes of U and Th are sufficiently stable to have survived since the formation of the Earth. All other heavy nuclei are produced either by the decay of these isotopes, or they have been produced artificially.

The energy liberated in α-decay, which is called the *Q-value* of the reaction, can be calculated by finding the mass loss in the process and converting the mass loss into an energy. We have

$$Q = [m(X) - m(X') - m(^4_2He)]c^2. \tag{15.14}$$

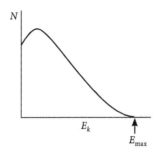

FIGURE 15.6 The kinetic energy spectrum of electrons emitted in β-decay.

The electron masses cancel as they did for calculating binding energies, and atomic masses may hence be used to calculate the Q-value.

Example 15.5

Calculate the Q-value of the decay of the $^{238}_{92}U$ into $^{234}_{90}Th$.

Solution

Using the masses of the $^{238}_{92}U$, $^{234}_{90}Th$, and $^{4}_{2}He$ nuclei given in Appendix B, we find that the mass loss in the α-decay is 0.004585 u. The Q-value of the α-decay is found by multiplying the mass loss by 931.5 MeV/u giving 4.27 MeV.

15.2.2   The β-stability valley

We now consider β-decay processes in which a nucleus decays by emitting an electron or a positron. The simplest example of a β-decay process, in which an electron is emitted, is the decay of the neutron into a proton, an electron, and an antineutrino described by the equation

$$n \rightarrow p + e^- + \bar{\nu}.$$

We recall that neutron decay occurs by the weak interaction and that the electron lepton number (L_e) is conserved since the electron has $L_e = +1$ and the antineutrino has $L_e = -1$.

Neutron decay can occur within a nucleus with Z protons and N neutrons leading to a nucleus with $Z + 1$ protons and $N - 1$ neutrons

$$^A_Z X \rightarrow ^A_{Z+1} X' + e^- + \bar{\nu_e}.$$

Since the daughter atom X' has one more electron than the parent atom X and the mass of the neutrino can be neglected, the Q-value for the reaction can be written

$$Q = [m(^A_Z X) - m(^A_{Z+1}X')]c^2, \qquad (15.15)$$

where $m(^A_Z X)$ and $m(^A_{Z+1}X')$ are the masses of the parent and daughter atoms. The energy release in the decay (the Q value) appears as the kinetic energy of the electron, the energy of the neutrino, and the recoil kinetic energy of the nucleus X'. The recoil energy of the daughter nucleus is usually negligible for β-decay processes, and, hence, the Q-value can be written

$$Q = K_e + E_\nu.$$

The kinetic energy spectrum of electrons emitted in β-decay is shown in Fig. 15.6. The electron has the maximum value of kinetic energy when the energy of the neutrino is equal to zero.

Example 15.6

The $^{77}_{32}Ge$ isotope decays by two successive β-decay processes to $^{77}_{34}Se$. Calculate the Q-values for each of these decay processes. What are the maximum kinetic energies of the emitted electrons?

Solution

The formulas for the two decay processes are

$$^{77}_{32}Ge \rightarrow\ ^{77}_{33}As + e^- + \overline{\nu}_e \,,$$

and

$$^{77}_{33}As \rightarrow\ ^{77}_{34}Se + e^- + \overline{\nu}_e \,.$$

$^{77}_{34}Se$ is the only stable isotope with $A = 77$.

The Q-value for each of these decay processes can be calculated using Eq. (15.15). For the first decay process, the Q-value is

$$Q = [76.923549\ \text{u} - 76.920648]931.5\ \text{MeV/u} = 2.70\ \text{MeV} \,,$$

while for the second decay process the Q-value is

$$Q = [76.920648\ \text{u} - 76.919915]931.5\ \text{MeV/u} = 0.68\ \text{MeV} \,.$$

The maximum values of the kinetic energy of the electrons emitted in these two decays is slightly less than the Q-values due to the small amount of kinetic energy carried off by the daughter nuclei.

Notice that for both of the β-decay processes in Example 15.6, the parent atom is more massive than the daughter atom. This condition must be satisfied for Q-value to be positive and for the process to occur. In a β-decay process, the atomic mass number A remains the same, while the value of Z changes by one unit. We can express the mass of an atomic isotope as a function of Z by replacing N by $A - Z$ in the semiempirical mass formula (15.9) to obtain

$$m(N, Z)c^2 = (Am_nc^2 - aA + bA^{2/3} + sA + \delta A^{-1/2}) - (4s + (m_n - m_p - m_e)c^2)Z + (4sA^{-1} + dA^{-1/3})Z^2 \,. \qquad (15.16)$$

This last equation can be written simply

$$m(N, Z)c^2 = \alpha - \beta Z + \gamma Z^2 \,, \qquad (15.17)$$

where

$$\begin{aligned}
\alpha &= Am_nc^2 - aA + bA^{2/3} + sA + \delta A^{-1/2} \\
\beta &= 4s + (m_n - m_p - m_e)c^2 \\
\gamma &= 4sA^{-1} + dA^{-1/3} \,.
\end{aligned} \qquad (15.18)$$

For a particular value of A, Eq. (15.17) can be used to plot the atomic mass as a function of Z. In Fig. 15.7, we show the atomic masses of atoms with $A = 64$ relative to the atomic mass of $^{64}_{28}Ni$. Open circles indicate odd-odd nuclei and filled circles indicate even-even nuclei. The two curves, which were drawn using the values of the parameters given by Eqs. (15.18), (15.7), and, (15.8), have minima near $Z = 29$. The position of the minima in the curves can be calculated by taking the derivative of $m(N, Z)c^2$ given by Eq. (15.17) and setting the derivative equal to zero. This gives $\beta/2\gamma$. The displacement of the curve for even-even nuclei relative to the curve for odd-odd nuclei is due to the pairing interaction, which according to Eq. (15.8) is positive for odd-odd nuclei and negative for even-even nuclei with the displacement of the two curves being equal to $2|\delta|A^{-1/2}/c^2$ for each value of A.

For a *beta*-decay process in which an electron is emitted, the value of Z increases by one unit. We can depict β-decay processes of this kind by giving the atomic mass number A and the atomic number Z of the initial and final states as follows

$$(A, Z) \rightarrow (A, Z+1) + e^- + \overline{\nu_e} \,.$$

Since the daughter atom has $Z + 1$ electrons and the mass of a neutrino can be neglected, the condition that must be satisfied for this process to occur is

$$m(A, Z) > m(A, Z+1),$$

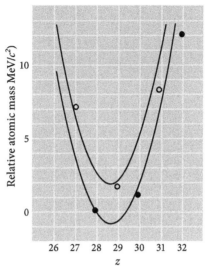

FIGURE 15.7 The atomic masses of atoms with $A = 64$ relative to the mass of $^{64}_{28}Ni$. Open circles indicate odd-odd nuclei and filled circles indicate even-even nuclei. (From W.N. Cottingham and D.A. Greenwood, 2001.)

where, as before, $m(A, Z)$ and $m(A, Z + 1)$ are atomic masses. Of the atoms depicted in Fig. 15.7, only $^{64}_{27}Co$, which lies to the left of the minimum can decay by β^- emission. The $^{64}_{27}Co$ atom, which has an odd-odd nucleus and is represented by an open circle in Fig. 15.7, decays into the $^{64}_{28}Ni$ atom with an even-even nucleus

$$^{64}_{27}Co \rightarrow ^{64}_{28}Ni + e^- + \overline{v_e} \,.$$

Beta decay processes for which a positron is emitted can be depicted

$$(A, Z) \rightarrow (A, Z - 1) + e^+ + v_e \,.$$

Since the daughter atom has $Z - 1$ electrons, the condition that must be satisfied for positron emission to occur is

$$m(A, Z) > m(A, Z - 1) + 2m_e c^2 \,.$$

The condition that the mass of the daughter atom be less than the mass of the parent atom minus twice the rest mass energy of the electron is somewhat more stringent for positron emission than the condition for electron emission.

Positron emission causes the atomic number Z to decrease and thus causes the atoms in Fig. 15.7 to move toward the left. The atom $^{64}_{27}Co$ cannot decay by positron emission since that would have the effect of reducing the atomic number Z and transforming $^{64}_{27}Co$ into an atom with greater mass. The atom $^{64}_{32}Ge$ with even-even nucleus can decay by positron emission into the atom $^{64}_{31}Ga$ with odd-odd nucleus according to the following formula

$$^{64}_{32}Ge \rightarrow ^{64}_{31}Ga + e^+ + v_e \,.$$

We also note that the atom $^{64}_{29}Cu$ with odd-odd nucleus, which lies near the minimum in Fig. 15.7, can emit an electron in decaying to $^{64}_{30}Zn$ or emit a positron in decaying to $^{64}_{28}Ni$. The two atoms, $^{64}_{28}Ni$ and $^{64}_{30}Zn$, which lie near the minimum of the curve shown in Fig. 15.7 are stable with respect to β-decay. As can be seen by considering Fig. 15.7, a β-decay of either of these two nuclei would result in an atom with greater atomic mass.

The atoms, whose nuclei are observed to be stable with respect to β-decay are denoted by dark squares in Fig. 15.8. To a very good approximation, the bottom of the β-decay valley indicated by dark squares in this figure corresponds to the value $Z = \beta/2\gamma$ obtained previously in conjunction with Eq. (15.17). Atoms with constant A, which are connected by β-decay processes, lie on straight lines with $N + Z = A$. These lines are perpendicular to the line $N = Z$ shown in Fig. 15.8.

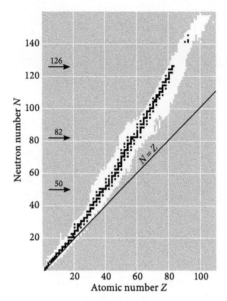

FIGURE 15.8 The nuclei (indicated by black squares) which are observed to be stable with respect to β-decay.

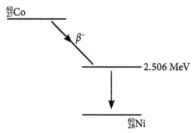

FIGURE 15.9 A β-decay process in which $^{60}_{27}Co$ decays into $^{60}_{28}Ni$ and the excited $^{60}_{28}Ni$ nucleus subsequently decays into the ground state by emitting a γ-ray.

15.2.3   Gamma decay

The processes of α and β-decay usually lead to nuclei which are in excited states. As for excited atomic states, the excited states of nuclei can decay into lower-lying states by emitting electromagnetic radiation. Since the separation of the energy levels of nuclei is much larger than for atoms, the radiation emitted by excited nuclei is much more energetic being in the form of γ-rays. Fig. 15.9 shows a typical β-decay process in which $^{60}_{27}Co$ decays into $^{60}_{28}Ni$ by emitting an electron. As shown in the figure, the excited $^{60}_{28}Ni$ nucleus subsequently decays into the ground state by emitting a γ-ray.

Example 15.7

Calculate the mass of the excited $^{60}_{28}Ni$ nucleus produced in the β-decay process shown in Fig. 15.9.

Solution

As can be seen in Fig. 15.9, the energy of the excited state of $^{60}_{28}Ni$ produced by the β-decay process lies 2.506 MeV above the ground state. The mass of the excited nucleus thus exceeds the mass of the unexcited nucleus by

$$\Delta m = \frac{2.506 \text{ MeV}}{931.5 \text{ MeV}/u} = 0.002690\, u.$$

The mass of the excited nucleus is thus

$$m = 59.930791\, u + 0.002690\, u = 59.933481\, u.$$

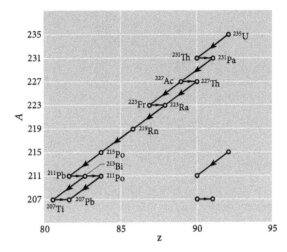

FIGURE 15.10 Members of a natural sequence of radioactive decay processes, which lead ultimately to stable isotopes.

Example 15.8

Calculate the Q-value of the initial β-decay process shown in Fig. 15.9.

Solution

The Q-value of the β-decay can be calculated using Eq. (15.15). We obtain

$$
\begin{aligned}
Q &= [m(^{60}_{27}Co) - m(^{60}_{28}Ni)]c^2 \\
&= (59.933822\,u - 59.930791\,u)\,(931.5\mathrm{MeV}/u) \\
&= 2.82\ \mathrm{MeV}
\end{aligned}
$$

Apart for a small correction for the kinetic energy of the recoiling nucleus, the Q-value of the reaction gives the total kinetic energy available to the emitted electron and neutrino.

15.2.4 Natural radioactivity

Of the elements in the periodic table, only hydrogen, helium, and a very small amount of lithium can be traced back to the early universe. All other elements were produced in nuclear fusion reactions in stars. The nuclear reactions in stars are the source of the immense amount of energy they radiate and produce all of the elements in the periodic table up to iron. As we have seen, iron has the most stable nucleus and does not fuse with other nuclei to form heavier nuclei. All of the elements heavier than iron were produced in the supernova explosions of massive stars and of binary stars. Supernovas generate a diverse variety of heavy elements and seed nuclear matter back into space where it can be incorporated into the formation of new stars.

The nuclear reactions in stars produce both stable and unstable nuclei. Unstable nuclei decay by α, β, and γ-emission into other nuclei which in turn can also decay. Since most decay processes have half-lives of a few days or a few years, most radioactive isotopes that were present when the Earth formed about 4.5 billion years ago have since decayed into stable elements. However, a few of the radioactive elements produced long ago have half lives that are comparable to the age of the Earth and still continue to decay.

The process of α-decay reduces the atomic mass number A of a radioactive isotope by four units and reduces the atomic number Z by two units, while β-decay leaves the atomic mass number unchanged and alters the atomic number by one unit. The radioactive decay processes that are observed in nature are members of decay sequences which lead ultimately to stable isotopes. An example of a radioactive sequence is shown in Fig. 15.10. In this sequence, the radioactive isotope $^{235}_{92}U$ decays by α-emission to the $^{231}_{90}Th$ isotope, which, in turn, decays by β-emission to $^{231}_{91}Pa$ isotope and then by α-emission to $^{227}_{89}Ac$. As can be seen in Fig. 15.10, the entire sequence terminates with the stable isotope $^{207}_{82}Pb$. Three radioactive sequences can be found in nature. Each of these sequences begins with a relatively long-lived isotope, proceeds through a considerable number of α and β-decays, and ends with a stable isotope.

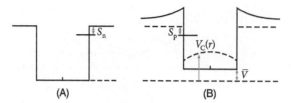

FIGURE 15.11 Potential energy curves for (A) neutrons and (B) protons.

15.3 The nuclear shell model

While the semiempirical theory for the binding energy and the mass of nuclei is very successful, experimental atomic masses show deviations from the semiempirical mass formula which are quantum mechanical in nature. Just as the atom can be described as electrons moving in an average central field due to the nucleus and the electrons, the nucleus itself can be described as protons and neutrons moving in a field due to both the strong and the electromagnetic forces. Because nucleons move in a finite region of space with definite values of the angular momentum, the table of nuclides shows recurring patterns that are very similar to the pattern of atomic elements described by the periodic table. As for atomic systems, the description of the nucleus in terms of the angular momentum of individual nucleons is called the *nuclear shell model*.

15.3.1 Nuclear potential wells

We begin our description of the shell model by considering the potential energy of nucleons. It is natural to expect that the form of the potential energy in which nucleons move is similar to the distribution of nuclear matter, which is approximately constant near the center of the nucleus and then falls off rapidly to zero. The decline of the nuclear potential energy in the surface region of the nucleus is actually more rapid than the decline of the nuclear density since tunneling of nucleons into the classically forbidden region causes the nuclear density near the surface to be more diffuse.

An illustration of the potential energy of neutrons and protons is given in Fig. 15.11(A) and (B). As shown in Fig. 15.11(A), the potential energy for neutrons is constant near the center of the nucleus and then rises to a finite value at the boundary. The potential energy for protons shown in Fig. 15.11(B) has a similar contribution due to the strong force, and it has a Coulomb part due to the Coulomb repulsion among the protons. The potential energy of protons also has a constant term stemming from the statistical larger effect of the neutron-proton interactions than the neutron-neutron and proton-proton interactions. Since stable nuclei generally have more neutrons than protons, the average number of neutrons seen by each proton is larger than the average number of protons seen by each neutron.

Each proton in a nucleus with atomic number Z interacts by the Coulomb interaction with $Z - 1$ other protons. The potential energy due to this interaction can be estimated by finding the electrostatic potential energy of a positive charge $+e$ due to a sphere of radius R in which a charge of $(Z - 1)e$ is uniformly distributed. This gives the following expression for the Coulomb energy of a proton

$$
V_C(r) = \begin{cases} \dfrac{(Z - 1)e^2}{4\pi\epsilon_0 R_0}\left[\dfrac{3}{2} - \dfrac{r^2}{2R_0^2}\right], & r \leq R_0 \\[2em] \dfrac{(Z - 1)e^2}{4\pi\epsilon_0 r}, & r > R_0, \end{cases}
$$

where R_0 is the nuclear radius. In Fig. 15.11(B), the contribution to the potential energy of a proton due to the Coulomb interaction $V_C(r)$ is shown and the statistical effect mentioned previously due to proton-neutron interactions is denoted by $\overline{V}$. The Coulomb potential V_C and the contribution $\overline{V}$ both make the potential well of the protons more shallow.

The highest occupied energy levels for neutrons and protons must be the same for the nucleus to be stable with respect to β-decay. If the height of the highest occupied neutron or proton level were to exceed the height of the other by more than about $m_e c^2$, the nucleus could decay by β-decay. The energy required to detach a neutron from the nucleus is called the *neutron separation energy* S_n. Similarly, the energy to detach a proton form the nucleus is called the *proton separation energy* S_p. The neutron and proton separation energies are shown in Figs. 15.11(A) and (B). The lowest energy levels of nucleons are not very sensitive to the precise shape of the potential well, and is similar for finite and infinite wells.

15.3.2 Nucleon states

Approximate wave functions and energies for nucleons can be obtained by solving the Schrödinger equation of a nucleon moving in an infinite potential well. Since nucleons cannot penetrate into the infinite wall of the well, the wave functions must vanish on the boundary. As for the states of the hydrogen atom described in Chapter 4, the wave functions of nucleons in spherical coordinates can be written as the product of a radial function and a spherical harmonic

$$\psi(r, \theta, \phi) = R(r)Y_{lm_l}(\theta, \phi),$$

where the orbital angular momentum quantum numbers, l and m_l, have the values, $l = 0, 1, 2, 3, \ldots$, and $m_l = l, l - 1, \ldots, -l + 1, -l$.

The function $Y_{lm_l}(\theta, \phi)$ is an eigenfunction of the angular momentum operator $\mathbf{l}^2$ corresponding to the eigenvalue $l(l + 1)\hbar^2$ and an eigenfunction of l_z corresponding to the eigenvalue $m_l\hbar$. The radial part of the wave function is denoted by $R(r)$. The spectroscopic notation is used in nuclear physics as in atomic physics with the letters, $s, p, d, f, g, \ldots$, being used to denote values, $l = 0, 1, 2, 3, 4, \ldots$. For each value of the angular momentum, though, the state with the lowest energy for nuclear physics is denoted by the principal quantum number, $n = 1$, and the next energy denoted by $n = 2$, and so forth. In this respect, the convention used in nuclear physics differs from the convention in atomic physics where the lowest state of a p electron with $l = 1$ is the $2p$ and the lowest state of a d electron with $l = 2$ is the $3d$.

Since the potential energy for neutrons is equal to zero inside the well, the Schrödinger for neutrons can be written

$$-\frac{\hbar^2}{2m}\nabla^2\psi = E\psi.$$

The Schrödinger equation may be expressed in terms of the radial and angular coordinates using the identity,

$$\nabla^2\psi = \frac{1}{r^2}\frac{d}{dr}\left(r^2\frac{dR}{dr}\right)Y_{lm_l}(\theta, \phi) - \frac{l(l + 1)}{r^2}RY_{lm_l}(\theta, \phi), \tag{15.19}$$

which is given in Appendix AA. The derivation of this equation depends upon the fact that the spherical harmonic, $Y_{lm_l}(\theta, \phi)$, is an eigenfunction of the angular momentum operator, $\mathbf{l}^2$, corresponding to the eigenvalue, $l(l + 1)\hbar^2$.

Using Eq. (15.19), the radial function $R(r)$, may be shown to satisfy the equation

$$-\frac{\hbar^2}{2mr^2}\frac{d}{dr}\left(r^2\frac{dR}{dr}\right) + \frac{l(l + 1)\hbar^2}{2mr^2}R = ER,$$

and this last equation may be simplified by dividing through with $\hbar^2/2m$ and by evaluating the derivatives. We thus obtain the following equation for the radial function of neutrons

$$-\frac{d^2R}{dr^2} - \frac{2}{r}\frac{dR}{dr} + \frac{l(l + 1)}{r^2} = k^2R, \tag{15.20}$$

where

$$k^2 = \frac{2mE}{\hbar^2}. \tag{15.21}$$

The possible energies of neutrons are thus related to the values of k^2 for which Eq. (15.20) has a solution satisfying the boundary conditions of a particle moving in a potential well. The fact that the potential well for protons is more shallow has the qualitative effect of shifting the proton levels up slightly above the neutron levels.

The solutions of Eq. (15.20), which are finite at the origin, may be expressed in terms of the *spherical Bessel functions* denoted $j_l(kr)$. The spherical Bessel functions for $l = 0, 1$, and 2 are given in Table 15.3 and illustrated in Fig. 15.12.

Knowing the nuclear wave functions, we can now determine the sequence of energy eigenvalues by imposing the boundary conditions. For $l = 0$, the radial wave functions can be written

$$R(r) = \frac{\sin x}{x},$$

where

$$x = kr.$$

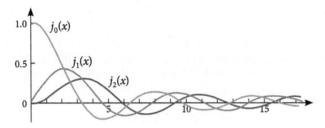

FIGURE 15.12 The spherical Bessel functions $j_l(kr)$ for $l = 0, 1$, and 2.

TABLE 15.3 The spherical Bessel functions $j_l(r)$ for $l = 0, 1, 2$.

l	$j_l(r)$
0	$j_0(x) = \dfrac{\sin x}{x}$
1	$j_1(x) = \dfrac{\sin x}{x^2} - \dfrac{\cos x}{x}$
2	$j_2(x) = \left(\dfrac{3}{x^3} - \dfrac{1}{x}\right)\sin x - \dfrac{3}{x^2}\cdot\cos x$

The condition that the wave function goes to zero on the surface of the well when $r = R_0$ is satisfied by requiring that $x = kR_0 = n\pi$, where $n = 1, 2, 3, \ldots$. The constant k must then be equal to $n\pi/R_0$, and Eq. (15.21) may be used to obtain the following sequence of eigenvalues for s-states

$$E(n, s) = \frac{\hbar^2 k^2}{2m} = \frac{n^2 h^2}{8m R_0^2}.$$

For $l = 0$, there is thus a sequence of energy eigenvalues, which can be labeled by the integer n or by the corresponding value of kR_0.

The procedure we have used to find the energy of $l = 0$-states can easily be generalized. For any value of l, the values of $x = kR_0$ for which the functions $j_l(x)$ are zero, have been tabulated. This leads to well-defined values of k, which determine the energy of higher-lying states in the well. Table 15.4 gives the lowest energies obtained in this way

TABLE 15.4 The lowest roots of the spherical Bessel functions $j_l(r)$.

State	x_{nl}	Number of states	Total number
$1s$	π	2	2
$1p$	4.49	6	8
$1d$	5.76	10	18
$2s$	2π	2	20
$1f$	6.99	14	34
$2p$	7.73	6	40
$1g$	8.18	18	58
$2d$	9.10	10	68
$1h$	9.36	22	90
$3s$	3π	2	92
$2f$	10.42	14	106
$1i$	10.51	26	132
$3p$	10.90	6	138
$2g$	11.70	18	156

The values of $x = kR_0$, for which the spherical Bessel functions $j_l(kr)$ are equal to zero, are given in the second column of Table 15.4. The first column of the table gives of values of l in spectroscopic notation. We note again that the first quantum number associated with each state gives the number of the solution for that particular value of l. For instance, the states, $1p$ and $2p$, are the first and second states for which $l = 1$. This convention differs from the convention used in atomic spectroscopy where the lowest two p states are the $2p$ and the $3p$ and the lowest d state is the $3d$. The third column in Table 15.4 gives the number of nucleons in each shell, while the fourth column gives the total number of nucleons up to that point.

15.3.3 Magic numbers

Nuclei with filled shells are important for our understanding of nuclear physics because they should enable us to identify values of the atomic number Z and the number of neutrons N for which nuclei are particularly stable. There are many ways of studying excited states of nuclei and for determining which nuclear isotopes are stable. As we shall consider in a following section, nuclei can be studied by bombarding them with energetic protons or with deuterons. Scattering experiments of this kind enable physicists to study the energies that can be absorbed by nuclei. There is a good deal of experimental evidence indicating that nuclei having certain numbers of protons and neutrons are especially stable. The values of Z and N for which nuclei are particularly stable are called *magic numbers*. The magic numbers are: 2, 8, 20, 28, 50, 82, and 126. Nuclei having Z or N equal to one of these numbers have properties reflecting the existence of an energy gap between occupied and unoccupied states. For example, tin with $Z = 50$ has ten stable isotopes, and there are seven stable isotopes with $N = 82$. The stable $_8^{16}O$ isotope has eight protons and eight neutrons, while the stable $_{82}^{208}Pb$ isotope has 82 protons and 126 neutrons.

There is not a close correspondence between the experimentally determined magic numbers and the numbers appearing in the last column of Table 15.4. The numbers, 18, 34, and 40, which correspond to filled shells are not magic numbers, and the magic number 50 corresponds to a partially filled $1g$ shell. The crucial step in obtaining a correspondence between the nuclear shell structure and the magic numbers is to include the *spin-orbit interaction*. The effects of the spin-orbit interaction are relatively more important for nuclei than for atoms.

15.3.4 The spin-orbit interaction

The spin and orbital angular momentum states of any particle with spin $s = 1/2$ and orbital angular momentum $l > 0$ can be combined to form states with the total angular momentum quantum number $j = l \pm 1/2$. As discussed in Chapter 4, the spin-orbit interaction causes a states according to the formula

$$h_{\text{s-o}}|(sl)j> = \begin{cases} \dfrac{\xi}{2}l\hbar^2, & j = l + \dfrac{1}{2} \\ -\dfrac{\xi}{2}(l+1)\hbar^2, & j = l - \dfrac{1}{2} \end{cases}$$

The splitting of levels by the spin-orbit interaction is illustrated in Fig. 4.19.

The spin-orbit interaction thus splits the $(4l + 2)$ states with the quantum numbers n and l into a sublevel corresponding to $2l + 2$ states with $j = l + 1/2$ and another sublevel corresponding to $2l$ states with $j = l - 1/2$. For example, the nd level, which corresponds to ten states are split into one sublevel with $j = 5/2$ corresponding to the six states and another sublevel with $j = 3/2$ corresponding to four states.

Experiment shows that the spin-orbit function $\xi(r)$ for nuclei is negative so that the states with $j = l + 1/2$ always have lower energy than the states with $j = l - 1/2$. The effect of the spin-orbit interaction upon the ordering of the energy levels of nuclei is most important for heavy nuclei for which the energy levels are closer together. A revised sequence of energy levels that includes the effect of the spin-orbit interaction is given in Table 15.5.

With the introduction of the spin-orbit interaction, the magic numbers all correspond to the filling of energy levels with particular values of the quantum numbers, s, l, and j. The energy gaps that occur between shells are represented in Table 15.5 by leaving a row vacant after each magic number.

The shell model successfully predicts the angular momentum of nuclei in the ground state. Nuclei with even numbers of protons and even numbers of neutrons (even-even nuclei) have angular momentum zero and even parity, while nuclei with an even number of protons and an odd number of neutrons or vice versa (even-odd nuclei) have angular momentum and parity equal to that of the odd nucleon in the shell being filled. As for atomic shells, the parity of a single nucleon with angular momentum l is even or odd depending upon whether $(-1)^l$ is even or odd. It is energetically favorable for

TABLE 15.5 A typical sequence for the filling of energy levels of nuclei. This table is taken from the book by W.N. Cottingham and D.A. Greenwood cited at the end of this chapter.

State	x_{nl}	Neutron		Proton	
1s	π	$1s_{1/2}$	2	$1s_{1/2}$	2
1p	4.49	$1p_{3/2}$	6	$1p_{3/2}$	6
		$1p_{1/2}$	8	$1p_{1/2}$	8
1d	5.76	$1d_{5/2}$	14	$1d_{5/2}$	14
		$2s_{1/2}$	16	$2s_{1/2}$	16
2s	2π	$1d_{3/2}$	20		
				$1d_{3/2}$	20
1f	6.99	$1f_{7/2}$	28		
		$2p_{3/2}$	32	$1f_{7/2}$	28
2p	7.73	$1f_{5/2}$	38		
		$2p_{1/2}$	40	$2p_{3/2}$	32
1g	8.18	$1g_{9/2}$	50	$1f_{5/2}$	38
		$2d_{5/2}$	56	$2p_{1/2}$	40
2d	9.10	$1g_{7/2}$	64		
		$1h_{11/2}$	76	$1g_{9/2}$	50
1h	3π	$3s_{1/2}$	78		
		$2d_{3/2}$	82	$1g_{7/2}$	58
3s	9.42	$2f_{7/2}$	90	$2d_{5/2}$	64
		$1h_{9/2}$	100		
2f	10.42	$3p_{3/2}$	104		
1i	10.51	$1i_{13/2}$	118	$1h_{11/2}$	76
		$2f_{5/2}$	124	$2d_{3/2}$	80
3p	10.90	$3p_{1/2}$	126	$3s_{1/2}$	82

pairs of protons and pairs of neutrons to form states having zero angular momentum and even parity, so that the angular momentum and parity of the nucleus is equal to the angular momentum and parity of the unpaired nucleon. There are very few exceptions to this rule.

No simple empirical rule has been found that gives the angular momentum and parity of odd-odd nuclei. Odd-odd nuclei, though, are energetically disfavored and very rare with only four being stable ($_1^2 H$, $_3^6 Li$, $_5^{10} B$, and $_7^{14} N$). All other odd-odd nuclei undergo β-decay to become even-even nuclei.

15.4   Excited states of nuclei

Thus far we have considered the energy levels of nuclei in their ground state. One method for studying the excited energy levels of nuclei is to scatter protons of known momentum from the nuclei and observe the momenta of scattered protons for different scattering angles. A scattering experiment of this kind is illustrated in Fig. 15.13. In this figure, the momenta of the incident and scattered protons are denoted by $\mathbf{p}_i$ and $\mathbf{p}_f$, respectively, and the scattering angle is denoted by θ. Before the collision, the target nucleus is supposed to be at rest. The momentum of the nucleus after the collision is denoted by $\mathbf{P}$.

In Fig. 15.13, the direction of the incoming proton is denoted by x, while the transverse direction in the plane of the collision process is denoted by y. The equations describing the conservation of the x- and y-components of the momentum are

$$p_i = p_f \cos\theta + P_x$$

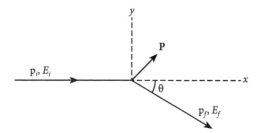

FIGURE 15.13 Illustration of an experiment in which a proton of known momentum is scattered by a nucleus.

$$0 = p_f \sin\theta + P_y$$

Solving these equations for the components of the momentum of the nucleus, we get

$$P_x = p_i - p_f \cos\theta$$
$$P_y = -p_f \sin\theta.$$

We now express the kinetic energy of the proton and the nucleus in terms of the incoming and outgoing momenta of the proton. Since the kinetic energy of the incident protons in the scattering experiments we will consider are around 10 MeV, nonrelativistic expressions may be used for the kinetic energy of the particles. The target nucleus is at rest before the collision. So, the total kinetic energy before the collision is equal to the kinetic energy of the proton

$$E_i = \frac{1}{2} m_p v_i^2 = \frac{1}{2m_p} p_i^2 .$$

After the collision, the total kinetic energy is

$$E_f = \frac{1}{2m_p} p_f^2 + \frac{1}{2m_A^*}(p_i - p_f \cos\theta)^2 + \frac{1}{2m_A^*}(p_f \sin\theta)^2 ,$$

where m_A^* is the mass of the nucleus in its final state.

Since the total energy is conserved in the collision process, the energy absorbed by the nucleus is equal to the loss of kinetic energy, which we denote by E. Collecting together terms, we obtain

$$E = E_i - E_f = \frac{1}{2m_p} p_i^2 - \frac{1}{2m_p} p_f^2 - \frac{1}{2m_A^*}(p_i^2 + p_f^2 - 2p_i p_f \cos\theta) .$$

The terms involving the square of the momentum of the proton can now be grouped together to give

$$E = E_i \left(1 - \frac{m_p}{m_A^*}\right) - E_f \left(1 + \frac{m_p}{m_A^*}\right) + \frac{2m_p}{m_A^*}(E_i E_f)^{1/2} \cos\theta . \tag{15.22}$$

Here E_i and E_f are the initial and final kinetic energies of the proton and m_A^* is the mass of the nucleus in its final state. When the kinetic energy of the incident proton is less than 100 MeV, the mass of the nucleus in its final state $m_A^* = m_A + E/c^2$ may be replaced by the mass in the ground state m_A with little error.

At a fixed scattering angle θ, the energy of the scattered protons are no longer monoenergetic, but have several well-defined values corresponding to the energy levels of excited nuclear states. As an example, we consider a scattering experiment performed by B. Armitage and R. Meads. In their experiment, protons with an energy 10.02 MeV were scattered from $^{10}_{5}B$. Fig. 15.14 shows the number of protons scattered within a small angular range at $\theta = 90°$ as a function of the final energy E_f. Using Eq. (15.22), one can show that the highest energy at 8.19 MeV corresponds to a kinetic energy loss E equal to zero and, hence, corresponds to elastic scattering. The peaks in Fig. 15.14 for lower energies correspond to excited states of $^{10}_{5}B$.

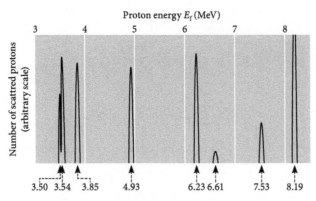

FIGURE 15.14 Protons scattered within a small angular range at $\theta = 90°$ as a function of the final energy E_f.

Example 15.9

Find the energy of the excited states of $^{10}_5 B$ corresponding to the peaks in Fig. 15.14.

Solution
Substituting the values, $E_i = 10.02$ MeV, $m_p = 1.007276$ u, and $m(^{10}_5 B) = 10.012937$ u, into Eq. (15.22), the equation becomes

$$E = 9.0120 \text{ MeV} - 1.1006 \text{ MeV} E_f.$$

We may now substitute the energies E_f given along the bottom of Fig. 15.14 to obtain the corresponding values of E given in Table 15.6.

TABLE 15.6 Excited states of $^{10}_5 B$ calculated by proton scattering.

E_f (MeV)	8.19	7.53	6.61	6.23	4.93	3.85	3.54	3.50
E (MeV)	0.0	0.72	1.74	2.16	3.59	4.77	5.12	5.16

The ground state and the lowest excited states of $^{10}_5 B$ are shown in Fig. 15.15.

Proton scattering experiments provide valuable information about excited nuclear states. Another technique, which can be used to determine the excited energy levels of nuclei is that of *deuteron stripping*, in which a nucleon of deuterium is taken away by the target nucleus. Consider for example the following reaction

$$^2_1 H + ^A_Z X \rightarrow ^{A+1}_Z X^* + p. \tag{15.23}$$

In this reaction, a neutron is stripped from an incident deuteron by the $^A_Z X$ nucleus. The products of this reaction include the nucleus $^{A+1}_Z X^*$ and a proton. The asterisk on the product nucleus indicates that it may be in an excited state. As for proton scattering experiments, the conservation of momentum and energy can be used to derive an expression for the energy of excited nuclear states in terms of the incident kinetic energy of the deuteron and the final kinetic energy of the proton. (See Problem 17.)

The number of excited states that nuclei have increases with atomic number. Deuterium, which is the lightest complex nucleus, has only one bound state and hence no excited spectrum. Very light nuclei have only a few excited states. However, the number of excited states increases rapidly as A increases. Fig. 15.16 shows the lowest energy levels of two *mirror nuclei*, $^{11}_5 B$ and $^{11}_6 C$. The nuclei are called mirror nuclei because the number of protons in one is equal to the number of neutrons in the other and vice versa. The energy levels of the two nuclei are very similar. For these two nuclei, the difference in the Coulomb energies is very small and the effects of the strong force are almost identical. The similarity of the energy levels of these two isotopes provides strong evidence for the isospin symmetry.

A qualitative understanding of the energy levels of $^{11}_5 B$ and $^{11}_6 C$ can be obtained from the shell model. Consider the $^{11}_6 C$ nucleus, which has six protons and five neutrons. The six protons fill the $1s_{1/2}$ and $1p_{3/2}$ shells. Of the five neutrons, two fill the $1s_{1/2}$ shell and the other three are in the $1p_{3/2}$ shell. As described earlier, two of the neutrons in the $1p_{3/2}$ level

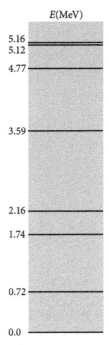

FIGURE 15.15 The ground state and the lowest excited states of $_5^{10}B$.

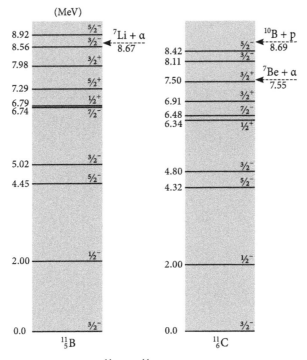

FIGURE 15.16 The lowest energy levels of two *mirror nuclei*, $_5^{11}B$ and $_6^{11}C$.

are coupled together to form a state having zero angular momentum and even parity with the remaining neutron giving the ground state angular momentum and parity. The parity of a single p-function with $l = 1$ is odd. This simple way of thinking correctly identifies the angular momentum and parity of the ground states of $_6^{11}C$ and $_5^{11}B$.

For the lowest excited state of $_6^{11}C$ nucleus, the odd neutron from the $1p_{3/2}$ shell is excited into the $1p_{1/2}$ shell resulting in a nucleus with angular momentum 1/2 and odd parity. One might be inclined to describe the next two excited states, which have angular momenta, 5/2 and 3/2, as excitations of the $1p_{3/2}$ neutron to the $1d_{5/2}$ and $1d_{3/2}$ states. We note,

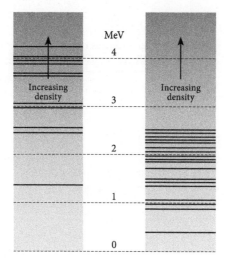

FIGURE 15.17 The energy levels of $^{46}_{20}Ca$ and $^{108}_{82}Pb$.

however, that the next two excited states have odd parity, while the parity of a d-state with $l = 2$ is even. The 5/2 and 3/2 states and many other excited states have more than one excited nucleon.

Fig. 15.17 shows the energy levels of $^{46}_{20}Ca$ and $^{108}_{82}Pd$. Heavier nuclei usually have a greater density of excited states and the density of excited states generally increases as the excitation energy increases.

Suggestions for further reading

W.N. Cottingham and D.A. Greenwood, *An Introduction to Nuclear Physics*, Second Edition (New York: Cambridge University Press, 2001).

Samuel S.M. Wong, *Introductory Nuclear Physics*, Second Edition (New York: Wiley, 1998).

Kenneth S. Krane, *Introductory Nuclear Physics* (New York: Wiley, 1988).

Basic equations

Binding energy

$$B(N, Z) = \left[Zm(^1_1H) + Nm_n - m(N, Z) \right] c^2 ,$$

where $m(^1_1H)$ is the mass of a hydrogen atom, m_n is the mass of a neutron, and $m(N, Z)$ is the mass of an atomic isotope.

The semiempirical formula

$$B(N, Z) = aA - bA^{2/3} - \frac{dZ^2}{A^{1/3}} - s\frac{(N - Z)^2}{A} - \frac{\delta}{A^{1/2}} ,$$

where A is the number of nucleons, N is the number of neutrons, and Z is the number of protons. The parameters a, b, d, s, and δ can be obtained by fitting the formula to the measured binding energies.

Magic numbers

The values of Z and N for which nuclei are particularly stable are

2, 8, 20, 28, 50, 82, and 126.

Summary

Information about the structure of nuclei can be obtained from high-energy electron scattering experiments. The results of these experiments show that the density of nuclear matter is roughly the same at the center of all nuclei. The nuclear density increases with A, but appears to approach a limiting value of about 0.17 nucleons per fm^{-3} for large A. All nuclei appear to have an inner region where the density is approximately uniform and a thin surface region where the distribution of nuclear matter falls off exponentially to zero.

The binding energy and mass of nuclei can be accurately described by empirical formulas. Superimposed upon the slowly varying properties of the nucleus described by the empirical formulas are deviations which are quantum mechanical in nature. The protons and neutrons in the nucleus move in a finite region of space with definite values of the angular momentum. The description of the nucleus in terms of the angular momentum of individual nucleons, which is known as the shell mode, makes it possible to understand the spin and parity of nuclei and to describe the spectra of excited states.

Questions

1. What kind of high-energy scattering experiments are used to determine the structure of the nucleus?
2. How is the scattering cross section related to the nuclear form factor?
3. Make a sketch showing how the density of nuclear matter varies with the distance r from the nuclear center.
4. Write down a formula expressing the nuclear binding energy $B(N, Z)$ in terms of the mass of the Z protons and N neutrons and the mass of the nucleus m_{nuc}.
5. To how much energy does an atomic mass unit u correspond?
6. To what physical effect does the term $-dZ^2/A^{1/3}$ in the empirical mass formula correspond?
7. The formation of which states will be discouraged by the term, $-s(N - Z)^2/A$ in the empirical mass formula?
8. The amount remaining of a radiative sample decreases by 50% over a certain period of time. By how much would the decay rate decline over the same period of time?
9. Is the atomic number of a daughter nucleus produced in the emission of an electron greater than or less than the atomic number of the parent nucleus?
10. Which particles are the primary recipients of the energy emitted in a beta decay process.
11. To which region of the spectrum does the electromagnetic radiation emitted by nuclei generally belong?
12. Why must the highest occupied energy levels of neutrons and protons in a nucleus be the same?
13. What condition can one impose on the wave functions of nucleons at the nuclear radius R_0?
14. Is the effects of the spin-orbit interaction relatively less important or more important for nuclei than for atoms?
15. Of the two nuclear states, $j = l + 1/2$ and $j = l - 1/2$, which has the lower energy?
16. To how many states does the $1d_{1/2}$ and $1d_{3/2}$ nuclear levels correspond?
17. What is the spin and parity of a nucleus having two protons and two neutrons?
18. What is the spin and parity of a nucleus having two protons and three neutrons?

Problems

1. How many protons and how many neutrons are there in the nuclei, $^{7}_{3}Li$, $^{63}_{29}Cu$, and $^{238}_{92}U$?
2. In a nuclear fission reaction, an atomic nucleus breaks up into two or more parts and also emits a number of neutrons. How many neutrons would be emitted in the fission reaction

$$^{236}_{92}U \rightarrow ^{90}_{36}Kr + ^{144}_{56}Ba?$$

3. Estimate the sizes of the nuclei $^{4}_{2}He$, $^{16}_{8}O$, $^{56}_{26}Fe$, $^{208}_{82}Pb$, and $^{93}_{237}Np$.
4. Using the mass of the neutron given in Appendix A and the atomic masses given in Appendix B, calculate the total binding energy and the binding energy per nucleon of the nuclei (a) $^{4}_{2}He$, (b) $^{16}_{8}O$, (c) $^{56}_{26}Fe$, (d) $^{208}_{82}Pb$, (e) $^{238}_{92}U$.
5. Find the energy needed to remove a proton from the nuclei (a) $^{4}_{2}He$, (b) $^{56}_{26}Fe$, (c) $^{208}_{82}Pb$.
6. Find the energy needed to remove a neutron from the nuclei (a) $^{4}_{2}He$, (b) $^{56}_{26}Fe$, (c) $^{208}_{82}Pb$.
7. Using the mass of the neutron given in Appendix A and the semiempirical formula (15.6), calculate the total binding energy and the binding energy per nucleon of the nuclei given in Problem 4.
8. Using the semiempirical formula (15.6) without the pairing term, derive an explicit expression for the binding energy per nucleon of a nucleus with atomic mass number A and atomic number $Z = N = A/2$. Show that the expression for the binding energy per nucleon you obtain has a maximum for $Z = A/2 = 26$.

9. A radioactive sample emits radiation 1200 times per second at $t = 0$. Its half-life is 2 min. What will its decay rate be after (a) 4 min, (b) 6 min, (c) (a) 8 min?

10. Tritium ($^3_1 H$) has a half-life of 12.3 years. What fraction of the tritium atoms would remain after 40 years?

11. The carbon isotope ($^{14}_6 C$) is continuously produced in the atmosphere by the reaction

$$n + {^{14}_7} N \rightarrow p + {^{14}_6} C,$$

where the neutron is due to cosmic rays. $^{14}_6 C$ decays back to $^{14}_7 N$ by the reaction

$${^{14}_6} C \rightarrow {^{14}_7} N + e^- + \overline{\nu}_e,$$

with a half-life of 5730 years. Since living organisms continually exchange carbon with the atmosphere, they have the same amount of the $^{14}_6 C$ isotope in a given sample of carbon as does the atmosphere.
(a) Using the fact that a gram of carbon in the atmosphere or in living organisms on the average emits 15.3 beta rays every minute, calculate the proportion of $^{14}_6 C$ in carbon. (b) What rate count would you expect from one gram of carbon extracted from a bone fragment that was 20,000 years old.

12. Complete the following decay reactions and find out how much energy is released in each reaction

$$(a) \qquad {^{209}_{83}} Bi \rightarrow {^{205}_{81}} Tl +$$
$$(b) \qquad {^{238}_{92}} U \rightarrow {^{234}_{90}} Th +$$
$$(c) \qquad {^{77}_{36}} Kr \rightarrow {^{77}_{35}} Br +$$
$$(d) \qquad {^{77}_{35}} Br \rightarrow {^{77}_{34}} Se +$$

13. $^7_4 Be$ decays by capturing an orbital electron in the reaction

$${^7_4} Be + e^- \rightarrow {^7_3} Li + \nu_e .$$

Find the energy released in the reaction.

14. (a) Give possible decay reactions for each of the following unstable nuclei $^6_2 He, {^8_4} B, {^{12}_4} B, {^{15}_8} O, {^{240}_{95}} Am$.
(b) How much energy would be released in each process?

15. (a) Find how much energy is released in the α-decay

$$^{234}_{92} U \rightarrow {^{230}_{90}} Th + {^4_2} He .$$

(b) How much of the total decay energy is carried off by the α-particle.

16. (a) Give the angular momentum and parity of the ground state of the following nuclei $^{17}_8 O, {^{17}_9} F, {^{31}_{15}} P, {^{31}_{16}} S, {^{32}_{16}} S, {^{40}_{20}} Ca, {^{45}_{21}} Sc$.
(b) For each nucleus give a possible single-particle excited state.

17. For the deuteron stripping reaction (15.23), use the conservation of energy and momentum to derive an expression for the energy of excited nuclear states analogous to Eq. (15.22) for proton-nuclear scattering.

18. In α-decay, the total kinetic energy of the α-particle and the daughter nucleus is

$$E = \frac{1}{2m_\alpha} \mathbf{p}_\alpha{}^2 + \frac{1}{2m_d} \mathbf{p}_d{}^2 .$$

Use the requirement that the total momentum is conserved in the decay process to show that the total kinetic energy of α-particle and the daughter nucleus can be written

$$E = \frac{1}{2m} \mathbf{p}_\alpha{}^2 ,$$

where the reduced mass m is defined by the equation

$$m = \frac{m_\alpha m_d}{m_\alpha + m_d} .$$

The total kinetic energy of the α-particle and the daughter nucleus thus has the form of the kinetic energy of a single particle with the reduced mass m.

19. Calculate the Q-value and mean lifetime of the double α-decay

$$^8_4Be \rightarrow 2(^4_2He).$$

The observed mean lifetime of this decay is 2.6×10^{-17} s.

20. Calculate the Q-value and mean lifetime of the α-decay

$$^{238}_{94}Pu \rightarrow ^{234}_{92}U + ^4_2He.$$

The observed mean lifetime of this decay is 128 years.

21. The isotope $^{194}_{79}Au$ decays by positron emission

$$^{194}_{79}Au \rightarrow ^{194}_{78}Pt + e^+ + \nu_e$$

and by α-decay

$$^{194}_{79}Au \rightarrow ^{190}_{77}Ir + ^4_2He.$$

Both the positron created in the first reaction and the α-particle created in the second reaction must tunnel through a Coulomb barrier to escape. Calculate the Q-value and the mean lifetimes for the two processes and explain the difference between the two lifetimes.

Index

A

Absorption, 96, 144, 271
  antiparticle, 365
  atomic, 146
  curve, 146
  maximum, 146–148, 151
  of light by atoms, 4
  particle, 365
  process, xxviii, 5, 6, 256, 257, 272
  radiation
    by atoms, 15
Acceptor atoms, 245, 251
Acceptors, 244
Additive quantum number, 348
Adjoining cesium atoms, 208
Adjoint, 322
  representation, 377
Affine connection, 330
Albert Einstein, xxvi, 4, 150
Alkali, 121
  atoms, 121, 147–149, 151, 193, 233
    energy levels, 148
  earth, 122
  metals, 234
Alkaline atoms, 146, 148
Allotment, 170
Alpha decay, 402
Alpha particles, xxix
Ammonia molecules beam, 143
Angular
  frequency, xvii
  momentum, xiii, 107
  wave number, xvii
Anode, 1
Anticolor states, 379
Anticommutation relations, 321
Antiparticle, xxx, 326, 338, 341, 343, 352, 353, 365
  absorption, 365
  creation, 365
Argon atom, 120, 127
Arsenic atoms, 244
Asymmetric colliding beam, 308, 309
Atom
  atomic masses, 404
  beam, 105
  daughter, 403–405
  energies, xxviii
  hydrogen, 6, 7, 81, 82, 95, 117, 119, 233, 355, 397

indium, 232, 245
iron, 133, 134
masses, 397
nitrogen, 143
oxygen, 234
parent, 403–405
silicon, 244
single, 96, 100, 207
Atomic
  absorption, 146
  centers, 233, 241
  electrons, xxix, 9, 105
  Hamiltonian, 130
  hydrogen, 164, 165
  isotope, 397, 404, 416
  magnetic moments, 104
  mass, 393, 404, 405, 408
    in atomic mass units, 393
    number, 393, 399, 402, 407
  mass unit, 397
  nucleus, xv, xxvi–xxx, 121, 148, 154, 345, 393
  number, 109, 127, 233, 393, 402, 407, 414
  orbitals, 133
  oxygen, 165
  physics, xxx, 85, 107, 137, 409
  shell model, 138
  shells, 241, 411
  species, 210, 232, 254, 393
  spectra, 5, 6
  spectroscopy, 411
  states, 107, 147
  structure, 10, 130
  transitions, xxviii, 12, 96, 141, 224, 270
  units, 85, 86, 100, 129, 136, 138, 153
Aufbau principle, 120
Average values, 68, 78, 89
Avogadro's number, 203
Azimuthal component of the angular momentum, 83
Azimuthal quantum number, 101, 107, 108, 120, 126, 355, 362, 363, 371

B

Band engineering, 262
Band gap, 226, 241
Band picture, 227
Band structure
  of semiconductors, 255

Bands
  heavy-hole, 228
  light-hole, 228
Bare nuclei, 397
Barrier potential, 246, 247
Baryon number, 350
Baryons
  light, 360
Base, 248
Basis, 207
Beam, xxiii, xxvii, 1, 11, 14, 17, 58, 105, 281, 308, 317
  atom, 105
  electromagnetic radiation, 1
  electron, xxiv, xxvi, 58
  laser, 142, 144, 148, 151
  molecules, 144
  radiation, 151
Beauty, 341
Beryllium
  atom, 94, 127
  nucleus, 127
Bias
  forward, 246
  reverse, 246
Biasing potentials, 245
Binding energy, 416
Binomial coefficient, 171, 199
Bipolar transistors, 248
Black body, 180
  radiation, 180, 184
Bloch function, 220
Bloch's theorem, 220, 235
Body-centered cubic, 206
Bohr
  magneton, 104
  model, 6–8, 85, 86
  radius, 8
Boltzmann constant, 173
Bond picture, 231
Bonding
  covalent, 231
  electrons, 232
  hydrogen, 231, 234
  ionic, 231, 233
  molecular, 231
Born-von Karman boundary conditions, 222
Boron atom, 131
Bose-Einstein
  condensation, 146, 193

distribution function, 192
distribution law, 200
statistics, 189
Bosons, 189
Bound states, 36, 44, 51, 353, 355
Bragg's law, 13, 16
Bravais lattice, 206, 207, 209, 211, 213, 215,
220, 235, 237, 254
Brillouin zone, 205, 213, 223, 225, 236, 243,
254–257
Building-up principle, 120

C

Cabibbo angle, 368
Carbon
atom, 123, 129, 210, 228, 237
nanotubes, 228, 231
Carriers
lifetime, 271
majority, 245
minority, 245
Cartesian coordinates, 111
Cathode, 1
Cavity, 2, 180, 181, 268, 269, 274
laser, 144, 266, 271–273
Central-field approximation, 119
Cesium atom, 208
Chemical potential, 194
Chlorine atoms, 208, 233
Christoffel symbol, 330
Cobalt nucleus, 357
Coefficient
binomial, 171, 199
reflection, 276
transmission, 276
Cohesive energy, 234, 399
Collective motions, xxx, 141
Collector, 248
Colliding beam, 292, 316, 317, 339
proton, 384
Colliding particles, 317, 340, 348
Collocation points, 51, 52, 86, 93, 156–158, 163
Color, 357, 363
confinement, 364
hypercharge, 363
isospin, 363
singlet states, 363
states, 363, 379
Complex conjugate representation, 376
Complex materials, 228
Conduction
band, 227, 242–244, 255, 277
for semiconductor, 259
electrons, 27, 80, 194, 196, 204, 226–228,
234, 247, 250, 255
energy, 255
states, 241
Confinement factor, 272
Constant
Boltzmann, 173
decay, 401
fine structure, 109
force, xiv
Rydberg, 5

spin-orbit, 108
velocity, 284, 286, 287, 298, 301, 307, 309
Contravariant vector, 304, 332, 333, 335
Coordination number, 206
Copper metal, 237
Core states, 241
Correlation energy, 135, 136
Cosmological constant, 336
Coupled states, 108
Covalent
bonding, 231
crystals, 233, 234
cohesive energy, 234
Covariant, 304
derivative, 334, 335, 380
vector, 304, 332
Cross product, xiii
Crystal
axes, 219, 220, 223
covalent, 233, 234
cubic, 206
diamond, 210
doped, 244
field, 226
germanium, 232, 244
lattice, 12, 205, 222, 236, 238
molecular, 233
planes, 217, 223
silicon, 243
structures, xxix, 205, 208–210, 232, 254,
258
Crystalline, 205
environment, 212
nature, 205
state, 205
Cubic Bravais lattices, 208, 209, 215
Cubic crystal, 206
Current gain, 249
Curvature scalar, 335
Cutoff frequency, 2

D

Dark energy, 336
Daughter
atom, 403–405
nucleus, 402, 403
De Broglie wavelength, 14, 16
Decay
alpha, 402
gamma, 406
Decay constant, 401
Decuplets, 360
Degrees of freedom, 180
Depletion zone, 245
Description
phenomenological, 270, 277
Deuteron stripping, 414
Diagram
direct, 136
exchange, 136
Diamond
crystals, 210
structure, 210

Diatomic
molecules, xxxi, 153, 154, 158, 163–165
particles, 203
Difference
energy, xvi, 6, 243, 278
mass, 348, 397
path lengths, xxiii
potential, xxviii, 2, 17, 246, 250
Diffracted electrons, 224, 225
Diffraction, xxiii
Diffraction phenomena, xxiii
Dirac
electron, 229
equation, 106, 321, 323, 324, 337, 338, 380,
381
matrices, 321, 337
particle, 323
spinors, 324
Direct
energy, 126
lattice, 213
term, 126
Displacement vector, 220
Dissipative processes, 271
Distance
atomic unit, 129, 138
separating
atoms, xxix
lattice planes, 235
Distinguishable particles, 170, 172, 190
Distribution, 170
Divalent metals, 195, 204
Donors, 244
atoms, 245, 251
impurity atoms, 245
Doping, 244
Doppler effect, 293
Drain, 249
Drift velocity, 228

E

Eigenfunction, 38, 39, 78, 84, 108, 109, 111,
117, 118, 409
superpositions, 39
Eigenstates, 70
mass, 342
Eigenvalues, 37–39, 47, 84, 95, 108, 111, 118,
264, 409
equation, 38–40, 50, 51, 77, 78, 86, 137,
158, 264
problem, 42, 51, 53, 86, 89, 154
Eigenvectors, 44, 45, 56, 326
Einstein, xxvi, xxviii, 2, 6, 11, 13, 31, 97, 106,
141, 151, 183, 187, 189, 279, 282,
295, 302, 305, 323, 329, 330, 336
coefficients, 96, 113
for stimulated emission, 98
field equations, 336
summation convention, 303
Elastic collision, 314
Electric dipole
operator, 98
transitions, 101
Electric potential, 250

Electromagnetic
  force, 341
  interaction, xxx, 329, 339, 343–346, 350,
      351, 356, 364, 365, 389
  radiation, xxv, xxvi, xxxi, 1, 12, 13, 17, 31,
      142, 265, 393, 406
    beam, 1
  waves, xxv, xxvi, 1, 181, 192, 265
Electron, xxix
  beam, xxiv, xxvi, 58
  charge, 233
    cloud, xxviii
    density, 220
  circulating, xxix, 102
  cloud, xxvii
  collide, 314, 327, 339
  collimated beam, 1
  configuration, 120
  density, 195
  diffraction, 19
  diffraction experiments, 19
  diffusing, 245, 248
  Dirac, 229
  emission, 405
  energies, 28, 120
  equivalent, 123, 124, 129
  free, 1, 145, 194, 205, 212–214, 221, 241,
      244, 251, 253, 365
    states, 327
  gas, 197
  interaction, 205, 236
  kinetic energy, 2, 15
  lepton number, 349, 353, 366, 391, 403
  magnetic moment, 112
  mass, 27, 28, 80, 243, 255, 256, 403
    for motion, 243
  motion, xiii–xv, 253, 262
  neutrino, 341, 348, 349
  orbiting, 6
  outgoing, 315, 316, 328
  pairs, 232, 233
  parity, 356
  potential energy, 277
  scattered, 11
  scattering, 56, 77, 265
    data, 396
    experiments, 394
  shells, 121, 232
  single, 81, 95, 107, 109, 121, 122, 128, 148,
      153, 354, 355, 371, 372
  states, 227, 231, 241, 256
  volts, xxviii, xxix, 11, 12, 28, 109, 124, 129,
      264, 394
  wave function, 221
  waves, 13
Electronegativity, 232, 233
Electronic
  bands, 227
  bonds, 232
  charge, 233
  configurations, 120, 134
  contribution, 397
  energy, 254
    levels, 144, 145

  states, 241
  structure, 227, 232
  transitions, 101, 165
Electrostatic energy, 234, 399
Electrostatic potential, 245, 246, 260, 261
  energy, 126, 408
Electroweak interactions, 381
Emission
  induced, 96
  of light by atoms, 4
  radiation
    by atoms, 15
  spontaneous, 96
  stimulated, 96, 141, 143, 144, 151, 270, 273
Emitter, 248
Energetic photons, xxviii
Energy
  atomic unit, 86, 129, 138
  bands, 227, 234, 255
  consumption, 247
  denominator, 136, 137
  density, 96, 98, 184, 193, 203, 247
  difference, xvi, 6, 243, 278
  eigenvalue equation, 119
  eigenvalues, 76, 94, 156, 158, 320, 409, 410
  electronic, 254
  gap, 11, 241, 243, 411
  ground state, 138
  in atomic units, 100
  in electron volts, 264
  in relativity theory, 339
  interaction, 172
  levels, 8, 34, 148, 151, 411, 414, 416
    for neutrons, 408
  loss rate, 274
  negative, 136, 277, 320, 325–327, 338
  nuclear, 402
  operator, 38, 77
  per photon, 274
  per volume, 180, 182
  photons, xxviii, xxix, xxxi, 2–4, 10, 11, 15,
      124, 141, 192, 203, 247
  positive, 136, 320, 325, 338
  profile, 260
  quanta, xxvi, 12
  range, 181, 195, 196, 201, 242, 260
  resolution, 260
  separation, 143
  states, 173, 190, 326
  surface, 229
  transition, 15
  vibrational, 257
Energy-momentum
  four-vector, 313
  relation, 313
  tensor, 336
Energy-time uncertainty relation, 77
Entropy, 185
Equilibrium positions, xiv, xv, 221, 257
Equipartition of energy, 180
Escape velocity, 178, 179, 202
Exchange
  energy, 126
  term, 126

Exoplanets, 163–165
  orbiting, 164
Extrinsic semiconductors, 241, 244

F

Fabry-Perot laser, 268
Fermi energy, 195, 196, 198, 241–243, 245
Fermi-Dirac
  distribution function, 194
  distribution law, 200
  statistics, 189
Fermion propagator, 365
Fermions, 189
Feynman
  diagram, 327, 328, 364–369, 384, 385, 388,
      389, 392, 395
  rules, 328
Field quanta, 329
Fine structure constant, 109
Finite potential well, 26
Finite well, 40
Fission, 402
Flavors, 341
Fluid
  normal, 193
Fluorine atoms, 131
Forbidden transitions, 101, 257
Force constant, xiv
Forward bias, 246
Four-momentum, 314
Fourier integral, 65, 77
Fourier theorem, xx
Fourier transform, 66, 77
Free
  atoms, 146
  electron, 1, 145, 194, 205, 212–214, 221,
      241, 244, 251, 253, 365
    mass, 27, 28, 40, 63, 263
    states, 327
    wave, 214
  particle, 1, 20, 30, 31, 33, 172, 205, 212,
      319, 324, 325, 380, 381
Frequency
  fundamental, xxi
  range, 98, 144, 180, 181
  shift, 293
  single, 181
  transition, 6, 98, 148
Fundamental representation, 375

G

G-value, 105
Galilean
  transformations, 279
  velocity transformation, 280, 291, 292
Gamma decay, 406
Gas
  electron, 197
  hydrogen, 4
  ideal, 172
  inert, 121
  molecules, 170
  particles, 172, 174–177, 180, 185, 188, 189
    states, 188

perfect quantum, 172, 189
photons, 192
planets, 165
rare, 121
Gate, 249
Gauss
    points, 48–51, 73–75, 90, 93, 156, 157
    quadrature points, 48, 49, 56, 73, 75, 76, 80, 89, 90, 93, 155, 156, 163
Generations, 341
Germanium
    atoms, 231, 232
    crystal, 232, 244
    semiconductor, 231
Gluon, xxx, 342
Gold
    atom, xxix
    nucleus, xxix, 394
Golden Rule, 364
Goldstone diagrams, 136
Graphene, 228, 230
Grating, xxiv
Graviton, xxx, 342
Gross structure, 81
Ground configuration, 120
Ground state, 17, 35, 127, 128, 134–136, 145, 151, 152, 165, 193, 406, 411, 412, 414, 415
    energy, 138
    hydrogen, 8
Group
    point, 254
    space, 254
Gyromagnetic ratio, 103

H

Hadrons, 341
Halogen, 122
    atoms, 233
Hamiltonian, 38
    of outer electron, 150
    operator, 117
Harmonic oscillator, xiv, xv, 28, 29, 34, 35, 40, 45, 47, 69, 75, 76, 155, 181, 328
    simple, 28
Harmonic waves, xvi, xviii, xx, xxi, 66
    superposition, 66
Hartree-Fock
    applet, 127
    equations, 126, 167
    method, 126
Heavier atoms states, 124
Heavier nuclei, 407, 416
Heisenberg Uncertainty Principle, 65, 77
Helicity, 357
Helium
    atom, xxvii, 107, 117, 125, 145, 151
    nucleus, 393, 398
Hermite polynomials, 30
Hermitian matrix, 322
Heterostructures, 258, 277
Hexagonal close-packed structure, 210
Higgs mass, 384

Holes, 147, 227, 230, 241, 244, 245, 256, 326
    heavy, 256
    light, 256
Hund's rule, 124
Hydrogen, xxvii, 4, 8, 81, 84, 102, 112
    atom, 6, 7, 81, 82, 95, 117, 119, 233, 355, 397
        beam, 105
        energy levels, 15
    atomic, 164, 165
    bonding, 231, 234
    energy eigenvalues, 120
    energy levels, 8, 19, 21, 82, 83, 85, 86, 102, 120, 320
    gas, 4
    ground state, 8
    in atomic units, 86, 112
    ion, 86
    ionization energy, 9, 233
    mass, 355
    molecular ion, 29, 153, 156, 158, 163
    nucleus, xxvii, 8, 344, 355
    wave function, 111
Hydrogen-bonded crystals, 233
Hypercharge, 358, 359, 388
    quantum number, 358

I

Ideal gas, 172
    law, 203
Incoming
    electrons, 328
    particles, 327, 329, 348, 351, 352
    photon, 348
    positron, 365
    proton, 412
Independent-particle model, 117
Induced emission, 96
Inelastic collision, 314
Insulator, 227
Interaction, 97, 102, 109–111, 113, 131, 368, 384, 389
    electromagnetic, xxx, 329, 339, 343–346, 350, 351, 356, 364, 365, 389
    electron, 205, 236
    electroweak, 381
    energy, 172
    particle, 338
    spin-orbit, 106, 411
    weak, xxx, 343, 351, 352, 366, 389
Interference, xviii, xxiii
Intrinsic semiconductor, 241
Invariant mass, 345–347
    distribution curve, 384
    spectrum, 346
Invariant space-time interval, 306
Ionic
    bonding, 233
    crystals, 209, 233, 234, 237
Ionization energy, 9–11, 121, 122, 137, 233
    hydrogen, 9, 233
Iron atom, 133, 134
Isospin, 357, 358, 362, 378, 380, 381, 388
    color, 363

matrices, 381
multiplet, 358, 388, 389
quantum numbers, 358–362, 381, 383
states, 362
symmetry, 358, 360, 396, 414
weak, 381

J

JK coupling, 151

K

Kinetic energy
    electron, 2, 15
    maximum value, 403
Klein-Gordon equation, 320

L

Laplacian, 81
Large Hadron Collider (LHC), 384
Larmor frequency, 104
Laser
    amplification, 142
    beam, 142, 144, 148, 151
        incident, 147
    cavity, 144, 266, 271–273
    cooling, 146
    threshold, 273
    transition, 141, 142, 145
Lattice
    planes, 215
    with a basis, 208
Law
    first
        thermodynamics, 187
    ideal gas, 203
    second
        thermodynamics, 188
    third
        thermodynamics, 188
    zeroth
        thermodynamics, 187
Length
    in atomic units, 100
Leptons, 341
    number, 348
Light
    baryons, 360
    cone, 300
    mesons, 359
Linear waves, xviii
Liquid hydrogen, 344
Lithium atom, 139
Longitudinal mass, 243
Lorentz
    condition, 305
    contraction, 287, 306
    invariant, 303
    transformations, 284, 286, 288, 291, 295, 298, 302, 306, 312, 323
LS terms, 122

M

Magic numbers, 411, 416

Magnetic
  field vector, 102, 110
  interactions, 102, 117
  moment, 102, 109, 150, 152
  potential energy, 109, 111, 112
  quantum number, 83
Magneto-optical traps, 147
Majority carriers, 245
Masers, 143
  ammonium, 143
Mass
  atomic, 393, 404, 405, 408
    number, 393, 399, 402, 407
  density, 336
  difference, 348, 397
  eigenstates, 342
  electron, 27, 28, 80, 243, 256, 403
  hydrogen, 355
  invariant, 345–347
  loss, 397, 398, 402, 403
  molar, 203
  neutrino, 342
  nuclei, 408, 417
  particle, 383
  single, 346
  spectra, 363
Matrices
  Dirac, 321, 337
  Hermitian, 322
  isospin, 381
  product, 62
  self-adjoint, 322
  spin, 326
Matrix product, 63, 269
Max Planck, xxvi, 180
Maximum
  absorption, 146–148, 151
  value, 26, 107, 123, 124, 177, 185, 186, 404
Maxwell
  distribution, 174
  speed distribution, 177
Maxwell-Boltzmann
  distribution law, 173, 199
  statistics, 174, 199
Mesons, 189, 341
  light, 359
Metal, 194, 200, 228, 234
  divalent, 195, 204
  electrode, 12
  gate, 250
  monovalent, 195
  surface, 1, 2, 17, 183
  transition, 234
  trivalent, 195
Metal-organic chemical vapor deposition, 258
Metallic
  bonding, 231
  crystals, 234
  nanotube, 231
Metric tensor, 303, 307, 329
Microscopic
  particle, 37
  states, 201

systems, xiii, xvi, xxvi, xxvii, xxxii, 13, 19, 21, 64, 150
  energy, 37
Miller indices, 218
Minority carriers, 245
Mirror nuclei, 414, 415
Mobility, 228
Mode, 181
Molar mass, 203
Mole, 203
Molecular-beam epitaxy, 258
Molecules
  beam, 144
  gas, 170
  ground states, 165
  oxygen, 165
Momentum, xiii, 314
  distribution law, 179
  interval, 174, 181
  operator, 37–39, 76, 78
  operator $\hat{p}$, 69
  range, 176
  space, 175
  vectors, xiii, 339
Momentum-position uncertainty relation, 77
Monovalent metal, 195, 204
Motion
  electron, xiii–xv, 253, 262
  orbital, 6, 103, 105, 106, 108, 110, 117
  particle, xiv, xvi, xxxii, 65, 295, 297
  relative, 106, 113, 294, 295, 311
Multilayered crystals, 258, 277
Multiplicity, 124
Muon, 289, 340–342, 349
  antiparticles, 341
  decay, 342, 366
  lepton number, 349, 366
  neutrino, 349

N
n-doped, 244
Natural radioactivity, 407
Natural units, 324
Nearest neighbors, 206
Nearly free electron model, 226
Negative
  energy, 136, 277, 320, 325–327, 338
    eigenvalues, 321
    states, 320, 321, 326, 365
  parity, 356, 357, 359
Neighboring
  atoms, 241, 258
  germanium atoms, 232
Neutral
  atoms, xxix, xxx, 147, 400
  neon atom, 140
  particles, 345–347, 360
    beam, 12
  titanium atom, 121
Neutrino, 314, 341, 342, 348
  electron, 341, 348, 349
  masses, 342
  muon, 349
  reactions, 348

Neutron
  decays, xxx, 403
  potential energy, 408
  separation energy, 408
n factorial, 171
Nitrogen atom, 143
Nodes, xix, 47
Nonequivalent electrons, 125
Nonlinear waves, xviii
Normal fluid, 193
Normalization condition, 24, 173
Nuclear
  binding energy, 397
  charge, xv, xxvii, 99, 109, 120, 121, 124, 126, 127, 131
  energy, 402
  force, xxx
  form factor, 395
  potential
    energy, 408
    wells, 408
  radiation, xxviii
  shell model, 408
  spin, 152, 357, 393
Nuclei
  atomic, xv, xxvi–xxx, 121, 148, 154, 345, 393
  daughter, 402, 403
  decays, 398, 401, 403
  energy levels, 406, 411, 412
  hydrogen, xxvii, 8, 344, 355
  mass, 408, 417
  mirror, 414, 415
  parent, 402
  parity, 417
  stable, 399, 401, 408
Nucleon, xxix, 393
  states, 346, 409

O
Observables, 37
Octets, 360
Operators, 37, 39, 78, 83, 157
  electric dipole, 98
Orbital
  angular momentum, 7, 83, 107, 123, 355, 357
    quantum number, 122, 125, 127, 355, 356, 409
    vector, 106, 110
  motion, 6, 103, 105, 106, 108, 110, 117
  quantum number, 109, 120, 123
Orthogonality conditions, 223
Oscillator, 45
  potential, xv
    energy, xxxi
Oxygen
  atom, 165, 234
  atomic, 165
  molecules, 165

P
p-doped, 244
p-n junction, 245

Packing fraction, 209
Parent
    atom, 403–405
    nucleus, 402
Parity, 101, 356
    conservation, 357
    electron, 356
    negative, 356, 357, 359
    nuclei, 417
    operator, 356
    opposite, 101, 356
    symmetry, 380
    transformation, 356, 357
    violation, 357
Partial derivative, 31, 32, 81, 157, 305, 321, 330
Particle, xiii
    absorption, 365
    beams, 14, 364
    density, 194
    Dirac, 323
    energy levels, 34
    free, 1, 20, 30, 31, 33, 172, 205, 212, 319,
        324, 325, 380, 381
    incoming, 327, 329, 348, 351, 352
    indistinguishable, 190, 192, 193
    interacting, 327
    interaction, 338
    mass, 383
    microscopic, 37
    model, 1, 12
    motion, xiv, xvi, xxxii, 65, 295, 297
    probability, 33
    single, 356
    spin, 189, 201
    velocity, 296
Particle/antiparticle pairs, 326, 350, 352
Partition function, 173, 174
Pauli exclusion principle, 118, 123, 141, 189,
        190, 326, 363, 380, 399
Pauli matrices, 321, 337, 372
Perfect quantum gas, 172, 189
Periodic
    boundary conditions, 222
    potential, 27, 253
    table, 120, 121
Phase-space spectra, 346
Phonons, 257
Photoelectric effect, xxvi, 1, 15
Photoelectrons, 3, 10, 11
Photons, xxvi, 342
    beams, 15
    density, 277
    energetic, xxviii
    energy, xxviii, xxix, xxxi, 2–4, 10, 11, 15,
        124, 141, 192, 203, 247
    gas, 192
    per, 192
Pion-nucleon scattering, 360
Planck, xxvi, xxviii, 2, 6, 31, 181–183, 192, 193
    black-body radiation formula, 200
    constant, 2
    distribution law, 203
    radiation law, 189
Plane waves, 65, 265

superposition, 68
Planetary atmospheres, 179
Poincaré algebra, 386
Point group, 254
Polar coordinates, 111
Polyatomic particles, 203
Position
    coordinate, 39
    relative, 221, 262
    vector, xiii, xxii, 82, 214, 219, 304
Positive
    energy, 136, 320, 325, 338
Positively charged
    holes, 248
    particles, 344
Positron
    emission, 405
    incoming, 365
    spinors, 327
    states, 327
Positronium, 355
    states, 355
Potential
    barrier, 60, 61, 63, 265, 269
    barriers
        for conduction electrons, 277
    difference, xxviii, 2, 17, 246, 250
    drop, 62
    energy
        for neutrons, 408, 409
        for protons, 408
    field, xvi, 129, 138, 261
    oscillator, xv
    stopping, 2
Primitive
    unit cell, 209
    vectors, 206
Principal quantum number, 7, 83, 120, 355
Principle of equivalence, 329
Principle quantum number, 122, 123, 125
Probability, xxvi, 23, 89
    current, 264, 319
    density, 21, 24, 319–323
    distribution, 177
    function, 177, 197, 198
    particle, 33
Propagator, 328, 365
Proper time, 291
Proportionality constant, 272
Proton
    colliding beam, 384
    incoming, 412
    isospin quantum numbers, 360
    potential energy, 408
    scattering experiments, 414
    separation energy, 408
    single, xxvii
    target, 342
Proxima Centauri, 168

Q
Q-value, 402
Quadrupole trap, 147
Quanta, xxvi, xxx, 2, 3, 257, 328, 329, 339

energy, xxvi, 12
Quantum barrier, 262, 265, 275
Quantum chromodynamics, 364
Quantum mechanics, 8
Quantum numbers, 23
    isospin, 358–362, 381, 383
    orbital, 109
    orbital angular momentum, 125, 127, 355,
        356, 409
    spin, 95, 106, 148, 355, 363
    strangeness, 351
Quantum wells, 27, 262
Quark states, 360, 363
Quark/antiquark pair, 341, 343, 350, 351, 355,
        356, 360, 364, 369, 372
Quarks, xxx, 341

R
Radial distances, 109, 130
Radiation
    absorption, 15
    beam, 151
    electromagnetic, xxv, xxvi, xxxi, 1, 12, 13,
        17, 31, 142, 265, 393, 406
    emission, 15
    field, xv, xxvi, 2, 4, 12, 96, 97, 106, 141,
        182, 185, 192, 201, 266, 270, 271
    nuclear, xxviii
    processes, 141, 273
    scattered, 1, 11, 12
    stimulated emission, 141, 143
Radiative transitions, xxvii, 94, 95, 100–102,
        141
Radio waves, xxv, xxvi
Rate equation, 270, 271
Reciprocal lattice, 205, 212, 213, 235
    lengths, 213
    vectors, 214, 216–219, 223, 225, 235, 238,
        239
Recoil energy, 403
Rectifiers, 247
Red giants, 184
Reflection, 267
    coefficient, 59
Relation
    energy-time uncertainty, 77
    momentum-position uncertainty, 77
Relative
    motion, 106, 113, 294, 295, 311
    position, 221, 262
    velocity, 282, 285
Relativistic kinetic energy, 313
Representation
    adjoint, 377
    complex conjugate, 376
Resonance
    cavity, 144
    states, 360
Rest
    energy, 313
    frame, 289
Reverse
    bias, 246
    breakdown voltage, 247

Ricci tensor, 335
Rocky planets, 165
Root vector, 373
Rubidium atoms, 148
    energy levels, 148
Ruby
    crystal, 144, 151
    laser, 145
Rydberg constant, 5

S
Scattered
    electron, 11
    photon, 11, 315
    protons, 412, 413
    radiation, 1, 11, 12
    waves, 62
Scattering
    pion-nucleon, 360
Schrödinger electron, 229
Schrödinger equation, 19, 81
    for electrons, 28
    for hydrogen, 82, 111, 119
    inside the barrier, 275
    inside the well, 275
Schrödinger time-dependent equation, 32, 34
Schrödinger time-independent equation, 21, 33, 34
Seekers
    high-field, 147
    low-field, 147
Selection rules, 101
Self-adjoint matrix, 322
Semiconductors, 228
    devices, 243, 245, 256, 258, 263, 267, 270
    energy bands, 255
    for electrons, 230
    intrinsic, 241
    lasers, 26, 27, 243, 267, 270, 272, 277
    material, 268
Semiempirical
    formula, 416
    mass formula, 398, 400, 404, 408
Separation of variables, 32
Sequence, 170
Shell, 120
    structure, 120
Shift
    blue, 293
    red, 293
Silicon
    atoms, 244
    crystal, 243
Single
    atom, 96, 100, 207
    electron, 81, 95, 107, 109, 121, 122, 128, 148, 153, 354, 355, 371, 372
        energy, 129
    frequency, 181, 182
    mass, 346
    orbital angular momentum, 355
    particle, 356
    photon, xxviii, 4, 141
    proton, xxvii

value, 124
variable, 47
Singlet, 124, 140, 360, 372, 379
    representation, 372, 379
    states, 124, 379
Sinusoidal waves, xx, xxi, 65
Slater
    determinant, 119
    integrals, 129
Sodium
    atom, 10, 11, 148, 226, 233
    chloride, 211, 212
    metal, 11, 17, 237
Solar
    cells, 247
    neutrinos, 342, 368
Source, 249
Space group, 254
Space quantization, 83
Space-charge region, 245
Space-time diagrams, 295
Spectroscopic notation, 83, 123, 128, 131, 134, 139, 409, 411
Spectrum, 4
Spherical Bessel functions, 409
Spherical harmonic, 84, 85, 101, 108, 113, 117, 119, 126, 137, 138, 371, 409
Spheroidal coordinates, 153
Spin, 356
    component, 95
    electron, 105
    function, 108, 119, 126
    magnetic moments, 111
    matrix, 326
    nuclear, 152, 357, 393
    orientation, 126, 194, 326, 328
    particle, 189, 201, 324
    quantum number, 95, 106, 148, 355, 363
    states, 98, 356, 363
    wave functions, 363
    zero, 357, 385
Spin-orbit
    constant, 108
    interaction, 106, 411
Spinning, 104, 357
Spline collocation, 47, 51, 80, 157, 158, 264
    method, 47, 51, 78, 86, 155, 156, 158, 163
Spontaneous emission, 96
Spontaneous transitions, 96, 98, 141
Stable nuclei, 399, 401, 408
Standing waves, xviii, xix, 175, 181
States
    atomic, 107, 147
    color, 363, 379
    conduction, 241
    coupled, 108
    electron, 227, 231, 241, 256
    electronic, 241
    energy, 173, 190, 326
    isospin, 362
    microscopic, 201
    positron, 327
    spin, 98, 356, 363
    stationary, 34, 97

superpositions, 39
valence, 241
vibrational, 143
Stationary
    neutrino states, 342
    states, 34, 97
    waves, xvii, xxii
Statistical weight, 170
Stefan-Boltzmann law, 184, 200
Stern-Gerlach experiment, 105
Stimulated emission, 96, 141, 143, 144, 151, 270, 273
    radiation, 141, 143
Stimulated transition, 113, 141
Stopping potential, 2
Stored optical energy, 274
Strange particles, 351, 352, 358
Strangeness quantum number, 351–354
Strong force, xxx
Subshell, 120
    closed, 120
    filled, 120
    open, 120
Summation convention, 302
Superfluid, 193
Supergravity, 387
Supermultiplets, 360
Superposition, xxiii, 66, 224
    principle, xviii
    waves, xxvii, 67
Supersymmetry, 385
    algebra, 387
Synchrotron radiation, 260

T
Target nucleus, 412–414
Tensor products, 157, 163
Terrestrial planet, 165
Thermal escape, 179
Thermodynamics
    first law, 187
    second law, 188
    third law, 188
    zeroth law, 187
Threshold frequency, 2, 3
Time dilation, 288
Transformation
    equations, 284
    matrix, 307
Transformations
    coordinate, 305
    Galilean, 305
    inverse, 306
    Lorentz, 286, 288, 291, 295, 298, 302, 306, 312, 323
    relativistic, 306
    velocity, 306
Transition
    amplitude, 269
    energy, 15
    frequency, 6, 98, 148
    laser, 141, 142, 145
    metal, 234
    optical, 256

probabilities, 91, 97, 112
radiative, 102
rate, 95, 98–100, 112, 141
    for absorption, 141
    in transitions, 98
spontaneous, 96, 98, 141
Transmission, 267
coefficient, 59
matrix, 77
Transpose operator, 73
Transverse mass, 243
Traveling waves, xvi, xxii, xxxi
Trivalent metals, 195
Truth, 341

U

Ultraviolet catastrophe, 182

V

Valence
band, 227, 255, 256
electrons, 228, 232, 244
    in crystals, 212
states, 241
Value
average, 23, 68, 78, 89
maximum, 26, 107, 123, 124, 177, 185, 186, 404
single, 124
zero, 8
Van der Waals force, xxx

Vector
boson, 316
equation, 311
momentum, xiii
notation, 302, 321
orbital angular momentum, 106, 110
pointing, xxii, 164, 216, 334
position, xiii, xxii, 82, 214, 219, 304
position-time, 306
primitive, 206
quantities, 311, 332
Velocity
constant, 284, 286, 287, 298, 301, 307, 309
distribution, 10
particle, 296
relative, 282, 285
vectors, xiii, 338
Vibrational
energy, 257
    levels, 143
states, 143
Volume element, 112

W

Wave function
antisymmetric, 118
hydrogen, 84, 111
polar coordinates, 112
Wavelengths, xxiii, xxv, 98
Waves, xiii
electromagnetic, xxv, xxvi, 1, 181, 192, 265
electron, 13

equation, 33
front, xxii
harmonic, xvi, xviii, xx, xxi, 66
light, 266
packets, 65
scattered, 62
standing, xviii, xix, 175, 181
stationary, xvii, xxii
superimposing, 65, 67
superposition, xxvii, 67
vector, xxii
Weak
force, xxx, 341, 343, 381, 382
interaction, xxx, 343, 351, 352, 366, 389
isospin, 381
    quantum number, 381
Wien displacement law, 200
Wigner-Seitz cell, 209
Work function, 11
Worldline, 295

X

X-ray scattering, 12

Z

Zeeman
effect, 109
splitting, 112
Zero point energy, 30
Zincblende structure, 210, 215, 233, 254, 256, 258